Review of Neuroscience

BEN PANSKY, Ph.D., M.D.
Professor of Anatomy,
Medical College of Ohio at Toledo,
Toledo, Ohio

DELMAS J. ALLEN, Ph.D.
Associate Professor of Anatomy,
Medical College of Ohio at Toledo,
Toledo, Ohio

Illustrations by Ben Pansky, Ph.D., M.D.

Review of Neuroscience

Macmillan Publishing Co., Inc.
New York

Collier Macmillan Canada, Ltd.
Toronto

Baillière Tindall
London

Macmillan Publishing Co., Inc.
866 Third Avenue, New York, New York 10022

Collier Macmillan Canada, Ltd.

Baillière Tindall · London

Library of Congress Cataloging in Publication Data

Pansky, Ben.
  Review of neuroscience.

  Bibliography: p.
  Includes index.
  1. Neurology. I. Allen, Delmas J., joint author. II. Title.
QP355.2.P36   612.8   80-11861
ISBN 0-02-390610-3

Printing: 1 2 3 4 5 6 7 8   Year: 0 1 2 3 4 5 6

Preface

This book has been written for the medical student, dental student, physician, surgeon, physical therapist, nurse, and others who require a visual comprehension and review of the fundamentals of neuroscience.

The beginning student is often overwhelmed by the amount of material necessary for a basic understanding of the nervous system. Those seeking to review are often astounded by progress in the field and the lack of time to study the mass of material that is available. Furthermore, both a mental gap and a real time lapse separate the average clinician, investigator, student, and teacher from his/her early studies of the nervous system. The functional and anatomic aspects of "the neurologic body" become buried as time and clinical material push basic knowledge into the background. Although the student of neuroscience continues to resurrect and expand his information because it is essential in his work, others in related fields are left behind, even though they frequently deal with subject matters that rely on information concerning control and function of the nervous system.

The material in *Review of Neuroscience* is presented in a simplified, concise, and outline form, but it is, nevertheless, complete. The text is functionally oriented and clinically informative to emphasize the need for understanding the fundamentals of neuroscience. The authors have made every effort to bring together a comprehensive review of the essential areas of neuroscience—namely, neuroembryology, neurophysiology, microscopic and gross neuroanatomy, neurology, and neuropathology. The basic aspects of each area have been diligently and carefully reviewed to present a clear and concise format for learning, an opportunity for review, and incentive for further study. Where repetition occurs, it does so to reemphasize particular points and to help the continuity of study. With only a few exceptions, illustrations appear on right-hand pages and corresponding text on opposite left-hand pages. The terminology used throughout is an anglicized version of the standard nomenclature (*Nomina Anatomica,* 4th ed.) approved at the Tenth International Congress of Anatomists, held at Tokyo in August, 1975.

The text is divided into three logical parts. Part I, "Structural and Functional Basis of Neuroscience," includes developmental neuroanatomy, gross and microscopic neuroanatomy, and neurophysiology. Part II, "Functional Mechanisms of Neuroscience," is subdivided to include material that is usually found in standard textbooks of neuroanatomy. Part III, "Neurology and Neuropathology," is also divided into relevant units. As each part is subdivided into essential units that comprise the specific subject matter of that part, an automatic outline is created that makes study easier, comprehension more reasonable, and learning more organized. Furthermore, the parts and even the units may be studied in sequence, as is often the case in formal coursework; however, since each is not dependent on the other, they may also be studied independently. Several appendixes have been added to the back matter of the text, not as an afterthought but to complete the student's overall concept of neuroscience. These include the following: Glossary; Disorders of Function (Migraine, Neuralgias, Vertigo, and Disturbances of Sleep); Examination of the

Nervous System; Clinical Testing of Cranial Nerve Function; Functional Classification of Systems and Reflexes; Language; Consciousness; and References.

Since students are often overwhelmed by excess explanations, references, and minute details and thereby lose sight of the real essentials, an outline format has been employed throughout that includes only the necessary words needed to convey basic thoughts and ideas. In addition, since illustrations play such an important role in the understanding and visualization of the biomedical sciences, over 1,000 original line-cut drawings were created for this text, some in color, as well as numerous photomicrographs and photographs. Since most readers of *Review of Neuroscience* would not have time to investigate multiple references to the research literature, but be more apt to consult standard textbooks, references to journal articles have been held to a minimum and the attention of the reader is directed to the textbooks and monographs listed in Appendix VIII.

The authors would like to express their appreciation to the many colleagues, teachers, and, above all, students at the Medical College of Ohio, whose comments and suggestions have helped to make this book possible. We are particularly grateful to Frank Redmond, of the Department of Pathology, who offered us help in his specialty of neuropathology, and who provided a collection of gross and microscopic specimens that enabled us to assemble many of the photographs found in Part III; to Dr. Hardy Waller, Associate Professor of Neurosciences and Physiology, who reviewed the unit on neurophysiology and was very supportive of the entire project; and to Dr. Stephen W. Corrie, Assistant Professor of Neurology, who read and reviewed Part III.

In addition, we would like to thank Dr. A. John Cristoforidas, Chairman and Professor of Radiology, and Dr. Edward Savolaine, Department of Radiology, for their help in the selection and interpretation of the CAT scans. Ms. Faye Keen, Library Media Technician Assistant in the Department of Radiology, assisted with the photographic work necessary to present the CAT scans and x-rays offered by the department. Of the many others whose services were essential for the completion of the text, we wish to mention George Gentry, Research Assistant in Anatomy; Verne Wright, Anatomy Technician; Jerry Lubinski, Photographer, Audiovisual Department; Irene Durivage and Joan Knighton, Anatomy Department secretaries who helped type the manuscript; Carol Young, for her work on the Glossary; Mary Horton, for her efforts with the Index; Ronald Watterson, Director of the MCO Library; and Dr. Liberato J. A. DiDio, Chairman and Professor of Anatomy, for giving us the time, place, and use of equipment, without which our efforts would have been formidable.

The authors cannot express too strongly their obligations to the publisher and particularly to the executive medical editor, Ms. Joan C. Zulch, for her continuous courtesy, enthusiasm, support in all matters, and infinite patience.

B. P.

D. J. A.

Contents

Unit Three. NEUROPHYSIOLOGY

PART II.   FUNCTIONAL MECHANISMS OF NEUROSCIENCE

Unit Four.   PATHWAYS, MUSCLE REGULATION, AND TYPICAL SPINAL CORD LESIONS

Unit Five. CRANIAL NERVES

Unit Six. THE DIENCEPHALON

Unit Eleven. THE RETICULAR FORMATION, LIMBIC AND OLFACTORY SYSTEMS

Unit Twelve. ORGANS OF SPECIAL SENSATION

PART III. NEUROLOGY AND NEUROPATHOLOGY

Unit Thirteen. INTRACRANIAL VASCULAR ACCIDENTS AND TRAUMA

Unit Fourteen. CONGENITAL MALFORMATIONS AND PERINATAL PATHOLOGY

Unit Fifteen. DEMYELINATING DISEASES

Unit Twenty-Two. TUMORS

Unit Twenty-Three. INFECTION AND INFLAMMATION

APPENDIXES

PART I

Structural and Functional Basis of Neuroscience

Unit One

DEVELOPMENTAL

NEUROANATOMY

1. EMBRYOLOGY OF THE CENTRAL NERVOUS SYSTEM: GENERAL DEVELOPMENT

I. Central nervous system (CNS) appears at the beginning of the 3rd week of development as an elongated plate of thickened ectoderm, the neural or medullary plate

A. THE PLATE is found in the middorsal region in front of the primitive pit occupying the anterior portion of the embryonic disk

1. The lateral edges of the plate soon elevate to form the neural folds; the depressed longitudinal furrow between the folds is the neural or medullary groove

B. THE NEURAL FOLDS elevate with development, approach each other in the midline, fuse, and form the neural tube which then loses connection to overlying ectoderm

1. Alongside the developing neural tube are strips of mesoderm which begin to show segmentation into somites near the end of the 3rd week of development
   a. The 1st somite appears behind the cephalic tip of the notochord
   b. Somites appear in a craniocaudal sequence, and 42–44 pairs are present by day 31
   c. Most of the axial skeleton and musculature develop from the somites
2. Closure of the neural tube begins near the 4th somite (future cervical level) and proceeds both cephalically and caudally
   a. At the cranial and caudal ends of the embryo, the fusion is delayed and the anterior and posterior neuropores temporarily form open connections between the lumen of the neural tube and the amniotic cavity
      i. The anterior neuropore indicates the lamina terminalis (most cephalic limit of developing brain) and closes at the 25th day (18–20 somite stage); the posterior neuropore closes in the 25 somite stage
3. The CNS then forms a closed tubular structure with a long caudal portion, the future spinal cord, and a broader cephalic portion, the future brain

C. DIVISIONS OF THE NEURAL TUBE: even before the neural tube is closed cephalically, the cephalic end shows 3 distinct dilatations, the primary brain vesicles:

1. The prosencephalon or forebrain is the most anterior vesicle
2. The mesencephalon or midbrain is the central vesicle
3. The rhombencephalon or hindbrain is the most posterior vesicle

D. UNEQUAL GROWTH RATES and cell migration result in flexures, constrictions, thickening, invaginations, and evaginations

1. Coincidental with the appearance of the vesicles (5-mm embryo), the neural tube bends ventrally to form 2 flexures: a cervical flexure at the junction of spinal cord and hindbrain; and a cephalic flexure, seen in the midbrain

E. AT 5 WEEKS OF AGE, 5 components are distinguished in the developing brain:

1. The prosencephalon which consists of 2 portions:
   a. An anterior telencephalon endbrain whose midposterior portion, the diencephalon is characterized by the outgrowth of the optic vesicles
      i. Dorsal and ventral evaginations from the diencephalon become the primordia of the pineal gland and posterior hypophysis, respectively
   b. Two anterolateral expansions called the primitive cerebral hemispheres.
2. The mesencephalon undergoes little change
3. The rhombencephalon consists of 2 parts:
   a. Metencephalon (anterior) which later forms the pons and cerebellum
   b. Myelencephalon (posterior) which later forms the medulla oblongata
   c. The boundary between the metencephalon and myelencephalon is marked by a 3rd flexure, the pontine flexure, compensatory to the other two

F. THE LUMEN OF THE SPINAL CORD, the central canal, is continuous with the brain vesicles permitting CSF to circulate freely between brain and cord

1. The cavity of the rhombencephalon is the 4th ventricle, of the diencephalon the 3rd ventricle, and of the cerebral hemispheres the lateral ventricles
2. The lumen between the 3rd and 4th ventricles is the narrow aqueduct of Sylvius
3. The interventricular foramina of Monro connect the 3rd and lateral ventricles
4. 4th ventricle opens into subarachnoid space via foramina of Luschka and Magendie

Notochordal process under ectoderm
Prochordal plate
Cut edge of amnion
Primitive knot (node)
Yolk sac
Primitive streak
Cloacal membrane

16 DAYS

Plate
Neural plate
Notochordal process
Primitive streak
Connecting stalk

Primitive groove
X-section
Yolk sac

19 DAYS

Neural plate
Cut edge of amnion
Primitive node
Primitive streak

Neural plate
Neuroecdodermal junction
X-section
Notochordal plate

18 DAYS

Notochord
Extraembryonic mesoderm
Entoderm
Intraembryonic mesoderm

Neural groove
Somatopleure
Neural fold
Somite
Splanchnopleure

20 DAYS
Neural fold
Somites
Neural groove
Primitive streak

X-SECTION
Neural groove
Somite
Notochord

22 DAYS
Neural fold
Pericardial bulge
Otic placode
Somite
Cut edge of amnion

Neural tube
Neural crest
Central canal
Surface ectoderm

23 DAYS
Ant. neuropore
Post. neuropore

25 DAYS
Ant. neuropore
1st + 2nd pharyngeal arches
Pericardial bulge
Cut edge of amnion
Vitelline canal
Post. neuropore
Lens placode
Heart bulge
Umbilical cord
Otic placode

Rhombencephalon
Mesencephalon
Cervical flexure
Cephalic flexure
Cord
Prosencephalon

Rhombencephalic isthmus
Metencephalon
Mes.
Pontine flexure
Myel.
Dien.
Optic vesicle
Tel.

Tel.
Dien.
Mes.
Met.
Myel.
Pontine flexure
Lamina terminalis
Intervent. foramen of Monro

28 DAYS

Forebrain (prosencephalon) → Telencephalon
Wall
Midbrain (mesencephalon) → Mesencephalon
Cavity
Hindbrain (rhombencephalon) → Metencephalon
→ Myelencephalon
Cord

3 PRIMARY VESICLES

Telencephalon → Cerebral hemispheres
Diencephalon → Thalami, etc.
Mesencephalon → Midbrain
Metencephalon → Pons / Cerebellum
Myelencephalon → Medulla oblongata
Cord

5 SECONDARY VESICLES

Lateral ventricle
IIIrd ventricle
Aqueduct
IVth ventricle

Pansky

2. SPINAL CORD: NORMAL DEVELOPMENT

I. Neuroepithelial, mantle, and marginal layers

A. THE WALL of a recently closed neural tube consists of only 1 cell type, the neuroepithelial cells, which extend over the entire thickness of the wall and form a thick pseudostratified epithelium
   1. The cells are connected to each other by terminal bars at the lumen
   2. During interphase, when DNA synthesis takes place, the cells are wedge-shaped with the broader portion containing the nucleus in the outer zone of the wall and a slender cytoplasmic part extending toward the lumen
   3. Just after DNA synthesis, the nucleus begins to move toward the lumen, while the cell contracts toward the terminal bars
   4. During metaphase, the cells are round and in broad contact with the lumen, squeezing the thin cytoplasmic processes of neighboring nondividing cells

B. DURING THE NEURAL GROOVE STAGE and just after tube closure, the neuroepithelial cells divide rapidly, resulting in the production of more cells, and we now refer to the thickened epithelium in the recently closed neural tube as the neuroepithelial layer or neuroepithelium

C. ONCE THE TUBE IS CLOSED, the neuroepithelial cells give rise to another cell type characterized by a round nucleus with pale cytoplasm and a dark-staining nucleolus, the primitive nerve cells or neuroblasts
   1. The neuroblasts form a zone that surrounds the neuroepithelial layer, called the mantle layer (the future gray matter of the spinal cord)
   2. The outermost layer of the cord contains the nerve fibers emerging from the neuroblasts in the mantle layer and is called the marginal layer
      a. As a result of myelination of the nerve fibers, the marginal layer takes on a "white" appearance and is called the white matter of the cord

II. Basal, alar, roof, and floor plates

A. WITH THE CONTINUAL ADDITION OF NEUROBLASTS to the mantle layer, each side of the neural tube shows a ventral and dorsal thickening
   1. The ventral thickenings, the basal or motor plates, contain the anterior motor horn cells and form the motor areas of the spinal cord
   2. The dorsal thickenings, the alar or sensory plates, form the sensory areas
   3. A longitudinal groove, the sulcus limitans, is found bilaterally on the inner surface of the tube. It marks the boundary between the anterior motor and posterior sensory areas and ends in the region of the mamillary recess in the ventral portion of the diencephalon
      a. One should therefore not expect any truly motor nerves emerging from the brain rostral to the mamillary recess
      b. Since all neural tissue rostral to this point is an extension of the alar plate, its function is regarded as sensory or associational

B. THE THIN DORSAL PORTION and ventral midline parts of the tube are the roof and floor plates, respectively. They contain no neuroblasts and serve primarily as pathways for nerve fibers crossing from one side of the cord to the other

C. THE BASAL PLATES bulge ventrally on each side of the midline as a result of the continuous enlargement of the neuroblasts, creating a deep longitudinal groove called the ventral fissure, which will later contain the anterior spinal artery

D. THE ALAR PLATES expand predominantly in a medial direction compressing the dorsal portion of the lumen of the neural tube
   1. The posterior median septum is formed where the 2 alar plates fuse in the midline

E. ACCUMULATION OF NEURONS between alar and basal plates causes the formation of the intermediate horn which contains motor neurons of the autonomic nervous system. The spinal cord acquires its definitive form: motor horns anteriorly, sensory horns posteriorly, intermediate horns laterally, and small lumen, the central canal

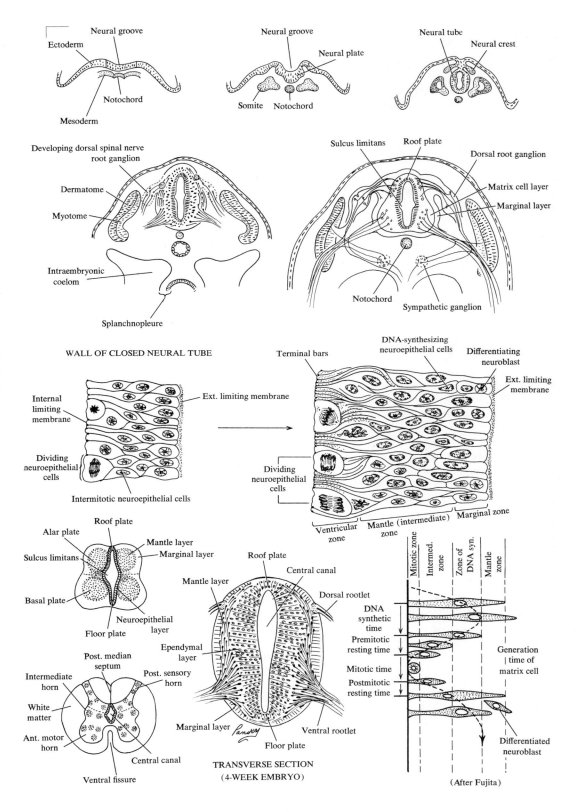

Ectoderm
Neural groove
Notochord
Mesoderm

Neural groove
Neural plate
Somite   Notochord

Neural tube
Neural crest

Developing dorsal spinal nerve root ganglion
Dermatome
Myotome
Intraembryonic coelom
Splanchnopleure

Sulcus limitans   Roof plate
Dorsal root ganglion
Matrix cell layer
Marginal layer
Notochord   Sympathetic ganglion

WALL OF CLOSED NEURAL TUBE

Internal limiting membrane
Dividing neuroepithelial cells
Intermitotic neuroepithelial cells
Ext. limiting membrane

Terminal bars
DNA-synthesizing neuroepithelial cells
Differentiating neuroblast
Ext. limiting membrane
Dividing neuroepithelial cells
Ventricular zone   Mantle (intermediate) zone   Marginal zone

Roof plate
Alar plate
Sulcus limitans
Basal plate
Floor plate
Mantle layer
Marginal layer
Neuroepithelial layer

Intermediate horn
White matter
Ant. motor horn
Ventral fissure
Post. median septum
Post. sensory horn
Central canal

Roof plate
Central canal
Dorsal rootlet
Mantle layer
Ependymal layer
Marginal layer
Ventral rootlet
Floor plate

TRANSVERSE SECTION
(4-WEEK EMBRYO)

Mitotic zone   Intermed. zone   Zone of DNA syn.   Mantle zone
DNA synthetic time
Premitotic resting time
Mitotic time
Postmitotic resting time
Generation time of matrix cell
Differentiated neuroblast

(After Fujita)

3. SPINAL CORD: NERVE AND GLIAL CELL DIFFERENTIATION

I. **Nerve cells:** the neuroblasts or primitive nerve cells arise exclusively by division of the neuroepithelial cells. Once the neuroblasts are formed, they lose their ability to divide

A. THE NEUROBLASTS OF THE ANTERIOR HORN are formed 1st, and only when most of these have migrated to the mantle layer does formation of nerve cells for the alar plate begin

B. THE NEUROBLASTS initially have a central process extending to the lumen, the transient dendrite, but it disappears when the cells migrate to the mantle zone. The neuroblasts are temporarily round, the so-called apolar neuroblasts

C. THERE IS FURTHER DIFFERENTIATION and 2 new cytoplasmic processes appear on opposite sides of the cell body. Thus, the bipolar neuroblasts are formed

D. THE PROCESS at one end elongates rapidly to form the primitive axon, whereas that at the other end develops a number of cytoplasmic arborizations which are known as the primitive dendrites

E. THE CELL is now a multipolar neuroblast and with further development becomes the adult nerve cell or neuron

F. THE AXONS of neurons in the posterior sensory horn behave differently than those in the anterior motor horn:
1. Those in the posterior horn penetrate the marginal layer of the cord and then ascend or descend to a higher or lower level (association neurons)
2. Those in the anterior horn break through the marginal zone and are seen on the ventral aspect of the cord where they form the anterior motor root of the spinal nerve (they conduct motor impulses from cord to muscles)

II. **Glia cells:** majority of primitive supporting cells are called gliablasts, formed by the neuroepithelial cells after the production of neuroblasts has ceased

A. THE GLIABLASTS migrate from the neuroepithelial layer to the mantle layer (some even to the marginal layer) where they differentiate into the fibrillar and protoplasmic astrocytes

B. ANOTHER TYPE OF SUPPORTING CELL, possibly of gliablast origin, is the oligodendroglia cell
1. This cell is mainly found in the marginal layer and forms the myelin sheath around the ascending and descending axons in that layer
2. They may be derived from mesenchyme cells which have penetrated into the CNS, thus their origin from neuroepithelial cells is in doubt

C. IN THE 2ND HALF OF DEVELOPMENT, a 3rd type of supporting cell, the microglia cell, appears in the CNS. It is believed that its origin is from mesoderm which surrounds the neural tube

D. WHEN THE NEUROEPITHELIAL CELLS cease to produce neuroblasts and gliablasts, they finally differentiate into the ependymal cells (as seen in the adult)

NEURAL TUBE

Roof plate

Central canal

Ependymal
(matrix cell) layer

Mantle layer

Sulcus limitans

Marginal layer

Mesenchyme

Mesenchymal cell

Microglial cell

Neuroepithelium
(neuroectoderm)

Ext. limiting membrane

Apolar neuroblast

Bipolar neuroblast

Unipolar
neuroblast

Dendrites

Axon

Multipolar neuroblast

Glioblast
(spongioblast)

Astroblast

Protoplasmic astrocyte

Oligodendroblast

Fibrous astrocyte

Ependyma

+

Epithelium of choroid plexus

Oligodendrocyte

Pansky

– 9 –

4. SPINAL CORD: NEURAL CREST CELLS AND MYELINATION

I. Neural crest cells

A. DURING INVAGINATION of the neural plate, a distinct group of cells appears along each edge of the neural groove, are ectodermal in origin, and are called the neural crest cells
  1. The cells temporarily form an intermediate zone between the tube and surface ectoderm
  2. The zone extends from the mesencephalon to the level of the caudal somite and in time divides into 2 parts, each of which migrates to the dorsolateral aspect of the neural tube
     a. Here the cells of the neural crest form a series of cell clusters that give rise to the sensory or dorsal root ganglia of the spinal and cranial nerves (5th, 7th, 9th, and 10th cranial nerves)
B. DURING DEVELOPMENT, the neuroblasts of the sensory ganglia form 2 processes:
  1. One grows centrally and penetrates the dorsal portion of the neural tube.
     a. In the spinal cord, they either end in the dorsal horn or ascend through the marginal layer to one of the higher brain centers
     b. Collectively, these processes are called the dorsal sensory root of the spinal nerve
  2. The other process grows peripherally and forms fibers of the ventral motor root and participates in the formation of the trunk of the spinal nerve.
     a. These processes eventually terminate in the sensory receptor organs
C. IN ADDITION TO FORMING THE SENSORY GANGLIA, the cells of the neural crest differentiate into sympathetic neuroblasts, Schwann cells, pigment cells, odontoblasts, meninges, and cartilage cells of the branchial arches
  a. Removal of neural crest cells of the trigeminal region results in facial abnormalities, including clefts of the primary palate

II. Myelination

A. MYELINATION OF THE PERIPHERAL NERVE is brought about by the neurilemma cells or cells of Schwann
  1. These cells originate from the neural crest, migrate peripherally, and wrap around the axons to form the neurilemma sheath
  2. Axons, varying in number from 1–20, can be enwrapped by one neurilemma cell
B. AT THE 4TH MONTH OF FETAL LIFE, the nerve fibers gradually obtain a whitish appearance as a result of myelin deposition between the axon and neurilemma
  1. This substance is formed by repeated coiling of the membrane around the axon
C. BOTH THE NEURILEMMA AND THE MYELIN SHEATH of the peripheral nerve fibers are formed by the cells of Schwann
D. THE MYELIN SHEATH of nerve fibers within the spinal cord is of different origin, being formed by the oligodendroglia cells
E. THOUGH MYELINATION OF NERVE FIBERS in the cord generally begins at about the 4th month of fetal life, some motor fibers that descend from higher brain centers to the cord do not become myelinated until the 1st year of postnatal life
  1. The tracts in the nervous system apparently become myelinated at about the time they begin to function

Neural groove

Neural plate

Neuroectodermal junction
(presumptive neural crest)

Extraembryonic
mesoderm

Notochordal plate

Intraembryonic mesoderm

Entoderm

Deepening neural groove

Intraembryonic
coelom

Developing notochord

Central canal

Neural crest

Somite

Migrating neural crest

Myotome
Dermatome

Somite

Neural tube

Notochord

Notochord

NEURAL CREST

Entire CNS (except vessels, choroid
plexus, and microglia)

Somatic and branchial motor axons

Autonomic preganglionic axons

Sensory and autonomic neurons + ganglia

Schwann cells and perineural epithelium
(satellite cells)

Meningocytes (pia, arachnoid, and villi)

Melanocytes

Chromaffin cells

Alar plate

Dorsal root
ganglion

Basal plate

Outgrowing motor axons

Dorsal sensory root

Dorsal horn

Ventral horn

Ventral motor
root

Spinal nerve trunk

Axon

Schwann cell

Interperiod lines

Internal (inner)
mesaxon

Outer
Inner

Membrane
lamina

Basement membrane

Axon

Schwann cytoplasm

MYELIN SHEATH

Major dense line

External (outer) mesaxon

Schwann cytoplasm

5. SPINAL CORD: LENGTH AND CONGENITAL MALFORMATIONS

I. Changes in cord position during development

A. WHEN THE CROWN-RUMP LENGTH of the embryo is about 30 mm (in the 3rd month of development), the spinal cord extends the entire length of the embryo and the spinal nerves pass through the intervertebral foramina at their level of origin

B. WITH INCREASING AGE, the vertebral column and dura mater lengthen more rapidly than the neural tube and the terminal end of the spinal cord gradually shifts to a higher level

C. AT BIRTH, the end of the cord is located at the level of the 3rd lumbar vertebra.
   1. Due to disproportionate growth, the spinal nerves run obliquely from their segment of cord origin to the corresponding level of the vertebral column
   2. The dura remains attached to the bony column at the coccygeal level

D. IN THE ADULT, the spinal cord terminates at the level of L2
   1. Below this point, the CNS is represented only by the filum terminale internum, marking the tract of spinal cord regression
   2. Nerve fibers below the terminal end of the cord are called, collectively, the cauda equina

E. WHEN CEREBROSPINAL FLUID is taken in a lumbar puncture, the needle is inserted at the lower lumbar level to avoid injuring the lower end of the spinal cord.

II. Congenital malformations of the cord and column

A. THE CNS forms a closed tubular structure which is detached from the overlying ectoderm by the end of the 4th week
   1. Occasionally, however, the neural groove fails to close, either as a result of faulty induction by the underlying notochord, or because of the action of environmental teratogenic factors on the neuroepithelial cells
      a. The neural tissue then remains exposed to the surface
      b. Such a defect may extend the entire length of the embryo or be restricted to a small area only—complete or partial rachischisis
      c. Failure of closure in the cephalic region is called anencephalis
      d. If localized in the cord region, the abnormality is called spina bifida
   2. Spina bifida refers to a wide range of defects
      a. In its most simple form, it is seen as a failure of the dorsal portions of the vertebrae to fuse with one another (bifid spine, literally)
         i. Usually localized in the sacrolumbar region, covered by skin and not noticeable on the surface except for the presence of a small tuft of hair over the affected area; called spina bifida occulta
            (a) Cord and nerves are normal; neurologic symptoms are absent
      b. If more than one or two vertebrae are involved in the defect, the meninges of the cord bulge through the opening and a sac covered with skin is seen on the surface; called meningocele
      c. If the sac is so large that it contains not only the meninges but also the spinal cord and its nerves, this abnormality is meningomyelocele
         i. Usually covered by a thin, easily torn membrane
         ii. Neurologic symptoms are usually present
      d. If the neural groove fails to close, the nervous tissue is widely exposed to the surface; called a myelocele or rachischisis
         i. Occasionally the neural tissue shows much overgrowth, but the excess tissue invariably becomes necrotic before or after birth
      e. Myelomeningoceles are usually associated with a caudal displacement of the medulla oblongata and a part of the cerebellum into the spinal canal
         i. The myelomeningocele is frequently combined with hydrocephaly; called Arnold-Chiari malformation

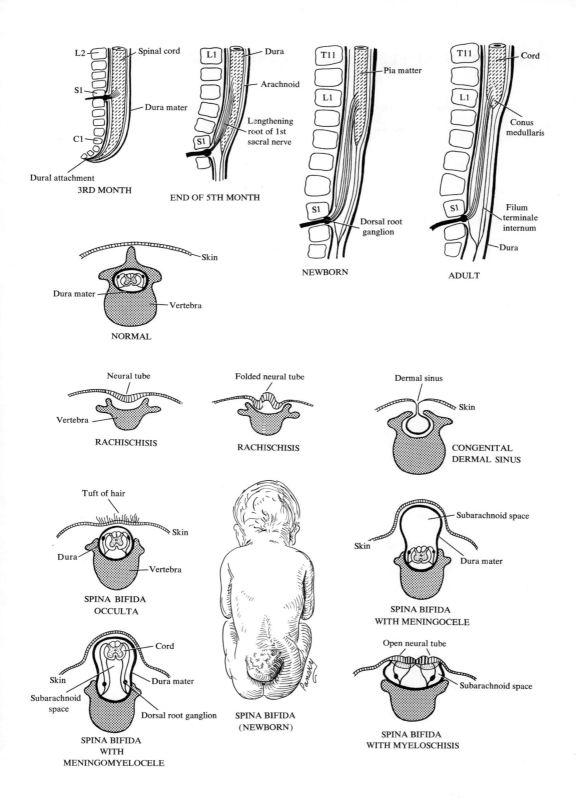

L2
Spinal cord
S1
Dura mater
C1
Dural attachment
3RD MONTH

L1
Dura
Arachnoid
Lengthening root of 1st sacral nerve
S1
END OF 5TH MONTH

T11
Pia matter
L1
S1
Dorsal root ganglion
NEWBORN

T11
Cord
L1
Conus medullaris
S1
Filum terminale internum
Dura
ADULT

Skin
Dura mater
Vertebra
NORMAL

Neural tube
Vertebra
RACHISCHISIS

Folded neural tube
RACHISCHISIS

Dermal sinus
Skin
CONGENITAL DERMAL SINUS

Tuft of hair
Skin
Dura
Vertebra
SPINA BIFIDA OCCULTA

Subarachnoid space
Skin
Dura mater
SPINA BIFIDA WITH MENINGOCELE

Cord
Skin
Dura mater
Subarachnoid space
Dorsal root ganglion
SPINA BIFIDA WITH MENINGOMYELOCELE

SPINA BIFIDA (NEWBORN)

Open neural tube
Subarachnoid space
SPINA BIFIDA WITH MYELOSCHISIS

6. BRAIN: MYELENCEPHALON—BASAL MOTOR PLATE

I. **Introduction:** the external shape of the cephalic part of the neural tube changes markedly with the appearance of brain vesicles and the development of cervical and cephalic flexures. Despite changes, some morphologic characteristics seen in the spinal cord are recognizable in most of the brain vesicles

A. DISTINCT BASAL AND ALAR PLATES, representing the motor and sensory areas, respectively, are seen on each side of the midline in most of the brain vesicles

B. THE SULCUS LIMITANS, which formed the boundary line between alar and basal plates in the cord, are present in the rhombencephalon and mesencephalon where it also forms the divider between sensory and motor areas

II. **The myelencephalon** is the most caudal brain compartment and extends from the 1st spinal nerve to the pontine flexure and gives rise to the medulla oblongata

A. THE MEDULLA differs from the cord in that its lateral walls rotate around an imaginary long axis in the floor plate (like opening a textbook); as a result, the roof plate stretches and becomes a single layer of cells

  1. The lateral wall structure is similar to that of the cord, and one sees alar and basal plates separated by the sulcus limitans

B. BASAL MOTOR PLATE OF THE MYELENCEPHALON, like the cord, contains the motor nuclei, but these are divided into 3 groups:

  1. A medial somatic efferent group: forms the cephalic continuation of the anterior horn cells containing the motor neurons which innervate striated muscle (derived from myotomes in the cephalic region)

    a. Since the somatic efferent group continues rostrally into the mesencephalon (through the metencephalon), it is referred to as the somatic efferent motor column

    b. It is represented in the myelencephalon by the neurons of the hypoglossal (XII) nerve which supplies the 4 occipital myotomes (tongue musculature)

    c. In the metencephalon and mesencephalon, the column is represented by the neurons of the abducens (VI), trochlear (IV), and oculomotor (III) cranial nerves, respectively, supplying eye muscles thought to be derived from preoptic myotomes

      i. The neurons of nerves III, IV, VI, and XII are located near the midline

  2. An intermediate special visceral efferent group which extends into the metencephalon and forms the special visceral efferent motor column which contains motor neurons supplying striated muscles derived from the mesenchyme of the pharyngeal or branchial arches

    a. In the myelencephalon, the column is represented by neurons of the accessory (XI), vagus (X), and glossopharyngeal (IX) cranial nerves

      i. In the adult, motor neurons of the above nerves are formed by the nucleus ambiguus and the bulbar portion of the accessory nerve

  3. A lateral general visceral efferent group contains the neurons whose axons grow out as preganglionic fibers to synapse in the parasympathetic ganglia supplying involuntary muscles of the heart, respiratory tract, and intestinal tract, as well as innervating the salivary glands

    a. In the myelencephalon, this group is represented by the dorsal nucleus of the vagus and the inferior salivatory nucleus (which, by way of cranial nerve IX innervates the parotid gland)

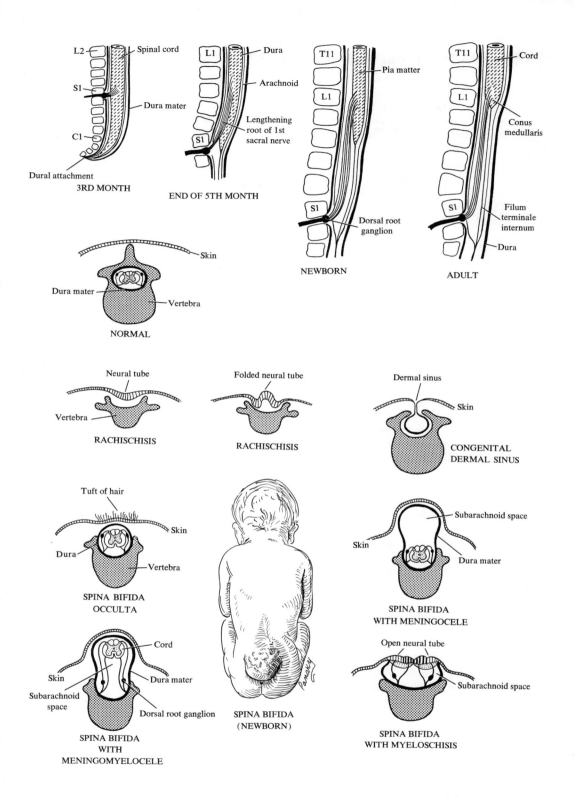

3RD MONTH

END OF 5TH MONTH

NEWBORN

ADULT

NORMAL

RACHISCHISIS

RACHISCHISIS

CONGENITAL DERMAL SINUS

SPINA BIFIDA OCCULTA

SPINA BIFIDA WITH MENINGOCELE

SPINA BIFIDA WITH MENINGOMYELOCELE

SPINA BIFIDA (NEWBORN)

SPINA BIFIDA WITH MYELOSCHISIS

6. BRAIN: MYELENCEPHALON—BASAL MOTOR PLATE

I. **Introduction:** the external shape of the cephalic part of the neural tube changes markedly with the appearance of brain vesicles and the development of cervical and cephalic flexures. Despite changes, some morphologic characteristics seen in the spinal cord are recognizable in most of the brain vesicles

A. DISTINCT BASAL AND ALAR PLATES, representing the motor and sensory areas, respectively, are seen on each side of the midline in most of the brain vesicles

B. THE SULCUS LIMITANS, which formed the boundary line between alar and basal plates in the cord, are present in the rhombencephalon and mesencephalon where it also forms the divider between sensory and motor areas

II. **The myelencephalon** is the most caudal brain compartment and extends from the 1st spinal nerve to the pontine flexure and gives rise to the medulla oblongata

A. THE MEDULLA differs from the cord in that its lateral walls rotate around an imaginary long axis in the floor plate (like opening a textbook); as a result, the roof plate stretches and becomes a single layer of cells

 1. The lateral wall structure is similar to that of the cord, and one sees alar and basal plates separated by the sulcus limitans

B. BASAL MOTOR PLATE OF THE MYELENCEPHALON, like the cord, contains the motor nuclei, but these are divided into 3 groups:

 1. A medial somatic efferent group: forms the cephalic continuation of the anterior horn cells containing the motor neurons which innervate striated muscle (derived from myotomes in the cephalic region)

  a. Since the somatic efferent group continues rostrally into the mesencephalon (through the metencephalon), it is referred to as the somatic efferent motor column

  b. It is represented in the myelencephalon by the neurons of the hypoglossal (XII) nerve which supplies the 4 occipital myotomes (tongue musculature)

  c. In the metencephalon and mesencephalon, the column is represented by the neurons of the abducens (VI), trochlear (IV), and oculomotor (III) cranial nerves, respectively, supplying eye muscles thought to be derived from preoptic myotomes

   i. The neurons of nerves III, IV, VI, and XII are located near the midline

 2. An intermediate special visceral efferent group which extends into the metencephalon and forms the special visceral efferent motor column which contains motor neurons supplying striated muscles derived from the mesenchyme of the pharyngeal or branchial arches

  a. In the myelencephalon, the column is represented by neurons of the accessory (XI), vagus (X), and glossopharyngeal (IX) cranial nerves

   i. In the adult, motor neurons of the above nerves are formed by the nucleus ambiguus and the bulbar portion of the accessory nerve

 3. A lateral general visceral efferent group contains the neurons whose axons grow out as preganglionic fibers to synapse in the parasympathetic ganglia supplying involuntary muscles of the heart, respiratory tract, and intestinal tract, as well as innervating the salivary glands

  a. In the myelencephalon, this group is represented by the dorsal nucleus of the vagus and the inferior salivatory nucleus (which, by way of cranial nerve IX innervates the parotid gland)

4 WEEKS

Forebrain
Midbrain
Hindbrain
Optic vesicle

5 1/3 WEEKS

Midbrain
Hindbrain
Otocyst
Diencephalon
Forebrain
Optic cup

6 WEEKS

Mes.
Met.
Dien.
Myel.
Tel.

7 WEEKS

Mes.
Met.
Dien.
Myel.
Tel.

8-WEEK EMBRYO (27-mm)

Mesencephalon
Cerebral hemisphere
Rhombencephalic isthmus
Pontine flexure
Myelencephalon
(medulla oblongata)
Diencephalon
V
VII, VIII
IX, X, XI
XII

Alar lamina
Midbrain
Isthmus
Forebrain
Basal lamina
Hindbrain
Sulcus limitans

DORSAL VIEW
(10-mm)

Mesencephalon
Metencephalon
Nerve VI
Rhombic lip
Basal
Roof plate
attachment
Plates
Alar
Myelencephalon

LOWER MYELENCEPHALON

Gracile nucleus
Cuneate nucleus
Central canal
Central gray
Pyramids

UPPER
MYELENCEPHALON

IV
Alar plate
Basal plate

Roof plate (ependymal roof)

IV
Alar plate
Basal plate

Choroid plexus
Special somatic afferent
(SSA)
General somatic afferent
(GSA)
Special somatic afferent
(SSA)
General visceral afferent
(GVA)
General visceral efferent (GVE)
Special visceral efferent (SVE)
Olivary nucleus
General somatic efferent (GSE)

Sulcus limitans
IVth ventricle
Alar plate
Basal plate

7. BRAIN: MYELENCEPHALON—ALAR SENSORY AND ROOF PLATE

I. The myelencephalon (cont.)

A. THE ALAR SENSORY PLATE contains the sensory relay nuclei, which like the basal plate are divided into 3 groups:
  1. The most lateral is the somatic afferent group, which receives impulses from the ear and surface of the head via the statoacoustic (VIII) and bulbospinal part of the trigeminal (V) nerves
  2. The intermediate is the special visceral afferent group, which receives impulses from the taste buds of the tongue and from the palate, oropharynx, and epiglottis
     a. These neurons later form the nucleus of the solitary tract
  3. The medial is the general visceral afferent group, represented by the dorsal sensory nucleus of the vagus with its neurons receiving interoceptive information from the heart and GI tract
  4. In addition to the sensory relay nuclei, other cells of the alar plate migrate downward to be ventrolateral to the basal plate and form a part of the olivary nuclear complex

B. ROOF PLATE, CHOROID PLEXUS, AND FORAMINA OF LUSCHKA AND MAGENDIE
  1. The roof plate of the myelencephalon consists of a single layer of ependymal cells which is later covered by vascular mesenchyme, the pia mater. Together, they make up the tela choroidea
     a. As a result of active proliferation of vascular mesenchyme, the tela forms a series of saclike invaginations that project into the underlying ventricular cavity in the region of the pontine flexure, forming the choroid plexus (tuftlike invaginations that produce the CSF of the CNS)
  2. At about 4 months of age, areas in the roof plate of the rhombencephalon thin out, bulge outward, and finally disappear, and the apertures formed are the 2 lateral foramina of Luschka and a median foramen of Magendie which allow the CSF to move freely between the ventricles and the surrounding subarachnoid space

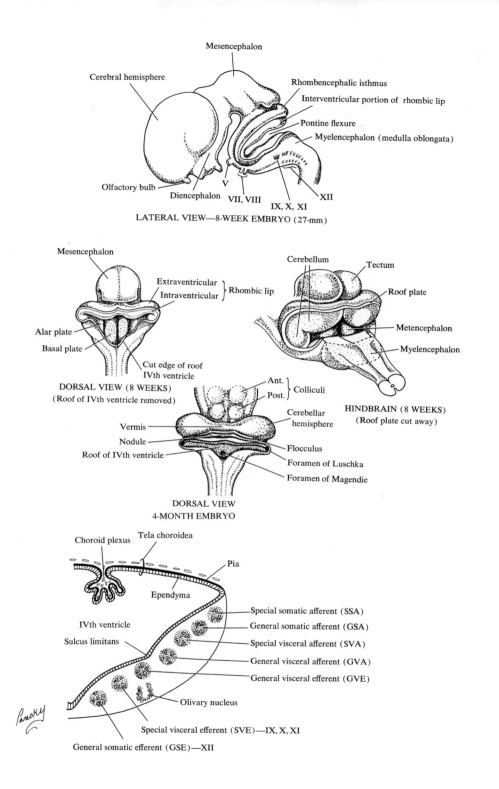

Mesencephalon

Cerebral hemisphere

Rhombencephalic isthmus

Interventricular portion of rhombic lip

Pontine flexure

Myelencephalon (medulla oblongata)

Olfactory bulb

Diencephalon

V

VII, VIII

IX, X, XI

XII

LATERAL VIEW—8-WEEK EMBRYO (27-mm)

Mesencephalon

Extraventricular
Intraventricular } Rhombic lip

Alar plate

Basal plate

Cut edge of roof
IVth ventricle

DORSAL VIEW (8 WEEKS)
(Roof of IVth ventricle removed)

Cerebellum

Tectum

Roof plate

Metencephalon

Myelencephalon

HINDBRAIN (8 WEEKS)
(Roof plate cut away)

Ant.
Post. } Colliculi

Cerebellar
hemisphere

Vermis

Nodule

Roof of IVth ventricle

Flocculus

Foramen of Luschka

Foramen of Magendie

DORSAL VIEW
4-MONTH EMBRYO

Choroid plexus

Tela choroidea

Pia

Ependyma

IVth ventricle

Sulcus limitans

Special somatic afferent (SSA)

General somatic afferent (GSA)

Special visceral afferent (SVA)

General visceral afferent (GVA)

General visceral efferent (GVE)

Olivary nucleus

Special visceral efferent (SVE)—IX, X, XI

General somatic efferent (GSE)—XII

Panoky

– 17 –

8. BRAIN: METENCEPHALON

I. **The metencephalon** develops from the anterior part of the rhombencephalon and extends from the pontine flexure to the rhombencephalic isthmus. It differs from the myelencephalon in forming 2 specialized components: the dorsal portion forms the cerebellum which functions as a coordination center for posture and movement and a ventral portion which becomes the pons which serves as the pathway for nerve fibers between the cord and cerebral and cerebellar cortices

A. BASAL PLATE AND PONS: the major morphologic features of the metencephalon do not change even though the lateral walls reapproach each other and the basal motor and sensory alar plates are easily seen. Each basal plate contains 3 groups of motor neurons:
   1. The medial somatic efferent group which gives rise to the abducens (VI) nerve
   2. The special visceral efferent group, containing the nuclei of the trigeminal (V) and facial (VII) nerves, which innervate the muscles of the 1st and 2nd branchial arches
   3. The general visceral efferent group which contains the superior salivatory nucleus, axons of which grow out in the facial nerve to supply the submandibular and sublingual glands as well as the nasal and lacrimal glands
   4. The marginal layer of the basal plates expands to serve as a bridge for the nerve fibers connecting the cerebral cortex and cerebellar cortex with the spinal cord and is known as the pons
      a. In addition to nerve fibers, the pons contains pontine nuclei which originate in the alar plates of the metencephalon and myelencephalon
      b. The axons of the nuclei grow toward the cerebellum and form the middle cerebellar peduncles

B. ALAR PLATE AND RHOMBIC LIP: development is complicated
   1. The ventromedial portion of the plate contains 3 groups of sensory nuclei:
      a. The lateral somatic afferent group contains neurons of the pontine portion of the trigeminal (V) nerve and a small part of the vestibulocochlear complex
      b. The special visceral afferent group: represented by the cranial portion of the nucleus of the solitary tract
      c. The general visceral afferent group: represented by the most cranial part of the dorsal sensory nucleus of the vagus (X) nerve
   2. The dorsolateral parts of the alar plates bend medially and form the rhombic lips which project partly into the lumen of the 4th ventricle and partly above the attachment of the roof plate (extraventricular part) to give rise to the cerebellum

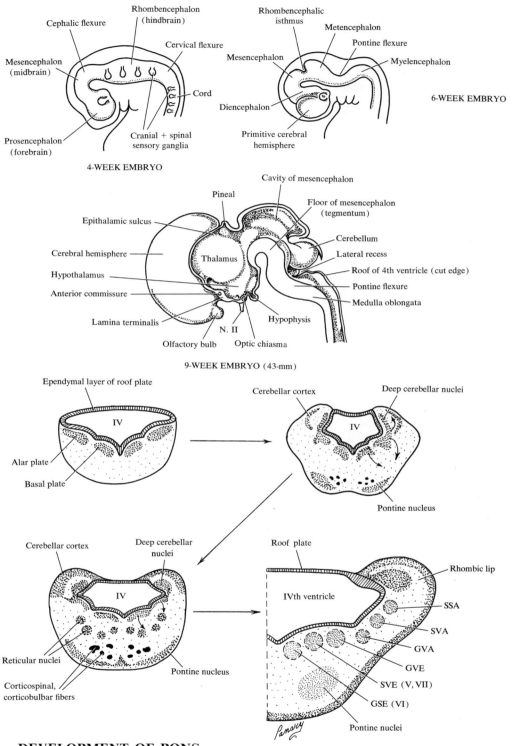

Cephalic flexure
Rhombencephalon (hindbrain)
Cervical flexure
Mesencephalon (midbrain)
Cord
Prosencephalon (forebrain)
Cranial + spinal sensory ganglia

4-WEEK EMBRYO

Rhombencephalic isthmus
Metencephalon
Pontine flexure
Mesencephalon
Myelencephalon
Diencephalon
Primitive cerebral hemisphere

6-WEEK EMBRYO

Cavity of mesencephalon
Pineal
Floor of mesencephalon (tegmentum)
Epithalamic sulcus
Cerebellum
Cerebral hemisphere
Lateral recess
Thalamus
Roof of 4th ventricle (cut edge)
Hypothalamus
Pontine flexure
Anterior commissure
Medulla oblongata
Lamina terminalis
Hypophysis
N. II
Olfactory bulb
Optic chiasma

9-WEEK EMBRYO (43-mm)

Ependymal layer of roof plate
IV
Alar plate
Basal plate

Cerebellar cortex
Deep cerebellar nuclei
IV
Pontine nucleus

Cerebellar cortex
Deep cerebellar nuclei
IV
Reticular nuclei
Corticospinal, corticobulbar fibers
Pontine nucleus

Roof plate
Rhombic lip
IVth ventricle
SSA
SVA
GVA
GVE
SVE (V, VII)
GSE (VI)
Pontine nuclei

Pansky

DEVELOPMENT OF PONS

CAUDAL METENCEPHALON

– 19 –

9. BRAIN: METENCEPHALON—CEREBELLUM

I. The metencephalon (cont.)

A. THE CEREBELLUM: the rhombic lips on the caudal portion of the metencephalon are widely separated, but just below the mesencephalon, they approach each other in the midline. With further deepening of the pontine flexure, the rhombic lips become compressed in a cephalocaudal direction to form the cerebellar plate

1. In the 12-week embryo, the plate shows a small midline portion, the vermis and 2 lateral portions, the hemispheres

2. A transverse fissure soon separates the nodule from the vermis; and the lateral flocculus from the hemispheres to form the flocculonodular lobe (the most primitive part of the cerebellum), which maintains connections with the vestibular septum

3. Many other transverse fissures appear, giving the cerebellum its adult appearance

4. Initially, the cerebellar plate consists of a neuroepithelial, a mantle, and a marginal layer

    a. With development, a number of cells formed by the neuroepithelium migrate through the marginal layer to the cerebellar surface to form the external granular layer

        i. The cells of this layer retain their ability to divide and form a proliferative zone on the cerebellar surface

5. In the 6th developmental month, the external granular layer begins to release various cell types, which migrate inward toward the differentiating Purkinje cells

    a. Thus, the external granular layer gives rise to the granule cells, the basket cells, and the stellate cells, in sequential order

        i. The production and migration of these small neurons continue for about $1\frac{1}{2}$ years after birth

6. Thus, the cerebellar cortex consists of Purkinje cells, Golgi II neurons, and the neurons produced by the external granular layer and reaches its definitive size a considerable time after birth

7. The deep cerebellar nuclei, such as the dentate, are formed by neuroblasts in the mantle layer and reach their definitive position long before birth

8. The greater portion of the original roof plate of the 4th ventricle forms the pia mater on the cerebellar surface

    a. The part located in front of, and behind, the cerebellum becomes specialized and forms the anterior and posterior medullary velum, respectively

DORSAL VIEW

Mesencephalon tectum

Rhombencephalic isthmus

Rhombic lip

Lateral recess

Primitive cerebellum

Sulcus limitans

Medulla

8-WEEK EMBRYO (28-mm)

Roof of 4th ventricle

Body of cerebellum

Posterolateral fissure

DORSAL VIEW

Lateral recess

Pons

Medulla

10-WEEK EMBRYO (56-mm)

Postclival fissure

Primary fissure

Anterior lobe

LATERAL VIEW

Body of cerebellum

Posterior lobe

Posterolateral fissure

Vermis Tonsil

DORSAL VIEW

Paraflocculus

Flocculus

Nodulus

Medulla

112-MM EMBRYO

Primary fissure

Uvula

Roof of IVth ventricle

Lateral recess

Mesencephalon

Culmen Pyramid

Nodule

Tonsil

Paraflocculus

Flocculus

Roof of IVth ventricle

Cerebellar rudiment

IV

(A)

120-MM EMBRYO DORSAL VIEW

(E)

Mesencephalon

Cerebellum

Rudiment of dentate nucleus

Inf. med. velum

Choroid plexus

Aqueduct

(D)

Mesencephalon

Cerebellum

Choroid plexus

Roof

(B)

White matter

Cerebellar cortex

Dentate nucleus

Aqueduct

Pons

(2)

Cerebellum

Inf. medullary velum

Roof

IV

Mesencephalon

(C)

Superficial cortex

Marginal layer

Mantle layer

Ependymal layer

(1)

Purkinje cells

Molecular layer

Granular layer

White matter

Granular cell

Golgi cell

Dentate nucleus

(3)

10. BRAIN: MESENCEPHALON

I. **The mesencephalon** is morphologically the most primitive of the brain vesicles. Its basal and alar plates, separated by the sulcus limitans (as seen in transverse sections), are easily identified

A. BASAL PLATE AND CRUS CEREBRI (BASIS PEDUNCULI)
   1. Each basal plate contains 2 groups of motor nuclei:
      a. A medial somatic efferent group, represented by the oculomotor (III) and trochlear (IV) nerves which innervate the preoptic (eye) muscles
      b. A small general visceral efferent group, represented by the Edinger-Westphal nucleus, which innervates the sphincter pupillary muscle
   2. The marginal layer of each basal plate enlarges and forms the basis pedunculi or crus cerebri
      a. The crura serve as pathways for the nerve fibers descending from the cerebral cortex to the lower centers in the pons and spinal cord
      b. In the adult, the crura contain the corticospinal, the corticobulbar, and the corticopontine tracts

B. ALAR PLATE AND COLLICULI
   1. The alar plates initially appear as 2 longitudinal elevations separated by a shallow midline depression
      a. With development, a transverse groove divides each longitudinal elevation into an anterior (superior) and a posterior (inferior) colliculus
         i. The nuclei of the posterior colliculus serve as synaptic relay stations for the auditory reflexes
         ii. The nuclei of the anterior colliculus serve as correlation and reflex centers for the visual impulses
         iii. The colliculi are formed by waves of neuroblasts produced by the neuroepithelial cells that migrate into the overlying marginal zone and become arranged in stratified layers
   2. Some feel that the cells of the alar plate also give rise to the nucleus ruber (red nucleus) and the substantia nigra (others feel they develop *in situ*)

C. THE CAVITY OF THE MESENCEPHALON is called the aqueduct of Sylvius (cerebral aqueduct) and connects the 3rd and 4th ventricles

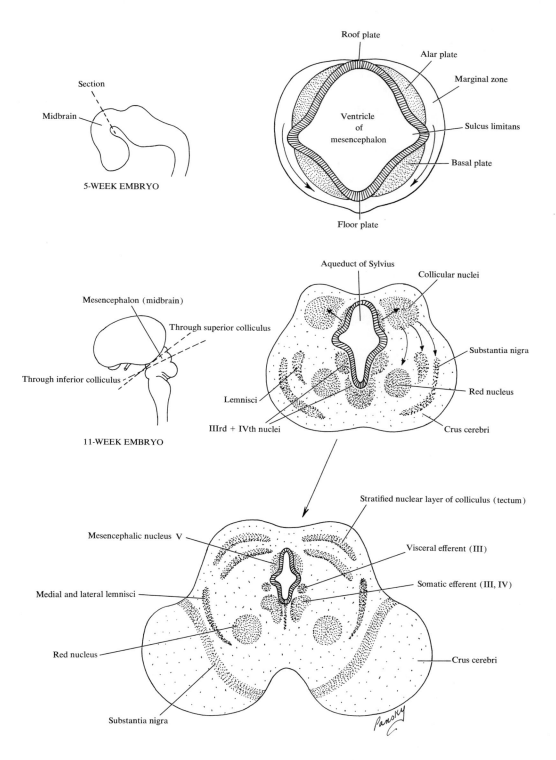

5-WEEK EMBRYO

Roof plate

Alar plate

Marginal zone

Ventricle
of
mesencephalon

Sulcus limitans

Basal plate

Floor plate

Section

Midbrain

Aqueduct of Sylvius

Collicular nuclei

Mesencephalon (midbrain)

Through superior colliculus

Through inferior colliculus

11-WEEK EMBRYO

Substantia nigra

Red nucleus

Lemnisci

IIIrd + IVth nuclei

Crus cerebri

Stratified nuclear layer of colliculus (tectum)

Mesencephalic nucleus V

Visceral efferent (III)

Somatic efferent (III, IV)

Medial and lateral lemnisci

Red nucleus

Crus cerebri

Substantia nigra

DEVELOPMENT OF MESENCEPHALON

11. BRAIN: DIENCEPHALON

I. **The diencephalon** develops from the median portion of the prosencephalon and consists of a roof plate, 2 alar plates, the 3rd ventricle, but has no floor or basal plates. It is bounded posteriorly by a plane passing behind the pineal gland and mamillary bodies and anteriorly by a plane passing just rostral to the optic chiasma and encircling the foramen of Monro (the lamina terminalis is considered to be a part of the telencephalon)

A. ROOF PLATE AND PINEAL BODY (EPIPHYSIS)

1. The roof plate consists of a single layer of ependymal cells covered by vascular mesenchyme
   a. The two later combine to form the choroid plexus of the 3rd ventricle
2. The most caudal part of the roof plate does not participate in the formation of the choroid plexus but develops into the pineal body or epiphysis
   a. The pineal initially appears as an epithelial thickening in the midline but begins to evaginate by the 7th week and eventually forms a solid organ located on the roof of the mesencephalon
3. Occasionally the roof plate forms another evagination near the interventricular foramen, called the paraphysis, which sometimes persists into postnatal life and may give rise to a small cyst
4. The roof plate is also thought to give rise to the epithalamus, a group of nuclei located on each side of the midline close to the pineal gland; however, the epithalamus may arise from the alar plate
   a. The epithalamic region is originally large but regresses to a small area where the habenular nuclei are found
      i. The habenular nuclei form a link in the olfactory conduction path and are connected to each other across the midline by a group of nerve fibers collectively called the habenular commissure (found rostral to the pineal stalk)
5. Another commissure, the posterior commissure, appears caudal to the pineal stalk and also connects 2 nuclear areas on each side of the midline

B. THALAMUS AND HYPOTHALAMUS

1. The alar plates form both the lateral walls and floor of the diencephalon
2. A distinct longitudinal groove is seen on the side facing the lumen, the hypothalamic sulcus, which divides the alar plate into a dorsal and ventral region, the thalamus and hypothalamus, respectively
   a. The sulcus is of a different nature than the sulcus limitans since it does not form a dividing line between motor and sensory areas
3. Following high proliferative activity, the thalami (right and left) gradually bulge into the lumen of the diencephalon and, frequently, the 2 may fuse in the midline to form the massa intermedia or interthalamic connexus (adhesion)
4. The thalamic nuclear areas eventually form 2 distinct nuclear groups:
   a. A dorsal thalamic group: important for the reception and transmission of visual and auditory impulses
   b. A ventral thalamic group: serving as a passage and relay station to higher centers
5. The hypothalamus which forms from the lower portion of the alar plate also differentiates into a number of nuclear groups:
   a. Nuclei that serve as regulation centers of visceral functions, i.e., sleep, digestion, body temperature, emotional behavior, etc.
   b. Two groups, the mamillary bodies, form the rounded elevations on the ventral surface of the hypothalamus on either side of the midline

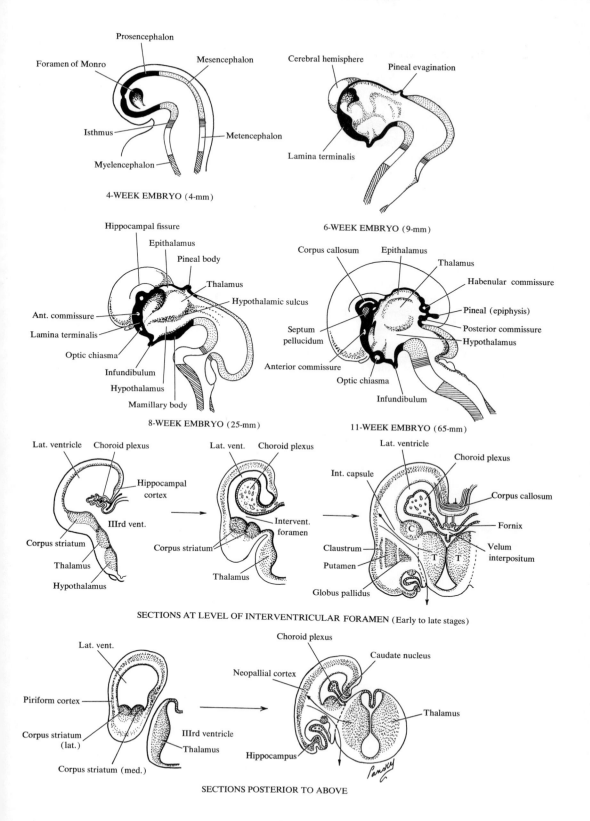

Prosencephalon

Foramen of Monro

Mesencephalon

Isthmus

Metencephalon

Myelencephalon

4-WEEK EMBRYO (4-mm)

Cerebral hemisphere

Pineal evagination

Lamina terminalis

6-WEEK EMBRYO (9-mm)

Hippocampal fissure

Epithalamus

Pineal body

Thalamus

Hypothalamic sulcus

Ant. commissure

Lamina terminalis

Optic chiasma

Infundibulum

Hypothalamus

Mamillary body

8-WEEK EMBRYO (25-mm)

Corpus callosum

Epithalamus

Thalamus

Habenular commissure

Pineal (epiphysis)

Posterior commissure

Septum
pellucidum

Hypothalamus

Anterior commissure

Optic chiasma

Infundibulum

11-WEEK EMBRYO (65-mm)

Lat. ventricle

Choroid plexus

Hippocampal
cortex

IIIrd vent.

Corpus striatum

Thalamus

Hypothalamus

Lat. vent.

Choroid plexus

Intervent.
foramen

Corpus striatum

Thalamus

Lat. ventricle

Choroid plexus

Int. capsule

Corpus callosum

C

Fornix

Claustrum

Putamen

T T

Velum
interpositum

Globus pallidus

SECTIONS AT LEVEL OF INTERVENTRICULAR FORAMEN (Early to late stages)

Lat. vent.

Choroid plexus

Caudate nucleus

Neopallial cortex

Piriform cortex

Corpus striatum
(lat.)

IIIrd ventricle

Thalamus

Corpus striatum (med.)

Thalamus

Hippocampus

SECTIONS POSTERIOR TO ABOVE

12. BRAIN: DIENCEPHALON—THE HYPOPHYSIS

I. The diencephalon (cont.)

A. THE HYPOPHYSIS (PITUITARY GLAND): develops from 2 completely different parts:

   1. An ectodermal outpocketing of the stomodeum just in front of the buccopharyngeal membrane called Rathke's pouch

      a. At 3 weeks of embryonic life, Rathke's pouch is seen as an invagination of the stomodeum and grows dorsally toward the infundibulum

      b. By the end of month 2, it loses its connection with the oral cavity and is in close contact with the infundibulum

      c. Occasionally, a small part of the pouch persists in the pharyngeal wall and is referred to as the pharyngeal hypophysis

      d. With development, the cells of the anterior wall of Rathke's pouch increase rapidly in number and form the anterior lobe of the hypophysis or the adenohypophysis

         i. A small extension of the adenohypophysis, the pars tuberalis, later grows along the stalk of the infundibulum and surrounds it

      e. The posterior wall of the pouch forms the pars intermedia of the hypophysis

      f. In man, the lumen of the pouch is usually obliterated, but may persist as a narrow cleft

      g. A not uncommon remnant of the pouch is the craniopharyngioma or Rathke's pouch tumor which is found intracranially and whose symptoms resemble those of anterior lobe tumors. The onset of symptoms is usually before the age of 15 years

   2. A downward extension of the diencephalon forms the infundibulum

      a. The infundibulum gives rise to the pituitary stalk and the pars nervosa or posterior lobe of the hypophysis (neurohypophysis)

         i. It is composed of neuroglia cells which later differentiate to form pituicytes

        ii. It also contains many nerve fibers coming from the hypothalamic area of the diencephalon

DEVELOPMENT OF
PITUITARY GLAND

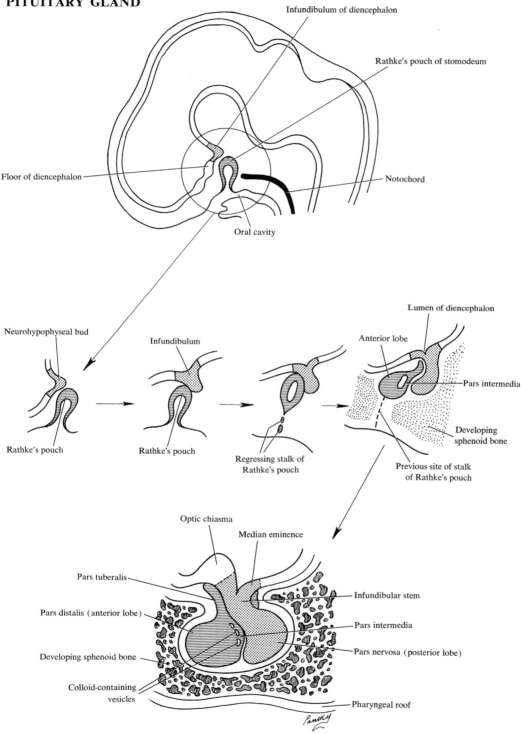

Infundibulum of diencephalon

Rathke's pouch of stomodeum

Floor of diencephalon

Notochord

Oral cavity

Neurohypophyseal bud

Rathke's pouch

Infundibulum

Rathke's pouch

Regressing stalk of
Rathke's pouch

Lumen of diencephalon

Anterior lobe

Pars intermedia

Developing
sphenoid bone

Previous site of stalk
of Rathke's pouch

Optic chiasma

Median eminence

Pars tuberalis

Pars distalis (anterior lobe)

Developing sphenoid bone

Colloid-containing
vesicles

Infundibular stem

Pars intermedia

Pars nervosa (posterior lobe)

Pharyngeal roof

13. BRAIN: TELENCEPHALON—CEREBRAL HEMISPHERES

I. **The telencephalon,** the most rostral portion of the brain vesicle, consists of 2 lateral outpocketings, the cerebral hemispheres, and a median portion, the lamina terminalis. The cavities of the hemispheres, the lateral ventricles communicate with the lumen of the diencephalon via the interventricular foramina of Monro

  A. THE CEREBRAL HEMISPHERES arise at the beginning of week 5 as bilateral evaginations of the lateral wall of the prosencephalon

    1. Cell proliferation occurs in the neuroepithelial layer and large numbers of neuroblasts are produced for the mantle layer

    2. By the middle of month 2, the mantle layer in the basal portions of the hemispheres (the part which initially formed the forward extension of the thalamus) increases in size

      a. This rapidly growing area bulges into the lumen of the lateral ventricle as well as into the floor of the foramen of Monro, has a striated appearance, and is called the corpus striatum

      b. The rest of the hemisphere wall temporarily stays thin and is called the pallium, the primordium of the cerebral cortex

    3. In the region where the wall of the hemisphere is attached to the roof of the diencephalon, it fails to develop neuroblasts and stays very thin

      a. Here the hemisphere wall consists of a single layer of ependymal cells covered by vascular mesenchyme, and together they form the choroid plexus

        i. The position of choroid plexuses on the medial surface of the hemisphere is a result of disproportionate growth of parts of the hemisphere

        ii. The plexus should have formed the roof of the hemisphere, but it protrudes into the lateral ventricle along a line, the choroidal fissure

      b. Just above the choroidal fissure, the wall of the pallium is thickened to form the hippocampus, which has an olfactory function and bulges out into the lateral ventricle

    4. As the hemispheres expand, they gradually cover the lateral aspect of the diencephalon, mesencephalon, and cephalic portion of the metencephalon

    5. The corpus striatum, a part of the hemisphere wall, also expands posteriorly and forms a longitudinal ridge in the floor of the lateral ventricle

      a. During this growth, the corpus striatum is divided into 2 portions (as a result of afferent and efferent axons passing to and from the cortex of the hemisphere and breaking through the nuclear mass of the corpus striatum)

        i. A dorsomedial portion which forms the caudate nucleus

        ii. A ventromedial portion which forms the lentiform nucleus

        iii. The fiber bundles between them known as the internal capsule

        iv. The lentiform nucleus is later divided into a lateral part, the putamen, and a medial, lightly staining part, the globus pallidus

    6. As a result of posterior expansion of the hemispheres, its medial surface approaches the lateral surface of the diencephalon, and they fuse

      a. With fusion, the caudate nucleus and the thalamus come into close contact, while the nerve fibers descending from the cortex of the hemisphere pass through the plane of fusion to enter the peduncles of the mesencephalon

    7. Growth of the hemispheres in an anterior, dorsal, and inferior direction results in the formation of the frontal, temporal, and occipital lobes

      a. A lag in growth in the region overlying the corpus striatum, between the frontal and temporal lobes, results in its depression and is called the insula. (This region is later overgrown by the adjacent lobes and at birth it is fully covered.)

    8. During late fetal life, the surface of the hemispheres grows so rapidly that many convolutions or gyri, separated by sulci and fissures, are seen on its surface

Prosencephalon
Mesenceph.
Rhombenceph.
Cord
4 WEEKS

Lamina terminalis
Foramen of Monro
Lateral ventricle
III
Optic cup
Aqueduct
IV
Central canal
6 WEEKS

Epiphysis
Cerebral hemisphere
Thalamus
Interventricular foramen (of Monro)
Lamina terminalis
Olfactory bulb
Infundibulum
Hypothalamus
8 WEEKS

Lateral ventricle
IIIrd ventricle
Aqueduct
IV
(no label)

Parietal lobe
Corpus callosum
Hippocampal fissure
Thalamus
Occipital lobe
Epiphysis
Anterior commissure
Frontal lobe
Optic chiasma
11 WEEKS

Interven. foramen
Ant. horn
Inferior horn
Body
IV
Post. horn

Neopallium cortex
Hippocampus
Lat. vent.
Choroid plexus
III
Palaeopallium cortex
Corpus striatum
Thalamus
Hypothalamus
TRANSVERSE SECTION: 8 WEEKS

Neopallium cortex
Lat. vent.
III
Corpus striatum
9 WEEKS

Choroid plexus
Caudate nucleus
Thalamus
III
Hippocampus
11 WEEKS

Neopallium cortex
Lat. vent.
Choroid plexus
Fornix
Intervent. foramen
III
Lat.
Med.
Corpus striatum
Thalamus
9 WEEKS

Corpus callosum
Neopallium cortex
Lateral ventricle
Choroid plexus
Caudate nucleus
Int. capsule
Globus pallidus
Putamen
Lentiform nucleus
Claustrum
Tail of caudate
Thalamus
III
Inf. horn
11 WEEKS

Smooth cerebral hemisphere
13 WEEKS

Lateral fissure
Insula
26 WEEKS

Central sulcus
Insula
Lat. fissure
35 WEEKS

Central sulcus
Lateral fissure
NEWBORN

14. BRAIN: TELENCEPHALON—CEREBRAL CORTEX AND COMMISSURES

I. The telencephalon (cont.)

A. CEREBRAL CORTEX DEVELOPMENT is from the pallium and is divided into 2 regions:

  1. Paleo- or archipallium: an area just lateral to the corpus striatum

    a. Appears in the 7th week of development and is formed by a mixture of cells migrating from the striated mantle layer into the marginal zone

    b. The cells here established form a thin nuclear layer near the surface that serves as a relay station for olfactory impulses

  2. Neopallium occupies the rest of the surface of the hippocampus and paleopallium

    a. The neuroepithelial cells of the neopallium begin to release a series of waves of neuroblasts shortly after the paleopallium appears

    b. When new neuroblasts arrive, they migrate through the earlier formed layer of cells to reach subpial position. Thus, the early formed neuroblasts get a deeper position in the cortex than those formed later

  3. At birth, the cortex appears stratified due to layers of neuroblasts and specific cell differentiation. In addition, different cortical areas acquire specific cell types; i.e., the motor cortex has many pyramidal cells, the sensory cortex is characterized by granular cells

B. COMMISSURES are fiber bundles, in the adult, which cross the midline and connect the right and left halves of the hemispheres

  1. The most important make use of the lamina terminalis, which extends from the roof plate of the diencephalon to the optic chiasma in the midportion of the telencephalon

  2. The 1st commissure to appear is the anterior commissure which is seen by the 3rd month of development and consists of fibers connecting the olfactory bulb and interconnects the left and right temporal lobes

  3. The 2nd one to appear is the hippocampal or fornix commissure

    a. Its fibers arise in the hippocampus, converge on the lamina terminalis near the roof plate of the diencephalon, and continue to form an arching system, just outside of the choroid fissure, to the mamillary bodies and hypothalamus

    b. It regresses greatly as the corpus callosum grows

  4. The most important commissure is the corpus callosum, which appears by week 10 of development and connects nonolfactory areas of the two hemispheres

    a. Initially it forms a small bundle in the lamina terminalis just rostral to the hippocampal commissure, but with continuous expansion of neopallium, it extends anteriorly and then posteriorly, arching over the thin roof of the diencephalon

    b. Growth of the corpus callosum in an anterior direction pulls the area of the lamina terminalis away from the fornix commissure and thins out the lamina terminalis locally, forming the septum pellucidum (may contain a small cavity which has no relation to the brain vesicles)

      i. It has been suggested that the septum represents the apposed walls of the 2 hemispheres anterior to the lamina terminalis

  5. Several other commissures appear in the lamina terminalis:

    a. Two appear just below and rostral to the stalk of the pineal gland, the posterior commissure, and the habenular commissure

    b. The third, the optic chiasma, seen in the rostral wall of the diencephalon contains fibers from the medial halves of the retinae which cross on their way to the lateral geniculate bodies and anterior colliculi

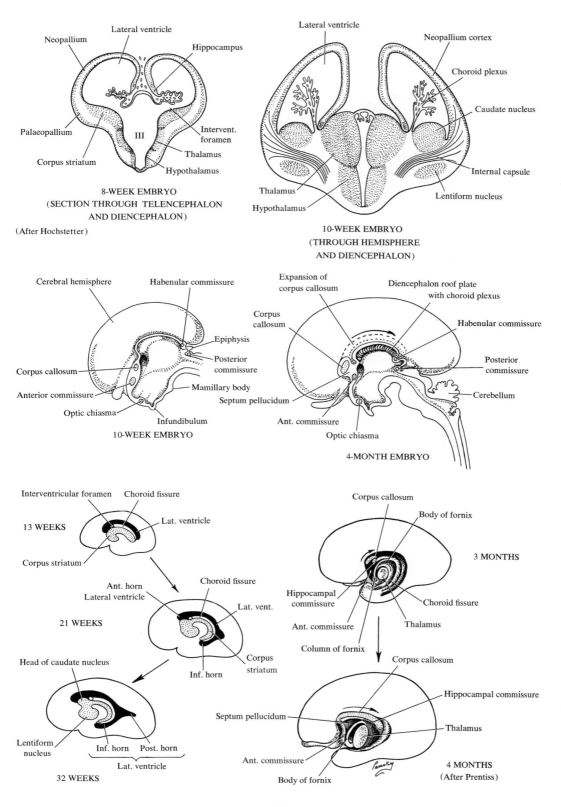

Neopallium

Lateral ventricle

Hippocampus

Palaeopallium

Corpus striatum

III

Intervent. foramen

Thalamus

Hypothalamus

8-WEEK EMBRYO
(SECTION THROUGH TELENCEPHALON
AND DIENCEPHALON)

(After Hochstetter)

Lateral ventricle

Neopallium cortex

Choroid plexus

Caudate nucleus

Internal capsule

Lentiform nucleus

Thalamus

Hypothalamus

10-WEEK EMBRYO
(THROUGH HEMISPHERE
AND DIENCEPHALON)

Cerebral hemisphere

Habenular commissure

Epiphysis

Posterior commissure

Corpus callosum

Anterior commissure

Mamillary body

Optic chiasma

Infundibulum

10-WEEK EMBRYO

Expansion of corpus callosum

Diencephalon roof plate with choroid plexus

Corpus callosum

Habenular commissure

Posterior commissure

Cerebellum

Septum pellucidum

Ant. commissure

Optic chiasma

4-MONTH EMBRYO

Interventricular foramen

Choroid fissure

13 WEEKS

Lat. ventricle

Corpus striatum

Ant. horn
Lateral ventricle

Choroid fissure

21 WEEKS

Lat. vent.

Corpus striatum

Inf. horn

Head of caudate nucleus

Lentiform nucleus

Inf. horn   Post. horn

Lat. ventricle

32 WEEKS

Corpus callosum

Body of fornix

3 MONTHS

Hippocampal commissure

Ant. commissure

Column of fornix

Choroid fissure

Thalamus

Corpus callosum

Hippocampal commissure

Septum pellucidum

Thalamus

Ant. commissure

Body of fornix

4 MONTHS
(After Prentiss)

15. BRAIN: CONGENITAL MALFORMATIONS

I. **Primary cause of many malformations:** ossification defect in the skull bones
A. THE MOST FREQUENTLY AFFECTED BONE is the squamous portion of the occipital bone, which may be partially or totally missing, and the resultant opening is often confluent with the foramen magnum
   1. If the opening of occipital bone is small, only the meninges bulge through it, creating a meningocele
   2. If the defect is large, part of the brain also bulges out; called meningoencephalocele
   3. If the defect is very large, part of the brain and part of the ventricle may penetrate through the opening into the meningeal sac, creating a meningohydroencephalocele

II. **Anencephalus:** characterized by failure of cephalic part of neural tube to close
A. AT BIRTH, brain consists of a mass of degenerated tissue exposed to the surface
   1. The defect is almost always continuous with an open cord in the cervical region
   2. The skull vault is gone, giving the head a characteristic appearance: the eyes bulge forward; neck absent; surfaces of face and chest form a continuous plane
   3. The last 2 months of pregnancy are characterized by hydramnios since the fetus lacks the control mechanism for swallowing
   4. The abnormality is recognized by x-ray since the skull vault is missing
B. THIS IS A COMMON ABNORMALITY (1:1,000): 4 times more frequently in females than males; 4 times more frequently in whites than blacks

III. **Hydrocephalus:** characterized by an abnormal accumulation of CSF in ventricular system (may be an external hydrocephalus with fluid between brain and dura)
A. HYDROCEPHALUS is thought to be due to an obstruction of the aqueduct which prevents CSF of the 3rd and lateral ventricles from passing into the 4th ventricle and from there into the subarachnoid space
   1. Malformation is often accompanied by a widening of skull sutures and the bones gradually thin out

IV. **Other malformations:** those above are serious and often incompatible with life. In addition, other defects of the CNS occur without external manifestations:
A. THE CORPUS CALLOSUM may be partially or completely absent without much functional disturbance
B. COMPLETE OR PARTIAL CEREBELLAR ABSENCE may show only a slight coordination disturbance
C. SEVERE IMBECILITY OR IDIOCY may show little morphologic brain abnormalities

V. **Chromosomal and environmental phenomena**
A. CONGENITAL MALFORMATION may be caused by chromosomal abnormalities, particularly mental retardation
   1. Mongolism is associated with 47 chromosomes, #21 being represented 3× instead of twice; condition called trisomy
   2. Klinefelter's syndrome also shows 47 chromosomes with an extra X chromosome and accounts for about 1% of mentally defective children
   3. Many sex chromosome abnormalities are associated with mental retardation
   4. Congenital metabolic defects, i.e., phenylketonurea, cause or accompany mental retardation
B. ENVIRONMENTAL TERATOGENS cause CNS malformations in offspring of experimental animals
C. CNS ABNORMALITIES can occur from fetal infection by Toxoplasma organism
D. RADIATION during early developmental stages may lead to microcephalus
E. VIRAL DISEASES and their effects are uncertain and variable

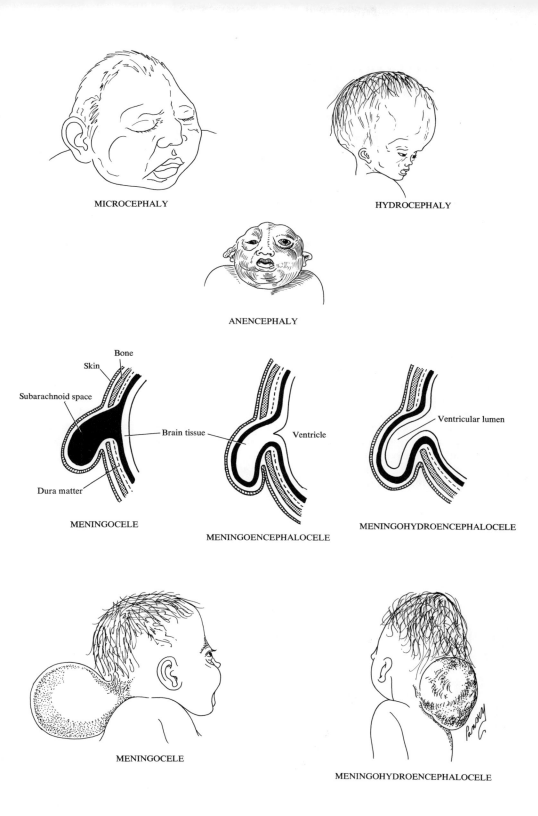

MICROCEPHALY

HYDROCEPHALY

ANENCEPHALY

Bone

Skin

Subarachnoid space

Brain tissue

Dura matter

MENINGOCELE

Ventricle

MENINGOENCEPHALOCELE

Ventricular lumen

MENINGOHYDROENCEPHALOCELE

MENINGOCELE

MENINGOHYDROENCEPHALOCELE

16. EMBRYOLOGY OF THE AUTONOMIC NERVOUS SYSTEM: GENERAL DEVELOPMENT

I. **Introduction:** functionally the system is divided into 2 parts: a sympathetic part, found in the thoracolumbar region, and a parasympathetic part, found in the cephalic (cranial) and sacral regions. Both divisions are composed of characteristic two-neuron efferent chains

II. **The sympathetic nervous system**
A. SYMPATHETIC NEUROBLASTS, originating in the neural crest of the thoracic region, in the 5th week of development migrate on each side of the spinal cord to the region just behind the dorsal aorta, where they form a bilateral chain of segmentally arranged sympathetic ganglia interconnected by longitudinal nerve fibers—the sympathetic chains found on each side of the vertebral column
  1. From their thoracic position, the neuroblasts migrate toward the cervical and lumbosacral regions extending the sympathetic chains to full length
  2. Though the ganglia are initially segmentally arranged, this arrangement is obscured, particularly in the cervical region, by fusion of the ganglia
  3. Some neuroblasts migrate in front of the aorta to form preaortic ganglia, i.e., celiac and mesenteric ganglia found at the roots of the main aortic branches
  4. Other neuroblasts migrate to the heart, lungs, and GI tract to give rise to sympathetic organ plexuses, i.e., submucosal plexus in the gut wall
B. WITH ESTABLISHMENT OF THE SYMPATHETIC CHAINS, nerve fibers originating in the visceroefferent column of the thoracolumbar segments of the spinal cord penetrate the ganglia of the chains and form synapses around the developing neuroblasts
  1. Some fibers either extend to higher or lower levels in the sympathetic chains or to preaortic or collateral ganglia before synapsing; called preganglionic fibers
    a. Have myelin sheaths
    b. Stimulate the sympathetic ganglion cells into action
    c. Passing from the spinal nerves to the ganglia, they form the so-called white rami communicantes (white communicating rami)
    d. Since the visceroefferent column extends only from T1 to L2 segments of the cord, the white rami are found only between these levels
  2. The axons of the sympathetic ganglion cells are called postganglionic fibers
    a. They have no myelin
    b. Pass either to other levels of the sympathetic chain or extend to the heart, lungs, and GI tract
    c. Other fibers called gray communicating rami pass from the chain to the spinal nerves and from there to peripheral blood vessels, hair, and sweat glands

III. **The parasympathetic nervous system**
A. THE ORIGIN of the parasympathetic ganglia found along the oculomotor (III), facial (VII), glossopharyngeal (IX), and vagus (X) nerves is controversial
  1. Some believe the cells of these ganglia migrate out of the CNS along the preganglionic fibers of the nerves
  2. Some feel the cells arise from neuroblasts originating in the sensory ganglia of the 5th, 7th, and 9th nerves
B. THE POSTGANGLIONIC FIBERS of the parasympathetic ganglia pass to the branchial arches and to the cardiac, pulmonary, and intestinal plexuses
C. THE ACTION OF THIS SYSTEM is said to be antagonistic to those of the sympathetic system

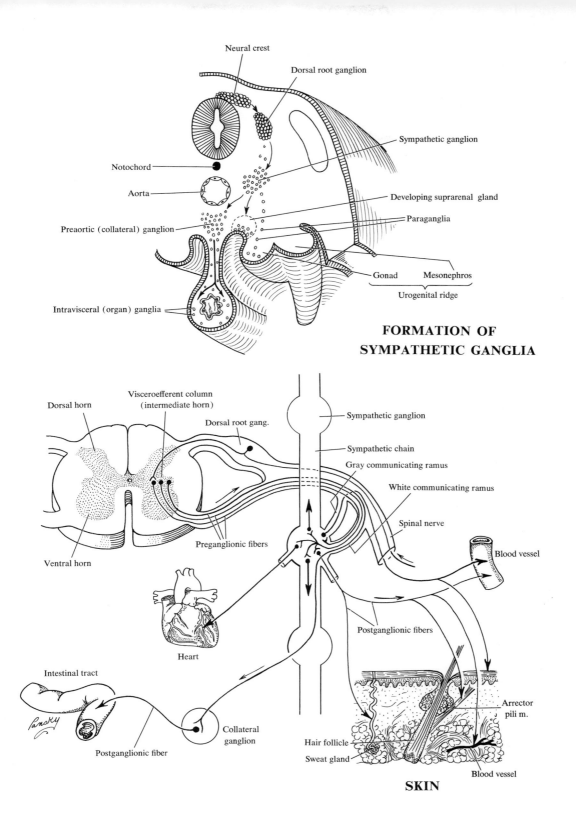

Neural crest

Dorsal root ganglion

Sympathetic ganglion

Notochord

Aorta

Developing suprarenal gland

Paraganglia

Preaortic (collateral) ganglion

Gonad   Mesonephros

Urogenital ridge

Intravisceral (organ) ganglia

FORMATION OF
SYMPATHETIC GANGLIA

Dorsal horn

Visceroefferent column
(intermediate horn)

Dorsal root gang.

Sympathetic ganglion

Sympathetic chain

Gray communicating ramus

White communicating ramus

Spinal nerve

Blood vessel

Preganglionic fibers

Ventral horn

Postganglionic fibers

Heart

Intestinal tract

Arrector
pili m.

Collateral
ganglion

Hair follicle

Postganglionic fiber

Sweat gland

Blood vessel

SKIN

17. AUTONOMIC NERVOUS SYSTEM: SUPRARENAL GLAND AND CAROTID BODY

I. **The suprarenal gland**
A. THE GLAND develops from 2 components:
   1. A mesodermal portion which forms the cortex
   2. An ectodermal portion which forms the medulla
B. THE CORTEX
   1. During the 5th week of development, mesothelial cells located between the root of the mesentery and developing gonad begin to proliferate and penetrate the underlying mesenchyme
   2. They differentiate into large acidophilic organs which form the fetal or primitive suprarenal cortex
   3. Shortly after, a 2nd wave of cells from the mesothelium penetrates the mesenchyme and surrounds the original acidophilic cell mass
      a. These cells are smaller than the 1st wave
      b. These smaller cells form the definitive cortex of the gland
   4. After birth, the fetal cortex regresses rapidly except for its outer layer which differentiates into the reticular zone
   5. The adult structure of the cortex is not achieved until near puberty
C. THE MEDULLA
   1. While the fetal cortex is forming, cells originating in the sympathetic system invade its medial part where they become arranged in clusters and cords
      a. These cells give rise to the adrenal medulla, but do not form nerve processes but rather stain yellow-brown with chrome salts and are thus called chromaffin cells
         i. The staining is probably due to epinephrine and norepinephrine in the cells
   2. During embryonic life, the chromaffin cells are widely scattered throughout the embryo, but in the adult, the only persisting group is found in the adrenal medulla.

II. **The carotid bodies**
A. THE CAROTID BODIES are formed by a mesodermal condensation around the origin of the internal carotid arteries and are supplied by branches of the glossopharyngeal (IX) nerve
   1. They are invaded by cells from nearby ganglia and then develop into chemoreceptor organs
B. THE CAROTID BODIES serve as reflex systems in the regulation of blood pressure.

SUPRARENAL GLAND

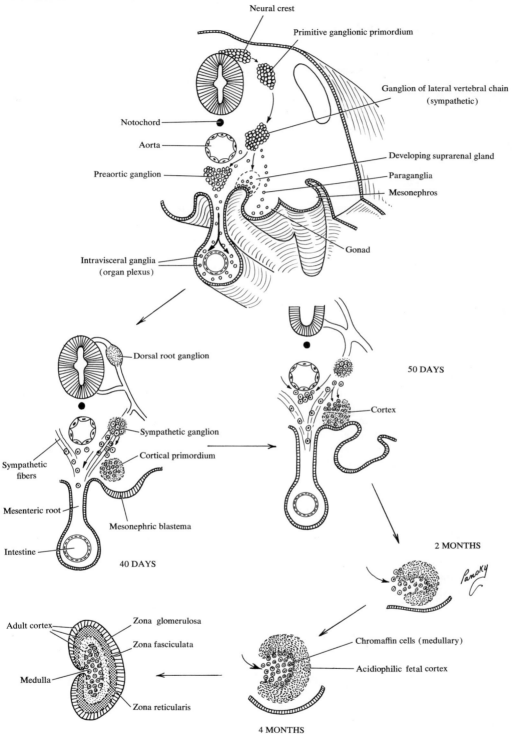

Neural crest

Primitive ganglionic primordium

Ganglion of lateral vertebral chain (sympathetic)

Notochord

Aorta

Developing suprarenal gland

Preaortic ganglion

Paraganglia

Mesonephros

Intravisceral ganglia (organ plexus)

Gonad

Dorsal root ganglion

50 DAYS

Cortex

Sympathetic ganglion

Cortical primordium

Sympathetic fibers

Mesenteric root

Mesonephric blastema

Intestine

40 DAYS

2 MONTHS

Chromaffin cells (medullary)

Acidiophilic fetal cortex

Adult cortex

Zona glomerulosa

Zona fasciculata

Medulla

Zona reticularis

4 MONTHS

18. THE EYE: OPTIC CUP, LENS VESICLE, RETINA, IRIS, AND CILIARY BODY

I. The optic cup and lens vesicle

A. THE EARLIEST SIGN of eye development is seen in the 22-day embryo as a pair of shallow grooves on each side of the invaginated forebrain

  1. With closure of the neural tube, the grooves form outpocketings of the forebrain, the optic vesicles, which are in contact with surface ectoderm

  2. In days, the vesicles cause chemical changes in surface ectodermal cells needed for lens formation, the optic vesicles invaginate and form the double-walled optic cup

    a. The inner and outer walls of the cup are initially separated by the intraretinal space, but with development, the lumen disappears and the walls oppose each other

  3. The invagination is not only centrally restricted in the cup but also involves its ventral rim where the choroid fissure is formed which extends along the undersurface of the optic stalk where it tapers off

    a. Fissure formation allows the hyaloid artery to reach the inner eye chamber

    b. During the 7th week, the lips of the choroid fissure fuse, and the mouth of the optic cup becomes a rounded opening, the future pupil

  4. In the meantime, the cells of the surface ectoderm, initially in contact with the optic vesicle, elongate and form the lens placode

    a. The placode subsequently invaginates and develops into the lens vesicle

    b. During week 5, the lens vesicle loses surface contact and is then located in the mouth of the optic cup

II. Retina, iris, and ciliary body

A. DEVELOPMENT OF THE OUTER LAYER of the optic cup is characterized by the appearance of small pigment granules during week 5 to form the pigment layer of the retina

B. THE POSTERIOR $\frac{4}{5}$ OF THE INNER LAYER of the optic cup (pars optica retinae):

  1. This part thickens and undergoes a series of changes similar to those seen in the wall of the brain vesicle.

  2. Bordering the intraretinal space, the ependymal layer differentiates here into the light-receptive elements, the rods and cones

  3. Adjacent to the photoreceptive layer, the mantle layer, as in the brain, gives rise to neurons and supporting cells

    a. In the adult, the outer nuclear layer, the inner nuclear layer, and the ganglion cell layer are distinguished

  4. On the surface of the mantle layer, the marginal zone contains the axons of the nerve cells of the deeper layers; and the nerve fibers in this marginal layer converge toward the optic stalk, which gradually forms the optic nerve

C. THE ANTERIOR $\frac{1}{5}$ of the inner layer of the optic cup (pars caeca retinae):

  1. Changes very little and remains one-cell layer thick but later divides into the pars iridica retinae, forming the inner layer of the iris, and the pars ciliaris retinae, which participates in the formation of the ciliary body

    a. The pars ciliaris retinae is recognized by its marked folding

      i. Externally it is covered by a layer of mesenchyme which forms the ciliary muscle

      ii. Internally, it is connected to the lens by a network of elastic fibers which form the suspensory ligament or zonula

      iii. Contraction of the ciliary muscle changes the tension in the ligament and controls the curvature of the lens

D. IN THE HUMAN, the sphincter and dilator pupillae muscles develop from ectoderm underlying the optic cup in the area between optic cup and overlying surface epithelium

E. THE IRIS in the adult is formed by the pigment-containing internal and external layers of the optic cup and by a layer of vascularized connective tissue which also contains the pupillary muscles

Surface ectoderm   Optic groove

Forebrain wall

21 DAYS

Ectoderm   Forebrain

Optic vesicle

Optic stalk

Lens placode

Invaginating
lens placode

27 DAYS

Invaginating optic
vesicle   Forebrain

Choroid fissure

29 DAYS

Intraretinal space

Lens vesicle

Pigmented
Nervous } Retinal layers

Wall of IIIrd
ventricle

Stalk

33 DAYS

Surface ectoderm

Choroid fissure

Hyaloid vessels

Optic cup

Optic stalk

Choroid fissure   **6 WEEKS**

Outer
Inner } Layer of
optic
stalk

Hyaloid a.

Choroid fissure

Intraretinal space

Brain wall

Lens vesicle

Lumen of optic stalk

Hyaloid a.

Surface ectoderm   Mesenchyme

Pigmented
Nervous } Retina

Sinus venosus
sclerae

Hyaloid a.

Lens

7 WEEKS

Sclera

Nervous layer
Pigment layer } Retina

Ciliary processes

Sphincter pupillae

Ciliary muscles

Ciliary processes

Dilator pupillae

**IRIS +
CILIARY BODY**

RETINA
(25 WEEKS)

Choroid

Conjunctival
sac

Eyelid

Cornea

Pupil

Ant. chamber

Iris

Cilary body

Vitreous body

Lens

Pigment
Nervous } Retinal layers

Hyaloid artery

Dura

Optic n.

5 MONTHS

Rod and cone layer

Outer nuclear layer

Inner nuclear layer

Ganglion cell layer

Optic nerve fibers

19. THE EYE: LENS, CHOROID, SCLERA, CORNEA, AND OPTIC NERVE

I. The lens
A. SHORTLY AFTER FORMATION of the lens vesicle, the cells of the posterior wall elongate anteriorly and form long fibers which gradually fill the vesicle lumen
B. THE PRIMARY LENS FIBERS reach the epithelium of the anterior wall of the vesicle by the end of the 7th week and form the nucleus of the lens
   1. Lens growth is not finished at this stage, but new, secondary lens fibers are continuously added to the central core
   2. The new fibers arise from cells in the equatorial zone

II. The choroid, sclera, and cornea
A. BY THE END OF THE 5TH WEEK and when the optic cup and lens vesicle have formed, the eye primordium is completely surrounded by loose mesenchyme
   1. The mesenchyme differentiates into an inner layer (comparable to the pia mater of the brain) and an outer layer (comparable to the dura mater)
      a. The inner layer forms a highly vascularized pigmented layer, the choroid
      b. The outer layer develops into the sclera and is continuous with the dura mater around the optic nerve
   2. Mesenchyme layers over the anterior aspect of the eye differentiate differently
      a. The cells arrange themselves so that a space, the anterior chamber, splits the mesenchyme into a thin inner layer just in front of the lens and iris (the iridopupillary membrane) and a thick outer layer which is continuous with the sclera
         i. The anterior chamber itself is lined by flattened mesenchymal cells which form the posterior lining of the cornea as well as the anterior covering of the iridopupillary membrane. The membrane in front of the lens normally disappears completely
B. THE CORNEA, from outside to inside, is formed by:
   1. An epithelial layer derived from surface ectoderm
   2. A layer of dense connective tissue, the substantia propria or stroma
   3. An epithelial layer that borders the anterior chamber
C. THE MESENCHYME, which surrounds the eye primordium, also invades the inside of the optic cup via the choroid fissure:
   1. It participates in formation of the hyaloid vessels (during intrauterine life supply the lens and form the vascular layer found on the retinal inner surface)
   2. Forms a delicate network of fibers between the lens and the retina, the interstitial spaces of which later fill with a transparent gelatinous substance, the vitreous body

III. The optic nerve
A. THE OPTIC CUP is initially connected to the brain by the optic stalk
   1. The choroid fissure is a groove on the stalk's ventral surface
   2. The nerve fibers of the retina returning to the brain are found among the cells of the inner wall of the stalk
B. DURING WEEK 7 the choroid fissure closes and a narrow tunnel is formed inside the optic stalk
C. WITH AN INCREASING NUMBER of nerve fibers growing toward the brain, the inner stalk walls increase in size and fuse with the outer walls
   1. The cells of the inner layer provide a network of neuroglia cells that support the optic nerve fibers
D. THUS, THE OPTIC STALK is transformed into the optic nerve and in its center is the hyaloid artery, later called the central artery of the retina

22 DAYS

Lens placode
Forebrain
Optic vesicle

4 WEEKS

Invaginating lens placode
Invaginating optic vesicle

6 WEEKS

Outer layer
Inner layer } of optic stalk
Hyaloid artery
Choroid fissure

33 DAYS

Intraretinal space
Internal } Retinal layers } Optic cup
External
Optic cup
Optic stalk
Wall of IIIrd ventricle
Lens vesicle
Surface ectoderm
Choroid fissure
Hyaloid vessels

6 WEEKS

Lens vesicle
Hyaloid artery
Lumen of optic stalk

7 WEEKS

Surface ectoderm
Mesenchyme
Hyaloid a.
Lens
Eyelid
Pars optica retinae
Pigmented layer of retina

15-WEEK EMBRYO

Conjunctival sac
Iridopupillary membrane
Sclera
Choroid
Pigment layer } Retina
Nervous layer
Vitreous body
Hyaloid a.
Optic nerve
Ant. chamber
Lens
Cornea
Eyelid
Iris
Ciliary body
Suspensory ligament

6 WEEKS

Lens
Lumen of stalk
Optic stalk
Outer layer
Inner layer
Optic stalk
Choroid fissure

7 WEEKS

Outer
Nerve fibers
Inner
Hyaloid artery

9 WEEKS

Optic nerve fibers
Central artery and vein of retina

OPTIC STALK DEVELOPMENT

20. THE EYE: CONGENITAL MALFORMATIONS

I. Coloboma iridis
A. NORMALLY, THE CHOROID FISSURE CLOSES during the 7th week of development, but when it fails to close, a cleft persists which is usually located in the iris only and is called coloboma iridis
  1. The cleft may extend into the ciliary body, the retina, the choroid, and the optic nerve
B. THIS MALFORMATION is frequently seen in combination with other abnormalities

II. Persistent iridopupillary membrane
A. THE MEMBRANE usually disappears during intrauterine life, but if reabsorption is incomplete, a network of connective tissue is seen suspended in front of the pupil
  1. This usually does not cause visual disturbance

III. Microphthalmia
A. THE OVERALL SIZE of the eye is too small, and the eyeball may be reduced to $\frac{2}{3}$ of its normal volume
B. THIS IS USUALLY ASSOCIATED with other ocular abnormalities
C. FREQUENTLY RESULTS from intrauterine infections such as cytomegalovirus or toxoplasmosis

IV. Anophthalmia
A. SOMETIMES THE EYE is grossly absent, and there is no trace of the eyeball except by histologic means
B. THIS CONDITION is usually accompanied by other serious craniocerebral abnormalities

V. Cyclopia
A. BOTH ORBITS are joined, and there exists only one median eye
B. THIS IS A RARE MALFORMATION and is often accompanied by a proboscis and other cerebral abnormalities

VI. Buphthalmos or congenital glaucoma
A. THE EYE IS GREATLY ENLARGED as a result of insufficient drainage of its aqueous humor
B. THERE IS HIGH INTRAOCULAR PRESSURE caused by an absence of the sinus venosus sclerae
C. THE CONDITION is frequently of genetic origin, but it is also seen after maternal rubella infection

VII. Congenital cataract
A. THE LENS has become opaque during intrauterine life
B. THE CONDITION is usually genetically determined, but it may be caused by environmental factors
  1. Seen in children of mothers who suffered from German measles between the 4th and 7th week of pregnancy
  2. If the mother is infected after the 7th week of pregnancy, the lens escapes damage, but the child may be deaf as a result of imperfect differentiation of the cochlea

VIII. Retinocele: herniation of the retina into the sclera, caused by failure of the choroid fissure to close. If severe, can cause eye protrusion

IX. Other conditions
A. A VIRUS may produce abnormalities, but the eye is only sensitive during certain developmental states
B. OTHER TERATOGENIC AGENTS related to eye abnormalities are hypoxia, vitamin deficiency, hypervitaminosis A, thyroid deficiency, and x-irradiation

COLOBOMA IRIDIS

PARTIALLY PERSISTENT
IRIDOPUPILLARY MEMBRANE

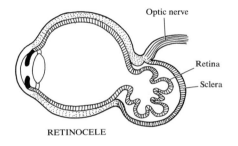

Optic nerve

Retina

Sclera

RETINOCELE

Central orbit

CYCLOPS

Central orbit

Oral opening

CYCLOPS HYPOGNATHUS
(nose absent; mouth rudimentary)

21. THE INTERNAL EAR

I. **Introduction:** the adult ear forms one unit serving both hearing and equilibrium. In the embryo, the ear develops from 3 different parts: the external ear, the sound-collecting organ, develops from the dorsal portion of the 1st pharyngeal cleft and 6 mesenchymal swellings; the middle ear, a sound conductor between the external and internal ear, arises from the 1st pharyngeal pouch; and the internal ear (converts sound waves into nerve impulses, registers equilibrium changes) formed from the ectodermal otic vesicle

II. **The internal ear**
A. THE OTIC PLACODES are the first indication of ear development seen in embryos of about 22 days as thickening of the surface ectoderm on each side of the rhombencephalon which invaginate rapidly and form the otic (or auditory) vesicles or otocysts
B. DURING LATER DEVELOPMENT, each vesicle divides into:
   1. A ventral component which gives rise to the saccule and the cochlear duct
   2. A dorsal component forms the utricle, semicircular canals, endolymphatic duct
   3. The structures are collectively known as the membranous labyrinth
C. THE MEMBRANOUS LABYRINTH is embedded in mesenchyme, but, with time, the surrounding mesenchyme converts into a cartilaginous shell, which in turn is ossified to form the bony labyrinth
   1. Thus the membranous labyrinth is entirely encased in a bony labyrinth with only narrow perilymphatic spaces separating the two
D. DURING OTIC VESICLE FORMATION, a small group of cells breaks away from its wall and forms the statoacoustic ganglion (other cells probably come from neural crest)
   1. The ganglion subsequently splits into cochlear and vestibular portions, with saccule, utricle, and semicircular canals (equilibrium)
E. THE SACCULE, COCHLEA, AND ORGAN OF CORTI
   1. In the 6th week of development, the saccular part of the otic vesicle forms a tubular-shaped outpocketing at its lower pole called the cochlear duct
      a. The duct penetrates the surrounding mesenchyme in a spiral fashion until, at the end of the 8th month, it has completed $2\frac{1}{2}$ turns
      b. Its connection with the remaining part of the saccule is confined to a narrow pathway, the ductus reuniens
   2. The mesenchyme around the cochlear duct differentiates into a thin, fibrous basement membrane lining the outside of the duct and a large shell of cartilage
   3. In week 10, the cartilage shell undergoes vacuolization and 2 perilymphatic spaces are formed, the scala vestibuli and the scala tympani
   4. The chochlear duct then is separated from the scala vestibuli by the vestibular membrane and from the scala tympani by the basilar membrane
   5. The lateral wall of the cochlear duct is attached to the surrounding cartilage by the spiral ligament. The duct is also connected and partly supported by a long cartilaginous process, the modiolus, the future axis of the bony cochlea
   6. The epithelial cells of the cochlear duct are initially alike, but with development they form 2 ridges:
      a. An inner ridge, the future spiral limbus, is the larger of the two and is located close to the cochlear center
      b. An outer ridge, which forms one row of inner and 3 to 4 rows of outer hair cells, the sensory cells of the auditory system
      c. The hair cells are covered by the tectorial membrane, initially a fibrillar gelatinous substance, which is carried by the spiral limbus and rests with its tip on the hair cells
   7. The neuroepithelial cells and the covering tectorial membrane together are known as the organ of Corti, the true organ of hearing
   8. The impulses received by the organ are transmitted to the spiral ganglion and then to the nervous system via auditory fibers of the 8th cranial nerve

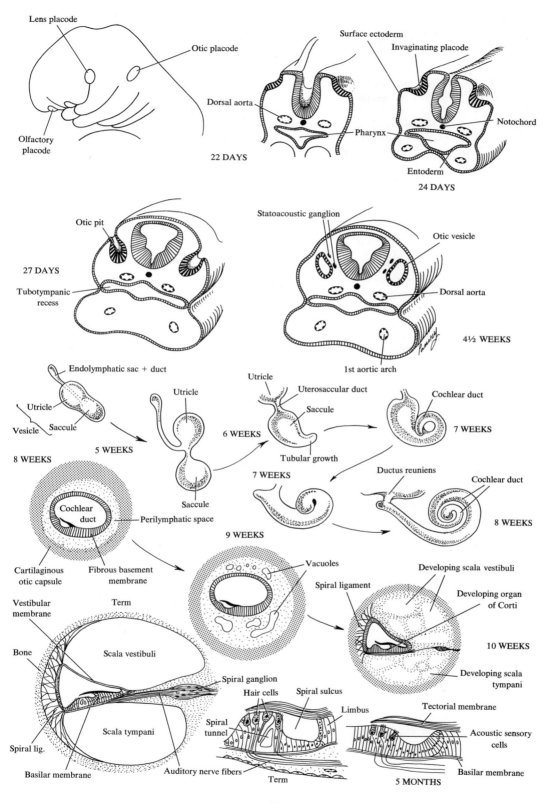

Lens placode

Otic placode

Olfactory placode

22 DAYS

Surface ectoderm

Invaginating placode

Dorsal aorta

Pharynx

Notochord

Entoderm

24 DAYS

Otic pit

27 DAYS

Tubotympanic recess

Statoacoustic ganglion

Otic vesicle

Dorsal aorta

1st aortic arch

4½ WEEKS

Endolymphatic sac + duct

Utricle

Vesicle

Saccule

8 WEEKS

5 WEEKS

Utricle

Saccule

7 WEEKS

Utricle

Uterosaccular duct

Saccule

Tubular growth

6 WEEKS

Cochlear duct

7 WEEKS

Ductus reuniens

Cochlear duct

8 WEEKS

9 WEEKS

Cochlear duct

Perilymphatic space

Cartilaginous otic capsule

Fibrous basement membrane

Vestibular membrane

Term

Bone

Scala vestibuli

Spiral lig.

Scala tympani

Basilar membrane

Auditory nerve fibers

Term

Vacuoles

Spiral ligament

Spiral ganglion

Hair cells

Spiral sulcus

Spiral tunnel

Limbus

Developing scala vestibuli

Developing organ of Corti

10 WEEKS

Developing scala tympani

Tectorial membrane

Acoustic sensory cells

Basilar membrane

5 MONTHS

22. THE INTERNAL AND EXTERNAL EAR

I. The internal ear (cont.)

A. UTRICLE AND SEMICIRCULAR CANALS
   1. During the 6th week of development, the semicircular canals are seen as flattened outpocketings of the utricular portion of the otic vesicle
   2. The central parts of the walls of the outpocketings eventually become apposed to each other and disappear, giving origin to the 3 semicircular canals
      a. One end of each canal dilates to form the crus ampullare, while the other end does not widen and is called the crus nonampullare
      b. Two of the latter type fuse, and thus only 5 crura enter the utricle, 3 with an ampulla and 2 without
   3. In week 7, the cells in the crus ampullare of each semicircular canal form a crest, the crista ampullaris, containing the sensory cells for maintenance of equilibrium
   4. Other sensory areas develop in the walls of the utricle and saccule to form the maculae acousticae
   5. Impulses generated in the sensory cells of the cristae and maculae, as a result of position changes of the body, are carried to the brain by the vestibular fibers of the 8th cranial nerve

II. The external ear

A. EXTERNAL AUDITORY MEATUS develops from the dorsal portion of the 1st pharyngeal cleft, which grows inward as a funnel-shaped tube to reach the entodermal lining of the tympanic cavity
   1. At the beginning of the 3rd month, the epithelial cells deep in the meatus proliferate to form a solid epithelial plate, the meatal plug
   2. In the 7th month, the plug dissolves, and the epithelial lining in the floor of the enlarged meatus then participates in the formation of the eardrum (tympanic membrane)
      a. If the meatal plug persists until birth, it may result in congenital deafness

B. THE EARDRUM OR TYMPANIC MEMBRANE
   1. The drum is made up of:
      a. The ectodermal epithelial lining at the bottom of the external auditory meatus
      b. The entodermal epithelial lining of the tympanic cavity
      c. An intermediate layer of loose connective tissue
   2. The major portion of the eardum is firmly attached to the malleus handle and is formed only after dissolution of the mesenchyme surrounding the ossicles
   3. The rest of the eardrum forms the separation between the external meatus and the original tubotympanic recess

C. THE AURICLE OR PINNA develops from a number of mesenchymal proliferations found at the dorsal ends of the 1st and 2nd pharyngeal arches and surrounding the 1st pharyngeal cleft
   1. These swellings, 3 on each side of the external meatus, are seen during the 6th week of development
   2. The hillocks fuse and are gradually formed into a definitive auricle
   3. Developmental abnormalities are not uncommon

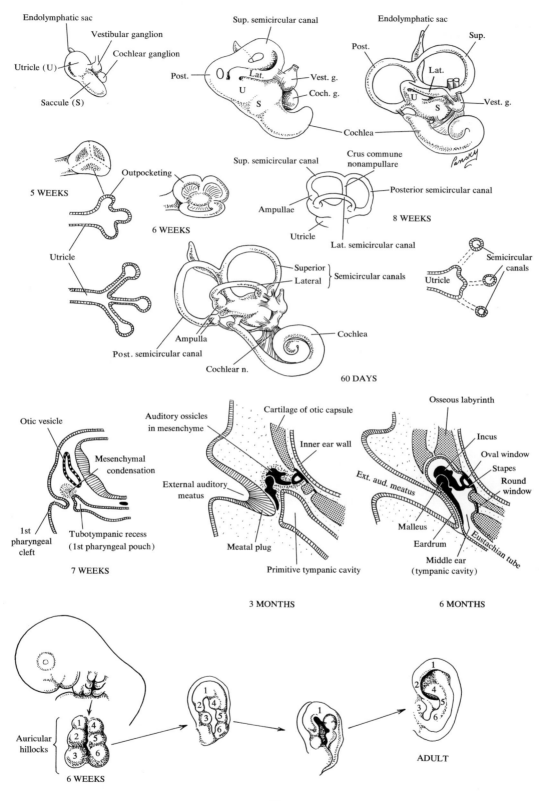

Endolymphatic sac
Vestibular ganglion
Cochlear ganglion
Utricle (U)
Saccule (S)

Sup. semicircular canal
Post.
Lat.
U
S
Vest. g.
Coch. g.

Endolymphatic sac
Post.
Sup.
Lat.
U
S
Vest. g.
Cochlea

5 WEEKS
Outpocketing
Utricle

6 WEEKS

Sup. semicircular canal
Crus commune nonampullare
Posterior semicircular canal
Ampullae
Utricle
Lat. semicircular canal

8 WEEKS

Superior
Lateral } Semicircular canals

Semicircular canals
Utricle

Ampulla
Post. semicircular canal
Cochlear n.
Cochlea

60 DAYS

Otic vesicle
Mesenchymal condensation
1st pharyngeal cleft
Tubotympanic recess (1st pharyngeal pouch)

7 WEEKS

Auditory ossicles in mesenchyme
Cartilage of otic capsule
Inner ear wall
External auditory meatus
Meatal plug
Primitive tympanic cavity

3 MONTHS

Osseous labyrinth
Incus
Oval window
Stapes
Round window
Ext. aud. meatus
Malleus
Eardrum
Middle ear (tympanic cavity)
Eustachian tube

6 MONTHS

Auricular hillocks

6 WEEKS

ADULT

-47-

23. THE MIDDLE EAR

I. The middle ear

A. TYMPANIC CAVITY AND EUSTACHIAN TUBE

1. The tympanic cavity is of entodermal origin and is derived from the 1st pharyngeal pouch, an outpocketing of the pharynx
   a. The pouch, lined with epithelium of entodermal origin, appears in embryos at about the 4th week
   b. The pouch grows rapidly in a lateral direction and temporarily comes in contact with the floor of the 1st ectodermal cleft
   c. The distal portion of the pouch, the tubotympanic recess, widens and gives rise to the primitive tympanic cavity; the proximal part remaining narrow and forming the auditory or eustachian tube
2. The eustachian tube connects the tympanic cavity with the nasopharynx
   a. The tube's pharyngeal orifice is surrounded by a considerable amount of lymphoid tissue, the tubal or pharyngeal tonsil (adenoids)
      i. Nasal inflammation associated with the tubal tonsillar swelling often results in occlusion of the tube and inflammation of the tympanic cavity resulting in otitis media (seen in young children)
3. During late fetal life, the tympanic cavity expands dorsally and posteriorly to form the tympanic antrum
   a. Its walls are covered with epithelium of entodermal origin
4. After birth, the bone of the developing mastoid process is invaded by epithelium of the tympanic cavity and epithelial-lined air sacs are created (pneumatization)
   a. Later, most of the mastoid air cells come in contact with the antrum and tympanic cavity

B. THE OSSICLES

1. By the end of the 7th week, the mesenchyme above the primitive tympanic cavity demonstrates a number of condensations caused by proliferation of the dorsal tips of the 1st and 2nd pharyngeal arches
   a. The condensations become the cartilaginous precursors of the auditory ossicles, the malleus, the incus, and the stapes
      i. The malleus and incus are derived from the 1st pharyngeal arch; the stapes is derived from the 2nd arch
2. The ossicles appear during the 1st half of fetal life but remain embedded in mesenchyme until the 8th month, when the surrounding tissue dissolves
3. The entodermal epithelial lining of the primitive tympanic cavity gradually extends along the wall of the newly developed space, the tympanic cavity being at least twice as large as before
4. When the ossicles are entirely free from the surrounding mesenchyme, the entodermal epithelium connects them in a mesentery-like manner to the cavity wall; and the supporting ligaments of the ossicles develop in these mesenteries
5. Since the malleus is derived from the 1st pharyngeal arch, its muscle, the tensor tympani, is innervated by the mandibular branch of the trigeminal nerve (V)
6. Since the stapes is of the 2nd pharyngeal arch, its muscle, the stapedius, is innervated by the facial (VII) nerve

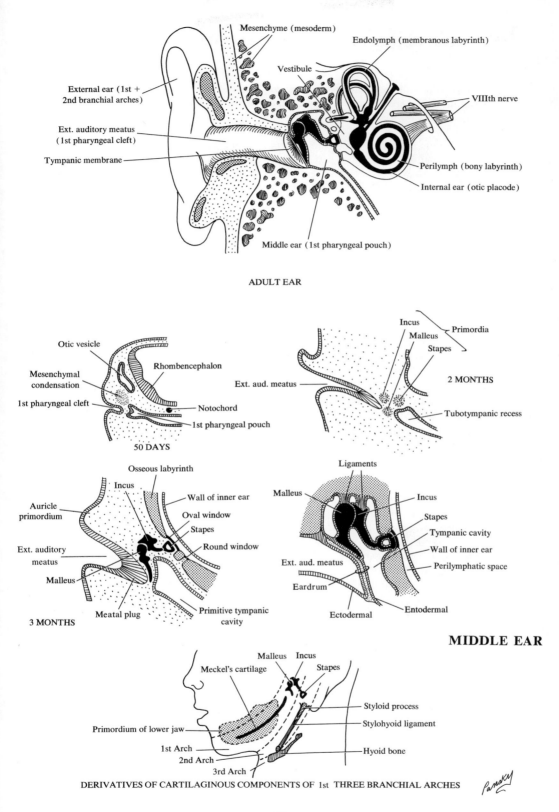

Mesenchyme (mesoderm)

Endolymph (membranous labyrinth)

Vestibule

External ear (1st + 2nd branchial arches)

VIIIth nerve

Ext. auditory meatus (1st pharyngeal cleft)

Tympanic membrane

Perilymph (bony labyrinth)

Internal ear (otic placode)

Middle ear (1st pharyngeal pouch)

ADULT EAR

Otic vesicle

Mesenchymal condensation

Rhombencephalon

1st pharyngeal cleft

Notochord

1st pharyngeal pouch

50 DAYS

Incus
Malleus
Stapes

Primordia

Ext. aud. meatus

2 MONTHS

Tubotympanic recess

Osseous labyrinth

Incus

Wall of inner ear

Auricle primordium

Oval window

Stapes

Round window

Ext. auditory meatus

Malleus

Meatal plug

Primitive tympanic cavity

3 MONTHS

Ligaments

Malleus

Incus

Stapes

Tympanic cavity

Wall of inner ear

Perilymphatic space

Ext. aud. meatus

Eardrum

Ectodermal

Entodermal

MIDDLE EAR

Malleus   Incus

Meckel's cartilage

Stapes

Styloid process

Stylohyoid ligament

Primordium of lower jaw

Hyoid bone

1st Arch
2nd Arch
3rd Arch

DERIVATIVES OF CARTILAGINOUS COMPONENTS OF 1st THREE BRANCHIAL ARCHES

Pansky

24. THE EAR: CONGENITAL MALFORMATIONS

I. **Malformations of the auditory system** can be classified in 3 different groups, corresponding to the specific embryologic characteristics of each of the ear's 3 parts:
A. LESIONS OF INTERNAL EAR
1. Cochlea
a. Rubella in 2nd month of pregnancy is often a cause of lesions since this is when internal ear differentiates
b. Epithelium of cochlea and vestibule is altered, but organ of Corti has some intact regions so children affected perceive some deep frequencies.
2. Canals: few malformations are known (seen with thalidomide)
B. LESIONS OF MIDDLE EAR: usually involve ossicles
1. After an infection, the mesenchymal plate separating the ossicles becomes sclerotic, impedes movement, and may cause total deafness
C. MALFORMATION OF THE EXTERNAL EAR
1. Results from absence or nonunion of tubercles
2. Major variations of the auricle have been associated with serious internal abnormalities such as kidney malformations
3. Abnormal position of the ear usually associated with abnormal mandibular development (agnathia or micrognathia)
a. Instead of moving to sides of head, ears develop at site of primordia—at level of 1st branchial groove (seen in newborn at the angle of the missing jaw—otocephalus and ingenetic)
4. Auricular appendages or tags—common and due to accessory auricular hillocks.
5. Absence and hypoplasia of auricles are rare—associated with 1st arch syndrome— failure of hillocks to develop (anotia) or suppressed development (microtia)
6. Auricular sinus and fistulas: sinuses are usually preauricular; fistulas connect exterior with tympanic cavity
7. Atresia of external auditory meatus: failure of meatal plug to canalize

II. CONGENITAL DEAFNESS IS USUALLY ASSOCIATED WITH DEAF-MUTISM
A. MAY BE CAUSED BY
1. Abnormal development of the membranous and bony labyrinths
2. Malformations of the eardrum and ossicles
3. In extreme cases, tympanic cavity and external meatus are completely absent
B. OTHER CAUSES OF DEAFNESS
1. Heredity
2. Environmental and other factors affecting the mother early in pregnancy
a. Rubella virus, affecting the embryo in the 7th and 8th week of development, will cause severe damage to the organ of Corti; diabetes; erythroblastosis fetalis; hypothyroidism; toxoplasmosis; x-radiation
C. MAJOR VARIATIONS of the auricle have been associated with serious internal abnormalities such as kidney malformations

AURICULAR APPENDAGES
OR TAGS

EXTERNAL EAR ANOMALY WITH
MAXILLOMANDIBULAR FISSURE

FISTULA AURIS
(PROBED)

MALFORMED AURICLE

SYNOTIA (WITH MANDIBULAR
HYPOSTOMIA AND MACROSTOMIA)

OTOCEPHALUS

HYPOPLASIA OF AURICLE
(WITH ABSENCE OF EXTERNAL
AUDITORY MEATUS)

-51-

Unit Two

GROSS AND

MICROSCOPIC

NEUROANATOMY

25. GENERAL HISTOLOGY OF NERVOUS SYSTEM AND NEURON CLASSIFICATION

I. **Introduction:** the nervous system may be defined as a specialized group of nerve cells (neurons) and supporting cells (glia) whose major functions are reception of information from external and internal environments; integration of incoming information; generation of new signals; and transformation and conduction of messages to special responding tissue (effectors)

A. NERVOUS TISSUE: characterized by its unique properties of excitation and conduction

B. NEURON is the anatomic and functional unit of the nervous system and chains of neurons comprise the nervous system
  1. Each neuron is contiguous with the next cell in the chain, but the cells are not continuous with one another
  2. Each neuron has a nerve cell body with all its protoplasmic processes, one axon, and usually one or more dendrites
    a. Dendrites are typically short, branching processes which form a major part of the receptor area of the neuron
    b. Each neuron has a single axon which varies greatly in length from one type of neuron to another and conducts impulses away from the soma or cell body

C. GANGLION is a collection of neurons outside the central nervous system

D. NUCLEUS is a cluster of neurons in the gray matter of the brain and spinal cord (nucleus in this sense refers to a functional aggregation of neurons and should not be confused with the nucleus of a cell)

II. **Classification of neurons**

A. ACCORDING TO FUNCTION
  1. Motor neurons include neurons of the autonomic nervous system (ANS) and of anterior horns of the spinal cord. They carry impulses that produce movement
  2. Sensory neurons carry impulses that produce sensation. Neurons of the dorsal root ganglia are of this type
  3. Association (intercalated or internuncial) neurons serve to connect motor and sensory neurons

B. ACCORDING TO NUMBER OF PROCESSES FROM CELL BODY
  1. Unipolar or pseudounipolar neurons have only 1 protoplasmic process
    a. This T-shaped process bifurcates and one branch functions as an axon, which enters the cord, the other as a dendrite, which extends to a peripheral tissue
    b. Neurons of this type are sensory and are found, e.g., in the dorsal root ganglia of the PNS and in the mesencephalic nucleus of the trigeminal (V) cranial nerve (the only example of unipolar neurons in the CNS)
  2. Bipolar neurons have 2 processes, one from each pole of the cell body; one process serves as an axon, the other as a dendrite
    a. Neurons of this type are found in the retina, olfactory epithelium, and in the acoustic ganglion of cranial nerve VIII
  3. Multipolar neurons have more than 2 processes from the cell body; only one process is an axon and the others are dendrites
    a. Neurons of the CNS are generally multipolar and include large motor neurons of the spinal cord, small association neurons of the cord, pyramidal cells of the cerebral cortex, Purkinje cells of the cerebellar cortex, and neurons of the ANS

C. ACCORDING TO LENGTH OF AXON
  1. Golgi type 1 neurons are found in the CNS and have long axons which leave their site of origin and extend for a considerable distance before ending
    a. They form tracts, commissures, and projection and association fibers
  2. Golgi type 2 neurons are found in the CNS and have short axons (0.5–5.0 mm)
    a. Their axons do not leave the gray matter where the cells are located
    b. Numerous in the cerebral and cerebellar cortices and spinal cord

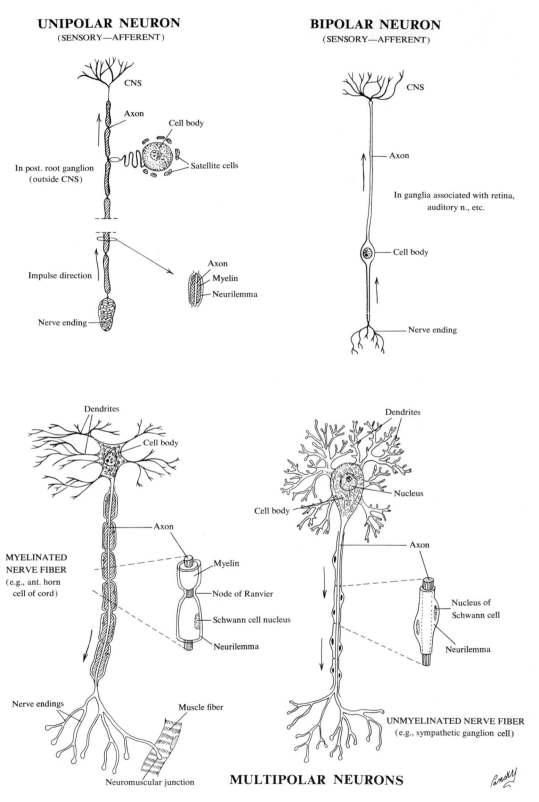

UNIPOLAR NEURON
(SENSORY—AFFERENT)

CNS

Axon

Cell body

Satellite cells

In post. root ganglion
(outside CNS)

Axon
Myelin
Neurilemma

Impulse direction

Nerve ending

BIPOLAR NEURON
(SENSORY—AFFERENT)

CNS

Axon

In ganglia associated with retina,
auditory n., etc.

Cell body

Nerve ending

Dendrites

Cell body

Axon

MYELINATED
NERVE FIBER
(e.g., ant. horn
cell of cord)

Myelin

Node of Ranvier

Schwann cell nucleus

Neurilemma

Nerve endings

Muscle fiber

Neuromuscular junction

Dendrites

Nucleus

Cell body

Axon

Nucleus of
Schwann cell

Neurilemma

UNMYELINATED NERVE FIBER
(e.g., sympathetic ganglion cell)

MULTIPOLAR NEURONS

26. GENERAL HISTOLOGY OF NERVOUS SYSTEM: NEUROGLIA—GENERAL

I. Neuroglia—general
A. SPECIALIZED NONNEURONAL SUPPORTING CELLS of the CNS are called neuroglia (nerve glue) or simply glia. These supporting cells are more numerous than neurons. They support neurons and provide axons with myelin sheaths that increase the speed of impulse conduction
B. UNLIKE NEURONS, glial cells have only 1 type process and do not form synapses
C. GLIAL CELLS can undergo cell division throughout adult life, particularly in response to damage to the CNS
D. CLASSIFICATION OF THE NEUROGLIA (three categories of glial cells can be recognized in the CNS)
   1. Macroglia can be subdivided into astrocytes and oligodendrocytes
   2. Microglia are not considered as true neuroglia by some authorities
   3. Ependymal cells are not always classified as true neuroglia
E. EMBRYOLOGICALLY, all supportive neuroglia except microglia are derivatives of the neural tube (neuroectoderm); microglia arise from mesoderm
F. MOST KNOWLEDGE about the shapes and disposition of glial cells has been gained by examination of tissues stained by metallic impregnation because ordinary stains such as hematoxylin and eosin (H & E) show only nuclei of glial cells

II. Neuroglia—specific
A. ASTROCYTES can be separated into 2 types—fibrous and protoplasmic—by special staining techniques (one being the gold sublimate method of Cajal). Both types of astrocytes are stellate (star-shaped) and have long processes
   1. Morphologic and functional characteristics of fibrous astrocytes
      a. Found primarily in white matter
      b. Processes are fewer in number, longer and straighter than those of protoplasmic astrocytes (resemble long-legged spiders)
      c. Some of their processes form expansions that are applied to the surfaces of blood vessels (perivascular end-feet or sucker feet)
         i. Other processes extend to surfaces of the CNS to form expansions known as the glial limiting membrane (secures pia mater to the CNS)
      d. Possess many cytoplasmic filaments which may occur in bundles (as seen in light microscopy) and provide rigidity and support
      e. They are the main supportive cells of the CNS and form firm links between capillaries
      f. They are important in the response of the CNS tissue to injury
         i. In the case of nonlethal injury to the astrocyte, the cell body swells, the processes elongate, the cell divides, and additional processes form
      g. They are often referred to as the scarring cells of the CNS because they fill in gaps after tissue is lost in disease processes
      h. When a lesion is extensive or when it is accompanied by liquefaction, the astrocytes are inadequate for the task and fibroblasts enter to assist in restoring supportive integrity
         i. Scarring results from the production of connective tissue fibers by fibroblasts (known as sclerosis) which differs from gliosis in which only astrocytes proliferate, as in response to chronic injury to the CNS
      i. Astrocytes are capable of limited phagocytosis (case of minimal CNS injury)

Sympathetic ganglion (low power)

Sympathetic ganglion (high power)

Dorsal root ganglion (low power)

Dorsal root ganglion (high power)

Spinal nerve in cross-section (0s04 preparation)

Myelinated nerve fibers in cross-section

Spinal nerve in longitudinal-section showing myelinated fibers and Schwann cell nuclei

Spinal nerve in oblique-section

27. GENERAL HISTOLOGY OF NERVOUS SYSTEM: NEUROGLIA—SPECIFIC

I. Neuroglia—specific (cont.)

A. ASTROCYTES (CONT.)

1. Morphologic and functional characteristics of protoplasmic astrocytes
   a. Found primarily in gray matter
   b. Processes are far more numerous, much shorter, and much more extensively branched than those of fibrous astrocytes (resemble bushy shrubs)
   c. Processes cover nonsynaptic neuronal surfaces and surround synaptic zones
      i. Their pattern of arrangement suggests they serve to isolate the receptive surfaces of neurons and prevent the haphazard flow of impulses
   d. Processes also terminate on capillaries as perivascular end-feet
   e. Possess fewer cytoplasmic filaments than fibrous astrocytes
2. Identification (differentiation) of the two types of astrocytes
   a. The 2 types are indistinguishable histologically without special stains
   b. Identification based on location alone is unreliable. (Protoplasmic astrocytes predominate in gray matter and fibrous astrocytes in white matter)
   c. Nuclei of both types are large, pale, and lack nucleoli
   d. Nuclei are often indented and usually more irregular in outline than those of other glial cells
   e. Nuclei are smaller than those of neurons but larger than oligodendrocytes.

B. OLIGODENDROGLIA (OLIGODENDROCYTES)

1. A special neurohistologic technique (del Rio-Hortega's silver method) shows
   a. Cells are small with round or polygonal bodies; most numerous cell in CNS
   b. Possess few protoplasmic processes (no perivascular end-feet)
   c. Differ from each other only in location; produce myelin in the CNS
2. Classification of oligodendrocytes
   a. Interfascicular oligodendrocytes
      i. Found in white matter of brain and spinal cord
      ii. Elaborate myelin sheaths that surround axons of the CNS; comparable to Schwann cells that produce myelin of PNS
      iii. Each process has a diameter of about 0.1 μm
      iv. On reaching the axonal surface, the process enlarges into a wide, flat tongue which then wraps around the axon 1–20 times
      v. This wrapping material is the lipoprotein plasma membrane and constitutes the myelin sheath; the thickness of the sheath is determined by the number of times that the tongue wraps around the axon
      vi. Processes of one oligodendrocyte may form internodal myelin for 3–50 fibers
   b. Perineuronal oligodendrocytes
      i. Located in gray matter close to nerve cell bodies and dendrites
      ii. Often compared to satellite cells that surround the perikarya of peripheral ganglia and appear to have a similar relationship
   c. Perivascular oligodendrocytes
      i. Lined up along blood vessels in both gray and white matter
      ii. Their function is still not known

C. DIFFERENTIATION OF OLIGODENDROCYTES AND ASTROCYTES WITH ROUTINE STAINS

1. Nuclei of oligodendrocytes are round to ovoid and usually appear smaller and darker than nuclei of astrocytes
2. Oligodendrocyte nuclei contain denser clumps of chromatin
3. Oligodendrocytes are smaller in size but appear in far greater abundance

Typical neurons and their cytoplasmic processes surrounded by numerous glial cells

Neuroglial cells and their processes

Fibrous astrocytes (low power)

A fibrous astrocyte (high power)

Protoplasmic astrocyte

Astrocytes, their processes and perivascular footplates

Oligodendrocytes

Microglia

28. SPINAL CORD: GENERAL GROSS APPEARANCE

I. Location and level
A. THE SPINAL CORD is that part of the CNS continuous with the medulla oblongata of the brainstem and located within the vertebral canal
B. IT EXTENDS, in the adult, from the foramen magnum cranially to the inferior border of the first lumbar vertebra caudally, where it tapers to a point (conus medullaris) and terminates
C. IT USUALLY MEASURES 42–45 cm in length, and its diameter varies at the different levels

II. Spinal cord and vertebral column relationship
A. BECAUSE OF DIFFERENCES IN GROWTH RATE between the spinal cord and vertebral column, the cord in the adult ends at the lower border of the first lumbar vertebra
B. SINCE THE SPINAL CORD IS APPROXIMATELY 25 CM SHORTER than the column, lumbar and sacral nerves possess very long intervertebral roots which extend from the cord to the appropriate intervertebral foramina where dorsal and ventral roots join to form spinal nerves
   1. In the third fetal month, the spinal cord fills the vertebral canal; at birth, the cord ends at a level between the first and second lumbar vertebrae (the lowest level is at the lower third of the third lumbar vertebra)
   2. Relationships in the adult are presented for various levels in the table

Spinal Nerve and Cord Segment	Vertebral Level	Spinous Process	Level of Nerve Exit
C1	C1	C1	Above C1
C8	C7	C6	Between C7–T1
T6	T5	T3	Between T6–T7
T12	T8	T8	Between T12–L1
L2	T10	T10	Between L1–L2
L5	T11	T12–L1	Below L5
S3	T12	L1–L2	Third sacral foramen

Quadriplegia

Complete cord
lesion

Paraplegia

**SPINAL CORD AND
NERVES RELATED TO
VERTEBRAL COLUMN**

C1
C1
C1
1

C7
T1
C8
T1
7 8
11
2
3
4
5
T5
6
7
8
9
10
11
12
T10
1
L1
2
3
4
5
L5
1
S1
2

C7
T1

T5

T10

L1

Cauda equina

L5

Sacrum

Coccyx

Spinous process C7
C7
Spinal cord segment T1

Spinous process T10
T10
Spinal cord segment L1

T12
T12
L1
Spinal cord segment
S1
L2

Termination of conus medullaris
at L1–L2 interspace

Subarachnoid space
(lumbar cistern)

Termination of subarachnoid
space at S2 or S3
S1
S2
Termination of filum
terminale on coccyx

Coccyx

29. SPINAL CORD: EXTERNAL MORPHOLOGY

I. External morphology

A. IT IS AN ELONGATED, UNSEGMENTED STRUCTURE intrinsically. Paired spinal nerves along its entire length produce an appearance of external segmentation

B. A SINGLE SPINAL CORD SEGMENT is defined as that part of the cord which receives or gives dorsal and ventral rootlets to a single pair of spinal nerves. There are 8 cervical segments, 12 thoracic segments, 5 lumbar segments, 5 sacral segments, and 1 coccygeal segment, or a total of 31 pairs of spinal nerves

C. INDIVIDUAL SPINAL SEGMENTS VARY IN LENGTH; those in the midthoracic region are usually much (at least 2×) longer than cervical and lumbar segments

D. GROSSLY, it has the appearance of a compressed cylinder with two enlargements
   1. Cervical enlargement corresponds to the nerves to the upper limb (brachial plexus) and is comprised of the lower four cervical and the first thoracic segments
   2. Lumbar enlargement corresponds to the area of nerves to the lower limbs (lumbosacral plexus) and is comprised of the five lumbar and the upper three sacral segments (i.e., acquires largest diameter at T12)

E. SINCE THERE IS NO SPINAL CORD WITHIN THE VERTEBRAL CANAL of most of the lumbar and all of the sacral vertebrae, the elongated dorsal and ventral roots are the only structures within the canal at these levels. The elongated mass of nerve roots suspended from the cord is known as the cauda equina (tail of a horse)

F. UP UNTIL THE THIRD FETAL MONTH, the length of the cord and vertebral canal are approximately equal. By the time of birth, the caudal extent of the cord is at the level between L1 and L2. This is one of the bases for the selection of the third or fourth lumbar interspace for the introduction of the needle in lumbar puncture procedures

G. THE CORD IS ATTACHED TO SURROUNDING STRUCTURES, by extensions of its innermost meningeal covering, the pia mater; it intimately invests the cord (and brain) like a sausage skin does a sausage

H. THROUGHOUT THE LENGTH OF THE CORD, extensions of the pia mater are pulled out from the lateral aspects of the cord between dorsal and ventral roots, to be "tacked" to the walls of the outer investing meningeal covering, the dura mater. This horizontal shelflike ligament on either side of the cord, with its numerous toothlike attachment points is termed the dentate (toothed) or denticulate ligament. These ligaments serve to protect it from excessive shock and provide useful handles for manipulating the cord safely during surgery

I. THE INFERIOR PIAL EXTENSION that serves to anchor the cord is the filum terminale. This connective tissue structure is formed caudal to the conus medullaris, where the nerve tissue terminates, but the pial covering continues. This terminal filament is continued caudally to its point of attachment at the coccyx where it is known as the coccygeal ligament after the additional meningeal coverings (arachnoid and dura) are added

J. ON THE CORD THE DURA CONTINUES out along the dorsal and ventral roots and spinal nerves to the point where they pass through the intervertebral foramina. Here the dura becomes attached to the periosteum. It extends down to the level of S2, caudal to the termination of the conus medullaris, to "bag" the cauda equina. A sheathlike extension continues further caudally, wrapping the filum terminale to form the coccygeal ligament

Lig. flavum

Vertebra

Epidural C. T. + fat

Internal vertebral venous plexus

Spinal dura

Subdural space

Arachnoid

Subarachnoid space

DORSAL VIEW

SPINAL CORD AND ITS COVERINGS

Denticulate ligament
Dorsal root gang.
Ventral
Dorsal } Rami
Dura

C1
C1
Cervical enlargement
C7
C8
T1
T1

Dorsal root gang.

T12
L1
Lumbar enlargement
T12
S1
L1
Conus medullaris
Cauda equina
L5
S1
Filum terminale
Co 1

DORSAL VIEW

Cord
Pia matter
Arachnoid
Dura mater
Denticulate lig.
Lumbar enlargement
Conus medullaris
Lower border L1
Subdural space
Filum terminale
Subarachnoid space
Vertebra S2
Coccygeal ligament
Base of coccyx

30. SPINAL CORD: MAJOR SUBDIVISIONS

I. Subdivisions of white and gray matter

A. TWO BASIC DIVISIONS of the spinal cord are seen in cross-section:

   1. Gray matter is located in the central part and shaped like a butterfly or in the shape of an H. It is comprised of glia cells and cell bodies of efferent and internuncial neurons.

      a. The narrow midline connection (crossbar of the H) of the gray matter is divided into dorsal and ventral gray commissures

      b. Dorsal projections of the gray are known as posterior gray horns

      c. Ventral projections of the letter H are the ventral gray horns

      d. Intermediolateral gray horns are found only at segmental levels T1–L2 or L3. This enlargement of the central gray matter is intermediate and lateral in position between the ventral and posterior gray horns

   2. White matter is arranged as the outer layer of the cord, and it is made up of nerve fibers and glia

      a. The surface reveals two longitudinally oriented grooves, located in the midline on the ventral and posterior surfaces

         i. The anterior median fissure is wide but shallow; it is found throughout the entire cord length

         ii. The posterior median sulcus is narrower and deeper and present at all levels

      b. Paired lateral sulci on the anterior and posterior surfaces

         i. Anterior lateral (ventrolateral) sulci are lines for the attachment of exiting ventral nerve rootlets

         ii. Posterior lateral (dorsolateral) sulci represent lines for attachment of entering dorsal nerve rootlets

B. IN CROSS-SECTIONAL VIEW it is observed that components from the ventral and posterior gray horns extend to the surface of the cord at these sulci, and divide each half of white matter into three bundles of cell processes (fibers) termed funiculi. These funiculi are named according to position

   1. The dorsal funiculi of the two sides are separated by the dorsal median sulcus and septum, and extend laterally to the dorsolateral sulcus

   2. The ventral funiculi are separated by the anterior median fissure and extend laterally to the ventrolateral sulcus

   3. The lateral funiculus of each half of white matter is bound medially by the gray horns and lies between the ventrolateral and dorsolateral sulci

C. IN THE CERVICAL AND UPPER THORACIC PORTIONS of the cord (above T7), the dorsal funiculi are each subdivided by the dorsal intermediate sulcus and septum into smaller lateral and medial bundles termed fasciculi

   1. Fasciculus cuneatus is the lateral portion of each dorsal funiculus and represents a tract carrying sensory impulses from upper half of body

D. IN REFERENCE TO THE CENTRAL CANAL of the spinal cord, the white matter (like the gray) crossing the midline is separated into dorsal and ventral commissures. The ventral commissure is more prominent than the dorsal, and the former is present throughout the spinal cord at all levels. Both commissural and decussating fibers cross in the white commissure

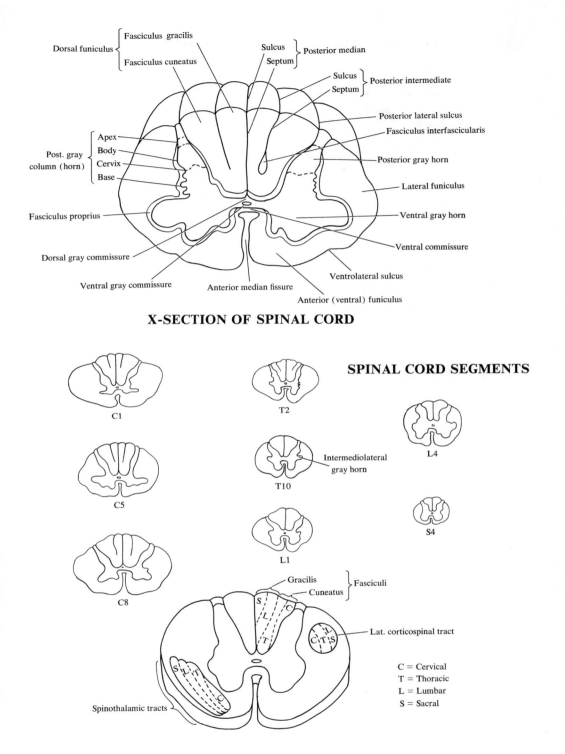

X-SECTION OF SPINAL CORD

Dorsal funiculus {
Fasciculus gracilis
Fasciculus cuneatus

Sulcus
Septum } Posterior median

Sulcus
Septum } Posterior intermediate

Posterior lateral sulcus

Fasciculus interfascicularis

Post. gray column (horn) {
Apex
Body
Cervix
Base

Posterior gray horn

Lateral funiculus

Fasciculus proprius

Ventral gray horn

Ventral commissure

Dorsal gray commissure

Ventrolateral sulcus

Ventral gray commissure

Anterior median fissure

Anterior (ventral) funiculus

SPINAL CORD SEGMENTS

C1

T2

L4

C5

T10

Intermediolateral gray horn

S4

C8

L1

Gracilis
Cuneatus } Fasciculi

Lat. corticospinal tract

C = Cervical
T = Thoracic
L = Lumbar
S = Sacral

Spinothalamic tracts

SOMATOTOPIC ARRANGEMENT OF FIBERS IN SPINAL CORD

31. SPINAL CORD: MICROSCOPIC ANATOMY

I. Nuclei with the gray matter: Neurons are not distributed uniformly with the gray substance of the spinal cord. Instead, their cell bodies are arranged somewhat into definitive columns or collections referred to as nuclei with the following distribution

A. NUCLEI OF THE POSTERIOR HORN

1. Intermediolateral nucleus is found at the levels of T1 to L3 (sympathetic). The axons of neurons in this nucleus pass via the ventral roots and the white communicating rami to the sympathetic chain ganglia or to more peripheral sympathetic ganglia along the aorta

2. Sacral autonomic nuclei are found on the lateral surface of the gray matter at the base of the anterior horn in sacral segments S2, S3, and S4. Axons of these neurons make up the parasympathetic pelvic plexus

3. Posteromarginal nucleus is a thin layer of cells that cover the tip of the posterior horn. They are intersegmental association neurons and their axons ascend or descend in the lateral funiculus for several segments

4. Substantia gelatinosa forms the outer "caplike" portion of the posterior horn. The neurons of this area form an important part of the conscious and reflex pathways for pain, temperature, and some tactile impulses

5. Nucleus proprius of the posterior horn (nucleus centrodorsalis) sends axons across the midline in the anterior white commissure to form the anterior spinothalamic tract (pain and temperature)

6. Nucleus dorsalis (Clarke's column) is found in the medial portion of the base of the posterior horn. It is found from about C8 to L3 and axons from neurons in this nucleus pass into the posterior spinocerebellar tract of the same side.

7. Secondary visceral gray is an indistinct cell column that extends through the thoracic and upper lumbar levels. It receives incoming visceral afferent fibers

B. NUCLEI OF THE ANTERIOR HORN: The neurons (anterior horn cells) are the largest found within the spinal cord and have the following characteristics

1. Multipolar neurons with large, round vesicular nuclei and clumps of coarse Nissl substance within the cytoplasm

2. Long axons with distinctive axonal hillocks

3. Large anterior horn neurons are grouped into medial and lateral groups, which can be further subdivided into
   a. Dorsomedial nucleus
   b. Ventromedial nucleus
   c. Ventral nucleus
   d. Central nucleus
   e. Ventrolateral nucleus
   f. Dorsolateral nucleus
   g. Retrodorsolateral nucleus

4. These nuclei are most apparent at the level of cervical and lumbar enlargements

5. They supply motor innervation to somatic effectors (skeletal muscle) and are known as lower motor neurons

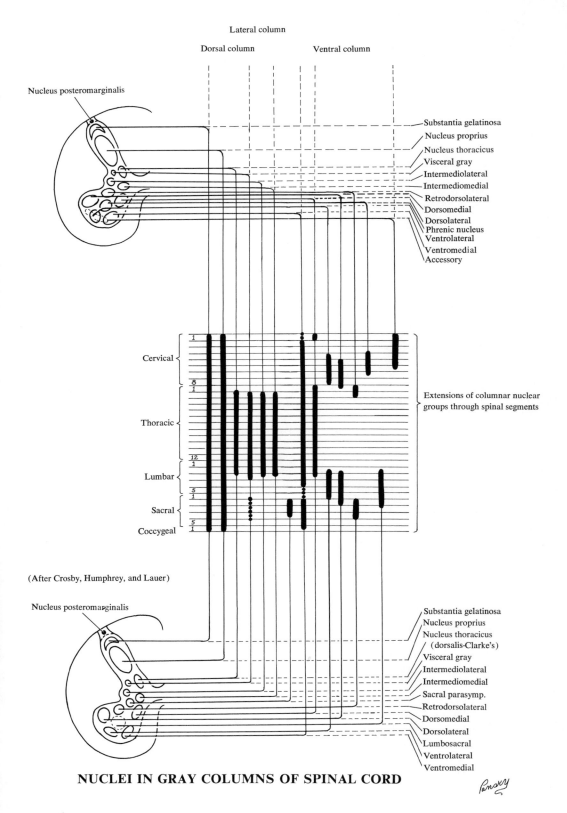

Lateral column

Dorsal column   Ventral column

Nucleus posteromarginalis

Substantia gelatinosa
Nucleus proprius
Nucleus thoracicus
Visceral gray
Intermediolateral
Intermediomedial
Retrodorsolateral
Dorsomedial
Dorsolateral
Phrenic nucleus
Ventrolateral
Ventromedial
Accessory

Cervical $\left\{ \begin{array}{c} \overline{1} \\ \\ \frac{8}{1} \end{array} \right.$

Thoracic $\left\{ \begin{array}{c} \\ \\ \frac{12}{1} \end{array} \right.$

Lumbar $\left\{ \begin{array}{c} \\ \frac{5}{1} \end{array} \right.$

Sacral $\left\{ \begin{array}{c} \\ \frac{5}{1} \end{array} \right.$

Coccygeal

Extensions of columnar nuclear
groups through spinal segments

(After Crosby, Humphrey, and Lauer)

Nucleus posteromarginalis

Substantia gelatinosa
Nucleus proprius
Nucleus thoracicus
  (dorsalis-Clarke's)
Visceral gray
Intermediolateral
Intermediomedial
Sacral parasymp.
Retrodorsolateral
Dorsomedial
Dorsolateral
Lumbosacral
Ventrolateral
Ventromedial

NUCLEI IN GRAY COLUMNS OF SPINAL CORD

Panoky

32. SPINAL CORD: REXED'S LAMINAE AND MAJOR TRACTS

I. **Rexed's laminae within gray matter of spinal cord:** The gray substance can be divided into nine layers or laminae of Rexed on the basis of the longitudinal arrangement of the neurons into columns and cytoarchitectonics. Rexed's laminae are more encompassing in area and are preferred by some authorities to the more traditional use of cell columns or nuclei because of problems with nomenclature and precise functions associated with the utilization of the latter. Basic facts essential for understanding laminae of Rexed include

A. SPINAL CORD is divided into nine cell laminae. Five are located in the posterior horn, two in the intermediate gray, and two in the anterior horn. An additional lamina surrounds the central canal

B. LAMINAE I THROUGH VI are found only in the posterior horn. They are concerned with sensory input to the spinal cord

C. LAMINA VI is found only in the cervical and lumbar enlargements

D. LAMINA VII is found in the intermediate gray area and extends into the anterior horn. It includes the nucleus dorsalis and the intermediolateral gray column

E. LAMINA VIII is located in the anterior horn. Many of the axons from the neurons located here cross to the opposite side (commissural)

F. LAMINA IX is found only in the anterior horn and is usually subdivided. It possesses alpha and gamma motor neurons whose axons leave the spinal cord via the anterior roots

G. THE FOLLOWING LAMINAE AND NUCLEI correspond in location
   1. Lamina I with the posteromarginal nucleus
   2. Lamina II with substantia gelatinosa
   3. Lamina III and IV with nucleus proprius
   4. Lamina IX with the subdivisions of the motor cell columns
   5. Lamina VII with the nucleus dorsalis and the intermediolateral gray

II. **Tracts within the white matter of the spinal cord:** The position of fiber tracts (a tract is comprised of numerous axons within the CNS with the same origin, termination, and function) has been determined on the basis of experimental work with various mammals, stimulations of various areas during surgical procedures, and pathologic observations. A knowledge of the relative positions of the following tracts will form the basis for an understanding of the details of long ascending and descending tracts presented later

A. ASCENDING TRACTS include the following long* and short† tracts
   1. Fasciculus gracilis*
   2. Fasciculus cuneatus*
   3. Anterior and posterior spinocerebellar*
   4. Anterior and lateral spinothalamic*
   5. Dorsolateral fasciculus†
   6. Fasciculus proprius†

B. DESCENDING TRACTS include the following long* and short† tracts
   1. Pyramidal (lateral and ventral corticospinal) tract*
   2. Extrapyramidal tracts are comprised of all long descending tracts and other than the lateral and ventral corticospinal tracts.*
   3. Dorsolateral fasciculus†
   4. Fasciculus proprius†
   5. Septomarginal fasciculus†
   6. Fasciculus interfascicularis†

REXED'S LAMINAE IN GRAY MATTER OF CORD

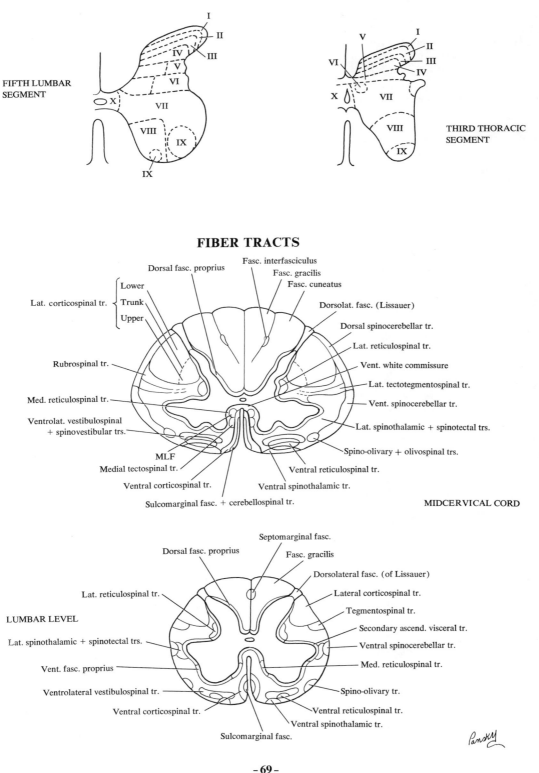

FIFTH LUMBAR
SEGMENT

I
II
IV
V
III
VI
X
VII
VIII
IX
IX

V
VI
I
II
III
IV
X
VII
VIII
IX

THIRD THORACIC
SEGMENT

FIBER TRACTS

Dorsal fasc. proprius

Fasc. interfasciculus
Fasc. gracilis
Fasc. cuneatus

Lat. corticospinal tr. { Lower / Trunk / Upper

Dorsolat. fasc. (Lissauer)

Dorsal spinocerebellar tr.

Lat. reticulospinal tr.

Rubrospinal tr.

Vent. white commissure

Lat. tectotegmentospinal tr.

Med. reticulospinal tr.

Vent. spinocerebellar tr.

Ventrolat. vestibulospinal
+ spinovestibular trs.

Lat. spinothalamic + spinotectal trs.

Spino-olivary + olivospinal trs.

MLF

Medial tectospinal tr.

Ventral reticulospinal tr.

Ventral corticospinal tr.

Ventral spinothalamic tr.

Sulcomarginal fasc. + cerebellospinal tr.

MIDCERVICAL CORD

Dorsal fasc. proprius

Septomarginal fasc.

Fasc. gracilis

Dorsolateral fasc. (of Lissauer)

Lat. reticulospinal tr.

Lateral corticospinal tr.

Tegmentospinal tr.

LUMBAR LEVEL

Secondary ascend. visceral tr.

Ventral spinocerebellar tr.

Lat. spinothalamic + spinotectal trs.

Med. reticulospinal tr.

Vent. fasc. proprius

Ventrolateral vestibulospinal tr.

Spino-olivary tr.

Ventral corticospinal tr.

Ventral reticulospinal tr.

Ventral spinothalamic tr.

Sulcomarginal fasc.

Pansky

– 69 –

33. SPINAL CORD: TYPICAL SPINAL NERVE

I. **Typical spinal nerve** is formed by a dorsal and a ventral root coming together inside the vertebral canal. Since the dorsal root, along with its dorsal root ganglion, is sensory and the ventral root carries motor fibers, the spinal nerve contains both modalities and is termed a mixed nerve

A. EACH RAMUS DIVIDES into two major subdivisions
   1. Dorsal ramus splits into a medial and a lateral branch (one of which supplies both skin as well as muscle, and the other muscle only)
   2. Ventral ramus gives off a lateral or perforating branch (which is cutaneous) and then continues on around supplying motor and sensory fibers to muscles of the body wall. It terminates as an anterior perforating branch (also cutaneous)

B. EACH CUTANEOUS BRANCH (ANTERIOR, POSTERIOR, AND LATERAL) DIVIDES into two branches
   1. Anterior and posterior cutaneous branches each has a medial and lateral branch
   2. Lateral cutaneous branch has an anterior and a posterior branch

C. THE SKIN OF ONE ENTIRE SIDE OF THE BODY is innervated in this fashion, and the opposite side has a comparable set of nerves

D. MOTOR INNERVATION TO AND SENSORY FIBERS FROM skeletal muscles are carried by the anterior and posterior rami. Thus motor fibers from the anterior horn cells within the cord are carried by both anterior and posterior rami, and sensory fibers, whose cell bodies are located in the dorsal root ganglia, are carried by both anterior and posterior rami. In addition, both rami carry fibers from the lateral sympathetic chain ganglia to glands of the skin and blood vessels in both the skin and muscles

E. EACH OF THE 31 PAIRS OF SPINAL NERVES develops along this general pattern as described. This typical pattern is clearly seen, however, only on the trunk and particularly in the thoracic region

F. SEGMENTAL INNERVATION: the fact that each of the 31 pairs of nerves developmentally follows the typical pattern gives rise to this concept
   1. Parallel segments of skin innervated by a single nerve form bands. This resembles conceptually a loaf of bread, with each slice representing a segment
   2. Slices or segments (known as dermatomes) do not retain their absolute horizontal arrangement. The ventral portion of each segment tends to be located at a more caudal level than the dorsal
   3. Within each band or segment, the nerve supply is not absolute. Each nerve supplies not only its own segment but half of each adjacent segment. As a result, destruction of a single segmental nerve will not cause a band of anesthesia (total loss of sensation) but hypoesthesia (partial loss of sensation)

G. THE SKIN OF THE NECK where it joins the trunk is innervated by C5 which is immediately adjacent to a segment innervated by T1. This does not mean that C6, C7, and C8 are omitted. Instead, this is merely the cutaneous representation of the fact that these intervening nerves take part in the formation of a special plexus to supply the superior extremity

X-SECTION THROUGH CORD AT ROOT ENTRY

Post. root
Dorsal root gang.
Nerve
Post. primary ramus
White
Gray
Rami communicantes
Lat. cutaneous
Ant. root
Sympathetic ganglion
Sympathetic chain
Ant. primary ramus

CORD SEGMENT

Spinal nerve
Post. (dorsal) root
Dorsal root gang.
Ant. (ventral) root

AN INTERCOSTAL SPINAL NERVE

Dorsal root gang.
Post. ramus
Med. br.
Lat. br.
Post. root
Post. br.
Rami communicantes
Sympathetic gang.
Ant. root
Ant. primary ramus
Lat. cutan. br.
Internal intercostal m.
Ant. br.
Ant. cutaneous
Ext. intercostal m.

DERMATOMES OF BODY
(After Reegan and Garrett)

C2
C3
C4
C5
T1
2
3
4
5
6
7
8
9
10
11
T12
L1
L2
L3
L4
L5
S1
C6
C7
C8
S2
C2
C3
C4
C5
C6
T12
L1
L4
C7
S3
S5
C1
C8
L1
L2
L5
L3
S2
S1
S2
L4
L5

-71-

34. SPINAL CORD: CERVICAL AND BRACHIAL PLEXI

The three major plexi (cervical, brachial, and lumbosacral) are formed from only the ventral rami of the spinal nerves. Dorsal rami in all cases are not involved and follow the typical segmental distribution just described

I. **Cervical plexus** is formed from the ventral primary rami of spinal nerves C1–C4. Main branches: phrenic nerve, motor branches to prevertebral and infrahyoid muscles of neck, and cutaneous branches to the neck and head

II. **Brachial plexus** is formed from the ventral rami of cervical nerves C5, C6, C7, C8, and most of T1 consists of
   A. ROOTS: used to identify the five spinal nerve components (ventral rami of C5–T1) until they combine to form distinct trunks
      1. Roots from C5 and C6 form the superior trunk
      2. Roots C8 and T1 form the inferior trunk
      3. Root from C7 continues alone to form the middle trunk
   B. THE THREE TRUNKS DIVIDE into an anterior and a posterior division which recombine to form three cords
      1. Posterior cord: formed by the posterior divisions of the three trunks
      2. Lateral cord: formed by anterior divisions of the upper and middle trunks
      3. Medial cord: formed by anterior division of the inferior trunk
   C. EACH CORD DIVIDES into two branches each of which either is a named nerve or contributes to a nerve as shown below
      1. Posterior cord gives rise to the axillary and the radial nerves which supply the extensor muscles of the upper extremity
      2. Lateral cord gives rise to musculocutaneous nerve and a branch to the median nerve
      3. Medial cord gives rise to the ulnar nerve and a branch to the median nerve
   D. TEN MORE SMALL NERVES take origin from the brachial plexus
      1. Dorsal scapular nerve arises directly from root C5
      2. Suprascapular nerve arises from a combination of C5 and C6
      3. Long thoracic nerve arises from C5, C6, and C7
      4. Lateral pectoral nerve arises from the lateral cord
      5. Medial pectoral nerve arises from the medial cord
      6. Medial brachial cutaneous nerve to arm from medial cord
      7. Medial antebrachial cutaneous nerve to forearm from medial cord
      8. Upper, middle, and lower subscapular nerves supply extensor muscles and arise from the posterior cord
   E. NERVES OF THE ARM, FOREARM, AND HAND
      1. Axillary nerve supplies the teres minor and deltoid muscles
      2. Radial nerve supplies all arm and forearm extensor muscles
      3. Musculocutaneous nerve supplies all the flexor muscles of the brachium or arm (coracobrachialis, biceps brachii, brachialis)
      4. Median nerve supplies nothing in the arm, but most of the flexor muscles of the forearm and part of the hand muscles
      5. Ulnar nerve supplies nothing in the arm, only two flexor muscles of the forearm (flexor carpi ulnaris and flexor digitorum profundus), but most of the muscles of the hand

CERVICAL PLEXUS
Muscle branches solid black

CUTANEOUS NERVES

BRACHIAL PLEXUS

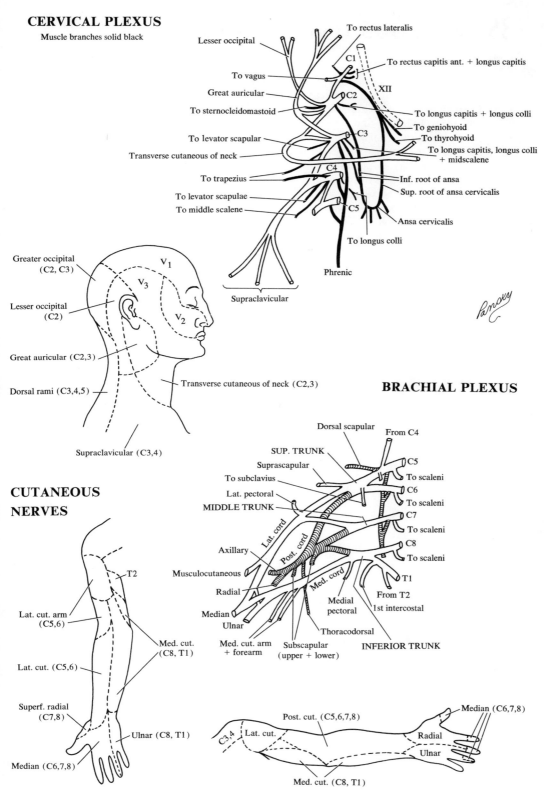

35. SPINAL CORD: LUMBOSACRAL PLEXUS

I. **Lumbosacral plexus** gives rise to the nerves that innervate the muscles of the hip, thigh, leg, and foot. In a very real sense, this plexus is actually comprised of two plexuses

A. LUMBAR PLEXUS is comprised of ventral rami of spinal nerves L1–L4 plus a contribution from T12. T12 and L1 supply cutaneous innervation primarily, therefore, it is simpler to think of the lumbar plexus as being the ventral rami of L2, L3, and L4. As in the case of the brachial plexus, muscles in both the original ventral (flexor) and original dorsal (extensor) groups are innervated by separate nerves

    1. Anterior divisions of the ventral rami of spinal nerves from L2–L4 combine to form the obturator nerve

    2. Posterior fibers of the anterior division of L2–L4 combine to form the femoral nerve

B. SACRAL PLEXUS is comprised of ventral rami of spinal nerves L4–L5 and S1–S3 which combine into a single dissectable nerve, the sciatic nerve, characterized by the following

    1. The sciatic nerve is actually two nerves in a common wrapper which separate into the tibial and common peroneal nerves which in turn supply muscles in the leg and foot

    2. As in the case of the lumbar plexus, two major nerves are formed from anterior and posterior divisions of the ventral rami of the spinal nerves

        a. Anterior divisions of L4–L5 and S1–S3 combine to form the tibial nerve which supplies the original ventral or flexor muscle group of the leg (located posteriorly) and the plantar surface of the foot

        b. Posterior divisions combine to form the common peroneal nerve which supplies muscles of the two original dorsal compartments of the leg which are located anteriorly and laterally (extensors and muscles on the dorsum of the foot)

CUTANEOUS NERVES

Subcostal (T12)

Femoral br. genitofemoral (L1,2)

Ilioinguinal (L1)

Lat. cutaneous of thigh (L2,3)

Obturator (L2,3,4)

Med. + intermed. cutaneous (L2,3)

Lat. cutaneous of leg (L5, S1,2)

Saphenous (L3,4)

Superf. peroneal (L4,5, S1)

Sural (S1,2)

Deep peroneal (L4,5, S1,2)

T12
L1
Iliohypogastric
Ilioinguinal
Genitofemoral
L2
Ventral rami
Lat. cutaneous of thigh
L3
L4
To psoas + iliacus
Femoral
Accessory obturator
Obturator
L5
Lumbosacral trunk

LUMBAR PLEXUS
DORSAL DIVISIONS—SHADED

Iliohypogastric (L1)

Dorsal rami (L1,2,3)

Dorsal rami (S1,2,3)

Lat. cut. thigh (L2,3)

Obturator (L2,3,4)

Med. cut. thigh (L2,3)

Post. cut. thigh (S1,2,3)

Lat. cut. leg (L4,5, S1)

Saphenous (L3,4)

Sural (L5, S1,2)

Med. calcaneal of tibial (S1,2)

L4
L5
S1
Sup. gluteal
Inf. gluteal
To piriformis
Post. femoral cut.
S2
Visceral
To obturator int. + sup. gemellus
S3
Perforat. cutaneous
S4
S5
Sciatic { Common peroneal / Tibial }
Co.
To quad. fem. + inf. gemellus
Pelvic splanchnics
Pudendal
Anococcygeal
To levator ani, coccygeus + sphincter ani ext.

SACRAL + COCCYGEAL PLEXI

36. SPINAL CORD: BLOOD SUPPLY

I. Arterial supply to the spinal cord is derived from two main sources

A. VERTEBRAL ARTERIES give rise to an anterior spinal artery and a pair of posterior spinal arteries

    1. Anterior spinal artery gives rise to a number of sulcal branches that enter the ventral median fissure. These branches supply the anterior and lateral funiculi and gray horns, central gray, and the basal portion of the posterior gray horn

    2. Posterior spinal arteries supply the posterior gray horns and the posterior funiculus (white matter). These paired spinal arteries traverse the posterior surface of the spinal cord, receive branches from the posterior radicular arteries, and then form a longitudinal plexiform channel near the point of entrance of the dorsal root

B. SEGMENTAL ARTERIES (ascending cervical, intercostal, lumbar, and sacral arteries) give rise to radicular arteries

    1. Each radicular artery travels along the spinal nerve, passes through the intervertebral foramina, and separates into small anterior and larger posterior radicular arteries. These radicular arteries provide the main blood supply to thoracic, lumbar, sacral, and coccygeal spinal segments

II. Venous drainage pattern is similar to and closely parallels the spinal arteries.* The following veins drain into an internal vertebral venous plexus (epidural in position) before draining into thoracic, abdominal, and intercostal veins

A. A single anterior median vein

B. A single posterior median vein

C. Paired anterior lateral veins

D. Paired posterior lateral veins

E. Vasocorona veins

*No valves are found in the spinal venous network.

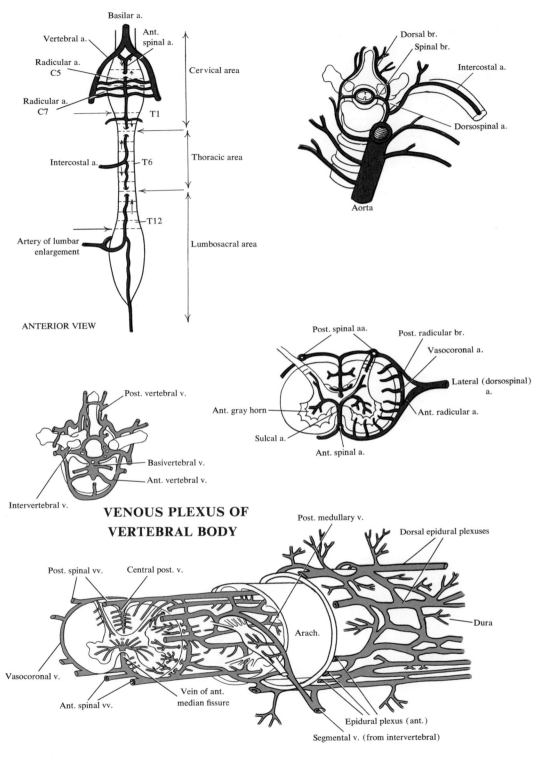

Basilar a.

Vertebral a.

Ant.
spinal a.

Radicular a.
C5

Radicular a.
C7

T1

Cervical area

Intercostal a.

T6

Thoracic area

T12

Artery of lumbar
enlargement

Lumbosacral area

ANTERIOR VIEW

Dorsal br.

Spinal br.

Intercostal a.

Dorsospinal a.

Aorta

Post. spinal aa.

Post. radicular br.

Vasocoronal a.

Ant. gray horn

Lateral (dorsospinal)
a.

Sulcal a.

Ant. radicular a.

Ant. spinal a.

Post. vertebral v.

Basivertebral v.

Ant. vertebral v.

Intervertebral v.

**VENOUS PLEXUS OF
VERTEBRAL BODY**

Post. medullary v.

Dorsal epidural plexuses

Post. spinal vv.

Central post. v.

Arach.

Dura

Vasocoronal v.

Vein of ant.
median fissure

Ant. spinal vv.

Epidural plexus (ant.)

Segmental v. (from intervertebral)

ARTERIES AND VEINS OF SPINAL CORD

Pansky

-77-

37. BRAINSTEM: DIENCEPHALON

I. **Introduction:** the brainstem is a collective term for the diencephalon, mesencephalon, metencephalon, and myelencephalon. It is that part of the brain remaining after the cerebral hemispheres and the cerebellum are removed. The definition of the term "brainstem" is somewhat variable. Some authors exclude the diencephalon, while others may include the deep structures of the telencephalon

II. **The diencephalon** lies between the cerebrum and midbrain and encloses the third ventricle. Its perimeter, viewed medially, includes the choroid plexus of the third ventricle; habenular commissure; pineal body (gland); posterior commissure; a line drawn through the posterior commissure and mamillary bodies (posterior boundary); mamillary bodies; tuber cinereum; hypophysis; optic chiasma; and a line drawn through the optic chiasma and the interventricular foramen (anterior boundary)

A. SUBDIVISIONS OF THE DIENCEPHALON (viewed in midsagittal plane)
   1. Epithalamus is the narrow band on the roof of the diencephalon and consists of
      a. Pineal body: a small mass lying in the depression between the superior colliculi
      b. Posterior commissure: comprised of myelinated fibers that cross the median plane on the dorsal aspect of the rostral end of the cerebral aqueduct. Some of its fibers interconnect the superior colliculi
      c. Habenula (habenular trigone) is a small triangular area anterior to the superior colliculus and contains the habenular nuclei (receives fibers from the stria medullaris)
      d. Habenular commissure: comprised of fibers interconnecting the nuclei
      e. Stria medullaris thalami runs parallel to the choroid plexus and is a fiber bundle from the septal area to the habenular nuclei
   2. Thalamus is the largest subdivision of the diencephalon. It is a large ovoid gray mass located above the hypothalamic sulcus and on either side of the third ventricle
      a. It is the major relay (integrative) station interposed between many subcortical structures and the cerebral cortex
      b. The massa intermedia (interthalamic adhesion) is a site of fusion of the dorsal thalami across the midline (absent in 25–30% cases)
   3. Hypothalamus is located below or ventral to the thalamus and forms the floor and the ventral half of the lateral walls of the third ventricle
      a. It contains the highest integrative centers of the ANS
      b. It is involved in the regulation of body temperature, endocrine gland functions, and the expression of emotions
      c. Hypothalamic structures visible on the ventral surface of the diencephalon include the optic chiasm and optic tracts, which delineate the floor of the diencephalon; mamillary bodies, which are paired white masses located inferior to the gray matter of the hypothalamic floor in front of the posterior perforated substance; tuber cinereum, which is a funnel-shaped eminence rostral to the mamillary bodies; median eminence, which is the most inferior extension of the tuber cinereum; and the infundibulum, a collective term for the median eminence and the infundibular stalk (a hollow process extending from the tuber cinereum to the posterior hypophysis)
   4. Subthalamus lies ventral to the thalamus and flanks the hypothalamus laterally
      a. It has no visible or prominent external structures but is an important subcortical station for voluntary muscle activities

Velum interpositum (cut)

Stria medullaris

Septum pellucidum

Hippocampal commissure

Choroid plexus

Interventricular foramen

Suprapineal recess

Body

Column of fornix

Crus of formix

Anterior commissure

Corpus callosum

Paraterminal body

Splenium

Habenular commissure

Genu

Longitudinal stria

Pineal body

Subcallosal gyrus

Post. commissure

Parolfactory area

Colliculi { Sup. / Inf. }

Lamina terminalis

Preoptic recess

Ant. medullary velum

Optic chiasma

Cerebral aqueduct

Pons

Infundibulum

Hypophysis

Medulla

Tuber cinereum

III

Uncus of hippocampal gyrus

Mamillary body

**MEDIAN SAGITTAL SECTION
THROUGH BRAINSTEM**

Hypothalamus + hypothalamic sulcus

Choroid plexus

Body of fornix

Tela choroidea

Corpus callosum

Lat. ventricle

Caudate nucleus

Thalamic nuclei { Ant. / Med. / Lat. }

IIIrd ventricle

Subthalamic nucleus

Putamen

Reticular stratum
(thalamus)

Globus pallidus

Lat. sulcus

Insula

Int. capsule

Claustrum

Inf. horn lat. vent.

Pons

Caudate nucleus (tail)

Substantia nigra

III

Crus cerebri   Mamillary body

Optic tract

CORONAL SECTION OF BRAIN—LEVEL OF MAMILLARY BODY

38. BRAINSTEM: MESENCEPHALON

I. **The mesencephalon or midbrain** is the smallest of the four major subdivisions of the brainstem and is positioned between the pons and diencephalon. Its anterior boundary is formed by a line drawn through the posterior commissure dorsally and a point just caudal to the mamillary bodies ventrally

A. PROMINENT EXTERNAL FEATURES ON THE MIDBRAIN

1. On the ventral surface, elevations on each side are formed by the cerebral peduncles (basis pedunculi, crus cerebri)

   a. The peduncles consist of nerve fibers of the pyramidal motor and corticopontine fibers

   b. They converge with one another on the pons as they emerge from the cerebrum in a position medial to the optic tracts

   c. The interpeduncular fossa lies between the cerebral peduncles

      i. Cranial nerve III (oculomotor) emerges from the side of this fossa

      ii. The floor of the fossa is known as the posterior perforated substance due to the many perforations produced by blood vessels that penetrate the midbrain

2. On the dorsal surface, four rounded eminences, the corpora quadrigemina, are arranged in pairs and called the superior and inferior colliculi

   a. The colliculi form the roof or tectum of the midbrain which measures about 1.5 cm long

      i. The superior colliculi are larger and darker and are associated with the optic system. The brachium (bridgelike structure) of the superior colliculus connects with the lateral geniculate body and contains fibers from the cerebral cortex and retina to the superior colliculus. The latter is primarily a reflex center for movements of the eyes and head in response to visual and other stimuli

      ii. The inferior colliculi are usually more prominent and are important relay nuclei on the auditory pathway to the thalamus. The brachium of the inferior colliculus connects with the medial geniculate body

   b. The cruciate sulcus separates the superior and inferior colliculi

   c. Cranial nerve IV (trochlear) emerges from the brainstem just inferior to the inferior colliculus and curves around the brainstem

MEDIAN SAGITTAL SECTION

Sup. + inf. colliculi

Aqueduct of Sylvius

Cerebellum

Pons

III

Mesencephalon

Mamillary body

Post. perforated area (substance)

IIIrd ventricle

Habenular trigone

Pineal gland

Pulvinar

Sup. + inf. colliculi

Cruciate sulcus

Med.
Lat. } Geniculate bodies

Cerebral peduncle (crus cerebri)

Taenia pontis

Pons

Brachium pontis

VII + VIII

VI

IV

V

IX

Facial colliculus

Obex

Gracilis tubercle

Cuneate tubercle

Post. median fissure

X

XII

XI

Tuberculum cinereum

C1

C2

DORSOLATERAL VIEW

I

II

Mamillary bodies

Ant. perforated substance

Hypophysis

III

Posterior perforated substance

Cerebral peduncle

IV

Interpeduncular fossa

V

VII

VIII

IX

X

XI

XII

VENTRAL VIEW

-81-

39. BRAINSTEM: INTERNAL GROSS ANATOMY OF THE MESENCEPHALON

I. The internal gross anatomy of the mesencephalon is well demonstrated in transverse sections at the following levels

A. LEVEL OF THE SUPERIOR COLLICULUS: since the posterior part of the thalamus bulges into the midbrain during development and maturation of the brain, certain thalamic nuclei, medial and lateral geniculate bodies, and the pulvinar may be identified. Major structures also identified at this level and associated with the midbrain include

1. Nucleus of cranial nerve III in the periaqueductal gray
2. Fibers of nerve II as they course to their exit
3. Medial longitudinal fasciculi (MLF) form a V at this level with the oculomotor nucleus cradled in the arms of the V
4. Red nucleus and substantia nigra extend the length of the mesencephalon and into the diencephalon
5. Cerebral aqueduct, medial lemniscus, brachium of the inferior colliculus, and the basis pedunculi are also present

B. LEVEL OF THE INFERIOR COLLICULUS

1. Lateral lemniscus and fibers from the lateral lemniscus form a covering for the inferior colliculus
2. Brachium of the inferior colliculus is formed by fibers that pass between the inferior colliculus and the medial geniculate body of the thalamus
3. Cerebral aqueduct connects the third and fourth ventricles
4. Periaqueductal gray, a rather extensive area of gray matter which surrounds the cerebral aqueduct
5. Nucleus of cranial nerve IV on each side including a group of neurons located on the ventromedian aspect of the periaqueductal gray; the nucleus itself is located on and actually invaginates the dorsal surface of a fiber tract known as the MLF
6. Other structures that can be seen and identified at this level include
   a. The medial lemniscus
   b. The central tegmental tract
   c. The decussation of the superior cerebellar peduncles

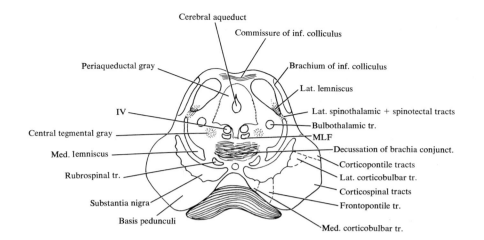

C. MIDBRAIN: LEVEL OF INFERIOR COLLICULUS

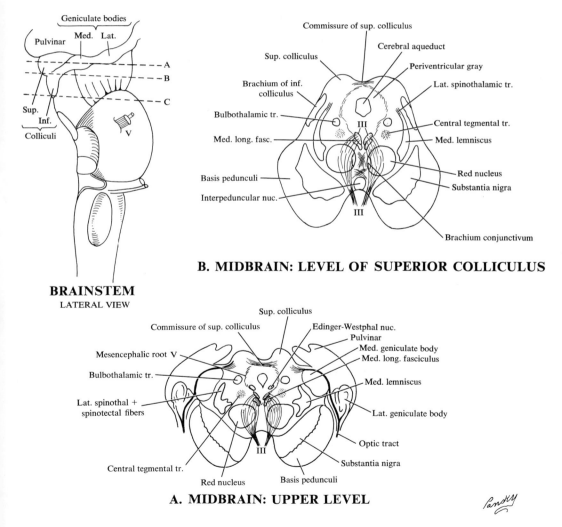

B. MIDBRAIN: LEVEL OF SUPERIOR COLLICULUS

BRAINSTEM
LATERAL VIEW

A. MIDBRAIN: UPPER LEVEL

40. BRAINSTEM: METENCEPHALON

I. **The metencephalon** is composed of both the pons and cerebellum in the adult. Since the latter is not defined as part of the brainstem proper, only the pons will be considered here

A. LOCATION OF THE PONS: ventral to the cerebellum and anterior to the medulla. The term pons means "bridge" and refers to the prominent ventral bulge or bridge between the cerebellar hemispheres

B. BOUNDARIES OF THE PONS: the rostral boundary is formed by a line through the superior pontine sulcus on the ventral surface and the root of cranial nerve IV on the dorsal surface; the caudal boundary is established by a line through the inferior pontine sulcus on the ventral surface and the stria medullares

C. EXTERNAL GROSS FEATURES OF THE PONS: the dorsal pons is concealed by the cerebellum, while the ventral pons is comprised of a ventral bulge which is continuous with the cerebellum via the large, paired brachium pontis (middle cerebellar peduncles)

1. Ventral surface of the pons has closely associated with it four cranial nerves
   a. Trigeminal (V) nerve is located on its lateral aspect
   b. Nerves VI, VII, and VIII are located in the inferior pontine sulcus which marks the caudal boundary of the pons
   c. Nerves V and VII are comprised of two roots of unequal size. The larger root for nerve V is the sensory root (portio major) and the smaller root is the motor root (portio minor). The larger root of VII is motor, while the smaller root is the intermediate nerve
   d. Nerve VIII is comprised of two separate and distinct nerves, the vestibular nerve (ventromedial) and the cochlear nerve (dorsolateral)

2. Dorsal surface of the pons is formed by the rostral half of the fourth ventricle. The lateral walls of this half of the fourth ventricle are comprised of the three pairs of cerebellar peduncles that form the lateral aspect of the pons and serve to connect the pons with the brainstem
   a. Median eminence lies on the floor of the fourth ventricle
   b. Facial colliculus is a prominent structure located on each side of the median eminence
   c. Locus ceruleus is a bluish area on each side of the median eminence positioned rostral to the vestibular area
   d. Median sulcus lies in the midsagittal plane
   e. Sulcus limitans is the more lateral sulcus separating the median eminence from the vestibular area

D. MAJOR INTERNAL GROSS FEATURES OF THE PONS: at the midpontine level, the following structures are present (along with other structures which originate at other levels)
   1. Superior medullary velum is located in the roof of the rostral portion of the fourth ventricle
   2. Superior cerebellar peduncles form the lateral walls of the fourth ventricle
   3. Corticospinal (pyramidal) fasciculi and pontine nuclei are positioned in the basilar portion of the pons
   4. Transverse pontine fibers pass to the cerebellum via the brachium pontis
   5. Main sensory nucleus, motor nucleus, and fibers of the trigeminal (V) nerve
   6. Medial lemniscus and trapezoid body are situated in the dorsal pons or tegmentum
   7. Other internal gross features depicted at the levels of the facial colliculus and cranial nerve IV are considered elsewhere in the brainstem

BRAINSTEM

DORSAL VIEW

Post. commissure

Habenular trigone

Pineal body

Pulvinar

Geniculate bodies — Lat. / Med.

Sup. / Inf. } Colliculi

Taenia pontis

Crus cerebri

Medial eminence

Sup. fovea (locus coeruleus)

Sup. / Middle / Inf. } Cerebellar peduncles

Sulcus limitans

Facial colliculus

Stria medullaris

Vestibular area

Hypoglossal trigone

Sulcus limitans

Vagal trigone

Cuneate tubercle

Obex

Tuberculum cinereum

Clava (gracile tubercle)

Fasciculi { Gracilis / Cuneatus

Nuc. lat. lemniscus

IV

Mesencephalic root V

MLF

Lat. lemniscus + lat. spinothalamic + spinotectal tracts

Med. lemniscus

Rubrospinal tr.

Decussation of brachia conj.

Corticopontile, corticobulbar + corticospinal tracts

A. UPPER PONS SECTION

Ant. medullary velum

Vent. spinocerebellar tr.

Lat. spinothat. + spinotectal tracts

Brachium conjunctivum

Mesenceph. nuc. V

Motor root V

Main sensory nuc. V

Brachium pontis

Sup. olivary nuc.

Trapezoid body + med. lemniscus

Rubrospinal tr.

Lat. lemniscus

V · V

B. MIDPONS SECTION

Sup. vestibular nuc. (VIII)

MLF

VI

VII

Spinal V tract

Lat. lemniscus

Nuc. spinal V tract

Rubrospinal + tectobulbar tracts

Lat. lemniscus

VIII

VII

Trapezoid body + med. lemniscus

VI

C. LOWER PONS SECTION

Panory

41. BRAINSTEM: MYELENCEPHALON

I. **The myelencephalon** refers to the medulla (medulla oblongata) in the adult. It is the pyramid-shaped portion of the brainstem which is located between the pons and the spinal cord. The rostral half of its dorsal portion forms the caudal half of the floor of the fourth ventricle.

A. BOUNDARIES OF THE MEDULLA are represented rostrally by a line through the stria medullares and the inferior pontine sulcus; caudally by a transverse plane through the rootlets of the first cervical spinal nerve

B. THE VENTRAL SURFACE OF THE MEDULLA reveals the pyramids, olives, pyramidal decussation, rootlets of cranial nerves IX, X, XI, and XII. Cranial nerves IX, X, and XI exit in the postolivary sulcus, while XII exits in the preolivary sulcus.

C. THE DORSAL SURFACE OF THE MEDULLA (open portion) consists of the caudal portion of the floor of the fourth ventricle. Structures found on this portion of the floor inferior to the stria medullares include
   1. Calamus scriptorius is the caudal portion (below the stria medullares) of the paired median eminences
   2. Hypoglossal and vagal trigones, which represent prominences caused by underlying nuclei, comprise each half of the calamus scriptorius

D. LATERAL SURFACE OF THE MEDULLA is comprised of the following named prominences
   1. Gracilis tubercle
   2. Cunate tubercle
   3. Tuberculum cinereum
   4. Olive
   5. Pyramid

E. MAJOR INTERNAL ANATOMIC FEATURES (at the level of the olive)
   1. Fasciculus and nucleus gracilis form the clava or gracilis tubercle on the dorsal surface of the medulla
   2. Fasciculus and nucleus cuneatus form the cuneate tubercle on the dorsal surface
   3. Spinal nucleus and tract of the trigeminal (V) nerve form the tuberculum cinereum
   4. Inferior olivary nuclear complex causes the bulges on the surface known as the olive
   5. Corticospinal tracts comprise the pyramids at the level of the medulla
   6. Other structures present include
      a. Nucleus of the hypoglossal (XII) nerve which forms the hypoglossal trigone
      b. Dorsal motor nucleus of the vagus (X) nerve which forms the vagal trigone
      c. Vestibular area
      d. Medial longitudinal fasciculus (MLF)
      e. Spinothalamic tract

B. MEDULLA: LEVEL OF LOWER IVth VENTRICLE

Nuc. prepositus (XII)
Inf. salivatory nuc.
Nuc. + tr. solitarius
Vest. bulbospinal tr.
Cochlear nuclei
Restiform body

Lat. vest. nuc.
Med. vest. nuc.
MLF
Vent. cochlear nuc.

VIII
Inf. olivary nuc.
Pyramid
Arcuate nuc.
Med. lemniscus

IX
Nuc. ambiguous

Dorsal nuc. X
Solitary nuc.
Lat. cuneate nuc.
Spinal tr. V
MLF
XII
Nuc. spinal tr.
X
Restiform body
Reticular form.
Inf. olivary nuc.
Dorsal + medial
access. oliv. nuc.
Pyramid
XII
Med. lemniscus
Arcuate nuc.

A. MEDULLA: LEVEL OF IXth NERVE

Colliculi
Collicular brachii
Geniculate bodies

Crus cerebri
IV
Medial eminence
Taenia pontis
Sup.
Inf. } Cerebellar peduncles
Mid.
Sup. fovea
(locus ceruleus)
Stria medullaris
Facial colliculus
Hypoglossal trigone
Vestibular area
Vagal trigone (ala cinerea)
Sulcus limitans
Tuberculum cinereum
Obex
Clava (gracile tubercle)
Gracile }
Cuneate } Fasciculi
Lat. column

DORSAL VIEW

A ----
B ----
C ----
D ----

IX
XII
X

XI

VENTRAL VIEW

Nucleus
Fasciculus } Gracilis
Cuneate fasciculus
Nuc. spinal V tr.
Spinal tr. V
XI
Lat. corticospinal tr.
Spinocerebellar
tracts
Pyramidal decussation

Nuclei { Gracile
Cuneate
Solitary nuc.
Nuc. spinal tr. V
Dorsal X nuc.
Spinothal. tr.
MLF
Ant. horn
Med. lemniscus

D. MEDULLA: LOWER PART

Med. access.
olivary nuc.
XII
Pyramid
Inf. olivary nuc.

Arcuate nuc.

**C. MEDULLA:
LEVEL OF SENSORY DECUSSATION**

42. GROSS MORPHOLOGY OF BRAIN

I. **Introduction:** the brain (encephalon) is the greatly enlarged and modified superior portion of the central nervous system. It is ensheathed by three protective, connective tissue membranes (meninges) and is contained within the cranial vault of the skull. Division into cerebrum, cerebellum, and brainstem provides a useful basis for the study of gross structures of the brain

II. **Basic subdivisions of the brain:** the brain is generally divided into five major divisions
   A. TELENCEPHALON (ENDBRAIN) gives rise to the bulk of the cerebrum in the adult and forms from the prosencephalon or forebrain
   B. DIENCEPHALON (BETWEEN BRAINS) originates as the second vesicle of the original prosencephalon
   C. MESENCEPHALON (MIDBRAIN) is the original middle vesicle and remains relatively unchanged. The term "mesencephalon" is retained
   D. METENCEPHALON (AFTERBRAIN) forms when the original rhombencephalon divides into two segments. It forms the pons and cerebellum in the adult
   E. MYELENCEPHALON (MEDULLA OBLONGATA) is the lowest division or region of the brainstem. It is the inferior segment of the original rhombencephalon

III. **Topography of the brain:** the two cerebral hemispheres, which make up the largest portion of the brain, are separated by the long, deep longitudinal fissure. The outer surface of each hemisphere is marked by a varied series of convolutions (elevated ridges) called gyri, and grooves termed sulci. Deeper furrows between some gyri are called fissures. The terms "sulci" and "fissures" are often used synonymously
   A. LOBES AND MAJOR FISSURES OR SULCI of the cerebral hemispheres; each hemisphere is subdivided into six lobes
      1. Frontal lobe is located rostral to the central sulcus (fissure of Rolando)
      2. Parietal lobe lies between the central sulcus and the parieto-occipital groove or sulcus
      3. Occipital lobe is positioned posterior to the parieto-occipital line and sulcus
      4. Temporal lobe lies inferior to the lateral (sylvian) fissure and rostral to the parieto-occipital line
      5. Central (insular) lobe lies deep to the temporal lobe and can be seen only when this lobe is retracted
      6. Limbic lobe is located on the medial aspect of each hemisphere along with medial portions of the frontal, parietal, and occipital lobes

Rhombencephalon (R)
Mesencephalon (M)
Prosencephalon (P)

Telencephalon
Diencephalon

R M D

T
M D
T

T
D
M T

Metencephalon
Myelencephalon (medulla)

T
D
Myel. Met. M

Tel.
Mes.
Cerebellum
Pons
Medulla
Cord

Forebrain (P)
Midbrain (mes.)
Hindbrain (R)

5
10
3 WEEKS
Cord

Mes. Met.
5
D
8
T
7 9 10
Cord
5 WEEKS
Myel.

Mes. Met.
D
5
T
Myel.
7 8 9 10
7 WEEKS

Tel.
Dien.
Sup.
Inf.
Colliculi
Cerebellum
Pons
Medulla

11 WEEKS

LOBES

Central sulcus

FRONTAL LOBE

PARIETAL LOBE

Frontal pole

Parieto-occipital sulcus

Lateral (sylvian) fissure
TEMPORAL LOBE
Temporal pole

OCCIPITAL LOBE

Occipital pole

LATERAL VIEW

Central sulcus

FRONTAL LOBE

PARIETAL LOBE

Corpus callosum

Parieto-occipital sulcus

OCCIPITAL LOBE

LIMBIC LOBE

Calcarine fissure

MEDIAL VIEW

TEMPORAL LOBE

Pansky

–89–

43. GYRI OF THE CEREBRAL HEMISPHERES— IMPORTANT LANDMARKS

I. Frontal lobe: recall that the central sulcus or fissure of Rolando separates the frontal and parietal lobes. Much of the frontal lobe is a vast unchartered ocean, but its function in higher cognitive processes, personality, etc., is likely

    A. PRECENTRAL GYRUS is referred to as the "motor cortex" and contains neurons whose axons give rise to corticospinal and corticobulbar (brainstem) tracts

    B. SUPERIOR AND MIDDLE FRONTAL GYRI are cortical areas in front of the precentral gyrus for the control of body and eye movements

    C. INFERIOR FRONTAL GYRUS has a region referred to as Broca's area (44, 45) which is important in the mechanics of speech production, especially in the "dominant" (usually left) hemisphere

    D. ORBITAL GYRI of the frontal lobe (inferior surface of the hemisphere) give rise to pathways that are important in the expression of emotion

II. Parietal lobe has two major sulci, the postcentral sulcus and the interparietal sulcus. These divide the lobe into

    A. POSTCENTRAL GYRUS is referred to as the "sensory cortex" and is the receiving area concerned with the appreciation of touch (somesthesis), the sense of position of the extremities (kinesthesis), and vibratory sense and other fine tactile discriminatory processes

    B. SUPERIOR AND INFERIOR PARIETAL LOBULES (LITTLE LOBES) are located posterior to the postcentral gyrus and are important in the synthetic aspects of multiple sensory experiences brought to consciousness

    C. A SUPRAMARGINAL AND AN ANGULAR GYRUS are subdivisions of the inferior parietal lobule. These gyri are important in the reception and organization of language functions, especially in the "dominant" hemisphere

III. Occipital lobe is divided by secondary occipital sulci into several unnamed gyri. On the medial surface of the hemisphere, the calcarine sulcus separates gyri that contain neurons specialized to receive visual information

IV. Temporal lobe has two sulci that form the boundaries of three gyri. These include the superior and inferior sulci and the superior, middle, and inferior gyri

    A. TRANSVERSE TEMPORAL GYRI (OF HESCHL) are found on the superior face of this lobe facing the lateral fissure and extending medially toward the insular (central) lobe

    B. THIS REGION RECEIVES AUDITORY INFORMATION, while the inferior temporal gyrus is important in processing visual information

V. Central (insular) lobe is buried deep in the lateral fissure and surrounded by the circular sulcus

    A. SHORT (IN THE ROSTRAL INSULA) AND LONG (IN THE POSTERIOR INSULA) OBLIQUE GYRI comprise most of the insular cortex

    B. THE LIMEN INSULA is a tonguelike extension of the insula that projects medially to the anterior perforated substance

VI. Limbic lobe and related gyri form a ring of cortex and associated structures surrounding the central core of the medial aspect of the cerebrum. These structures are important in mechanisms of emotional expressions (see "Limbic System")

Transverse sinus

SURFACE ANATOMY OF CEREBRUM

Precentral sulcus

Ant. central gyrus

Parts of inferior frontal gyrus { Triangular / Opercular / Orbital

Operculum

Central sulcus

Post. central gyrus

Interparietal sulcus

Supramarginal gyrus

Interparietal sulcus

Superior frontal gyrus

Angular gyrus

Inferior parietal lobule

Middle frontal gyrus

Parieto-occipital sulcus

Lateral occipital gyri

Orbital gyrus

Lateral cerebral fissure

Superior temporal gyrus

Superior temporal sulcus

Transverse occipital sulcus

Middle temporal gyrus

Middle temporal sulcus

Inferior temporal gyrus

Post. limb of lateral (sylvian) fissure (sulcus)

Circular sulcus

CENTRAL (INSULAR) LOBE

Short gyri

Limen insula

Long gyrus

Central sulcus

44. GROSS STRUCTURES OF BRAIN: MIDSAGITTAL ASPECT

I. The medial surface of the hemisphere consists of many named gyri and sulci or fissures

A. CINGULATE GYRUS is a crescentric convolution between the cingulate sulcus and the corpus callosum

B. THE UNCUS is wedge-shaped and lies between calcarine and parieto-occipital fissures

C. PARAHIPPOCAMPAL GYRUS lies between the hippocampal and collateral fissures

D. OCCIPITOTEMPORAL GYRUS (FUSIFORM) is medial to the inferior temporal sulcus

E. LINGUAL GYRUS is between the calcarine and collateral fissures

F. CUNEUS is the curved (hooked) anterior part of the parahippocampal gyrus

G. PRECUNEUS is separated from the cuneus gyrus by the parieto-occipital fissure

H. PARACENTRAL LOBULE is the quadrilateral gyrus around the end of the central sulcus

I. SUPERIOR FRONTAL GYRUS

II. The corpus callosum is a huge bundle of nerve fibers from neurons located entirely in the cortex. It is a major commissural (interhemispheric) pathway connecting the cerebral cortex on one side with its corresponding or homologous area on the other side.

A. THE ROSTRUM, GENU, BODY, AND SPLENIUM are divisions of the corpus callosum

B. THE FORCEPS MAJOR is formed by the splenium and the callosal fibers (U-shaped bundles) radiating to the occipital lobe

C. THE FORCEPS MINOR is formed by the genu and the U-shaped bundles of callosal fibers radiating to the frontal lobe

D. THE GENU blends with the rostrum, which is continuous with the lamina terminalis, a plate of neural tissue which forms the anterior boundary of the third cerebral ventricle

E. THE TAPETUM is a thin sheet of callosal fibers that radiate to line the outer border of the inferior and posterior horns of the lateral ventricles

F. THE INDUSIUM GRISEUM (SUPRACALLOSAL GYRUS) is a thin vestigial layer of gray matter found on the superior surface of the corpus callosum and parallel to its long axis. It contains two bilateral pairs of fiber bundles called the lateral and medial longitudinal striae. The striae are continuous with the septal region rostrally and with the fascicular gyrus caudally

III. The anterior commissure is a bundle of white fibers that crosses the midline to interconnect the cerebral hemispheres. The rostral portion joins both olfactory bulbs; the remainder connects the piriform areas of both hemispheres

IV. The interventricular foramina (of Monro) serve for the passage of cerebrospinal fluid from the lateral ventricles to the third ventricle

V. The median telencephalon includes the following structures: anterior commissure; lamina terminalis; and neural tissue anterior to a line drawn through the foramen of Monro and the optic chiasma

VI. The fornix is a large bundle of axons forming an arc from the hippocampal formation to the mamillary body and part of the limbic system

VII. The septum pellucidum is a thin-walled paired structure (midplane) separating the lateral ventricles and extending between the fornix and corpus callosum. They are usually separated in the midline by a small cavity (cavum septum pellucidum)

VIII. The transverse cerebral fissure separates thalamus, mesencephalon, and cerebellum from the cerebral hemispheres

IX. The subcallosal (paraterminal) gyrus is cortical gray matter that covers the inferior aspect of the rostrum of the corpus callosum and is continuous above the genu with the indusium griseum

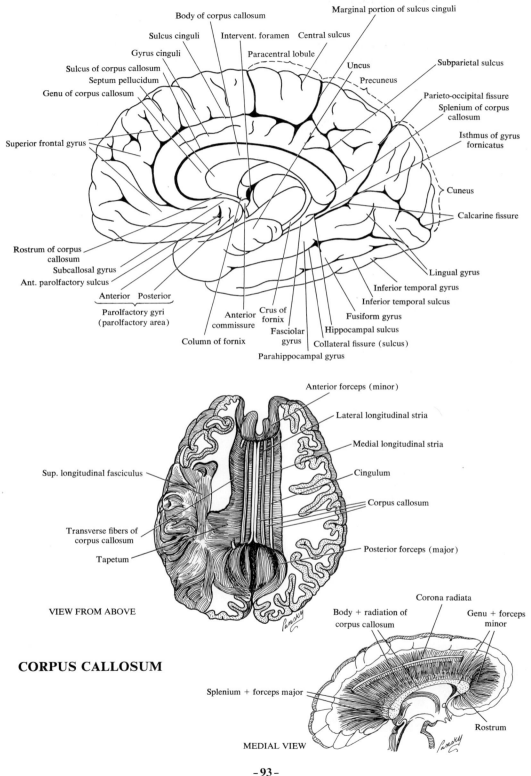

Body of corpus callosum

Marginal portion of sulcus cinguli

Sulcus cinguli   Intervent. foramen   Central sulcus

Gyrus cinguli   Paracentral lobule

Sulcus of corpus callosum   Uncus

Septum pellucidum   Precuneus

Genu of corpus callosum

Subparietal sulcus

Parieto-occipital fissure

Splenium of corpus callosum

Superior frontal gyrus

Isthmus of gyrus fornicatus

Rostrum of corpus callosum

Cuneus

Calcarine fissure

Subcallosal gyrus

Ant. parolfactory sulcus

Lingual gyrus

Anterior   Posterior

Inferior temporal gyrus

Inferior temporal sulcus

Parolfactory gyri (parolfactory area)

Anterior commissure

Crus of fornix

Fusiform gyrus

Column of fornix

Fasciolar gyrus

Hippocampal sulcus

Collateral fissure (sulcus)

Parahippocampal gyrus

Anterior forceps (minor)

Lateral longitudinal stria

Medial longitudinal stria

Sup. longitudinal fasciculus

Cingulum

Corpus callosum

Transverse fibers of corpus callosum

Posterior forceps (major)

Tapetum

VIEW FROM ABOVE

Corona radiata

Body + radiation of corpus callosum

Genu + forceps minor

CORPUS CALLOSUM

Splenium + forceps major

Rostrum

MEDIAL VIEW

–93–

45. BASAL ASPECT OF THE BRAIN

I. The basal aspect of the brain includes both telencephalic and brainstem structures

A. TELENCEPHALIC STRUCTURES include portions of the frontal, temporal, and occipital lobes
1. The frontal lobe is demarcated by the olfactory sulcus (parallel to the medial border) and the orbital sulci (an irregular group) and gyri
2. The gyrus rectus lies medial to the olfactory sulcus
3. The rhinal sulcus and collateral sulcus serve to delineate the limbic lobe
4. The rhinencephalon is a phylogenetically old part of the cerebral hemisphere and includes portions associated with olfaction: olfactory bulb and tract, olfactory striae (medial, intermediate, and lateral), olfactory trigone, anterior perforated substance, and piriform area

B. DIENCEPHALIC STRUCTURES
1. Optic nerves
2. Optic tracts
3. Hypophysis
4. Tuber cinereum
5. Mamillary bodies
6. Anterior and posterior perforated substance

C. MESENCEPHALIC STRUCTURES
1. Interpeduncular fossa
2. Crura cerebri
3. Oculomotor nerve (III)
4. Trochlear nerve (IV)

D. METENCEPHALIC STRUCTURES
1. Pons
2. Trigeminal nerve (V)

E. MYELENCEPHALIC STRUCTURES
1. Olives (protuberance of the inferior olivary nucleus)
2. Pyramids and pyramidal decussation
3. Abducent nerve (VI)
4. Facial nerve (VII)
5. Vestibulocochlear nerve (VIII)
6. Glossopharyngeal nerve (IX)
7. Vagus nerve (X)
8. Spinal accessory nerve (XI)
9. Hypoglossal nerve (XII)
10. Posterolateral (postolivary) sulcus
11. Preolivary sulcus

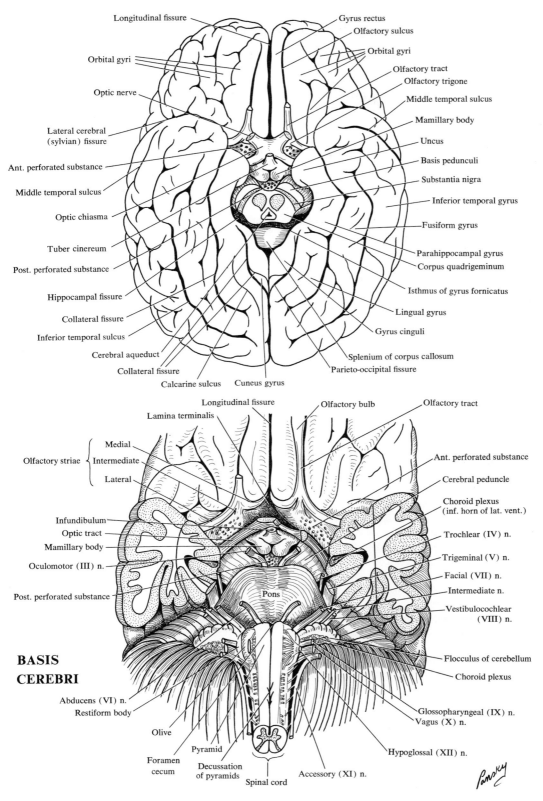

Longitudinal fissure
Gyrus rectus
Olfactory sulcus
Orbital gyri
Orbital gyri
Olfactory tract
Olfactory trigone
Optic nerve
Middle temporal sulcus
Mamillary body
Lateral cerebral (sylvian) fissure
Uncus
Basis pedunculi
Ant. perforated substance
Substantia nigra
Middle temporal sulcus
Inferior temporal gyrus
Optic chiasma
Fusiform gyrus
Tuber cinereum
Parahippocampal gyrus
Post. perforated substance
Corpus quadrigeminum
Hippocampal fissure
Isthmus of gyrus fornicatus
Collateral fissure
Lingual gyrus
Inferior temporal sulcus
Gyrus cinguli
Cerebral aqueduct
Splenium of corpus callosum
Collateral fissure
Parieto-occipital fissure
Calcarine sulcus
Cuneus gyrus

Longitudinal fissure
Lamina terminalis
Olfactory bulb
Olfactory tract
Medial
Olfactory striae
Intermediate
Lateral
Ant. perforated substance
Cerebral peduncle
Choroid plexus (inf. horn of lat. vent.)
Infundibulum
Trochlear (IV) n.
Optic tract
Trigeminal (V) n.
Mamillary body
Facial (VII) n.
Oculomotor (III) n.
Intermediate n.
Post. perforated substance
Vestibulocochlear (VIII) n.
Pons
BASIS CEREBRI
Flocculus of cerebellum
Choroid plexus
Abducens (VI) n.
Restiform body
Glossopharyngeal (IX) n.
Vagus (X) n.
Olive
Pyramid
Hypoglossal (XII) n.
Foramen cecum
Decussation of pyramids
Accessory (XI) n.
Spinal cord

46. MAJOR ASSOCIATION FIBERS OF THE CEREBRAL HEMISPHERES

I. **Definitions:** Association fibers are nerve fibers that interconnect cortical regions of the same cerebral hemisphere. Projection fibers connect cortical areas of the cerebrum with subcortical regions. Commissural fibers cross the midline and interconnect similar cortical regions in the two cerebral hemispheres

II. **Association fibers on the lateral aspect of the cerebral hemispheres**
A. UNCINATE FASCICULUS (uncinate means hook-shaped) interconnects the cortex of the uncus and temporal pole with the cortex of the inferior frontal region
B. INFERIOR OCCIPITOFRONTAL FASCICULUS is located along the inferior portion of the extreme capsule, dorsal to the uncinate fasciculus. It interconnects the cortex of the lateral or inferolateral portion of the frontal lobe and cortex of the occipital lobe, with connections along the way, including the inferior temporal and fusiform gyri of the temporal lobe
C. SUPERIOR LONGITUDINAL FASCICULUS is located along the dorsolateral border of the putamen, lateral to the internal capsule. It underlies and interconnects the cortices of the frontal, parietal, and occipital lobes and arches inferiorly and anteriorly with connections in the temporal lobe cortex
D. ARCUATE FASCICULUS curves over and around the posterior part of the insula to pass into the temporal lobe. It is a continuation of the superior longitudinal fasciculus (synonym for superior longitudinal fasciculus)
E. LATERAL OCCIPITAL FASCICULUS (also known as vertical or perpendicular occipital fasciculus and as the fasciculus of Wernicke) passes vertically through the occipital lobe and interconnects the fusiform gyrus of the temporal lobe and the posterior part of the parietal lobe
F. INFERIOR LONGITUDINAL FASCICULUS interconnects occipital lobe cortex and temporal lobe cortex in the inferior and lateral portion of the hemisphere

III. **Association fibers on the medial aspect of the cerebral hemispheres**
A. STRATUM CALCARIUM refers to a well-developed sheet of fibers curving around the bottom of the calcarine fissure from the cuneus above to the lingual gyrus
B. CINGULUM (means girdle) is an association bundle of the cerebrum located within the cingulate gyrus. It has connections all along its course with adjacent frontal, parietal, and temporal lobe cortex
C. SUPERIOR OCCIPITOFRONTAL FASCICULUS is located along the caudate nucleus medial to the interdigitating fibers of the internal capsule and corpus callosum. Its fibers interconnect the cortex of the occipital and temporal lobes with those of the frontal lobe and insula (synonym for subcallosal fasciculus)

IV. **Association and commissural fibers in coronal section of cerebral hemispheres**
A. CORPUS CALLOSUM (means hard body) is the thick band of commissural fibers interconnecting areas of the neopallium (cerebral cortex and underlying white matter)
B. CINGULUM
C. SUPERIOR AND INFERIOR OCCIPITOFRONTAL FASCICULI
D. SUPERIOR LONGITUDINAL FASCICULUS
E. ARCUATE FASCICULUS
F. UNCINATE FASCICULUS
G. ANTERIOR COMMISSURE

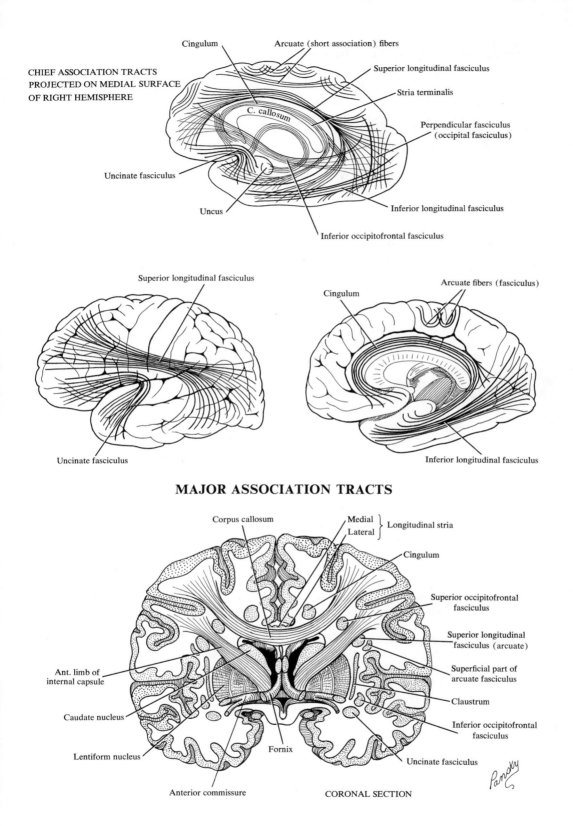

CHIEF ASSOCIATION TRACTS
PROJECTED ON MEDIAL SURFACE
OF RIGHT HEMISPHERE

Cingulum

Arcuate (short association) fibers

Superior longitudinal fasciculus

Stria terminalis

Perpendicular fasciculus
(occipital fasciculus)

C. callosum

Uncinate fasciculus

Uncus

Inferior longitudinal fasciculus

Inferior occipitofrontal fasciculus

Superior longitudinal fasciculus

Cingulum

Arcuate fibers (fasciculus)

Uncinate fasciculus

Inferior longitudinal fasciculus

MAJOR ASSOCIATION TRACTS

Corpus callosum

Medial } Longitudinal stria
Lateral }

Cingulum

Superior occipitofrontal
fasciculus

Superior longitudinal
fasciculus (arcuate)

Superficial part of
arcuate fasciculus

Claustrum

Inferior occipitofrontal
fasciculus

Uncinate fasciculus

Ant. limb of
internal capsule

Caudate nucleus

Lentiform nucleus

Fornix

Anterior commissure

CORONAL SECTION

-97-

47. MENINGES: THE DURA MATER

I. **Introduction:** the meninges are the three layers of nonneural, connective tissue coverings, which enclose the brain and spinal cord, and provide a shock absorber filled with cerebrospinal fluid (CSF). The CSF is found in the ventricular system and the subarachnoid space

II. **The three meningeal layers** consist of the dura mater, arachnoid, and pia mater, each of which forms a separate continuous sheet to surround and protect the soft brain and the spinal cord
  A. THE DURA MATER (PACHYMENINX) is the outer layer of the meninges and is a dense, tough, nonstretchable membrane composed of two layers around the brain and one layer around the spinal cord
    1. The outer layer of dura is the fibrous or periosteal layer and is actually the connective tissue membrane of the skull bone
    2. The inner layer of dura is the meningeal layer, which folds into a doubled layer or partition in several regions of the skull. Four membranes formed by these folds or reflections are seen grossly in a sagittal plane
      a. The falx cerebri is the double fold of dura between the cerebral hemispheres in the midline; its free edge follows the corpus callosum
      b. The falx cerebelli is a similar midline partition between the cerebellar hemispheres
      c. The tentorium cerebelli is a doubled partition located within the transverse fissure separating the occipital lobe (cerebrum) and the cerebellum
      d. The diaphragma sellae is a double layer of dura that forms the roof of the hypophysis fossa (sella turcica). The infundibular stalk perforates this dural sheet
    3. Branches of the trigeminal (V) nerve, vagus (X) nerve, and sympathetic nerves supply the dura mater
      a. Ethmoidal nerves of the ophthalmic division of V supply part of the falx cerebri and the anterior cranial fossa
      b. Recurrent tentorial branches of the ophthalmic division of V innervate the tentorium cerebelli and the posterior falx cerebri
      c. Branches of the maxillary and mandibular divisions of V innervate the middle cranial fossa
      d. Recurrent meningeal branches of the vagus (X) nerve innervate the posterior cranial fossa
      e. Sympathetic autonomic fibers follow the middle meningeal arteries to innervate the dura mater, but not the cerebral blood vessels
      f. Recurrent spinal nerve branches supply the dura mater of the spinal cord

III. **Clinical application:** dura mater has been used as a homograft in repair of thoracic wall and diaphragm defects, for the correction of transposition of the great arteries, and as a tracheal prosthesis. Since 1971, more than 2000 bioprosthetic cardiac valves constructed of homologous dura mater have been implanted for the correction of acquired or congenital valvular disease

Sup. cerebral vv.

Middle meningeal vessels

Hypophysis

Ophthalmic a. + n.

Sphenoparietal sinus

Diaphragmatic sinus

Cavernous sinus

Sup. petrosal sinus

Inf. petrosal sinus

9,10,11 nn.

3

4

5

6

12

Sup. sagittal sinus

Inf. sagittal sinus

Transverse sinus

Falx cerebri

Tentorium cerebelli

Straight sinus

Confluence of sinuses

Great cerebral vein

Occipital sinus

Falx cerebelli

Foramen magnum

Skin

Pericranium

Bone

Dura mater

Arachnoid

Pia mater

Sup. sagittal sinus

Falx cerebri

Subdural space

Subarachnoid space

Falx cerebri

Diaphragma sellae

Tentorial notch

Tentorium cerebelli

Falx cerebelli

Pericranium

Sup. sagittal sinus

Dura mater

Arachnoid mater

Brain
(cerebrum)

Pia mater

Falx cerebri

Tentorium cerebelli

Transverse sinus

Inf. sagittal sinus

Falx cerebelli

CORONAL SECTION

Pansky

– 99 –

48. MENINGES: ARACHNOID AND PIA MATER

I. The arachnoid is the thin, delicate middle layer of meninges which, like the dura mater, bridges the sulci and extends into only the major fissures. Arachnoid tends to follow dura

A. NUMEROUS FINE TRABECULAE extend from the arachnoid membrane to the pia mater to provide its spider-web appearance

B. THE SUBARACHNOID SPACE is the trabeculated space between the arachnoid and pia mater membranes and contains CSF

C. THE ARACHNOID GRANULATIONS are pia-arachnoid projections which extend into the superior sagittal sinus. The granulations (pacchionian bodies) are each comprised of a number of arachnoid villi which appear to function as one-way valves allowing passage of substances from the CSF to venous blood in the sinuses

D. THE ARACHNOID, like the dura, sends prolongations over the roots of the spinal nerves

II. The pia mater is the inner layer of meninges which is intimately attached to the brain and spinal cord, following every contour (sulci and fissure), carrying with it small blood vessels that nourish the underlying nervous tissue

A. THE PIA-GLIAL MEMBRANE is formed by astrocyte end-feet which end in the pia

B. THE PIA MATER appears to serve as a barrier to the entrance of harmful materials and organisms

C. THE LEPTOMENINGES is a name given to the combination of the pia mater and the arachnoid membrane

D. MENINGITIS is an infectious disease of the nervous system that involves the meninges

Arachnoid granulation
Skin (scalp)
Emissary v.
Galea aponeurotica
Diploic v.
Dura { Endosteal / Meningeal }
Diploe (parietal bone)
a. } In subarachnoid space
v.
Vein entering perivascular space
Subarachnoid space
Sup. sagittal sinus
Arachnoid
Pia mater
Dura (falx cerebri)

Arachnoid granulation
Endothelial covering
Arachnoid
Dura matter
Subarachnoid space
Pia

Arachnoid
Subarachnoid space
Arachnoid trabeculae
Perivascular channel
Pia mater
Brain
Blood vessel

Pia
Covering of arachnoid
CNS
Fused pia + arachnoid (reticular perivascular sheath)
Pervascular space
Penetrating vessel

Epineurium (mesodermal)
Perineurium (neural crest)
Peripheral nerve
Nerve root
Arachnoid
CNS
Pia

**MENINGEAL RELATIONS
TO VESSELS + AND NERVES
PENETRATING BRAIN**

Venous lacuna
Venous sinus
Dura
Arachnoid membrane
Subarachnoid space
Pillars (trabeculae)
Pia mater
Fenestrated glial limiting membrane
Cerebral cortex

Cranial dura
Arachnoid + granulation
Diploe
Inner table
Outer table
Cranial flat bone

49. MENINGES: VENOUS DURAL SINUSES

I. **The venous dural sinuses** are epithelially lined spaces or channels between the two layers of dura of the brain, and drainage of blood from the brain is chiefly into these sinuses. The dural sinuses do not have valves and are usually triangular in shape

A. THE SINUSES RECEIVE BLOOD from three major sources
1. The cerebral veins are the major source of blood drainage
   a. The great anastomotic vein drains into the superior longitudinal sinus
   b. The small anastomotic vein drains into the transverse sinus
   c. The deep cerebral vein drains the basal ganglia
   d. The internal cerebral vein drains the basal vein
   e. The great cerebral vein (vein of Galen) is formed at the junction of the two internal cerebral veins, is a short midline vein, and drains into the straight sinus
2. The diploic veins lie between the layers of cranial bone. They anastomose freely with each other and also communicate with the meningeal veins internally and the superficial veins externally
3. The emissary veins connect extracranial and intracranial veins

B. THE VENOUS SINUSES have little tendency to collapse as do most other veins because of the fibrous consistency of the dura mater. Because they have a tendency to remain filled even against gravity, they are subjected to profuse bleeding when ruptured by a bone fragment

C. THE DURAL SINUSES are both paired and unpaired
1. The unpaired dural venous sinuses include
   a. Superior longitudinal (sagittal) sinus begins as a continuation of the nasal vein at the foramen cecum and receives blood from cerebral, diploic, and emissary veins. It also receives CSF from arachnoid granulations
   b. Inferior longitudinal (sagittal) sinus receives blood from veins on the medial surface of the brain
   c. Straight sinus occupies the area between the falx cerebri and tentorium cerebelli. It is the posterior continuation of the great cerebral vein of Galen and joins the superior sagittal sinus to form the confluence of sinuses
   d. Occipital sinus: smallest of the sinuses. Begins near the foramen magnum and ascends in the attached margin of the falx cerebelli to end in the confluence of sinuses. It also receives the inferior cerebellar veins
2. The paired dural venous sinuses include
   a. Transverse sinus receives blood from the superior sagittal sinus and straight sinus and drains into the internal jugular veins
   b. Cavernous sinuses are located lateral to the body of the sphenoid bone, and receive blood from the middle cerebral veins and superior ophthalmic veins. They communicate with each other via intercavernous sinuses as well as with the petrosal sinuses and pterygoid plexus of veins. These sinuses are the largest and most clinically important of the deep interconnecting sinuses and drain into the internal jugular veins and transverse sinuses
   c. Sigmoid sinuses are a continuation of the transverse sinuses. In their S-shaped course toward the jugular foramen, they receive blood from the inferior cerebrum, cerebellum, and emissary veins
   d. Petrosal sinuses: the superior and inferior petrosal sinuses drain from the cavernous sinuses into the transverse sinuses and internal jugular veins, respectively
   e. Sphenoparietal sinuses extend along the crest of the lesser wing of the sphenoid bone and drain into the cavernous sinuses

CEREBRAL VENOUS SINUSES

Basilar plexus

Middle cerebral v.

Sup. } Longitudinal (sagittal)
Inf. } sinuses

Falx cerebri

Sphenoparietal s.

Great vein of Galen + st. sinus

Cavernous sinus

Confluence

Tentorium cerebelli

Sup. } Petrosal ss.
Inf. }

Int. carotid a.

Sigmoid s.

Jugulars

III
IV

Ophthalmic (V) n.

VI

Cavernous s.

Sphenoid sinus

Sagittal sinuses { Sup.
Inf.

Diploic +
emissary vv.

Sup. orbital v.

Cavernous s.

Emissary v.

Inf. } Petrosal sinuses
Sup. }

Sigmoid s.

Confluence

Occipital s.

Jugular v.

Transverse s.

Post. terminal v.

Choroidal v.

Septal v.

Int. cerebral v.

Basilar v.

Grt. cerebral v.

Transverse sinus

Int. jugular v.

Intercavernous sinuses

Cavernous sinus

Ophthalmic vv.

Septal v.

Post. terminal v.

Sphenoparietal sinus

Choroidal v.

Sup. } Petrosal
Inf. } sinuses

Int. cerebral v.

Basal v.

Great cerebral v.

Sigmoid sinus

St. sinus

Basilar plexus

Transverse sinus

Transverse sinus

Occipital sinus

FLOOR OF SKULL

DEEP CEREBRAL VEINS

FIGURE 1. **Scanning electron microscopy of cranial meninges.** **A.** A cross section through two parallel cerebral gyri showing relationships of pia mater (*PM*), subarachnoid space (arrow), and arachnoid mater (*AM*) after the removal of the dura mater. **B.** The outer surface of the meningeal layer of dura mater is less cellular and the main characteristic of this artificially created surface is the tangle of collagen fibers as shown in this micrograph. A thin attenuated layer of fibrocyte-like cells (arrows) covers this collagen network. **C.** A tangled network (retiform) of arachnoid trabeculae spans the subarachnoid space overlying the convexities of the cerebellum. In the middle of the field, three oval-shaped macrophages rest on a flattened portion of this branching network. **D.** This relatively smooth macrophage appears to be intimately attached to the pia mater (foreground), and it gives rise to a long, thick cytoplasmic process on its opposite side. **E.** Several overlapping pia mater cells with attenuated processes (*AP*) and microappendages (*MA*) are shown in this micrograph. **F.** This view of the base of a rather complex arachnoid trabecula shows its continuity with the pia mater (*PM*).

FIGURE 2. **Electron microscopy of suprapial macrophages and associated subpial structures.**
A. Pia mater cells appear to be able to round up and become free macrophages. A short cytoplasmic process extends from this rounded-up cell to make contact to the adjacent pial surface. **B.** This free macrophage rests upon the surface lining of the pia mater. The long, thin cytoplasmic process (arrow) measuring 1,000 Å in diameter extends over portions of three pia mater cells before passing out of view. **C.** This macrophage rests on the pial surface facing onto the subarachnoid space (*SAS*). Pial cell processes overlap extensively and make contact with this macrophage. Beneath the pial layers, a connective tissue space (*CTS*) containing unit collagen fibrils exists. **D.** These three macrophages appear to be engulfing a red blood cell. A cytoplasmic process (arrow) from one travels a short distance over the pial surface before disappearing into a fenestration. **E.** Removal of the pia mater after critical point drying reveals a tangle of subpial connective tissue cells (*CT*) and fibers and macrophages (*M*). **F.** Removal of these connective tissue components reveals the glia limitans and portions of two large severed blood vessels and many smaller blood vessels penetrating the substance of the cerebrum.

50. VENTRICULAR SYSTEM

I. **The ventricular system** is a series of four communicating cavities within the brain, lined by ependyma and filled with CSF which is elaborated from blood by the choroid plexuses

A. THE CAVITIES OF THE SYSTEM consist of a pair of lateral ventricles (right and left), the third ventricle, and the fourth ventricle with the latter two connected by the cerebral aqueduct (of Sylvius)

  1. The two lateral ventricles are contained within the cerebral hemispheres, each of which is connected with the third ventricle through the interventricular foramen of Monro. Each lateral ventricle has four distinct parts

    a. The anterior horn lies anterior to the interventricular foramina and is located in the frontal lobe

      i. Its anterior border and roof are surrounded by the corpus callosum

      ii. Its medial wall is formed by the septum pellucidum

      iii. Its floor and lateral wall are formed by the caudate nucleus (head)

    b. The body (corpus) lies posterior to the foramina of Monro and is located in the parietal lobe

      i. It extends posteriorly to a point adjacent to the splenium

      ii. Its roof is also surrounded by the corpus callosum

      iii. Its medial wall is formed by the septum pellucidum

      iv. Structures found on the floor of the body include (medial to lateral): fornix, choroid plexus, dorsal thalamus, stria terminalis, vena terminalis, and caudate nucleus

    c. The inferior (temporal) horn is contained in the temporal lobe

      i. Its roof is comprised of white matter of the cerebral hemispheres

      ii. Its medial wall and floor are formed by the fimbria of the fornix and the hippocampus; the stria terminalis and caudate nucleus (tail) are found along its medial wall

      iii. The rostral end is partially surrounded by the amygdala

    d. The posterior horn is located in the occipital lobe

      i. Its roof is formed by the corpus callosum

      ii. The calcar avis, an eminence produced by the calcarine fissure, is found on its medial wall

    e. The trigone (atrium) of the lateral ventricle is the cavity or chamber at the junction of the body, posterior horn, and inferior horn

  2. The third ventricle is a thin vertical chamber or cavity located in the midline between the paired lateral ventricles

    a. Its roof is formed by a thin layer of ependyma

    b. Its lateral walls are formed primarily by the paired thalami

    c. Its floor and lower lateral walls are formed by the hypothalamus and subthalamus.

    d. Its anterior boundary is formed by the lamina terminalis and anterior commissure

  3. The fourth ventricle is a broad, rhomboid-shaped cavity overlying the pons and medulla and is bound dorsally by the cerebellum and medullary veli

    a. It is continuous with the central canal of the upper cervical spinal cord below and the cerebral aqueduct (of Sylvius) of the mesencephalon above. The latter connects the third and fourth ventricles

    b. The lateral boundaries of the floor are the cerebellar peduncles, the cuneate tubercles, and the clava

    c. The calamus scriptorius is the most inferior portion of the rhomboid fossa or floor of the ventricle

    d. This ventricle also communicates with the subarachnoid space via the two foramina of Luschka and the foramen of Magendi

VENTRICULAR SYSTEM

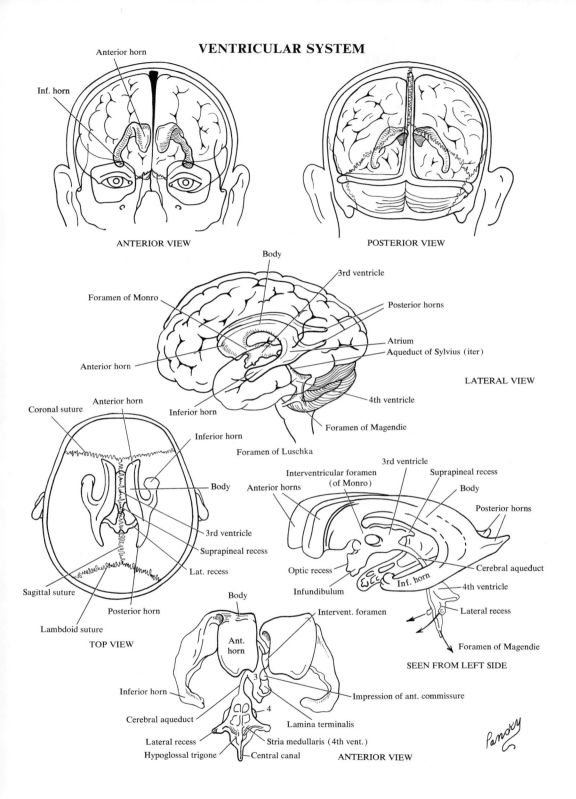

ANTERIOR VIEW

POSTERIOR VIEW

Body

3rd ventricle

Foramen of Monro

Posterior horns

Anterior horn

Atrium

Aqueduct of Sylvius (iter)

LATERAL VIEW

4th ventricle

Foramen of Magendie

Foramen of Luschka

Inferior horn

Coronal suture

Anterior horn

Inferior horn

Body

3rd ventricle

Suprapineal recess

Lat. recess

Sagittal suture

Posterior horn

Lambdoid suture

TOP VIEW

3rd ventricle

Interventricular foramen
(of Monro)

Suprapineal recess

Anterior horns

Body

Posterior horns

Optic recess

Cerebral aqueduct

Infundibulum

Inf. horn

4th ventricle

Lateral recess

Foramen of Magendie

SEEN FROM LEFT SIDE

Body

Ant.
horn

Intervent. foramen

Inferior horn

Impression of ant. commissure

Cerebral aqueduct

Lamina terminalis

Lateral recess

Stria medullaris (4th vent.)

Hypoglossal trigone

Central canal

ANTERIOR VIEW

Inf. horn

Anterior horn

Pansky

51. CHOROID PLEXUS AND CEREBROSPINAL FLUID

I. **The choroid plexus** is a rich network of blood vessels of the pia mater which projects into each ventricular cavity to form a semipermeable filter between arterial blood and the CSF

A. AN EPITHELIAL LAYER OF EPENDYMA covers each choroid plexus

B. THE TELA CHOROIDEA of the lateral ventricle is a thin membrane resembling a web which is continuous with the choroid plexus of the third ventricle through the interventricular foramina. It is formed by vascular folds invaginating the ependyma. Thus, the tela choroidea is pia plus ependyma

C. THE CHOROID PLEXUS EXTENDS from the interventricular foramen to the rostral end of the inferior horn of the ventricular system

   1. The anterior and posterior choroidal arteries, which are branches of the internal carotids and posterior cerebral arteries, respectively, supply the plexuses of the lateral ventricles

   2. The choroid plexus of the 3rd ventricle is found in its roof and extends on each side from the interventricular foramina to the caudal extent of the roof

   3. The choroid plexus of the fourth ventricle is found in the roof of the medulla and is T-shaped. Right and left halves extend in the direction of the lateral recesses

II. **Cerebrospinal fluid** is a clear, colorless, almost protein-free filtrate of blood which forms in the ventricles and circulates through the subarachnoid space

A. FUNCTION: serves to support and cushion the CNS against injury

   1. May serve nutritive functions and remove waste products of neuronal metabolism

   2. Its pressure measured in the lumbar cistern is 100–150 mm H_2O when the subject is lying down and 200–300 mm H_2O in the sitting position

   3. It contains small amounts of protein, glucose, and potassium and relatively large amounts of sodium chloride

   4. Very few cellular components are found in CSF (1–5 cells/mm^3 is normal)

   5. Bears close resemblance to blood plasma ultrafiltrate except for great differences in protein concentration (CSF, 25 mg/100 gm; plasma, 6500 mg/100 gm)

   6. It has higher Mg^{3+}, Na^+, and Cl^- concentrations and lower K^+, Ca^{2+}, and glucose concentrations than plasma

   7. It is a secretory product produced by active transport mechanisms; Na^+, K^+, and Cl^- are actively transported from choroid plexi into the ventricles, while water follows passively to maintain equilibrium

   8. There is considerable evidence that the choroid plexi are the chief sites for production of CSF, but small amounts are believed to be produced from the ependyma, glial elements, and pia-arachnoid capillary beds

   9. It is constantly being formed at the rate of 600–700 ml/day

   10. The total volume in the ventricles and subarachnoid space is about 140 ml. with an average ventricular capacity of about 20-25 ml

B. CIRCULATION OF CEREBROSPINAL FLUID is through the ventricles and subarachnoid spaces to literally float the brain and spinal cord

   1. From the lateral ventricles it flows through the interventricular foramina (of Monro) into the third ventricle; then it flows through the aqueduct of the midbrain (of Sylvius) into the fourth ventricle; from the latter it flows through the single (Magendie) and two lateral (Luschka) foramina and recesses to circulate in the subarachnoid space around the brain and spinal cord

   2. It returns to the venous system through small membranous villi, the arachnoid granulations, located in the superior sagittal sinus

   3. The flow of CSF is unidirectional because of the pressure gradient between its source, arterial blood and venous blood

Dura mater { Endosteal layer / Meningeal layer }

Cerebral v.

Sup. sagittal sinus

Arachnoid granulations

Choroid plexus of lateral ventricle

Great cerebral v.

Interventricular foramen

Superior cistern

Choroid plexus of third ventricle

Aqueduct of Sylvius

Interpeduncular cistern

Choroid plexus of fourth ventricle

CEREBROSPINAL FLUID
CIRCULATION

Pontine cistern

Subdural space

Cerebellomedullary cistern
(cisterna magna)

Obex

Subarachnoid space

Central canal

Arachnoid

Dura mater

Pia mater

Dura mater

Arachnoid

Pia mater

Tip of conus

Filum terminale

Subarachnoid septum

Extradural space

Subdural space

Dura mater

Arachnoid

Dorsal root

Spinal cord

Spinal nerve

Body of vertebra

Prolongation of subarachnoid space

Root sleeve (dura + arachnoid)

Denticulate ligament

Subarachnoid space

Ventral root

Pia mater

Panoky

RELATIONS OF SPINAL CORD TO MENINGES AND VERTEBRA

FIGURE 3. **Scanning electron microscopy of the ventricular system.** **A.** Ciliated (*C*) and nonciliated (*NC*) ependymal cells viewed from the luminal surface of the lateral ventricle. **B.** Hexagonal ependymal cell boundaries outlined by microvilli and covered by scattered tufts of cilia. **C.** Rounded convolutions of choroid plexus in the fourth ventricle with barely discernible epiplexus cells (arrows) on its surface. **D.** A population of polymorphic supra-ependymal (Kolmer cell macrophages) cells on the posterior medullary velum. **E.** Bleb formation on the surface of choroid plexus ependymal cells. **F.** High power of portions of four choroid plexus cells showing their luminal surface covered with microvilli.

FIGURE 4. **Electron microscopy of supraependymal cells to demonstrate their polymorphism.** **A.** An oval-shaped macrophage with few, short delicate processes found resting on the luminal surface of the temporal horn of the lateral ventricle. **B.** A triangular-shaped supraependymal cell whose processes flatten distally to blend with the ependymal surface. The ependymal surface possesses microvilli and/or cilia of varying dimensions. **C.** An ovoid-shaped Kolmer cell macrophage with several surface indentations. This epiplexus cell measured 11.0 μm through its thickest dimension, and its two processes varied in diameter along their lengths. **D.** A flattened supraependymal cell with numerous processes highly variable in shape and length. **E.** A section through the ovoid cell in *C* reveals an indented, irregularly shaped nucleus, many vacuoles, and few cytoplasmic organelles. Surface indentations shown here overlie internal vacuoles. The metallic coat (arrows) as required for prior scanning electron microscopy appears on the cell surface. **F.** Neuron-like cell observed sparingly on nonciliated ventricular surfaces. The cell in the foreground sends a process through an opening in the ependymal surface. A terminal swelling (synapse) at the arrow and neuropil underlying this cell type were infrequently observed.

52. SUBARACHNOID SPACE CISTERNS

I. **The subarachnoid space** surrounding the brain and spinal cord shows local variations. Over the convexities of the cerebral hemispheres, this space is small except in the depths of the sulci. At the base of the brain, around the brainstem, and at the termination of the spinal cord, the pia and arachnoid are separated, creating several large spaces, called subarachnoid cisternae

A. THREE NAMED CISTERNAE (CISTERNS) LOCATED ON VENTRAL ASPECT OF BRAINSTEM
   1. The chiasmatic cistern is located in the region of the optic chiasma
   2. The interpeduncular cistern is located in the interpeduncular fossa of the mesencephalon
   3. The pontine cistern is located at the pontomedullary junction

B. TWO NAMED CISTERNAE (CISTERNS) POSITIONED ON POSTERIOR ASPECT OF BRAINSTEM
   1. The cerebellomedullary cistern (cisterna magna), one of the largest cisterns, is located between the choroid plexus of the medulla and the cerebellum. The foramina of the fourth ventricle open into this cistern
   2. The superior cistern (cisterna ambiens) surrounds the superior and lateral surfaces of the mesencephalon. This cistern contains the great cerebral vein of Galen and the posterior cerebral and superior cerebellar arteries, which is of great clinical importance

C. The lumbar (spinal) cistern is located between lumbar and upper sacral levels of the spinal column (L1 to S2)
   1. This cistern contains the filum terminale and dorsal and ventral nerve roots of the cauda equina
   2. CSF is withdrawn here in a lumbar spinal tap

Supracallosal cistern   Foramen of Monro   3rd ventricle

Aqueduct of Sylvius

Superior cistern
(cisterna ambiens)

C.C.

Tentorium cerebelli

Cistern of lamina terminalis

Cistern of optic chiasma (chiasmatic)

Pons

Cerebellum

Med.

4th ventricle

Interpeduncular cistern

Foramen of Magendie

Pontine cistern

Cerebellomedullary cistern
(cisterna magna)

SAGITTAL VIEW

Supracallosal cistern

Superior cistern
(cisterna ambiens)

Central canal

Pia mater

Arachnoid

Dura

Chiasmatic cistern

Tip of conus

Interpeduncular cistern

Lumbar (spinal)
cistern

Filum terminale

Cerebellomedullary cistern
(cisterna magna)

- 113 -

53. BRAIN BARRIERS

I. **Ultrastructural studies on the brain, choroid plexus, and meninges** have helped to modify the concept of the hematoencephalic (blood-brain) barrier demonstrated in 1887 by Ehrlich. This original concept was based on the fact that certain aniline dyes, which stain most tissues of the body readily, when injected intravascularly, failed to stain the brain and spinal cord. Thus, a barrier was said to exist between the CNS and the vascular system. The concept has been altered to include the various morphologic features as well as functional systems which serve to subdivide the CNS and associated structures into named compartments separated by specific brain barriers

A. THE BLOOD-BRAIN BARRIER exists between the brain and vascular compartments. The perivascular components of this barrier consist of two continuous layers, the endothelial cells of the blood vessels and the basement membrane. The barrier functions as a differential filter that controls the exchange of various substances from the blood to interstitial fluid

B. THE BLOOD-CSF BARRIER exists between the brain and the CSF compartments. It is an effective one-way entry into the CSF compartment. The epithelium (ependyma) and adnexa of the choroid plexi are responsible for the actively secreted CSF

C. THE BRAIN-CSF BARRIER exists between the brain and the CSF comparments. The ependyma and the subjacent glial cells form a potential barrier between CSF and the interstitial fluid of the brain

D. COMPARTMENTS: central nervous system is usually divided into two compartments
   1. The intracellular fluid compartment of the brain is comprised of neurons and glial cells. Substances pass into and out of neurons and glial cells from the extracellular space and through plasma membranes instead of by direct exchange from the plasma
   2. The intercellular (interstitial or extracellular) compartment is comprised of the space between nerve cells (neurons and glia) and capillaries. Electron microscopic studies of the brain suggest a total extracellular space of about 4%. Neurochemical studies, using brain chloride as a measure of the extent of the extracellular volume, give values between 25–40%

E. ABSENCE OF BLOOD-BRAIN BARRIER: certain areas of the brain appear to be devoid of a blood-brain barrier. These areas are highly vascular and some are secretory in function. These areas include area postrema, hypophysis, pineal body, hypothalamic regions, subfornical organ, supraoptic crest, and choroid plexi
   1. The structure of the blood vessels in these areas differs in that their endothelial cells have fenestrations or pores
   2. There may be an extensive perivascular space containing basement membrane material and other connective tissue elements (connective tissue stroma of the perivascular space in choroid plexi may be 50–100 μm thick)
   3. The boundary between such an extensive perivascular space and the brain parenchyma may be lined with a definitive basement membrane as well as the basal part of the endothelial cells
   4. On the brain parenchymal side of the basement membrane, astrocytic end-feet fit close together up against the membrane

F. BREAKDOWN OF THE BLOOD-BRAIN BARRIER
   1. When large molecules pass from the blood to the brain parenchyma during induced cerebral edema (anoxia, cold-induced, etc.), gross structural damage to small blood vessels has been demonstrated
   2. Histamine, bacterial endotoxins, and other agents that increase extravasation in most tissues do not appear to affect the vascular endothelium of the brain
   3. Systemically induced inflammation does not appear to affect the blood-brain barrier

ECF diffusion across pia

Intracranial plasma – 70–90 ml

CSF

ECF

PLASMA

ECF

CSF

Ventricular CSF
20–24 ml

Choroid plexus

ECF

ECF

CSF

ECF diffusion
across ependyma

CSF
110 ml

CSF

Extracellular fluid in
ultrastructural spaces
185–250 ml

Subarachnoid
CSF 85–90 ml

Arachnoid villi

BLOOD-BRAIN BARRIER

VENTRICULAR CSF

Choroid plexus epithelium

Nonfenestrated capillary
endothelium

PLASMA

Astrocyte foot
processes

Tight junction

BM

N   N   N   N   N

Endothelial
Epithelial

Basement membrane (BM)

Spotty tight junction

Space occupied
by cell processes
+ ECF

Lateral interdigitations

PLASMA

Endothelial cell

Meningeal cell process

Closed fenestration

SUBARACHNOID CSF

Meningeal cell

BM

N   N   N   N

Astrocyte foot processes

ECF

Ependymal cell

ECF

Subependymal glia

INTERFACES OF EXCHANGE

54. BLOOD SUPPLY OF THE BRAIN:
ARTERIAL SUPPLY

I. **The arterial supply to intracranial structures** is basically derived from branches of the two internal carotid arteries and two vertebral arteries

A. THE INTERNAL CAROTID ARTERIES arise at the bifurcation of the common carotids in the neck below the angle of the mandible. They ascend to the base of the skull, extend anteromedially through the carotid canal, curve as a sigmoid-shaped vessel within the cavernous sinus, and at the lesser wing of the sphenoid bone turn between the anterior and middle clinoid processes to reach the brain. The internal carotid arteries do not give off any branches in the neck; all branches are within the cranial cavity, and consist of

1. Ophthalmic artery is given off just as the internal carotid leaves the cavernous sinus. It has an important branch, the central retinal artery, which runs in the center of the optic nerve to end in the retina

2. Anterior choroidal artery is given off just before the carotid divides into its terminal branches
   a. It runs along the optic tract, around the cerebral peduncle to the level of the lateral geniculate body, and then becomes part of the choroid plexus of the lateral ventricle
   b. It sends branches to cerebral peduncles, internal capsule, caudate nucleus, hippocampus, and optic tract

3. Anterior and middle cerebral arteries are the terminal divisions of the internal carotid as it reaches the ventral surface of the brain lateral to the optic chiasma
   a. Anterior cerebral artery runs medially over the optic tract to the longitudinal cerebral fissure to supply blood to the frontal lobes
      i. Anterior communicating artery is located in the fissure
      ii. Branches pass to the medial aspect of the frontal and parietal lobes; anterior perforated substance, septum pellucidum, and part of the corpus callosum
      iii. Medial striate artery, when present, supplies the caudate nucleus, putamen, and internal capsule (anterior limb)
   b. Middle cerebral artery extends laterally (in lateral fissure) over the insula, giving off branches to the lateral aspect of temporal and parietal lobes
      i. Lateral striate artery supplies basal ganglia and internal capsule
      ii. Posterior communicating artery runs dorsally to join posterior cerebral branch of basilar artery. In its course, it gives off branches to the internal capsule and thalamus

B. VERTEBRAL ARTERIES are branches of the subclavian artery and ascend in the neck through the transverse foramina of cervical vertebrae. They enter the foramen magnum and join to form the basilar artery at the caudal end of pons. Branches include

1. Anterior spinal artery descends on ventral aspect of cord to supply it

2. Posterior spinal artery descends on dorsal aspect of cord to supply it

3. Posterior inferior cerebellar arteries are given off laterally to supply part of the cerebellum and the choroid plexus of the fourth ventricle

Prerolandic br. (MC)

Rolandic br. (MC)

Ant. }
Post. } Parietal brs. (MC)

Terminal twigs of ant. cerebral a.

Terminal twigs of post. cerebral a.

Frontal br. (MC)

Orbitofrontal br. (MC)

Parieto-occipital (angular) br. (MC)

Middle cerebral artery

Post. }
Middle } Temporal brs. (MC)
Anterior }

ARTERIAL SUPPLY— LATERAL VIEW

CEREBRAL CORTEX

Anterior cerebral

Anterior cerebral

Posterior cerebral

Middle cerebral

Middle cerebral

Posterior cerebral

Precallosal br. (AC)

Paracentral br. (AC)

Precuneal br. (AC)

Internal frontal brs. (AC)
{ Post.
Middle
Ant. }

Parieto-occipital br. (PC)

Callosomarginal br. (AC)

Calcarine (PC) br.

Frontopolar br. (AC)

Orbital br. (AC)

Posterior }
Anterior } Temporal branches (PC)

Anterior cerebral artery

Post. cerebral artery

ARTERIAL SUPPLY— MEDIAL VIEW

-117-

55. BLOOD SUPPLY OF THE BRAIN: BASILAR ARTERY AND CIRCLE OF WILLIS

I. Arterial supply (cont.)

A. BASILAR ARTERY is formed by the union of the two vertebral arteries, and it courses along the ventral aspect of the pons. Its branches include

1. Pontine arteries supply the pons
2. Anterior inferior cerebellar arteries originate at the level of the abducens, facial, and vestibulocochlear nerves, between the pons and medulla to supply the cerebellum
3. Labyrinthine arteries follow the facial and vestibulocochlear nerves through the internal acoustic meatus to the inner ear
4. Superior cerebellar arteries branch off the basilar artery near its termination at the level of the upper pons. They are in close proximity to the oculomotor and trochlear nerves and extend superiorly around the pons. They supply the upper cerebellum, corpora quadrigemina, choroid plexus of the third ventricle, and the pineal body
5. Posterior cerebral arteries are the terminal branches of the basilar artery. They receive the posterior communicating arteries from the internal carotid.
   a. Supply the medial and inferior aspects of the occipital and temporal lobes
   b. Give off posterior choroidal branches to the choroid plexi of the third and lateral ventricles

II. The circle of Willis is formed by the anterior and posterior communicating arteries and proximal portions of the anterior, middle (internal carotid), and posterior cerebral arteries

A. FUNCTION OF CIRCLE OF WILLIS makes possible an adequate blood supply to the brain in case of occlusion of either the carotid or vertebral arteries

1. The circle surrounds the optic chiasma and pituitary stalk
2. The circle of Willis is subject to frequent anatomic variation and a "normal" circle is seen in only about 50% of the population
   a. Common variations occur when the posterior cerebral arteries arise directly from the internal carotid via an enlarged posterior communicating artery, when one or both posterior communicating arteries are absent, and when there are multiple small anterior communicating arteries
3. Numerous arteries arise from the circle, penetrate the brain substance, and are very important because not only are they small and easily plugged but they are also end arteries (no collateral circulation) and supply vital areas

Recurrent br. (Heubner) (ant. cerebral)

Ant. cerebral a.

Ant. communicating a.

Middle cerebral a.

Internal carotid a.

Medial + lateral striate (lenticulostriate) aa.

Ant. choroidal a.

Post. communicating a

Pontine brs.

Basilar a.

Sup. cerebellar a.

Post. cerebral a.

Vertebral a.

Int. auditory a.

Ant. spinal a.

Post. inf. cerebellar a.

Ant. sup. cerebellar a.

Post. spinal a.

ARTERIES AT BASE OF BRAIN

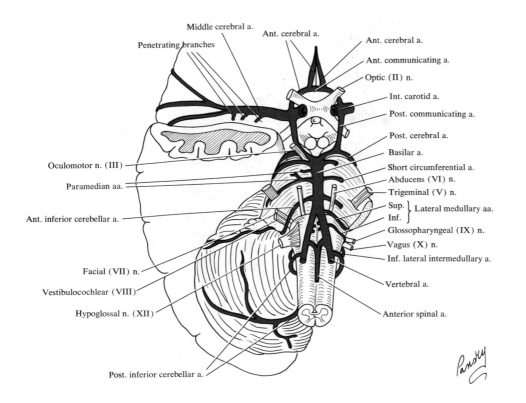

Middle cerebral a.

Ant. cerebral a.

Ant. cerebral a.

Penetrating branches

Ant. communicating a.

Optic (II) n.

Int. carotid a.

Post. communicating a.

Post. cerebral a.

Basilar a.

Oculomotor n. (III)

Short circumferential a.

Paramedian aa.

Abducens (VI) n.

Trigeminal (V) n.

Sup. } Lateral medullary aa.
Inf. }

Ant. inferior cerebellar a.

Glossopharyngeal (IX) n.

Vagus (X) n.

Inf. lateral intermedullary a.

Facial (VII) n.

Vertebral a.

Vestibulocochlear (VIII)

Hypoglossal n. (XII)

Anterior spinal a.

Post. inferior cerebellar a.

Pansky

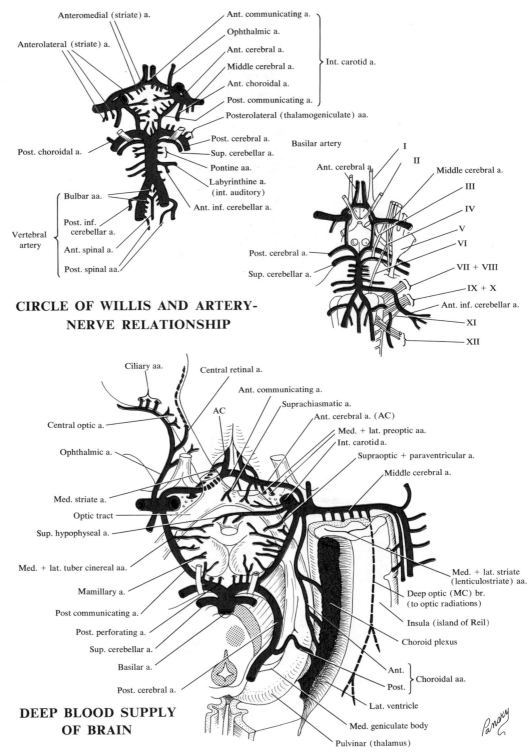

Anteromedial (striate) a.

Anterolateral (striate) a.

Post. choroidal a.

Bulbar aa.

Post. inf. cerebellar a.

Vertebral artery

Ant. spinal a.

Post. spinal aa.

Ant. communicating a.

Ophthalmic a.

Ant. cerebral a.

Middle cerebral a.

Ant. choroidal a.

Post. communicating a.

Int. carotid a.

Posterolateral (thalamogeniculate) aa.

Post. cerebral a.

Sup. cerebellar a.

Pontine aa.

Labyrinthine a. (int. auditory)

Ant. inf. cerebellar a.

Basilar artery

Ant. cerebral a.

Post. cerebral a.

Sup. cerebellar a.

I

II

III

IV

V

VI

VII + VIII

IX + X

Ant. inf. cerebellar a.

XI

XII

Middle cerebral a.

CIRCLE OF WILLIS AND ARTERY-NERVE RELATIONSHIP

Ciliary aa.

Central optic a.

Ophthalmic a.

Med. striate a.

Optic tract

Sup. hypophyseal a.

Med. + lat. tuber cinereal aa.

Mamillary a.

Post communicating a.

Post. perforating a.

Sup. cerebellar a.

Basilar a.

Post. cerebral a.

Central retinal a.

AC

Ant. communicating a.

Suprachiasmatic a.

Ant. cerebral a. (AC)

Med. + lat. preoptic aa.

Int. carotid a.

Supraoptic + paraventricular a.

Middle cerebral a.

Med. + lat. striate (lenticulostriate) aa.

Deep optic (MC) br. (to optic radiations)

Insula (island of Reil)

Choroid plexus

Ant.
Post.
Choroidal aa.

Lat. ventricle

Med. geniculate body

Pulvinar (thalamus)

DEEP BLOOD SUPPLY OF BRAIN

FIGURE 5. **Circle of Willis.**

MIDDLE MENINGEAL ARTERY

Ant. meningeal a. (br. ant. ethmoidal of ophthalmic)

Middle meningeal a.

Maxillary a.

Carotid aa.

Ext. } Int. }

Common carotid a.

Transverse sinus

Post. meningeal a. (below tentorium)

Vertebral a.

Subclavian a.

Brachiocephalic a.

Pansky

FRONTAL SECTION

Caudate n.

Corticomedullary br.

Central or basal brs. to basal ganglia

Middle cerebral a.

Int. carotid a.

Cerebral peduncle

Ant. cerebral a.

Caudate n.

Penetrating brs. middle cerebral a.

Int. capsule

Middle cerebral a.

Putamen

Claustrum

Globus pallidus

Ant. choroidal a.

Uncus

Post. cerebral a.

Th.

BASAL GANGLIA SUPPLY

Middle cerebral (via lateral striate) a.

Putamen

Caudate nucleus

Ant. cerebral (via medial striate a.)

Post. choroid a.

Post. cerebral a.

Post. communicating a.

Ant. choroid a.

Globus pallidus

Amygdaloid complex

Inf. horn lat. ventricle

Choroidal a.

Hippocampal a.

Dentate gyrus

Subiculum

Post. cerebral a.

Inf. temporal a.

Tentorium

FIGURE 6. **Middle meningeal artery and basal ganglia arterial supply.**

–121–

56. BLOOD SUPPLY OF THE BRAIN: VENOUS DRAINAGE

I. **Veins of the brain:** all terminate in the dural sinuses and are classified as external and internal cerebral veins

A. EXTERNAL CEREBRAL VEINS
1. Superior cerebral (8–12) drain the superior, lateral, and medial surface of the hemisphere and terminate in the superior sagittal sinus
2. Middle cerebral veins begin on the lateral surface of the hemispheres and end directly or indirectly in the cavernous sinuses (there are also intercommunications with the superior and transverse sinuses)
3. Medial cerebral veins drain the medial surface of the hemispheres and terminate in the inferior sagittal sinus
4. Inferior cerebral veins are small and drain the inferior surface of the hemispheres
   a. Those on the frontal lobe connect via the superior cerebral tributaries with the superior sagittal sinus
   b. Those on the temporal lobe join either the middle cerebral or basilar venous system and end in the cavernous or superior petrosal sinuses
5. Basal veins are formed by the confluence of 3 small veins: anterior cerebral draining the frontal lobe; deep middle cerebral, draining the insula and adjoining gyri; and inferior striate veins, draining the corpus striatum
   a. Receive tributaries from the interpeduncular fossa, hippocampal gyrus, inferior horn of the lateral ventricle, and mesencephalon
   b. Terminate in the great cerebral vein

B. INTERNAL CEREBRAL VEINS: each formed by a confluence of two veins
1. The terminal vein lies in the groove between the thalamus and corpus striatum and receives tributaries from both
2. The choroid vein runs the entire length of the choroid plexus and drains the hippocampus, fornix, and corpus callosum
3. When the internal cerebral veins reach the splenium of the corpus callosum, they are joined by the basal veins to form the great cerebral vein (of Galen)

C. CEREBELLAR VEINS are of two groups
1. Superior pass rostrally and medially to end in the straight sinus, internal cerebral veins, transverse sinus, and superior petrosal sinus
2. Inferior ends in the transverse, superior petrosal, and occipital sinuses

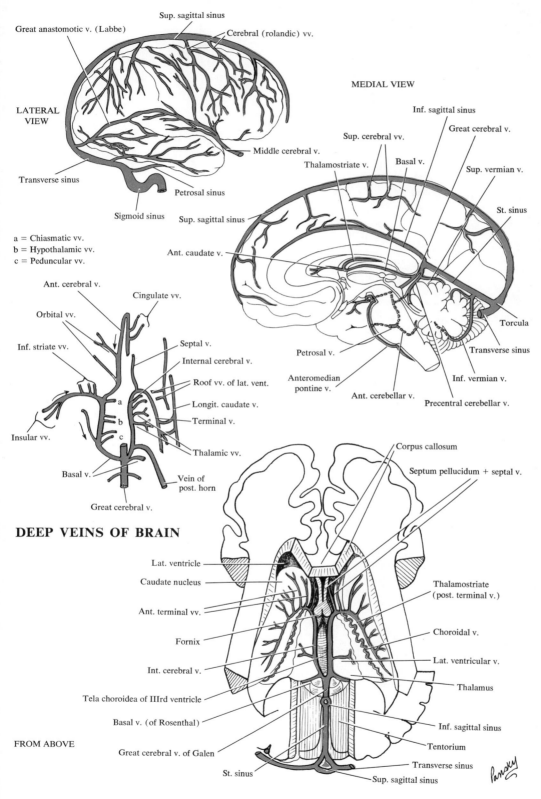

Sup. sagittal sinus

Great anastomotic v. (Labbe)

Cerebral (rolandic) vv.

LATERAL VIEW

MEDIAL VIEW

Inf. sagittal sinus

Great cerebral v.

Sup. cerebral vv.

Middle cerebral v.

Thalamostriate v.

Basal v.

Sup. vermian v.

Transverse sinus

St. sinus

Petrosal sinus

Sigmoid sinus

Sup. sagittal sinus

Ant. caudate v.

a = Chiasmatic vv.
b = Hypothalamic vv.
c = Peduncular vv.

Torcula

Transverse sinus

Ant. cerebral v.

Cingulate vv.

Petrosal v.

Inf. vermian v.

Orbital vv.

Septal v.

Anteromedian pontine v.

Ant. cerebellar v.

Precentral cerebellar v.

Internal cerebral v.

Inf. striate vv.

Roof vv. of lat. vent.

Longit. caudate v.

Terminal v.

Insular vv.

Thalamic vv.

Basal v.

Vein of post. horn

Great cerebral v.

DEEP VEINS OF BRAIN

Corpus callosum

Septum pellucidum + septal v.

Lat. ventricle

Caudate nucleus

Thalamostriate (post. terminal v.)

Ant. terminal vv.

Choroidal v.

Fornix

Lat. ventricular v.

Int. cerebral v.

Thalamus

Tela choroidea of IIIrd ventricle

Basal v. (of Rosenthal)

Inf. sagittal sinus

FROM ABOVE

Tentorium

Great cerebral v. of Galen

Transverse sinus

St. sinus

Sup. sagittal sinus

Pansky

Unit Three

NEUROPHYSIOLOGY

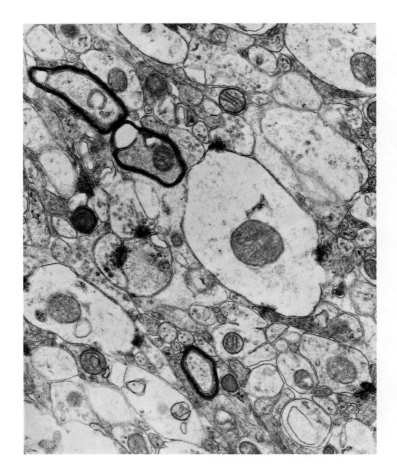

57. MEMBRANE POTENTIALS: PRINCIPLES OF ELECTRICITY—CHEMICAL REACTIONS AND ELECTRIC CHARGES

I. **Introduction:** a difference in potential, the membrane potential, usually exists between the inside of a cell and the extracellular fluid around it

A. THE FUNCTION OF A CELL, in many cells (muscle cells, glandular cells, etc.), can be controlled by the size of the membrane potential

B. THE SPECIALIZED ROLE OF THE NERVOUS SYSTEM is to propagate changes in membrane potential in its cells and to transmit them to other cells

II. **Chemical reactions** are basically electric in nature since they involve exchanging or sharing of negatively charged electrons between atoms to form ions or bonds

A. MOST CHEMICAL REACTIONS RESULT IN NEUTRAL MOLECULES, containing equal numbers of electrons and protons, but in some cases ions are formed which have a net electric charge

   1. Organic molecules: negative carboxyl group $RCOO^-$ or positive amino group RNH_3^+

   2. Inorganic ions: Na^+, K^+, and Cl^-

B. ELECTRICAL PHENOMENA: the cell environment contains many charged particles and the interaction of these plays an important role in cell function

   1. Except for water, the major chemical components of the extracellular fluid are Na^+ and Cl^- ions; the intercellular fluid contains high concentrations of K^+ ions and organic molecules with ionized groups, particularly proteins and phosphate compounds (RPO_4^{2-})

C. ALL PHYSICAL AND CHEMICAL PHENOMENA can be described using the fundamental units of measurements: length, time, mass, temperature, and electric charge

   1. Each of these units is independent in that none can be defined in terms of a combination of the others; e.g., electric charge is a fundamental unit of measurement and cannot be defined in terms of length, time, mass, or temperature

III. **Electric charges:** two types of charge that behave symmetrically are positive $(+)$ charges and negative $(-)$ charges

A. LIKE CHARGES repel each other; unlike charges attract each other

   1. When electrons were discovered, they behaved like the charged particles which we called negative. Protons, on the other hand, behaved like positive charges. Thus, electrons are negative and protons are positive

     a. All electrical phenomena result from the interaction of negative electrons and positive protons

     b. The total number of $(+)$ and $(-)$ particles in the universe are said to be equal, thus, the universe is electrically neutral, as are atoms, which contain an equal number of electrons and protons

B. AN ELECTRICAL FORCE draws the opposite charges together. When $(+)$ and $(-)$ charges are separated, the force is measured and the relation between the amount of force, quantity of charge, and the distance separating the charges can be observed

   1. The amount of force acting between electric charges increases when the charged particles are moving closer together and with increasing quantity of charge

   2. Energy (E) is the ability to do work (W) and work is the product of force (F) and distance (D): $W = FD$

     a. If opposite charges come together as a result of attracting forces between them, work is done and energy released by the moving particles since a force is exerted over a distance between them opposite to their electric attracting force. Thus, energy must be added to the system, in the form of work, in order to separate the charges

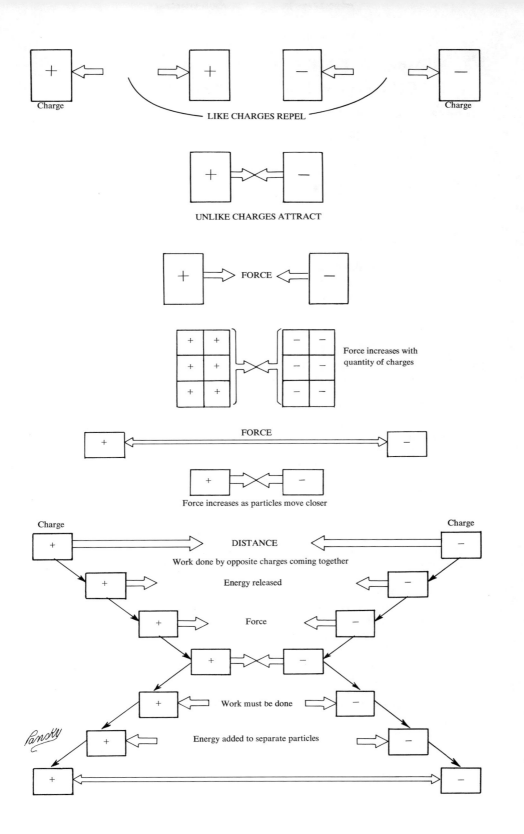

Charge   LIKE CHARGES REPEL   Charge

UNLIKE CHARGES ATTRACT

+   FORCE   −

Force increases with quantity of charges

FORCE

Force increases as particles move closer

Charge   Charge

DISTANCE

Work done by opposite charges coming together

Energy released

Force

Work must be done

Energy added to separate particles

58. MEMBRANE POTENTIALS: PRINCIPLES OF ELECTRICITY—VOLTAGE, CURRENT, RESISTANCE

I. Electric charges (cont.)

A. AN ELECTRIC FIELD is produced when charges tend to separate because of diffusion, which opposes further separation
   1. The electric field is a vector which has both magnitude and direction
   2. The potential difference is the sum of all the electric field vectors
B. WHEN ELECTRIC CHARGES ARE SEPARATED, they have a potential of doing work if they are allowed to come together again
   1. Voltage is a measure of the potential of separated electric charge to do work and is defined as the amount of work done by an electric charge when moving from one point in a system to another
      a. Voltage is always measured with respect to 2 points in a system so one refers to the potential difference between the 2 points and the units of measurements are known as volts (the terms "voltage" and "potential difference" are synonymous)
   2. The total amount of charge that can be separated in most biological systems is very small and thus the potential difference is very small
      a. The voltage across the nerve cell membrane is about 70 millivolts (mV) or $\frac{1}{10}$ V

II. Current is the movement of electric charge per unit of time

A. IF AN ELECTRIC CHARGE IS SEPARATED BETWEEN 2 POINTS, there is a potential difference between these points, and the electric force of attraction between the opposite charges tends to make the charges flow, producing a current
   1. The amount of charge that does move, the current, depends on the nature of the material lying between the separated charges which may be either a conductor or insulator
      a. The space may be occupied by a copper wire, a solution of water and ions, glass, rubber, or it may even be empty (a vacuum)
   2. The ability of an electric charge to move through these media varies
      a. The flow depends on the number of charged particles in the material that are able to move and carry current
      b. The amount of current also depends on the interactions between the moving charges and the materials
   3. The ability of a material to store charges is determined by its dielectric constant
B. RESISTANCE is the hindrance of the movement of electric charge through a particular material
   1. The higher the resistance, the lower the amount of current flow for any given voltage
      a. Glass and rubber have high electric resistance and are used as insulators
      b. Materials of low resistance, such as copper wire, are used as conductors
   2. Pure water is a relatively poor conductor because it contains very few charged particles, but when sodium chloride (NaCl) is added, the solution becomes a relatively good conductor with a low resistance since sodium (Na) and chloride (Cl) ions provide charges that can carry the current
C. THE WATER COMPARTMENTS inside and outside the cells in the body contain numerous charged particles (ions) which are capable of moving between areas of charge separation
   1. Lipid contains very few charged particles and thus has a high electric resistance
      a. The lipid component of the cell membrane provides a region of high electric resistance separating two water compartments (inside and outside of the cell) of low resistance

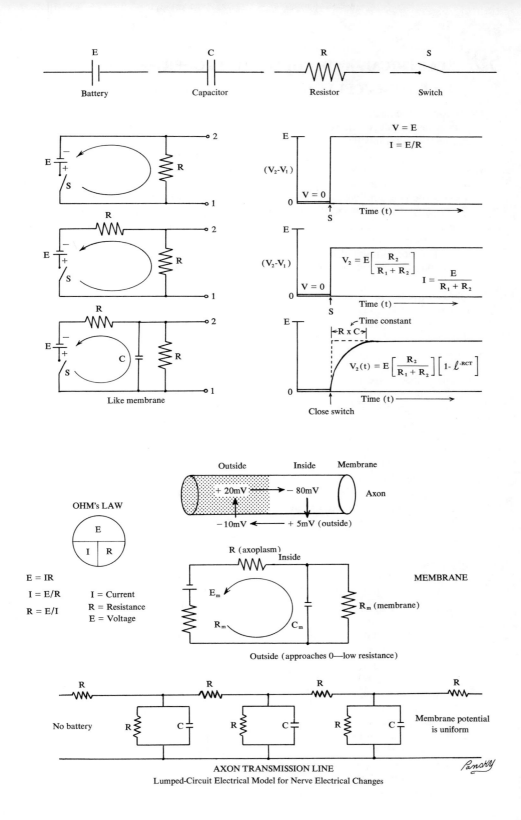

E
Battery

C
Capacitor

R
Resistor

S
Switch

$V = E$
$I = E/R$

$(V_2 - V_1)$

$V = 0$

$V = 0$

Time (t)

S

$V_2 = E\left[\dfrac{R_2}{R_1 + R_2}\right]$ $I = \dfrac{E}{R_1 + R_2}$

$V = 0$

Time (t)

S

Time constant

R x C

$V_2(t) = E\left[\dfrac{R_2}{R_1 + R_2}\right]\left[1 - \ell^{-RCT}\right]$

Time (t)

Close switch

Like membrane

Outside   Inside   Membrane

+ 20mV   − 80mV   Axon

− 10mV   + 5mV (outside)

OHM's LAW

E

I   R

$E = IR$

$I = E/R$ $I = $ Current

$R = E/I$ $R = $ Resistance

$E = $ Voltage

R (axoplasm)
                    Inside

E_m

R_m C_m

MEMBRANE

R_m (membrane)

Outside (approaches 0—low resistance)

R   R   R   R

No battery   R   C   R   C   R   C   Membrane potential
                                              is uniform

AXON TRANSMISSION LINE
Lumped-Circuit Electrical Model for Nerve Electrical Changes

Pansky

– 129 –

59. ELECTRICAL AND CHEMICAL PROPERTIES OF A RESTING CELL: DIFFUSION POTENTIAL

I. **Measuring the membrane potential:** the inside and extracellular space of a cell are filled with aqueous salt solutions in which most of the molecules dissociate into (+) or (−) charged ions (atoms or molecules)

A. CATIONS are positively (+) charged atoms; anions are negatively (−) charged atoms

B. AN ELECTRIC POTENTIAL DIFFERENCE (VOLTAGE) across a cell membrane can be determined by inserting a fine electrode into the cell, another into the extracellular fluid around the cell, and connecting the two by means of a voltmeter

1. All cells of the body exhibit a membrane potential (true for nerve and muscle cells *only* when they are not stimulated). The inside of the cell is (−) charged with respect to the outside which is relatively (+) charged

   a. The nonstimulated potential is called the resting potential and varies from 5–100 mV, depending on the cell type and its chemical environment

II. **Diffusion potential** is generated across a membrane largely due to the tendency for movement of K^+ down its concentration chemical gradient* so that the inside of the cell is negative with respect to the outside of the cell (outward K^+ and inward Cl^- diffusion)

A. INTRODUCTION: the different ionic concentrations of either side of the cell would soon cancel each other out by simple diffusion of the mobile particles were it not prevented or compensated for by

1. K^+ ions move through the membrane in both directions but because of higher concentrations inside the cell, the ions pass through the membrane pores from the inside $30\times$ more often than from the outside, resulting in a net efflux of K^+ ions. The motive force is the higher osmotic pressure of K^+ inside the cell

2. The osmotic pressure difference for K^+ would soon equalize except for an opposing force applied by an electric field which is the membrane potential

   a. When a K^+ ion leaves the cell, since it is (+) charged, an excess positive charge appears on the outer surface of the membrane (capacitor) which corresponds to an excess of (−) charges now present on the membrane's inside Thus, the membrane potential serves to oppose the efflux of more ions and the efflux of (+) charges in itself builds up an electric potential that impedes the efflux of more (+) charges from the cell

B. PERMEABLE MEMBRANE: If we have 2 solutions of different ionic composition and ignore the nature of the membrane between them, they would react as follows

1. With solution 1 of dilute NaCl at a concentration of 0.1 M and that of solution 2 also dilute NaCl but at a concentration of 0.01 M, and if the membrane between is permeable to all ion species

   a. Both Na and Cl on side 1 are more concentrated and they will diffuse down their concentration gradients and move from side 1 to side 2

   b. Mobility and movement of Cl, however, are 50% greater than those of Na^+ so Cl^- will move to side 2 faster and side 2 will transiently become slightly (−) charged with respect to side 1. This electric gradient, due to differential diffusion of charged particles in solution, is called the diffusion potential and disappears in time as the concentrations on both sides equalize

2. If we change solutions to resemble extracellular fluid (0.15 M NaCl), and the intracellular fluid is 0.15 M KCl, and again ignore the separation membrane

   a. Cl^- concentrations on the 2 sides are equal, but Na^+ and K^+ are not

   b. The mobility of K^+, like Cl^-, is about 50% greater than Na^+, and thus K^+ will diffuse down its concentration gradient faster than Na^+, and positive charge will initially leave side 2 faster than it enters and side 2 will become electronegative with respect to side 1. Again, in time, equilibrium will prevail

*The magnitude of the diffusing tendency from one area to another is proportionate to the difference in concentration of the substance in 2 areas.

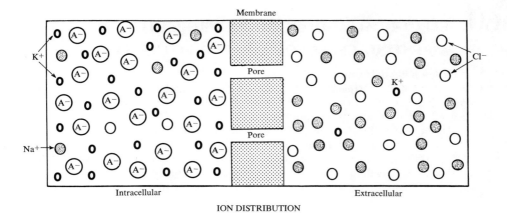

Membrane

K+

Na+

Pore

Pore

Cl−

K+

Intracellular

Extracellular

ION DISTRIBUTION

IONIC CONCENTRATIONS INSIDE AND OUTSIDE A NERVE CELL
(mmol/liter of water)

Ion	Extracellular	Intracellular
Na+	150	15
K+	5	150
Cl−	125	10
HCO$_3$−	27	8
A− (Protein)	——	155
Resting potential: −70 to −90mv		

Voltmeter

Cell

Electrode

Electrode

Cell

+20
0
−40
−80

Extracellular potential

Intracellular potential

Time

Time

① NaCl 0.1 M ② NaCl 0.01 M → ① ② Cl− Na+

① NaCl 0.15 M ② KCl 0.15 M → ① K+ ② Na+

Extracellular   Intracellular

Extracellular   Intracellular

DIFFUSION POTENTIALS THROUGH A COMPLETELY PERMEABLE MEMBRANE

60. ELECTRICAL AND CHEMICAL PROPERTIES OF A RESTING CELL: EQUILIBRIUM POTENTIAL

I. Diffusion potential (cont.)

A. EQUILIBRIUM POTENTIAL deals with intra- and extracellular concentrations, but now involves the addition of a selectively permeable membrane between compartments, so that K^+ can pass through but not Na^+ or Cl^-

1. All the Na^+ will remain on side 1 and some of the K^+, diffusing down its gradient, will be added to it; and side 1 will become relatively positive

2. This diffusion potential, however, will not disappear or equalize in time, but will be of greater magnitude, depending on the concentration and gradient of K^+

3. The concentration gradient, which causes net diffusion of particles from a region of higher to a region of lower concentration, is called the concentration force

   a. As this force moves K^+ from side 2 to side 1 and the latter becomes more ($+$), the electric potential difference begins to influence the movement of the ($+$)-charged K particles [they are attracted by the relatively negative charge of side 2 and repulsed by the ($+$) charge of side 1]

   b. This attraction caused by a difference in electric charge (or repulsion due to a similar charge) is called the electric force

   c. As long as the concentration force driving K^+ from side 2 to 1 is greater than the electric force driving the opposite direction, there will be net movement of K^+ from side 2 to side 1 and the potential difference will increase

   d. Side 1 will become more and more positive until the electric force opposing the entry of K^+ equals the concentration force favoring its entry. The membrane potential at which the latter occurs is called the equilibrium potential and, at that time, there is no net movement of the ion because the forces acting on it are exactly balanced

      i. The value of the equilibrium potential for any ion depends on the concentration gradient for that ion across the membrane (if the concentrations on the 2 sides are equal, the concentration force would be zero and the electric potential needed to oppose it is also zero)

      ii. The K^+ concentrations of neurons and extracellular fluid typically show the equilibrium potential for K^+ to be about 90 mV, the inside of the cell being ($-$) with respect to the outside

      iii. For neurons, the Na^+ equilibrium potential is 60 mV, the inside being positive

4. *The direction of the diffusion potential across a membrane is determined both by the permeability properties of the membrane and the orientation of the concentration gradients. The diffusion potential for each ion species is different from those for other ion species because the concentration gradients are different*

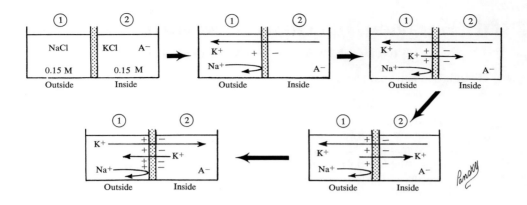

Cell permeable only to K⁺—diffusion down concentration gradient leaves inside of cell negatively charged

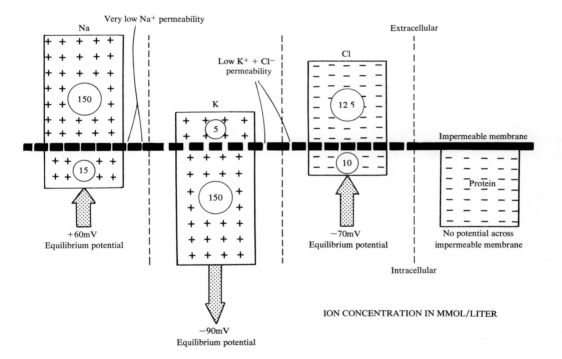

ION CONCENTRATION IN MMOL/LITER

Arrows are the magnitude of transmembrane (equilibrium) potentials needed to maintain internal + external ion concentration across a permeable membrane   (Arrow direction: effect of equilibrium potentials needed to maintain with regard to ion exchange)

(After Lewis)

61. ELECTRICAL AND CHEMICAL PROPERTIES OF A RESTING CELL: MEMBRANE POTENTIAL

I. Diffusion potential (cont.)

A. THE RESTING CELL MEMBRANE POTENTIAL: why is there a negative resting potential in a nerve or muscle cell?

1. The source of the resting potential is the unequal distribution of diffusible ions, particularly K^+, inside and outside the cell; i.e., K^+ are about 20–100× greater in concentration inside a muscle cell than outside it

2. The distribution of Cl^- ions is the reverse of the K^+ ions; generally, about 20–100× lower inside the cell than outside

3. The majority of intracellular anions are large protein anions (A^-), not Cl^-; Na^+ concentration is about 5–15× lower inside the cell than outside

4. Extracellular salt solution (salty like blood) is essentially a NaCl solution with an NaCl content of about 9 g/l, equal to a physiologic solution

5. In a nerve cell at rest, K^+ ion concentration is greater inside the cell than outside, and the Na^+ concentration is the opposite, with the cell membrane being about 50–75× more permeable to K^+ than to Na^+

6. Resting membrane potential (about -80 mV) is close to K^+ equilibrium potential because the membrane is not perfectly impermeable to Na^+ which continually diffuses down its electric and concentration gradients, adding a small amount to the inside of the cell. Measured membrane potential of a neuron (cell body) is closer to -70 mV than to the K^+ equilibrium potential of -90 mV, resulting in a membrane not at the K^+ equilibrium potential and a continual net diffusion of K^+ out of the cell

B. THE MEMBRANE ($Na^+ - K^+$) PUMP: active membrane transport mechanisms use forces from cellular metabolism to pump Na^+ back out of the cell and K^+ back in

1. For each 3 Na^+ ions pumped out of the cell, 2 K^+ ions are pumped into it

2. The pump makes a small direct contribution to the membrane potential because of the net outward pumping of cations

3. The pump makes an essential indirect contribution to membrane potential by maintaining the concentration gradients down which the ions diffuse. The membrane potential is due primarily to the diffusion of these ions

C. SUMMARY

1. Some K^+ ions diffuse out of the cell down their concentration gradient and some move in down the electric gradient, because the membrane potential is not as negative as the K^+ equilibrium potential (-70 mV rather -90 mV). Less K^+ enters the cell passively than leaves it

2. The difference is relatively small and is made up by active transport via the membrane pump such that the total K^+ entering equals that leaving—no gain or loss

3. Na^+ is driven passively into the cell by both electric and concentration forces, but because the membrane permeability of a resting cell to Na^+ is so low, the amount entering is low

4. There is no passive force to remove Na^+ from the cell. That which enters must be actively transported out by the membrane pump. The amount of Na^+ pumped out equals, in most cases, the amount of K^+ pumped in

5. When more than 1 ion species can diffuse across the membrane, the membrane permeability properties, as well as the concentration gradient of each species, must be considered when evaluating membrane potential

   a. If the membrane is impermeable to a given ion species, no ion of that species can cross it and contribute to diffusion potential, regardless of the electric and concentration gradients that are present

   b. The greater the membrane permeability to an ion species, the greater the influence of that ion on the diffusion potential

6. Since the resting membrane is more permeable to K^+ than Na^+ (and the concentration differences are about the same), the resting membrane potential is much closer to the K^+ equilibrium potential than to that of Na^+

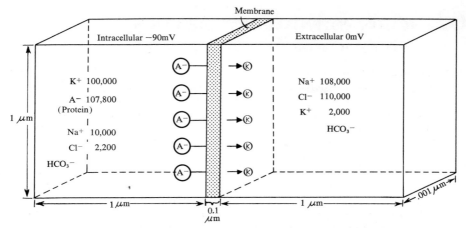

Membrane

Intracellular −90mV

Extracellular 0mV

K+ 100,000

A− 107,800
(Protein)

Na+ 10,000

Cl− 2,200

HCO₃⁻

Na+ 108,000

Cl− 110,000

K+ 2,000

HCO₃⁻

1 μm

1 μm

0.1 μm

1 μm

.001 μm

MEMBRANE CHARGE DURING RESTING POTENTIAL (K+ diffused through
membrane but stays fixed at its outside due to A− charge inside)

Extraneural

Neural membrane

Intraneural

$E_K - E = -11$mV

K+

Diffusion fluxes

K+

Pump fluxes

Resting potential:
−80 mV

$Na^+ - K^+$
pump

Metabolic
drive of
pump

Na+

Driving potential $E_{Na} - E = 155$ mV

Na+

Exterior

Membrane

Interior

Na+ ← NaY ← NaY ← Na+

Y

Y

1/3 ADP

Enzyme converts Y to X

Enzyme converts X to Y with
energy expenditure

X

X

1/3 ATP

X + Y: CARRIER MOLECULES

K+ → KX → KX → K+

Na+ − K+ TRANSPORT THROUGH MEMBRANE VIA CARRIER MOLECULES (after Glynn)

62. ACTION POTENTIAL: INTRODUCTION AND IONIC BASIS

I. **Introduction:** resting potential is needed for nerve cells and muscle fibers to fulfill their body functions. The nerve cell, however, must pick up information and transmit it in order to coordinate and integrate body function. When the cells work and are "active," brief (+) charges in the membrane potential occur and are called "action potentials," nerve impulses, or spikes. The latter are initiated in 3 ways: by activation of receptors; by synaptic input from other neurons; or spontaneously by intrinsic changes in the cell membrane (pacemaker neurons)

A. THE MEMBRANE POTENTIAL UNDERGOES RAPID CHANGE when nerve and muscle cells seem physiologically active, suddenly changing from -90 mV to $+30$ mV and then rapidly returning to its original value. This rapid change of membrane potential or sequence of depolarization and repolarization of the membrane that occurs spontaneously when the membrane is depolarized beyond the threshold potential is called an action potential and may only last about $\frac{1}{1000}$ second

   1. Only nerve and muscle cells can produce action potentials, a property called excitability. Electrical excitability is seen in axonal membranes; chemical excitability is seen at the synaptic membranes
   2. Excitable membranes can transmit action potentials along their surfaces, transmitting a signal from one portion of the nerve or muscle to another

B. PHASES OF THE ACTION POTENTIAL (a.p.)

   1. The action potential begins with a rapid (+) change in potential, the upstroke or rising phase, which lasts about 0.2–5 milliseconds in nerves and muscles
      a. During this rising phase, the cell loses its negative resting charge of polarization and this phase is also called the depolarization phase
         i. The membrane is said to be depolarized or stimulated when its potential is less negative than the resting membrane potential (closer to zero)
   2. The depolarization phase sets processes in motion to restore the resting membrane charge. The excitation-induced depolarization phase is followed by a spontaneous repolarization (moving from zero back to the resting potential)
   3. When the potential moves beyond its resting level, it is hyperpolarizing

C. THE ACTION POTENTIAL AND EXCITATION ARE TRIGGERED when the membrane, starting from the resting potential, is depolarized to about -50 mV. The potential from which it starts is called the threshold. The membrane charge is unstable at threshold

   1. The charge dissipates rapidly and automatically and usually even reverses polarity, resulting in a rapid rise of the action potential that goes beyond the zero potential and is called the overshoot phase
   2. The condition of a spontaneous progressive discharge of the membrane, triggered at threshold, is called excitation and lasts only a brief time, less than 1 millisecond

II. **Ionic basis**

A. HIGH INTRACELLULAR K^+ CONCENTRATION is a prerequisite for resting potential; high extracellular Na^+ concentration is necessary for action potential. Excitability depends on low intracellular Na^+ concentration so that Na^+ can flow into the cell

B. THE ACTION POTENTIAL MUST RESULT FROM A TRANSIENT CHANGE in either the concentration gradients or the membrane permeabilities, the latter being the case

   1. In the resting state, the membrane is 50–75× more permeable to K^+ than Na^+ ions; thus, the magnitude and polarity of the resting potential are due almost entirely to the movement of K^+ ions out of the cell

C. DURING THE ACTION POTENTIAL membrane permeability to Na^+ and K^+ is altered

   1. In the rising phase of the action potential, membrane permeability to Na^+ increases 600-fold and Na^+ ions rush into the cell. There is little change in K^+ ion permeability of the membrane. More (+) charge enters the cell in the form of Na^+ ions than leaves it in the form of K^+ ions; the membrane potential decreases and eventually reverses polarity, becoming (+) inside and (−) outside the membrane

AXON

Overshoot

Depolarization (rising) phase

Repolarization

Threshold

Hyperpolarizing   Depolarizing } After potentials

Resting

Membrane potential (mV)

Time (msec) ⟶

Resting membrane potential

Action potential (1 msec)

Pansky

CHANGES IN MEMBRANE POTENTIAL DURING AN ACTION POTENTIAL

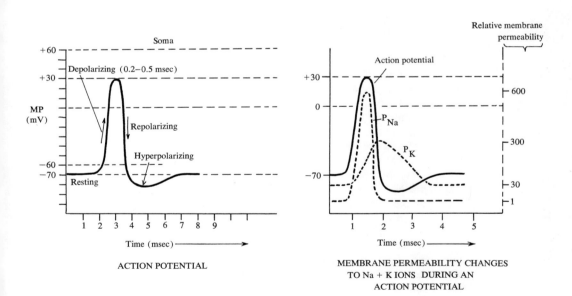

Soma

Depolarizing (0.2–0.5 msec)

MP (mV)

Repolarizing

Hyperpolarizing

Resting

Time (msec) ⟶

ACTION POTENTIAL

Relative membrane permeability

Action potential

P_{Na}

P_K

Time (msec) ⟶

MEMBRANE PERMEABILITY CHANGES
TO Na + K IONS DURING AN
ACTION POTENTIAL

63. ACTION POTENTIAL: IONIC BASIS AND SUMMARY EVENTS

I. Ionic basis (cont.)

A. ACTION POTENTIAL IN NEURONS lasts about 1 ms (0.001 sec) and the membrane quickly returns to resting level because
   1. Permeability to Na^+ is altered when the membrane potential is changed; e.g., hyperpolarization causes a decrease in Na^+ permeability whereas depolarization results in an increase in Na^+ permeability
   2. Increased Na^+ permeability (activation) is rapidly turned off (Na^+ inactivation)
   3. The membrane permeability to K^+ increases over its resting stage
   4. As the membrane becomes more ($+$) inside and the drive for Na^+ entry is reduced, Na^+ permeability decreases toward its resting value and Na^+ entry quickly decreases (this alone would restore the potential to its resting level)
   5. The entire process is further speeded up by a simultaneous increase in K^+ permeability, causing more K^+ to move out of the cell down their concentration gradient, allowing K^+ diffusion to regain predominance over Na^+ diffusion and the membrane to return to its resting level
   6. Massive movements are not necessary; e.g., only 1 of every 100,000 K^+ ions in the cell need diffuse out to change the membrane to its resting value, and very few Na^+ ions need enter the cell to cause depolarization during an action potential, resulting in virtually no change in the concentration gradients during that action potential

B. SUMMARY OF IONIC MOVEMENTS DURING ACTION POTENTIAL
   1. As a result of depolarization beyond the threshold, the Na^+ conductance increases rapidly (increased permeability to Na^+)
   2. The K^+ conductance also increases, but with a time lag
   3. Therefore, initially, Na^+ ions flow rapidly into the cell, creating further depolarization of the membrane, and the membrane potential moves in the direction of the Na^+ equilibrium potential at $+60\,mV$
   4. Then K^+ ions flow out of the cell, restoring the resting potential charge and repolarizing the membrane to its resting potential
   5. Despite large changes in the conductance of the membrane during the action potential, the ionic shifts through the membrane are small in relation to the number of ions surrounding the membrane
   6. The Na^+ ions that flowed into the cell with the action potential are expelled again in the course of time by the Na^+ pump
      a. The active Na^+ transport compensates not only for the resting Na^+ influx but also for the Na^+ influx during excitation
      b. The active Na^+ transport is of no importance for the individual action potential; if the pump is blocked (poisoned with DNP), despite active transport elimination, thousands of action potentials can occur before intracellular Na^+ concentration becomes too high to render the cell inexcitable
      c. Energy-consuming processes such as the Na^+ pump are only necessary to maintain the concentration gradients
      d. We have no explanation for the physiochemistry of how Na^+ permeability of the membrane is altered when the level of the membrane potential changes
      e. Whatever the responsible mechanisms, they are found only in the cell membranes of nerve and muscle cells; changing the membrane potential in other types of cells does not result in a change in membrane permeability

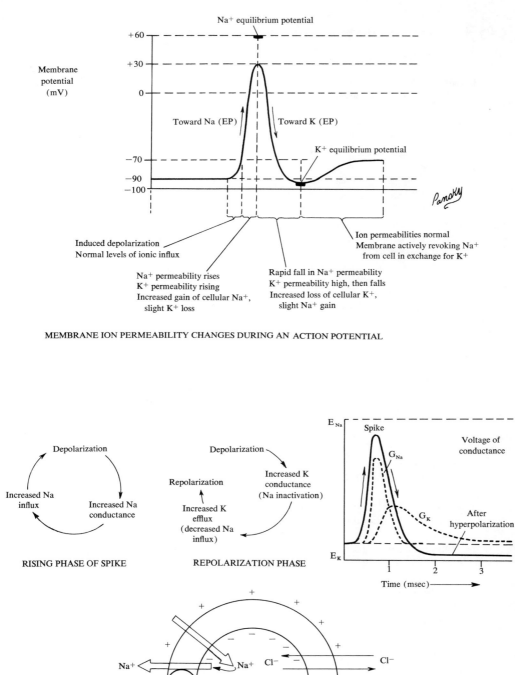

Na+ equilibrium potential

+60

+30

Membrane
potential
(mV)

0

Toward Na (EP)   Toward K (EP)

K+ equilibrium potential

−70

−90
−100

Induced depolarization
Normal levels of ionic influx

Na+ permeability rises
K+ permeability rising
Increased gain of cellular Na+,
    slight K+ loss

Rapid fall in Na+ permeability
K+ permeability high, then falls
Increased loss of cellular K+,
    slight Na+ gain

Ion permeabilities normal
Membrane actively revoking Na+
    from cell in exchange for K+

MEMBRANE ION PERMEABILITY CHANGES DURING AN ACTION POTENTIAL

Depolarization

Increased Na
influx

Increased Na
conductance

RISING PHASE OF SPIKE

Depolarization

Repolarization

Increased K
conductance
(Na inactivation)

Increased K
efflux
(decreased Na
influx)

REPOLARIZATION PHASE

E_{Na}

Spike

G_{Na}

Voltage of
conductance

G_K

After
hyperpolarization

E_K

1   2   3

Time (msec)

Na+ − K+ PUMP
(ACTIVE TRANSPORT)

Na+

Na+

Cl−

Cl−

Inside

K+

Outside

K+

Membrane

64. ACTION POTENTIAL: THRESHOLD AND REFRACTORY PERIOD

I. Threshold: partially depolarizing an excitable membrane initiates an action potential only when the strength of the stimulus is sufficient to depolarize the membrane potential to a critical level, known as the threshold potential. This is a threshold stimulus

A. SUBTHRESHOLD STIMULI (WEAKER STIMULI) do not initiate an action potential. Stronger stimuli, suprathreshold stimuli, elicit action potentials, but the response is no different from that following a threshold stimulus

B. THRESHOLD POTENTIAL FOR MOST EXCITABLE MEMBRANES is 5–15 mV more depolarized than the resting membrane's potential (if resting potential of a neuron is -70 mV, the threshold's potential may be -60 mV). To initiate an action potential in such a membrane, the potential must be decreased by at least 10 mV

C. EXPLANATION OF THRESHOLD: at any potential between resting and threshold, Na^+ movement into the cell is less than K^+ ion movement out of the cell despite the increased Na^+ permeability which prevents further depolarization beyond that induced directly by the stimulus and drives the potential rapidly back to resting levels as soon as the stimulus is removed

  1. Once threshold is reached, the process is no longer dependent on stimulus strength, and the depolarization continues on to become an action potential solely because the membrane permeability changes allow Na^+ ions to diffuse down the electric and concentration gradients which exist across the membrane

  2. Action potentials triggered either by stimuli just strong enough to depolarize the membrane a bit above threshold or by very strong stimuli are identical

  3. Action potentials occur either maximally or not at all (all-or-none law). A single action potential cannot convey any information about the magnitude of the stimulus that initiated it

  4. The ability to distinguish between levels of stimuli, e.g., a loud sound or whisper, depends on the number of action potentials transmitted per unit of time (depends on frequency of action potentials and not on their size)

  5. Most nerve cells respond at frequencies between 0–100 action potentials per second (some even higher for brief periods of time)

II. Refractory periods: If one applies a threshold strength stimulus to a membrane and then stimulates the membrane with a 2nd threshold stimulus, the membrane does not always respond to the 2nd stimulus. The membrane appears unresponsive for a period of time even though identical stimuli are applied. The membrane, during this period, is said to be refractory to the 2nd stimulus

A. TWO SEPARATE REFRACTORY PERIODS ASSOCIATED WITH AN ACTION POTENTIAL: if the 2nd stimulus is increased to suprathreshold levels can be distinguished

  1. During the 1-millisecond period of the action-potential spike, a 2nd stimulus will not produce a 2nd action potential response no matter how strong it is; the membrane is said to be in its absolute refractory period

  2. Following the above, there is an interval during which a 2nd action-potential response can be produced, only if the stimulus strength is greater than threshold—the relative refractory period (can last 10–15 milliseconds or longer)

B. REFRACTORY PERIOD MECHANISMS are related to the membrane mechanisms that alter the Na^+ and K^+ permeability

  1. The absolute refractory period corresponds with the period of inactivation of Na^+ ion channels

  2. The relative refractory period corresponds with the period of recovery of Na^+ channels. K^+ permeability becomes high during repolarization and is continuous throughout the negative and positive afterpotentials

C. AFTER AN ACTION POTENTIAL, time is needed to return the membrane structure to its original resting state. Refractory periods limit the number of action potentials an excitable membrane can produce in a given period of time

Action potentials
(generated at threshold potential)

Membrane potential (mV)

+30

0

Threshold potential

−70

Resting potential

Increasing stimulus above
threshold does not alter action
potential response

Stimulates
strength
(volts)

+

0

−

Time (msec) ⟶

Subthreshold
stimuli

Threshold
stimulus

Suprathreshold
stimuli

+30

0

MP
(mV)

|◄──Refractory period──►|

−70

1st stimulus
(threshold)

Stimulus
strength
(volts)

|◄────2nd stimulus────►|

+

0

−

0   5   10

Time (msec) ⟶

After action potential, membrane is
refractive for several msec

+30

0

MP
(mV)

Threshold
potential

−70

Relative refractory period
Absolute refractory period

Stimulus
strength
(volts)

+

0

−   1st stimulus   2nd stimulus

0   5   10

Time (msec) ⟶

With time, stimulus size for 2nd action
potential decreases

+30
+20
0
−20
−40
−60
−80
−90
−100

Membrane
potential
(mV)

Threshold

A   B   C   D

REFRACTORY PHASE
AFTER EXCITATION

0   1   2   3   4   5   6   msec

|◄──Absolute──►|◄─Relative─►|
refractory phase

A = Action potential
B, C, D, = Action potentials triggered
        at periods after 1st

Dotted lines: Depolarization of fiber
        to threshold

65. PROPAGATION OF AXON DEPOLARIZATION

I. **Introduction:** If the action potential of a nerve fiber is measured at 2 points and the nerve is stimulated at one end, an action potential 1st appears at the measuring point closest to the site of stimulation and a little later at the 2nd measuring point. Thus, the action potential is propagated or conducted from the site of stimulation past the 2nd electrode

A. VELOCITY OF CONDUCTION is determined by dividing the distance between the 2 points that the action potential passes by the time it takes to travel that distance

    1. Conduction velocities in nerve fibers range from 1 m/s to greater than 100 m/s; depend on characteristics of fiber; and are typical for each type

B. ONE ACTION POTENTIAL does not itself travel along the membrane, but each action potential triggers, by local current flow, a new one at an adjacent area of the membrane

    1. The old action potential provides the electric stimulus that depolarizes the new membrane site to just past its threshold potential

    2. When this happens, the Na^+ activation cycle at the new membrane site takes over and an action potential occurs there

    3. Once the new site is depolarized to threshold, the action potential generated there is solely dependent on the electrochemical gradients and membrane permeability properties at the new site

    4. Since the above factors are the same as those involved in the generation of the old one, the new action potential is identical to the old. No distortion takes place as the signal passes along the membrane and the signal at the end of the membrane is the same as that at the beginning

C. PROCESS OF PROPAGATION

    1. When an action potential is initiated at a site, Na^+ permeability changes so that Na^+ rushes in across the membrane, making the inside of the cell relatively more ($+$) and leaving the outside more ($-$). At this time, the remainder of the cell membrane is at its normal resting potential

    2. Like charges attract and unlike repel and thus current (the flow of positive charges) flows away from the activated membrane region through the cytoplasm and toward the activated region through the extracellular fluid

    3. The addition of ($+$) charge to the inside of the cell and removal of ($+$) charge from the outside decrease the potential difference across the membrane and this initial depolarization, due to local current flow, acts as a stimulus to trigger an action potential

    4. Meanwhile, at the original membrane site, Na^+ inactivation is occurring and K^+ permeability is increasing so that the membrane is repolarizing

    5. These processes repeat themselves until the end of the membrane is reached

    6. By convention, the direction of movement of the ($+$) ions is labeled the direction of current flow (negatively charged particles, however, can and do move in the opposite direction)

    7. The change in potential or electric charge moves from one point to another much faster than the ions themselves move between the points because of the influence of the molecules themselves. However, local current flow is decremental—its amplitude decreases with increasing distance

    8. Even though local current flow is decremental because membranes depend on local current flow, it is over a short distance and doesn't matter because only the amount of local current flow which is necessary to depolarize action potential occurs at the new site and is transmitted without a decrement because the new action potentials are continually generated along the membrane

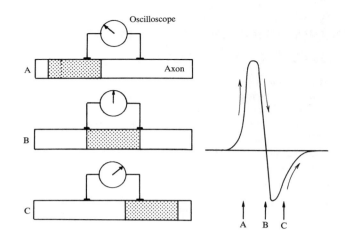

DIPHASIC RECORDING OF ACTION POTENTIAL WITH EXTERNAL ELECTRODES

PROPAGATION OF A NERVE IMPULSE

66. PROPAGATION OF ACTION POTENTIAL: DIRECTION AND VELOCITY

I. Propagation of the action potential (cont.)

A. DIRECTION OF ACTION-POTENTIAL PROPAGATION
   1. The only direction of action-potential propagation is away from the stimulation site
   2. The action potentials in skeletal muscle membrane are initiated near the middle of the cell and propagate from this region toward the 2 ends, but in most nerve-cell membranes, action potentials are initiated at one end of the cell; this unidirection propagation is determined by the stimulus location and not the inability to conduct in the other direction

B. VELOCITY OF ACTION-POTENTIAL PROPAGATION
   1. Velocity depends on the fiber diameter and whether or not the fiber is myelinated
      a. The larger the fiber diameter, the faster is the action-potential propagation because a large fiber offers less resistance to local current flow and adjacent regions are brought to threshold faster
      b. Myelin electrically insulates the membrane, making it more difficult for current to flow between intra- and extracellular fluid compartments
      c. Action potentials do not occur along the sections of membrane protected by myelin but occur only where the myelin coating is interrupted at the nodes of Ranvier where the membrane is exposed to the extracellular fluid. Thus, the action potential appears to jump from node to node as it propagates along the myelinated fiber; called saltatory conduction
      d. The membrane of nodes adjacent to the active node is brought to threshold faster and undergoes an action potential sooner than if no myelin were present
   2. The velocity of action-potential propagation in large myelinated fibers can exceed 250 miles per hour

Fiber Type	Fiber Diameter (μm)	Conduction Velocity (m/sec)	Spike Duration (msec)	Absolute Refractory Period (msec)	Functions
A (α)	12–22	70–120	0.4–0.5	0.2–1.0	Somatic motor, primary muscle–spindle afferents Propioceptors
A (β)	5–13	30–70	0.4–0.5	0.2–1.0	Touch, kinesthesia (muscle sense), pressure
A (γ)	3–8	15–40	0.4–0.7	0.2–1.0	Motor (to mm spindle) Touch and pressure
A (δ)	1–5	12–30	0.4–1.0	0.2–1.0	Pain and temperature (heat and cold) Pressure
B	1–3	3.0–5.0	1.2	0.6–1.2	Automatic preganglionic fibers
C (dorsal root)	0.2–1.2	0.2–2.0	2	2	Pain, reflexes
C (sympathetic nerves)	0.3–1.3	0.7–2.3	2	2	Postganglionic sympathetic

Chronaxie = Excitation time
       constant for nerve (<1 msec for normal nerve)
       may be 100X normal in injured nerve

Current strength (volts)

Utilization time

Rheobase

Chronaxie   Duration of current

STRENGTH—DURATION RELATIONSHIPS (EXCITATION TIME)
(Rheobase: Least strength which an applied current could have and still excite the membrane)

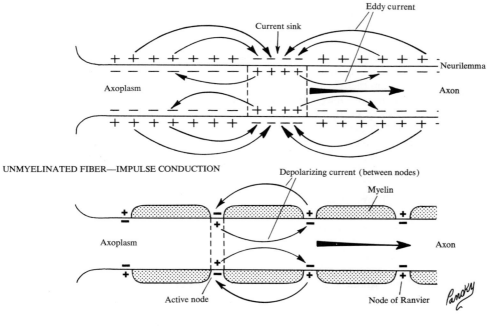

Eddy current

Current sink

Neurilemma

Axoplasm

Axon

UNMYELINATED FIBER—IMPULSE CONDUCTION

Depolarizing current (between nodes)

Myelin

Axoplasm

Axon

Active node

Node of Ranvier

MYELINATED FIBER—IMPULSE CONDUCTION
(SALATORY CONDUCTION)

67. PHYSIOLOGIC ACTIVITIES OF NEURONS: FUNCTIONAL ANATOMY OF THE NEURON

I. **The human nervous system** consists of neurons and glial cells which make up the brain and spinal cord. It also includes the many nerve processes that run between them as well as the receptors, muscles, and glands they innervate

A. THE BRAIN AND SPINAL CORD form the central nervous system (CNS) and are protected by the skull and vertebral column, respectively

B. ALL NERVE CELLS (or parts of them) that lie outside of the skull and vertebral column are part of the peripheral nervous system (PNS)

II. **The neuron, the basic unit of the nervous system:** the human brain has about 25 billion cells

A. EACH NEURON is bounded by a cell membrane that encloses the cell contents

    1. Only about 10% of the cells in the nervous system are neurons

        a. The remaining cells are glial cells which probably support the neurons physically and sustain them metabolically

    2. The neuron consists of 3 parts, each associated with a particular function

        a. The dendrites (usually several) and cell body or soma

            i. Form a series of branched cell outgrowths connected to the cell body and serve as an extension of the cell membrane of the body

            ii. There are great variations in dendritic formations: profusely branched; ratio of soma surface to dendritic surface may be larger in others; some neurons have no dendrites; and dendrites can be several hundred micrometers long

            iii. The dendrites and soma are the site of the specialized junctions with other neurons through which signals are passed to the cell

            iv. The soma contains the nucleus and other organelles involved in metabolic processes and maintains the metabolism of the neuron, including its growth and repair (diameter of neuronal cell bodies are in the order of 5–100 mm)

        b. The axon (neurite or nerve fiber) is a long single process extending from the cell body (usually longer than the dendrite)

            i. Its first portion plus the part of the cell body where it is joined to the cell is called the initial segment

            ii. Can give off collaterals (branches) along its course, and near its end undergoes branching into many axon terminals, the last part of which is enlarged and is responsible for transmitting a signal from the neuron to the cell contacted by the terminal. Axons and collaterals vary in length from a few μm to well over a meter

            iii. Every axon in a peripheral nerve lies in a "tube" formed by special glial cells, the Schwann cell. Together, the axon and Schwann sheath (neurolemma) around it are called the nerve fiber

                (a) The number of nerve fibers bundled together to form a nerve varies from dozens or hundreds to tens of thousands

                (b) About $\frac{1}{3}$ of all nerve fibers are encased, during growth, by the Schwann cell winding around the axon so that between the axon and the Schwann-cell body, a sheath is formed consisting of a lipid-protein mixture called myelin; nerve fibers "insulated" in this manner are called myelinated nerve fibers, A or B fibers

                (c) The myelin sheath is discontinuous, being interrupted at regular intervals which, under the microscope, look like constrictions or nodes and are called the nodes of Ranvier (occur about every 1–2 mm)

            iv. Nerve fibers without the myelin sheath are called unmyelinated or C fibers

                (a) The unmyelinated fibers are also unsheathed by Schwann cells; often a Schwann cell may enclose several unmyelinated axons

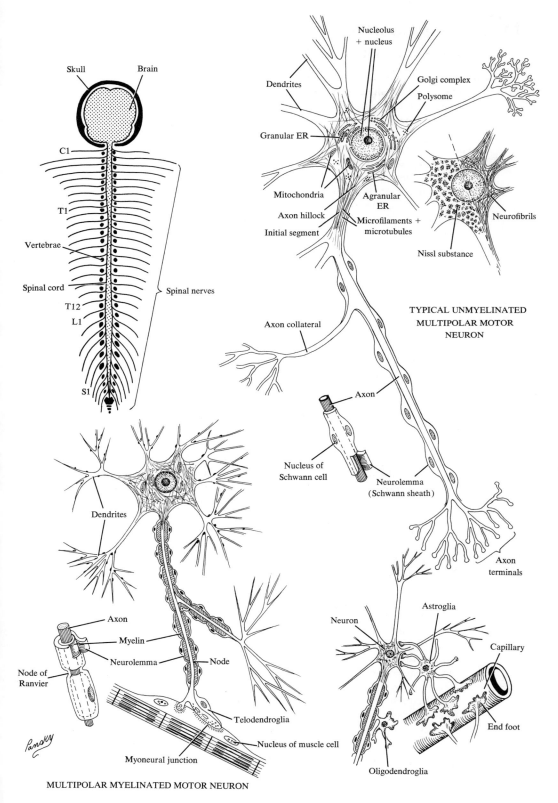

Skull Brain

C1

T1

Vertebrae

Spinal cord

T12
L1

S1

Spinal nerves

Nucleolus + nucleus

Dendrites

Golgi complex

Polysome

Granular ER

Mitochondria

Agranular ER

Axon hillock

Microfilaments + microtubules

Initial segment

Neurofibrils

Nissl substance

TYPICAL UNMYELINATED
MULTIPOLAR MOTOR
NEURON

Axon collateral

Axon

Nucleus of
Schwann cell

Neurolemma
(Schwann sheath)

Axon
terminals

Dendrites

Astroglia

Neuron

Capillary

Axon

Myelin

Neurolemma

Node

Node of
Ranvier

End foot

Telodendroglia

Nucleus of muscle cell

Myoneural junction

Oligodendroglia

MULTIPOLAR MYELINATED MOTOR NEURON

Pansky

68. PHYSIOLOGIC ACTIVITIES OF NEURONS:
CLASSIFICATION AND RECEPTORS

I. Classification of neurons: regardless of shape they can be placed into 3 classes

A. AFFERENT NEURONS lie largely outside of the CNS

   1. At their peripheral endings have receptors, which, in response to physical or chemical environmental changes, cause action potentials to be generated in the afferent neuron

   2. These neurons carry information from receptors into the brain or spinal cord. After transmission to the CNS, some of the afferent information may be perceived as a conscious sensation

   3. Afferent nerve fibers from the viscera are called visceral afferents, all others are called somatic afferents

B. EFFERENT NEURONS also lie largely outside of the CNS and transmit the final integrated information from the CNS out to the effector organs (muscles or glands)

   1. Efferent neurons to skeletal muscles are called motor neurons; all the rest belong to the autonomic nervous system and are called autonomic efferents

C. THE INTERNEURONS lie within the CNS where they originate and terminate

   1. 99% of all nerve cells belong to this group

   2. The interneurons and their connections in large part account for thoughts, feelings, learning, language, etc.

   3. The number of interneurons in the pathway between the afferent and efferent neurons varies with the complexity of the action

   4. One type of reflex, the stretch reflex, has no interneurons, whereas stimuli invoking memory or language may involve thousands of interneurons

II. Receptors: different energy forms (sensory elements) relay information about our external and internal environment: pressure, temperature, pain, light, etc.

A. TELERECEPTORS: for our distant environment via our eyes and ears

B. EXTEROCEPTORS: for our immediate environment via skin receptors

C. PROPRIOCEPTORS: about the attitude and position of the body in space, with receptors in the labyrinth and those of muscles, tendons, and joints

D. INTEROCEPTORS AND VISCEROCEPTORS: for information on events in the viscera

E. ALONE can deal with the variety of energy forms; the rest of the system can use only action potentials or small subthreshold changes in membrane potentials. Thus, all energy forms must be translated into the language of action potential

   1. The receptors are either specialized peripheral endings of neurons or separate cells intimately connected to them

     a. Not all neurons have specialized receptor regions—seen only on afferent neurons

   2. There are many types of receptors, each of which is specific, namely, responds to one form of energy or stimulus more than another, although virtually all receptors can be activated by several different forms of energy

     a. The receptors of the eye normally respond to light (electromagnetic radiation with wavelengths between 400 and 800 nm, violet to blue) but can be activated by intense mechanical stimuli such as a "poke" in the eye and we see "stars"

     b. Receptors of the ear, on the other hand, are specialized to transduce sound waves with frequencies from 20 to 20,000 Hz (cycles per second)—sensory modalities (include vision, audition, etc.)

     c. Usually much more energy is needed to excite a receptor by energy forms to which it is not specific, the receptor being highly sensitive to its own special energy form which is an adequate stimulus

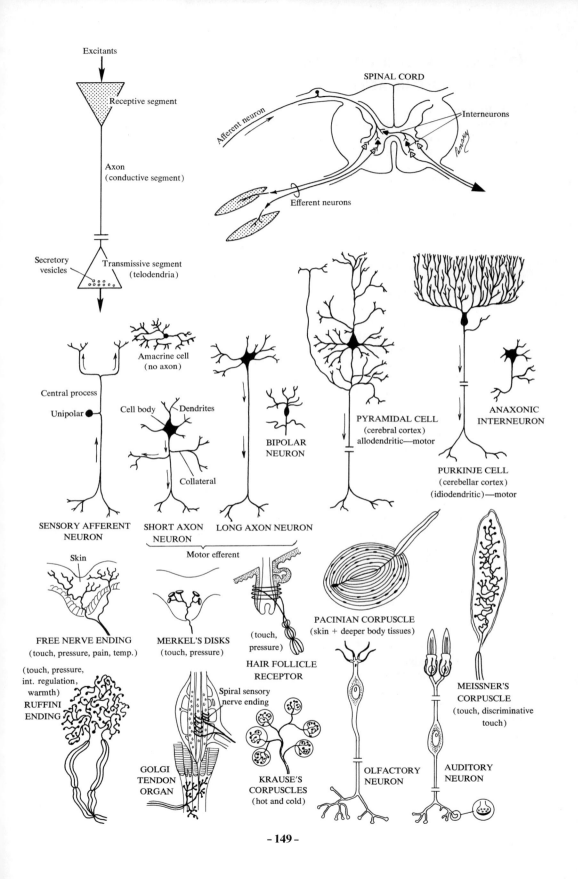

Excitants

Receptive segment

Axon
(conductive segment)

Secretory
vesicles

Transmissive segment
(telodendria)

SPINAL CORD

Afferent neuron

Interneurons

Efferent neurons

Amacrine cell
(no axon)

Central process

Unipolar

Cell body

Dendrites

BIPOLAR
NEURON

PYRAMIDAL CELL
(cerebral cortex)
allodendritic—motor

ANAXONIC
INTERNEURON

PURKINJE CELL
(cerebellar cortex)
(idiodendritic)—motor

Collateral

SENSORY AFFERENT
NEURON

SHORT AXON
NEURON

LONG AXON NEURON

Motor efferent

Skin

FREE NERVE ENDING
(touch, pressure, pain, temp.)

MERKEL'S DISKS
(touch, pressure)

(touch,
pressure)

HAIR FOLLICLE
RECEPTOR

PACINIAN CORPUSCLE
(skin + deeper body tissues)

(touch, pressure,
int. regulation,
warmth)
RUFFINI
ENDING

Spiral sensory
nerve ending

GOLGI
TENDON
ORGAN

KRAUSE'S
CORPUSCLES
(hot and cold)

OLFACTORY
NEURON

AUDITORY
NEURON

MEISSNER'S
CORPUSCLE
(touch, discriminative
touch)

69. GENERATOR POTENTIAL: RECEPTOR ACTIVATION

I. Receptors (cont.)

A. RECEPTOR ACTIVATION OR THE GENERATOR POTENTIAL

1. The following refers only to the general mechanisms for receptor activation in which the receptor is the peripheral ending of the afferent neuron and responds to pressure or mechanical changes, namely, the transformation of mechanical energy into electrochemical energy of action potentials which take place at the very end of the afferent neuron (in the part lacking a myelin sheath)

    a. Mechanical stimuli (e.g., pressure) bend, stretch, or press on the receptor membrane and, somehow, perhaps by opening pores in the membrane, increase membrane permeability

    b. With increased permeability, ions move across the membrane down their electric and concentration gradients (the permeability change is to Na^+ and one or more other ions)

    c. Since the intracellular fluid of all nerve cells has a higher concentration of K^+ and a lower concentration of Na^+ than the extracellular fluid, and the inside of the reacting neuron is about 70 mV negative with respect to the outside, change in membrane permeability at the receptor results in a net outward diffusion of a small number of K^+ ions and a simultaneous movement of a larger number of Na^+ ions inward

    d. The result is a net movement of $(+)$ charge into the cell leading to a decrease in membrane potential (depolarization) with the movement of Na^+ into the nerve fiber playing the major role in depolarizing the nerve ending (K^+ and other ions such as Cl^- have less effect). The initial depolarization of the receptor is called the generator potential (g.p.)

    e. The g.p. occurs at the unmyelinated nerve ending where the cell membrane is specialized for excitation by an "adequate" or appropriate stimulus

    f. The g.p. cannot by itself produce action potentials but the depolarization is conducted by local current flow a short distance from the nerve ending to the first node in the myelin sheath. If the amount of depolarization that reaches the first node is large enough to bring the membrane there to threshold, an action potential is initiated; it is the action potential and not the generator potential that propagates along the nerve fiber

    g. If after one action potential is fired, local current flow through the first node remains sufficient to depolarize the node again to threshold, another action potential occurs and as long as the first node is depolarized to threshold, action potentials continue to fire and propagate along the membrane of the afferent nerve

    h. The action potential is all-or-none and its amplitude is always the same, regardless of the size of the stimulus. The generator potentials, however, are not all-or-none but their amplitude and duration vary with the stimulus and other factors which might influence action-potential frequency in the afferent neuron

    i. Summary: generator potential is a local response whose only function is to trigger an action potential; the latter traveling along the nerve fiber to its terminations

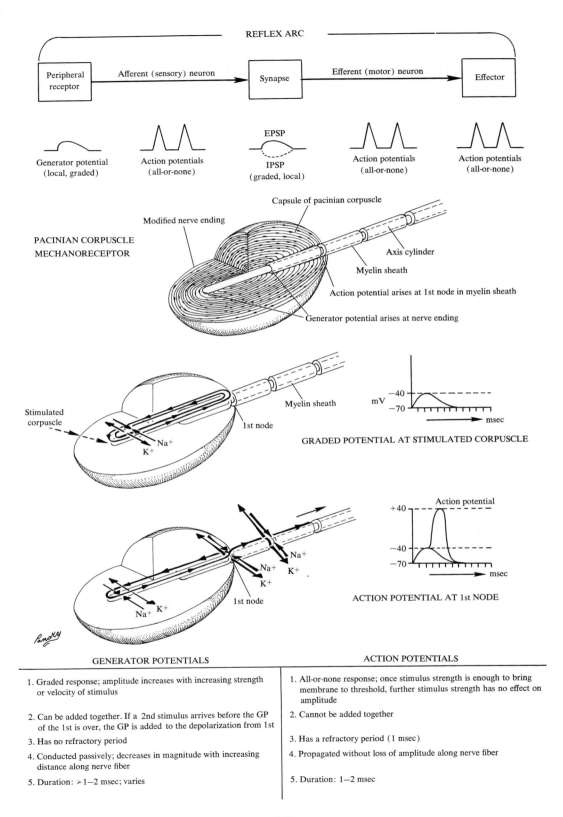

REFLEX ARC

| Peripheral receptor | — Afferent (sensory) neuron → | Synapse | — Efferent (motor) neuron → | Effector |

Generator potential
(local, graded)

Action potentials
(all-or-none)

EPSP

IPSP
(graded, local)

Action potentials
(all-or-none)

Action potentials
(all-or-none)

**PACINIAN CORPUSCLE
MECHANORECEPTOR**

Capsule of pacinian corpuscle

Modified nerve ending

Axis cylinder

Myelin sheath

Action potential arises at 1st node in myelin sheath

Generator potential arises at nerve ending

Stimulated corpuscle

Myelin sheath

1st node

Na^+

K^+

−40

−70

mV

msec

GRADED POTENTIAL AT STIMULATED CORPUSCLE

Action potential

+40

−40

−70

msec

Na^+

Na^+ K^+

K^+

1st node

Na^+ K^+

ACTION POTENTIAL AT 1st NODE

GENERATOR POTENTIALS	ACTION POTENTIALS
1. Graded response; amplitude increases with increasing strength or velocity of stimulus	1. All-or-none response; once stimulus strength is enough to bring membrane to threshold, further stimulus strength has no effect on amplitude
2. Can be added together. If a 2nd stimulus arrives before the GP of the 1st is over, the GP is added to the depolarization from 1st	2. Cannot be added together
3. Has no refractory period	3. Has a refractory period (1 msec)
4. Conducted passively; decreases in magnitude with increasing distance along nerve fiber	4. Propagated without loss of amplitude along nerve fiber
5. Duration: >1–2 msec; varies	5. Duration: 1–2 msec

70. RECEPTORS: GENERAL POTENTIAL AMPLITUDES AND INTENSITY CODING

I. Amplitude and duration of the generator potential (g.p.)

A. AMPLITUDE determines action potential frequency (number of action potentials fired per unit of time); it does not determine action potential size

B. AMPLITUDE VARIATIONS are due to the stimulus intensity, rate of change of the stimulus application, summation of successive g.p.*s*, and adaptation.

1. G.p.*s* become larger with greater intensity of the stimulus because the permeability changes increase and the transmembrane ion movements are greater (possible increase in number of pore size allowing transmembrane ion flow)

2. Amplitude of the g.p. also rises with a greater rate of change of stimulus application as well as rate of removal of stimulus (so-called off-response which further depolarizes the membrane leading to a burst of action potentials)

3. Summation of generator potentials
   a. Amplitude can be varied by adding 2 or more g.p.*s*, possible because they are a graded phenomenon and last 5 to 10 msec; thus, if the nerve ending is stimulated again before the g.p. from a preceding stimulus has died, the 2 summate and make a large single g.p.

4. Adaptation is a decrease in frequency of action potentials in the afferent neuron despite a constant stimulus energy and is not due to nerve fatigue (which does not occur in nerve membrane). It occurs because
   a. Stimulus energy can be dissipated in the tissues as it passes through them to reach the receptor
   b. Receptor membrane responsiveness can decrease with time and the g.p. amplitude drops even though the energy reaching the receptor stays the same
   c. Even if the g.p. amplitude remains unchanged, there can be a decreased frequency of action potentials in response to a constant stimulus because the membrane changes at the 1st node

C. SUMMARY: the magnitude of a given g.p. varies with stimulus intensity, rate of change of stimulus application, summation, adaptation, and cessation of the stimulus. Amplitude of the g.p. does not determine the amplitude of the action potentials in the afferent neuron but determines how many, if any, a.p.*s* take place

II. Intensity coding: there are different stimulus intensities

A. INCREASED STIMULUS STRENGTH means a larger g.p. and higher frequency of firing action potentials

1. There is an upper limit to the positive correlation between stimulus intensity and action-potential frequency
   a. When stimulus strength becomes great, the g.p. reaches a maximum; further increase in rate of firing of a.p.*s* by that receptor cannot take place
   b. Though a certain receptor cannot generate a higher frequency of action potentials, receptors at other branches of the same neuron can be stimulated

2. Most afferent neurons have many branches, each with a receptor at its ending

3. Receptors at different branches do not all respond with equal ease to a given stimulus
   a. As stimulus strength increases, more and more receptors begin to respond and action potentials generated by these receptors propagate along the branch to the main afferent nerve fiber and, if the membrane is not refractory, they increase the frequency of action potential there

4. In addition to the increased frequency of firing in a single neuron, similar receptors on the nerve endings of other afferent neurons are also activated as stimulus strength increases because a larger area is affected. This calling in of receptors on additional nerve cells is called recruitment

① Stimulus

② Rapidly adapting receptor potential

Spike train

③ Slowly adapting receptor potential

Spike train

RECEPTOR ADAPTATION

ADAPTATION

Nerve

Touch

Pressure

Muscle spindle

Stimulus

0 1 2 3 4 5 6 7 8 Time Sec.

Dendrites (receptive)

Stimulus

Soma

Receptor potential

2
0

50
mV
80

Generator potentials

A.P.

0
50 mV
80

Axon

Action potentials

+50
0
−80

RECEPTOR POTENTIAL = GENERATOR POTENTIAL
(acts as an electric stimulus to axon)

Intensity (arbitrary units)

Time

Duration

S   S   S

Amplitude (mV)

25
20
15
10
5
0

Generator receptor potential

Duration

Amplitude (mV)

120

60

0

Action potentials

Inadequate to evoke all-or-none response

All-or-none response

Increased firing rate with greater stimulus

RELATIONSHIP OF STIMULUS INTENSITY, AMPLITUDE
OF RECEPTOR POTENTIAL, AND FREQUENCY OF
DISCHARGE IN A NERVE CONTINUOUS WITH THE RECEPTOR

71. SYNAPSES: ANATOMY AND DEFINITIONS

I. **Introduction:** A synapse is an automatically specialized junction between 2 neurons (usually axonal endings on another cell) where the electric activity in 1 neuron influences the excitability of the 2nd

A. MOST SYNAPSES OCCUR between the axon terminals of one neuron and the cell body or dendrites of a second—the axodendritic synapse; there may, however, be axosomatic, axoaxonal somatosomatic, somatoaxonic, dendrodendritic synapses as well

B. THE NEURONS CONDUCTING TOWARD THE SYNAPSES are called presynaptic and those conducting information away are postsynaptic neurons
   1. Every postsynaptic neuron has thousands of synaptic junctions on the surface of its dendrites or cell body so that information from hundreds of presynaptic nerve cells converges on them

C. IF AN AXON ENDS ON A SKELETAL MUSCLE FIBER, this synapse is called a neuromuscular plate or junction [synapses on muscle fibers of the viscera (smooth muscle) and on glandular cells have no special names]

D. EACH ACTIVATED SYNAPSE produces a small electric signal, either excitatory or inhibitory, in the postsynaptic cell and for a short time visualizes thousands of synapses from many different presynaptic cells converging on a single postsynaptic cell. The level of excitability of this cell at any moment in time depends on the number of synapses that are active at any one time and how many are excitatory or inhibitory
   1. If the postsynaptic neuron reaches threshold and generates a response, action potentials are transmitted out along its axon to the terminal branches which diverge to influence the excitability of many other cells
   2. Postsynaptic neurons thus function as neural integrators and their output reflects the sum of the incoming pieces of information arriving as excitatory and inhibitory inputs

II. **Functional anatomy of a synapse**

A. THE AXON TERMINAL of the presynaptic neuron is a slight swelling at the end of the synaptic knob

B. THE SYNAPTIC CLEFT separates the pre- and postsynaptic neurons and prevents direct propagation of the action potential from the pre- to the postsynaptic cell

C. MEMBRANE-ENCLOSED VESICLES found in the synaptic knob transmit information across the cleft by means of a chemical agent stored in the small vesicles
   1. When the action potential in the presynaptic neuron reaches the axon terminal and depolarizes the knob, small amounts of the chemical transmitter are released from the knob into the cleft. Once released from the vesicles, the transmitter diffuses across the cleft and combines with receptor sites (a membrane molecular configuration with which chemical transmitters combine) on the part of the postsynaptic cell (the subsynaptic membrane) lying under the knob

D. THE COMBINATION OF THE TRANSMITTER with the receptor sites causes changes in the permeability properties of the postsynaptic membrane and in the membrane potential of the postsynaptic cell
   1. There is a synaptic delay (less than 0.001 second) between excitation of the nerve terminal of the presynaptic neuron and membrane-potential changes in the postsynaptic cell, which is a function of the release mechanism that frees transmitter substance from the knob
   2. Synaptic activity is terminated when the transmitter is chemically transformed into an ineffective substance and simply diffuses away from the receptor sites or is taken back up by the synaptic knob

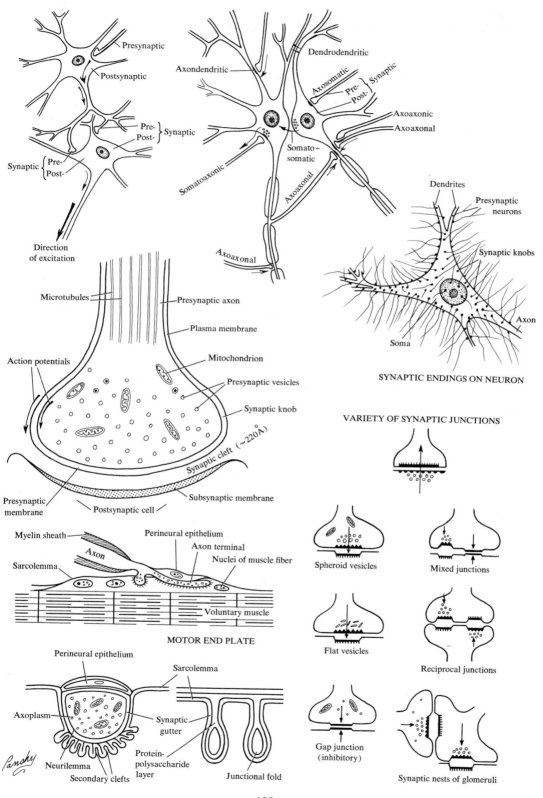

Presynaptic

Postsynaptic

Pre- } Synaptic
Post-

Synaptic { Pre-
Post-

Direction
of excitation

Dendrodendritic

Axondendritic

Axosomatic

Pre- } Synaptic
Post-

Axoaxonic
Axoaxonal

Somato-
somatic

Somatoaxonic

Axoaxonal

Axoaxonal

Dendrites

Presynaptic
neurons

Synaptic knobs

Axon

Soma

SYNAPTIC ENDINGS ON NEURON

Microtubules

Presynaptic axon

Plasma membrane

Action potentials

Mitochondrion

Presynaptic vesicles

Synaptic knob

Synaptic cleft (~220Å)

Presynaptic
membrane

Postsynaptic cell

Subsynaptic membrane

VARIETY OF SYNAPTIC JUNCTIONS

Spheroid vesicles

Mixed junctions

Flat vesicles

Reciprocal junctions

Myelin sheath

Perineural epithelium

Axon terminal

Nuclei of muscle fiber

Axon

Sarcolemma

Voluntary muscle

MOTOR END PLATE

Perineural epithelium

Sarcolemma

Axoplasm

Synaptic
gutter

Protein-
polysaccharide
layer

Neurilemma

Secondary clefts

Junctional fold

Gap junction
(inhibitory)

Synaptic nests of glomeruli

Pansky

– 155 –

72. SYNAPSES: EXCITATORY AND INHIBITORY

I. **Functional anatomy of a synapse (cont.)**
A. EXCITATORY SYNAPSES
   1. When an excitatory synapse is activated, it increases the likelihood that the membrane potential of the postsynaptic cell will reach threshold and that the cell will undergo an action potential
   2. The chemical-transmitter receptor site combination increases the permeability of the subsynaptic membrane to $(+)$ charged ions so that they are free to move according to the electric and chemical forces acting on them
   3. At the synaptic membrane of excitatory synapses, there is a simultaneous movement of a relatively small number of K^+ out of the cell and a larger number of N^+ into it. The net effect is the movement of $(+)$ ions into the neuron which slightly depolarized the postsynaptic cell; referred to as the excitatory postsynaptic potential (EPSP), and its only function is to help trigger an action potential
B. INHIBITORY SYNAPSES
   1. When an inhibitory synapse is activated, changes are produced that lessen the likelihood that the cell will undergo an action potential
   2. At inhibitory synapses, the combination of the chemical transmitter with the receptor sites on the subsynaptic membrane also changes the permeability of the membrane, but only the permeability to Cl^- is increased; Na^+ and K^+ permeabilities are not
      a. The greater permeability to K^+ is responsible for the changes in membrane potential associated with activity of an inhibitory synapse
      b. The resting membrane potential approximates the Cl^- equilibrium potential
   3. The increased negativity or hyperpolarization is called an inhibitory postsynaptic potential (IPSP)
      a. When a neuron is acted on by an inhibitory synapse, its membrane potential is moved farther from the threshold level

E = Excitatory
I = Inhibitory

SYNAPTIC POTENTIALS

Membrane potential (mV)

S = Stimulation of presynaptic neuron

SUM OF I + E DETERMINES IF IMPULSE IS INITIATED AT INITIAL AXON SEGMENT
(After Curtis and Eccles)

Action potential

EPSP

Excitatory synapse

Na^+

K^+

Cl^-

Na^+

Nerve impulse

Action potential

No action potential

Inhibitory synapse

IPSP

SEQUENCE IN EXCITATION (A); AND IN INHIBITION (B) FROM PRE- TO POSTSYNAPTIC NEURON

5 msec

Action potential

E_{Na}

EPSP

E_{EPSP}

EPSP minus IPSP

EPSP

E_{Cl}

E_{IPSP}

IPSP

E_K

mV

EFFECT OF IPSP ON ACTION POTENTIAL: EQUILIBRIUM POTENTIALS OF Na^+, K^+, Cl^-, EPSP, AND IPSP ARE SEEN

Panaky

73. SYNAPSES: ACTIVATION OF POSTSYNAPTIC CELL

I. Functional anatomy of a synapse (cont.)

A. ACTIVATION OF THE POSTSYNAPTIC CELL: In most neurons, postsynaptic integration is possible because one excitatory synaptic event is not sufficient by itself to change the membrane potential of the postsynaptic neuron from resting to threshold level; an EDSP produced by a single presynaptic fiber is smaller than 1 mV
   1. Whereas changes of up to 25 mV are needed to depolarize the membrane from resting to threshold
   2. It requires the combined effect of many fibers to bring the postsynaptic membrane to its threshold level
B. THERE IS A GENERAL DEPOLARIZATION OF THE MEMBRANE toward threshold when excitatory synaptic activity predominates (known as facilitation) and a hyperpolarization when inhibition predominates
C. WHEN 2 SYNAPTIC POTENTIALS SUMMATE as a result of successive stimulation of the same presynaptic fiber, it is called temporal summation
D. IF 2 EPSPS SUMMATE BUT THEY ORIGINATE AT DIFFERENT PLACES on the postsynaptic neuron, this is called spatial summation
E. INHIBITORY POTENTIALS can show both temporal and spatial summation
F. DIFFERENT PARTS OF THE NEURON have different thresholds
   1. The neuronal cell body and larger dendritic branches reach threshold when their membrane is depolarized at about 25 mV from the resting level
   2. In many cells, the initial segment of the neuron has a threshold that is less than half of the body
G. THE SUBSYNAPTIC MEMBRANE is depolarized at an activated excitatory synapse and hyperpolarized at an activated inhibitory synapse
   1. By the mechanism of local current flow, current flows through the cytoplasm from an excitatory synapse and toward the inhibitory one
   2. Thus, the entire body, including the initial segment, becomes slightly depolarized during activation of an excitatory synapse and slightly hyperpolarized during activation of an inhibitory synapse
H. SYNAPTIC EVENTS last more than 10 × as long as do action potentials
I. THE GREATER THE DEPOLARIZATION due to synaptic events, the greater is the number of action potentials fired (up to the limit imposed by the duration of the absolute refractory period)
J. NEURONAL RESPONSES are almost always in the form of so-called "bursts" or "trains" of action potentials
K. THE BIAS OF THE CELL'S ACTIVITY by synapse grouping is greatly lessened by the fact that the initial segment acts to average all the synaptic input as a result of its lower threshold
   1. However, those synapses nearer the initial segment have a greater influence on cell activity than those at the ends of the dendrites, and this synaptic placement provides a mechanism for giving different inputs a greater or lesser influence on the postsynaptic cell's output. In motoneurons, inhibitory synapses are predominantly on the soma and therefore to the initial segment than are excitatory synapses, which are predominantly on the dendrites
L. IN SOME CELLS, action potentials can be initiated in regions other than the initial segment

Time →

Membrane potential (mV)

+40

0

−70

EPSP

IPSP

Excitation

Inhibition

Pattern at axon hillock

All-or-none APs with transient polarity reversal

SYNAPTIC INPUT TO A NEURON

Excitatory synapse

Na⁺

Inhibitory synapse

Cl⁻

K⁺

mV Membrane potential

−70

Threshold

Excitatory

Time (msec) →

−70

Threshold

Inhibitory

Inhibitory synapse

Excitatory synapses

3 2 1

Time

0

Trans-membrane potential (mV)

−70

Stimulate

① ① ① ① ② ① ③ ① ③
                 +②   +②

Temporal summation due to successive stimulation of same neuron

Spatial summations: summate 2 different presynaptic neurons

A   B
Independent firing A + B stimulate 1,2,5,6

Fired together stimulate all 6

Spatial facilitation   Summation

1 2 3 4 5 6

1 2 3 4 5 6

Subthreshold

Convergence

A   B

1 2 3 4 5 6

4   4

6

Occlusion: when fired together (8 < 6)

Synaptic knob

Post.

Presynaptic

Divergence

Pre.

Post.   Pre.

Post.

– 159 –

74. SYNAPSES: CHEMICAL TRANSMITTERS AND PRESYNAPTIC INHIBITION

I. Chemical transmitters

A. PRESYNAPTIC NEURONS influence postsynaptic neurons only by means of chemical transmitters

B. THERE ARE MANY DIFFERENT SYNAPTIC TRANSMITTERS, but all synaptic endings from a single presynaptic cell probably liberate the same transmitter

C. SUBSTANCES THOUGHT TO BE TRANSMITTERS are listed on the opposite page

D. SYNAPSES OPERATE IN ONLY ONE DIRECTION because all the transmitter is stored on one side of the synaptic cleft. In contrast, action potentials travel along a nerve fiber in either direction but, due to one-way conduction across synapses, action potentials pass through the nervous system in only one direction

E. THE CHEMICAL TRANSMITTER continues to combine with the receptor sites on the sub-synaptic membrane until it is inactivated by combining chemically with another substance and is then taken up again by the presynaptic endings or just diffuses away from the synaptic region

    1. If the transmitter substance is not removed quickly from the receptor sites, a single volley of impulses in the presynaptic fibers may cause a prolongation of firing in the postsynaptic neuron

F. SYNAPSES ARE VULNERABLE to many drugs and toxins that can modify the synthesis, storage, release, inactivation, or uptake of the transmitter substance, or block the receptor sites on the subsynaptic membrane to prevent combination with the transmitter

    1. The toxin tetanus bacillus acts at the inhibitory synapses on the motor neuron by blocking the receptor sites. This eliminates inhibitory input to the motor neurons and allows the unchecked excitatory inputs to have full influence, leading to muscle spasticity and seizures (i.e., lockjaw)

G. PRESYNAPTIC INHIBITION: an inhibitory influence can be exerted on the postsynaptic neuron (the activation of inhibitory synapse) which results in hyperpolarization of the postsynaptic cell and depresses all information fed into the cell

    1. Presynaptic inhibition provides the means by which certain inputs to the postsynaptic cell can be selectively altered. Here the postsynaptic cell is not necessarily depressed but, rather, its membrane potential is determined by a different group of synaptic inputs and different bits of information

    2. Presynaptic inhibition operates by affecting the transmission of a single excitatory synapse

        a. At excitatory synapses, the amount of chemical transmitter released from the synaptic knob is directly related to the amplitude of the action potential in the axon terminal. The magnitude of the resulting postsynaptic EPSP is directly related to the amount of chemical transmitter released

          i. The excitatory synapse is less effective if the amplitude of the action potential in the presynaptic terminal is depressed (action potentials usually have an amplitude of about 100 mV or -70 to $+30$ mV)

          ii. This is accomplished in presynaptic inhibition by means of a synaptic junction between one neuron and the synaptic knob of a 2nd. When activated, the synapse slightly depolarizes the membrane of the knob of the 2nd neuron, but the depolarization is not great enough to cause an AP potential in the 2nd neuron but only brings the membrane potential of the synaptic knob closer to threshold and reduces the amplitude of the action potentials when neuron 2 is fired. The amount of chemical transmitter released from the knob of neuron 2 is decreased, the permeability changes in the postsynaptic cell are smaller, the size of the EPSP is decreased, and the postsynaptic cell is less influenced by inputs from neuron 2

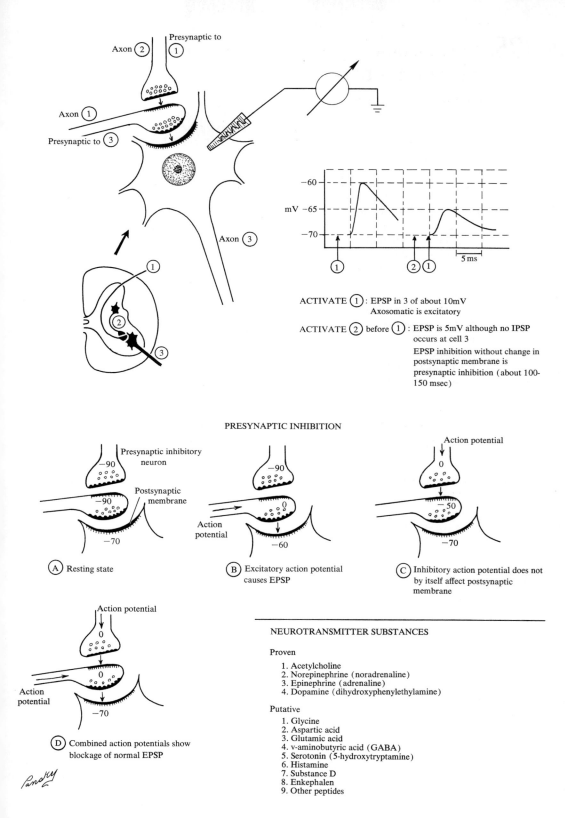

Axon ② Presynaptic to ①

Axon ①

Presynaptic to ③

Axon ③

mV −60
−65
−70

① ② ①

5 ms

ACTIVATE ① : EPSP in 3 of about 10mV
Axosomatic is excitatory

ACTIVATE ② before ① : EPSP is 5mV although no IPSP
occurs at cell 3

EPSP inhibition without change in
postsynaptic membrane is
presynaptic inhibition (about 100-
150 msec)

PRESYNAPTIC INHIBITION

Presynaptic inhibitory
neuron

Postsynaptic
membrane

−90
−90
−70

Ⓐ Resting state

Action
potential

−90
0
−60

Ⓑ Excitatory action potential
causes EPSP

Action potential

0
−50
−70

Ⓒ Inhibitory action potential does not
by itself affect postsynaptic
membrane

Action potential

0
0
−70

Action
potential

Ⓓ Combined action potentials show
blockage of normal EPSP

NEUROTRANSMITTER SUBSTANCES

Proven

1. Acetylcholine
2. Norepinephrine (noradrenaline)
3. Epinephrine (adrenaline)
4. Dopamine (dihydroxyphenylethylamine)

Putative

1. Glycine
2. Aspartic acid
3. Glutamic acid
4. v-aminobutyric acid (GABA)
5. Serotonin (5-hydroxytryptamine)
6. Histamine
7. Substance D
8. Enkephalen
9. Other peptides

75. PHYSIOLOGIC ACTIVITIES OF NEURONS: NEUROEFFECTORS

I. Neuroeffector communication (effectors)

A. EFFERENT NEURONS innervate muscle cells or glandular cells
   1. The striated skeletal muscles, the smooth muscles of the viscera and blood vessels, and the glands are the "executive organs" or effectors of the nervous system
B. CHEMICAL TRANSMITTERS transmit information from axons to the effector cells
   1. When an action potential reaches the terminal portions of the axon, it causes the release of transmitter which diffuses to the effector cell and changes its activity
C. THE STRUCTURE OF THE AXON TERMINALS and their anatomic relationships to the effector cells vary, depending on the effector cell type described later
D. THE NEUROEFFECTOR TRANSMITTERS have been well characterized and are either acetylcholine or norepinephrine
   1. The somatic division of the PNS is made up of all the fibers going from the CNS to skeletal muscle cells
      a. The cell bodies of these neurons are found in groups within the brain or spinal cord; their large diameter, myelinated axons leave the CNS and pass directly to skeletal muscle cells
      b. The transmitter substance released by these neurons is acetylcholine
   2. Fibers of the autonomic division of the PNS innervate cardiac and smooth muscle cells and glands
      a. Anatomic and physiologic differences in the ANS are the basis for further subdivision into sympathetic and parasympathetic components
      b. In both the sympathetic and parasympathetic divisions, the chemical transmitter for the ganglionic synapse between pre- and postganglionic fibers is acetylcholine
      c. The chemical transmitter at the junction between the parasympathetic postganglionic fiber and the effector organ is also acetylcholine
      d. The transmitter between the sympathetic postganglionic fiber and the effector organ is norepinephrine (a member of the chemical family of catecholamines)
E. CHOLINERGIC FIBERS release acetylcholine; those that release norepinephrine are called adrenergic

METABOLIC CYCLE OF ACh AT NEUROMUSCULAR JUNCTION

Presynaptic axon

Presynaptic action potential

Presynaptic knob

Synthesis Storage Release

Uptake into presynaptic ending

Choline

Cholin-esterase

ACh

Acetic acid

ACh

Synaptic cleft

Postsynaptic

Receptor (conductance change)

CHOLINERGIC NEURON

ACh

ACh Choline

Receptors ⟶ AChE

AChE = Acetylcholinesterase

ADRENERGIC NEURON

Dopa ⟵ Tyrosine

Dopamine

NE ⟶ MAO ⤏ Inactive metabolites

NE

NE

Receptors ⤏ COMT ⤏ Norepinephrine

COMT = Catechol-O-methyltransferase

Impulse depolarizes knob ①

K⁺ ⟶ Na⁺ ②

Ca⁺

Transmitter substance released ③

Reaction on postsynaptic membrane ④

SYNAPTIC TRANSMISSION

Impulse frequency

① ② ③

Low Moderate High

Amplitude

Temporal Summation
(frequency modulation or signal averaging)

SPATIAL SUMMATION

① ② ③

S1
S2
S3

Stimulus (s) 1: Gets fiber 1
2: Gets fibers 1 + 2
3: Gets fibers 1 + 2 + 3

76. SOME PATTERNS OF NEURAL ACTIVITY: FLEXION REFLEX

I. Flexion reflex is a coordinated functional operation of a reflex arc, but not the entire nervous system which is to respond in neural control systems

A. A PAINFUL STIMULUS to the toe leads to foot withdrawal reflex
1. Toe receptors are stimulated and transform the energy of the stimulus into the electrochemical energy of action potentials in afferent neurons
2. Information regarding the intensity of the stimulus is coded both by the frequency of action potentials in single afferent fibers and by the number of different afferent fibers stimulated
3. The afferent fibers branch after entering the spinal cord, each branch terminating at a synaptic junction with another neuron
4. In flexion reflex, the 2nd neurons in the pathway are interneurons (between the afferent and efferent limbs of the reflex arc), thus this reflex is one of the large classes of polysynaptic reflexes
5. The afferent neuron branches have the following function
    a. Some synapse with interneurons that carry information to higher centers, resulting in a conscious correlate of the stimulus
    b. Other branches synapse with different interneurons that synapse on efferent neurons innervating flexor muscles which, when activated, cause flexion of the ankle and withdrawal of the foot from the stimulus. With an intense stimulus, the afferent discharging zone is large, so that muscles of joints such as the knee and thigh also flex
    c. As the injured leg is flexed away from the stimulus, the opposite leg is extended more strongly to support the added share of the body's weight—the so-called cross-extensor reflex
    d. Other afferents activate interneurons that inhibit motor neurons of antagonistic muscles that oppose the reflex flexion (reciprocal innervation)

B. FLEXION REFLEX NEURAL NETWORKS are simple, yet coordinated and purposeful
1. Foot flexion away from the stimulus removes the receptors and prevents further damage (negative feedback)
2. Purposefulness of the reflex response is independent of pain sensation, e.g., withdrawal occurs before the sensation of pain is experienced
3. The observed response varies with the stimulus strength: low levels of stimulus lead to responses localized to the area of the stimulus, whereas higher levels of stimulation cause more widespread responses

II. Swallowing involves the coordinated interaction of about 20 muscles whose efferent neuron's cell bodies are found in the lower portion of the brain

A. SWALLOWING REQUIRES AN ORDERLY SEQUENCE of excitation of many neurons
1. A single brief stimulus (10 sec) triggers off a chain of responses and, once initiated, its course is largely unaffected by further sensory input
2. A more complex network of interacting neurons is built into the swallowing reflex to ensure the timing of the response than is in the flexion reflex where the stimulus induces a brief and nearly simultaneous response
3. Swallowing, however, is a less complex reflex than flexion reflex because it elicits a stereotyped all-or-none response whose pattern and magnitude are not altered by the stimulus strength

III. Respiratory movement is mediated by interconnected groups of neurons resulting in a rhythmic firing of phrenic motor neurons, probably not of "pacemaker" neurons

IV. Spontaneously depolarizing neurons, "pacemaker" neurons in some animals, provide a so-called biologic clock for the nervous system and may also function in the sleep-wake cycle, the reproductive cycle, hormone secretion, etc.

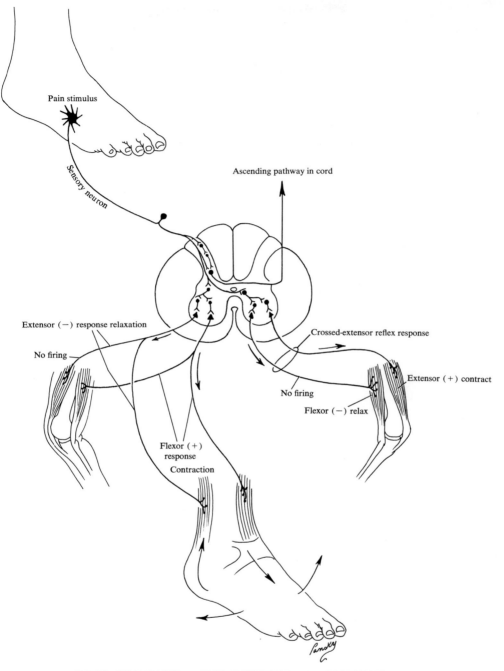

Pain stimulus

Sensory neuron

Ascending pathway in cord

Extensor (−) response relaxation

No firing

Crossed-extensor reflex response

Extensor (+) contract

No firing

Flexor (−) relax

Flexor (+)
response

Contraction

FLEXOR (WITHDRAWAL) + CROSSED-EXTENSOR REFLEX PATHWAYS

PART II

Functional Mechanisms of Neuroscience

Unit Four

PATHWAYS, MUSCLE REGULATION, AND TYPICAL SPINAL CORD LESIONS

77. GENERAL SOMATIC AFFERENT PATHWAYS: PAIN AND TEMPERATURE—FROM BODY

I. **Introduction:** the general somatic afferent (GSA) system includes those pathways that conduct pain, temperature, touch, pressure, and proprioception. Axons conveying pain and temperature take essentially the same course through the central nervous system and will be considered together

II. **Peripheral receptors** for pain are naked, free ends of nerve fibers that spread out among epithelial (epidermis) cells; those for temperature remain in question (probably include free nerve terminals, end-bulbs of Krause, and Ruffini's corpuscles)

III. **First-order neurons (neuron I)** are small, unipolar (pseudounipolar) nerve cells
   A. PERIPHERAL PROCESS (DISTAL PART OF AXON) brings sensory information from receptors toward the cell body
   B. CELL BODIES (GANGLION CELLS) of the sensory neurons are located in dorsal root ganglia (no synapses in the ganglia)
   C. PROXIMAL (CENTRAL) BRANCH OF AXON extends to spinal cord in the dorsolateral sulcus at the dorsal root zone
      1. Smallest, finely myelinated fibers (pain and temperature) enter the spinal cord occupying a lateral position in the entry zone
      2. Largest, most heavily myelinated fibers are medial in position (proprioceptive)
      3. Medium-sized fibers (tactile) are intermediate in position
   D. DORSOLATERAL TRACT (ZONE OF LISSAUER) is where afferent fibers enter the cord
      1. Entering fibers divide into short ascending and descending branches that run longitudinally for 1–3 cord segments
      2. Collaterals are given off to posterior horn along course of dorsolateral tract (fasciculus)
      3. Parent fibers and collaterals terminate in the dorsal gray horn (substantia gelatinosa or lamina II)
      4. Short connector neurons may extend from the substantia gelatinosa to more ventral nuclei (laminae) for reflex activity

IV. **Axons of second-order neurons (neuron II):** nerve cell bodies are located in the substantia gelatinosa
   A. CROSS TO THE OPPOSITE (CONTRALATERAL) SIDE OF THE SPINAL CORD in the anterior white commissure in close proximity to the central canal
   B. AFTER CROSSING, THE AXONS TURN CRANIALLY to enter the lateral spinothalamic tract (and spinotectal tract) in the anterior portion of the lateral funiculus
   C. THIS TRACT EXTENDS WITHOUT INTERRUPTION through the cord to the medulla
   D. SOME FIBERS TERMINATE IN OR SEND COLLATERALS to the brainstem reticular formation in the medulla and pons
   E. IN THE MIDBRAIN, this tract is located at the posterolateral aspect of the medial lemniscus
   F. BEFORE THESE SECOND-ORDER NEURONS END in a specific thalamic nucleus, some few fibers enter the reticular system (diencephalon) as an integral part of the reticular activating mechanism of the cerebrum

V. **Third-order neurons (neuron III):** cell bodies are located in the ventral posterolateral nucleus (VPL) of the thalamus
   A. SYNAPSES OCCUR WITH AXON TERMINALS of second-order neurons
   B. AXONS OF THE THIRD-ORDER NEURONS pass through the internal capsule (posterior limb) as a part of the thalamocortical fiber system
   C. THESE AXONS CONTINUE AS THE CORONA RADIATA and finally end in the postcentral gyrus of the parietal lobe (areas 3, 1, 2 of the somesthetic area)

Arm
Head
Leg

Post. part of internal capsule
(post. limb)

Leg

Arm

Leg
Parietal cortex

VPM

Central fissure
Postcentral gyrus
areas: 3–1–2

L
A
H

RN   RN   VPL thalamus

Section

Ventral
posterolateral nucleus of thalamus
(VPL)

MIDBRAIN

Reticular formation

MEDULLA

Spinal nucleus of nerve V

Lateral spinothalamic tract

LOWER MEDULLA

**PAIN AND TEMPERATURE—
FROM BODY**

Substantia gelatinosa

III

Dorsal root ganglion

I

II

CORD

Substantia gelatinosa

Spinothalamic nuclei

CORD

Superficial end
organs

Ruffini corp.

Free ending

Vsl. wall

Free
ending

Krause's end bulb

Deep visceral

Pansky

– 171 –

78. GENERAL SOMATIC AFFERENT PATHWAYS:
PAIN AND TEMPERATURE—FROM FACE

I. **Peripheral receptors in the face:** same as those of body

II. **First-order neurons (neuron I)**
A. PERIPHERAL PROCESSES (DENDRITE IN FUNCTION) are thinly and intermediately myelinated fibers found in
1. Trigeminal (V) nerve: all three divisions
2. Facial (VII) nerve: with receptors for some GSA fibers in the deep face
3. Glossopharyngeal (IX) nerve: with receptors in the middle ear and pharynx
4. Vagus (X) nerve: with receptors in the external auditory meatus
B. CELL BODIES OF FIRST-ORDER NEURONS are small and medium-sized pseudounipolar cells located in
1. Semilunar (gasserian) ganglion of the trigeminal (V) nerve
2. Geniculate ganglion of the facial (VII) nerve
3. Superior ganglion of the glossopharyngeal (IX) nerve
4. Superior ganglion of the vagus (X) nerve
C. CENTRAL OR PROXIMAL PROCESSES (AXONAL IN FUNCTION)
1. Enter the brainstem at the levels of nerves V, VII, IX, and X
2. Pain and temperature fibers bend inferiorly to form a distinct bundle, the spinal tract of the trigeminal nerve (ipsilaterally)
3. Spinal tract of V descends through the pons and medulla to end in the third or fourth cervical segment where it merges with the dorsolateral fasciculus of the cord
   a. Pain fibers in the spinal tract (V) are medial to those for temperature
   b. Many collaterals arise from fibers within the spinal tract (V)
D. TERMINATION (SYNAPSE) OF NEURON I: collaterals and axon terminals enter the elongated nucleus of the spinal tract of V (descending nucleus or spinal trigeminal nucleus)
1. This nucleus parallels and adjoins the spinal tract on its medial side
2. This nucleus is continuous with the substantia gelatinosa of the spinal cord

III. **Second-order neurons (neuron II)**
A. NERVE CELL BODIES OF SECOND-ORDER NEURONS make up the descending nucleus of V
B. AXONS OF SECOND-ORDER NEURONS
1. Cross ventromedially to the opposite side of the brainstem at about the same level of origin in the spinal nucleus of V
2. Approach the lateral spinothalamic tract of the contralateral side
3. Turn rostrally to enter the ventral trigeminal lemniscus or tract (next to the medial lemniscus)
4. Continue through the brain stem in this tract in close association with the medial lemniscus
5. Synapse mainly in the ventral posteromedial (VPM) nucleus of the thalamus
6. Contribute some fibers to the reticular formation (?)

IV. **Third-order neurons (neuron III)**
A. AXONS ARISE FROM CELL BODIES in the VPM nucleus of the thalamus
B. AXONS PASS THROUGH THE INTERNAL LIMB of the internal capsule and corona radiata
C. TERMINATE IN THE INFERIOR PORTION (FACE AREA) of the postcentral gyrus of the parietal lobe (areas 3, 2, and 1)

Head

Arm

Leg

Leg

Arm

Head

III

Postcentral gyrus
areas: 3–2–1

Central fissure

L
A
H

Ventral posteromedial nucleus (VPM)

II

Section

MIDBRAIN

Ventral trigeminal lemniscus or tract

Ventral posteromedial nucleus
(VPM)

Ventral trigeminal
lemniscus (tract)

Reticular
formation

Semilunar gang.

Ophthalmic V

Maxillary V

Mandibular V

V

VII

PONS

MEDULLA

IX

X

Dorsolateral fasciculus

Spinal tract V (I)

Substantia gelatinosa

Nucleus of V

SPINAL CORD

PAIN AND TEMPERATURE— FROM FACE

Pansky

VISCERAL AND
SOMATIC PAIN

To cortex

To cortex via thalamus

Solitary tract

Reticular
formation

V

VII

IX

X

C1

Spinal tract +
nucleus of V

2° trigeminal fibers

C4

Spinothalamic tract (somatic pain)

Spinotectal tract (somatic pain)

Visceral pain fibers in fasciculi proprii

Somatic
Visceral } Pain

Collaterals to adjacent segments

2

To nearest segment

3

1

SPREAD OF PAIN

Pain + temperature

Opposite to spinothalamic tr.

C T L S

Lat. corticospinal

L

S

T

C

Lat. spinothal. tract

Cervical (C)

Thoracic (T)

Lumbar (L)

Sacral (S)

Spinothalamic nucleus

FIGURE 7.

-174-

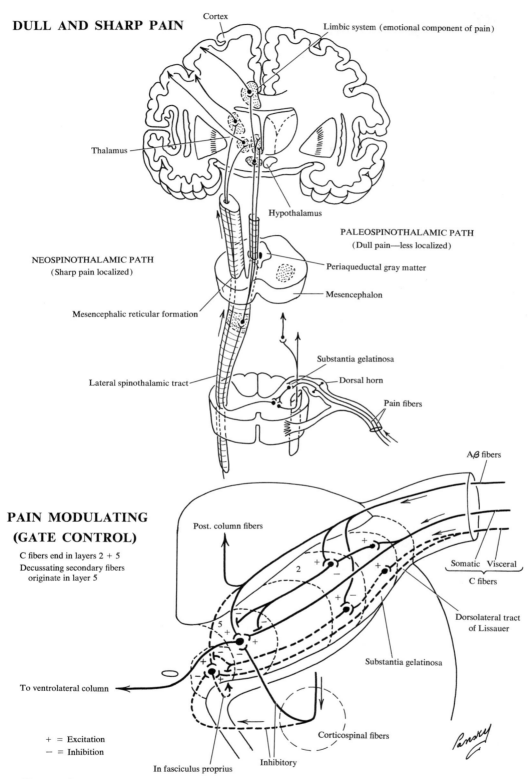

DULL AND SHARP PAIN

Cortex

Limbic system (emotional component of pain)

Thalamus

Hypothalamus

PALEOSPINOTHALAMIC PATH
(Dull pain—less localized)

NEOSPINOTHALAMIC PATH
(Sharp pain localized)

Periaqueductal gray matter

Mesencephalon

Mesencephalic reticular formation

Substantia gelatinosa

Dorsal horn

Lateral spinothalamic tract

Pain fibers

Aβ fibers

PAIN MODULATING
(GATE CONTROL)

C fibers end in layers 2 + 5
Decussating secondary fibers
originate in layer 5

Post. column fibers

Somatic Visceral

C fibers

Dorsolateral tract
of Lissauer

Substantia gelatinosa

To ventrolateral column

Corticospinal fibers

+ = Excitation
− = Inhibition

In fasciculus proprius

Inhibitory

FIGURE 8.

- 175 -

79. GENERAL SOMATIC AFFERENT PATHWAYS: DISCRIMINATIVE (FINE) TOUCH—FROM BODY

I. **Definition:** tactile discrimination is concerned with the sense of deep pressure (pressure touch) spatial localization (two-point discrimination), stereognosis (perception of size, shape, and texture), awareness of movement of body parts, position sense, and vibratory sense. The term "tactile localization" is often used as a collective word to refer to the last four senses (above) which also require the involvement of proprioceptive pathways

II. **Receptors for tactile discrimination, conscious proprioception (kinesthesis), stereognosis, etc.**
   A. MEISSNER'S CORPUSCLES (FOR TOUCH) are located in dermal papillae just adjacent to surface epithelium, They are widely dispersed in areas of thin skin, but are close together in palmar and plantar areas
   B. PERITRICHIAL ARBORIZATIONS (FOR TOUCH) are fine nerve endings around hair follicles
   C. TACTILE DISKS OF MERKEL (FOR TOUCH) are platelike expansions of nerve fibers associated with epidermis
   D. FREE NERVE ENDINGS are for touch and conscious proprioception
   E. PACINIAN CORPUSCLES are for pressure and conscious proprioception

III. **Cell bodies of first-order neurons (neuron I)** are large, unipolar (pseudounipolar) and are located in the dorsal root (spinal) ganglia
   A. PERIPHERAL PROCESSES (FUNCTION AS DENDRITES) of the unipolar neurons are heavily myelinated
   B. CENTRAL PROCESSES (AXONS) enter the cord through the medial division of the dorsal root

IV. **Course of fibers (axons) of first-order neurons** enter the posterior funiculus and branch into
   A. SHORT BRANCHES, which end in laminae III and IV of the posterior gray horn (then associated with the anterior spinothalamic pathway)
   B. LONG ASCENDING BRANCHES which form the essential part of first-order fibers and continue rostrally in the posterior white columns (fasciculus gracilis and cuneatus) of the same side
      1. Fibers from the lower six thoracic, lumbar, and sacral levels form the fasciculus gracilis; from T6 and above, the fasciculus cuneatus
   C. FIBERS IN THE ABOVE TWO FASCICULI enter the lower medulla to synapse in the nucleus gracilis and cuneatus, respectively

V. **Course of axons of second-order neurons (neuron II):** arc anteriorly as the internal arcuate fibers, decussate in the lower medulla as the lemniscal decussation (sensory decussation), and turn rostrally to form the medial lemniscus
   A. IN THE MEDULLA: the medial lemniscus forms a dorsoventrally oriented column just dorsal to the pyramids (cuneate fibers are dorsal to those of the gracilis)
   B. IN THE PONS: the medial lemniscus has its long axis parallel to the horizontal plane with gracilis fibers (from lowest body levels) most lateral in position
   C. IN THE MESENCEPHALON: the lemniscus bends outward to lie along the surface of the brainstem (along with the more superficial spinothalamic tract)
   D. THE SECOND-ORDER FIBERS synapse in the ventral posterolateral (VPL) nucleus of the thalamus on neurons positioned more rostral and dorsal than those receiving simple (crude) touch, pain, and temperature

VI. **Course of axons of third-order neurons (neuron III):** from the VPL nucleus, fibers pass laterally through the posterior limb of the internal capsule and corona radiata to end primarily in the postcentral gyrus and adjacent parietal cortex

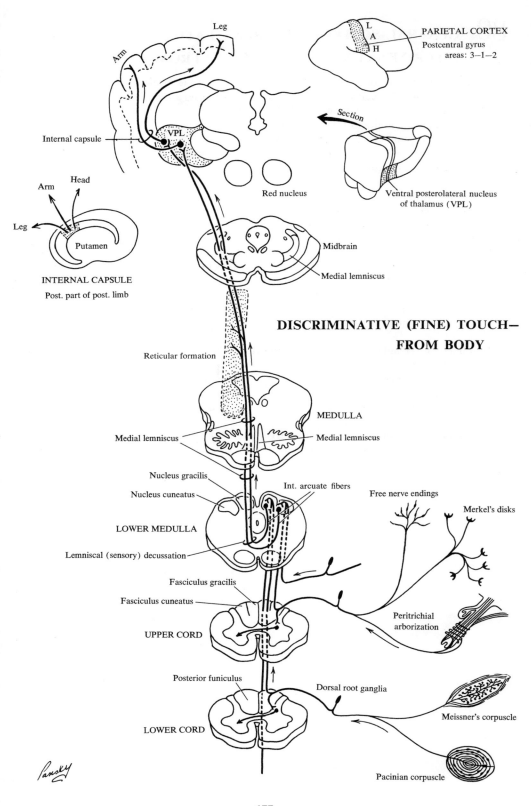

PARIETAL CORTEX
Postcentral gyrus
areas: 3—1—2

L
A
H

Leg

Arm

Internal capsule

VPL

Section

Red nucleus

Ventral posterolateral nucleus
of thalamus (VPL)

Head
Arm
Leg
Putamen

INTERNAL CAPSULE
Post. part of post. limb

Midbrain

Medial lemniscus

DISCRIMINATIVE (FINE) TOUCH—
FROM BODY

Reticular formation

MEDULLA

Medial lemniscus

Medial lemniscus

Nucleus gracilis

Nucleus cuneatus

Int. arcuate fibers

Free nerve endings

Merkel's disks

LOWER MEDULLA

Lemniscal (sensory) decussation

Fasciculus gracilis

Fasciculus cuneatus

Peritrichial
arborization

UPPER CORD

Posterior funiculus

Dorsal root ganglia

Meissner's corpuscle

LOWER CORD

Pacinian corpuscle

Pansky

– 177 –

80. GENERAL SOMATIC AFFERENT PATHWAYS: DISCRIMINATIVE (FINE) TOUCH—FROM FACE

I. **Receptors:** for the discriminative general senses from the face and head (anterior to a coronal plane through the ears) are primarily Meissner's corpuscles
   A. THE RECEPTORS are numerous and close together at the tip of the tongue, and two points only 1.1 mm apart can be distinguished here
      1. The receptors are also numerous on the lips
   B. THE RECEPTORS TRANSMIT THEIR SENSATION via the ophthalmic, maxillary, and mandibular divisions of the trigeminal (V) nerve

II. **Course of first-order neurons (neuron I)**
   A. THE PERIPHERAL PROCESSES follow the ophthalmic, maxillary, and mandibular divisions of the trigeminal (V) nerve
   B. CELL BODIES are located in the semilunar (gasserian) ganglion
   C. CENTRAL PROCESSES form part of the sensory root of the trigeminal nerve on the lateral aspect of the pons and follow the same course as outlined for crude (light or simple) touch from the head
   D. SYNAPSE OF FIRST-ORDER NEURONS takes place in the main (principal) sensory trigeminal nucleus. This nucleus is often referred to as the cranial equivalent of the nucleus gracilis and nucleus cuneatus

III. **Course of second-order neurons (neuron II)** on their ascent to the thalamus is unclear and disputed
   A. THE CELL BODIES are located in the main sensory nucleus of V
   B. THE AXONS ascend in or near the medial lemniscus
   C. THEY SYNAPSE in the VPL nucleus of the thalamus

IV. **Course of third-order axons (neuron III)** passes through the posterior limb of the internal capsule and corona radiata to end in the postcentral gyrus and adjacent parietal cortex

Arm Leg

Central fissure Post. central gyrus
areas: 3–1–2

L
A
H

Head

VPL

Section

Head Arm

Leg

Putamen

Ventral posterolateral nucleus
of thalamus (VPL)

INTERNAL CAPSULE
Post. part of post. limb

MIDBRAIN

Medial lemniscus

Main sensory nucleus of V

Mesencephalic tr. + nuc. V
Main sensory nucleus of V
Motor nucleus of V
Spinal tr. + nuc. V

V

Semilunar ganglion (V)   Medial lemniscus   PONS

Ophthalmic V

Maxillary V

Mandibular V

MEDULLA

X

XII

Gracile ⎫
Cuneate ⎬ Nuclei

Int. acruate fibers

LOWER MEDULLA

DISCRIMINATIVE (FINE) TOUCH— FROM FACE

Pansky

81. GENERAL SOMATIC AFFERENT PATHWAYS: LIGHT (CRUDE) TOUCH—FROM BODY

I. **Definition:** light touch is concerned with the sense of simple touch, light pressure, and a very crude sense of tactile localization. This level of tactile sense is stimulated, for example, by the gentle stroking of the skin with cotton

II. **Receptors for light (protopathic) touch: probably include**
   A. FREE NERVE ENDINGS
   B. MERKEL'S (TACTILE) DISKS or platelike expansions of nerve fibers associated with certain epithelial cells in the epidermis
   C. PERITRICHIAL ARBORIZATIONS of fine nerve fibrils around hair follicles which detect any movements of the hair
   D. MEISSNER'S CORPUSCLES in dermal papillae beneath the volar and plantar surfaces
   E. PACINIAN CORPUSCLES which lie deep in the dermis and in tendons and joints
   F. GENITAL CORPUSCLES found in the connective tissue of the external genitalia

III. **Cell bodies of first-order neurons (neuron I)** are medium-sized, pseudounipolar and are located in the dorsal root ganglia of spinal nerves
   A. THE PERIPHERAL PROCESSES (FUNCTION AS DENDRITES) are large and well myelinated (intermediate type)
   B. CENTRAL PROCESSES (FUNCTION AS AXONS) group into the medial division of the dorsal root and enter the spinal cord along the dorsolateral sulcus
      1. Entering fibers become part of the posterior funiculus on the ipsilateral side (posterolateral tract of Lissauer)
         a. Fibers from the upper half of the body enter the posterior funiculus laterally (fasciculus cuneatus)
         b. Fibers from the lower half of the body enter the posterior funiculus medially (fasciculus gracilis)
      2. In the appropriate fasciculus, each nerve fiber bifurcates into a long ascending and a short descending branch, both of which give off collaterals
      3. All descending, all collaterals, and some ascending fibers synapse with internuncial neurons of the posterior horn (laminae VI and VII)
      4. Some primary ascending fibers continue superiorly into the lower medulla to synapse (nucleus gracilis and nucleus cuneatus)

IV. **Course of axons of second-order neurons (neuron II)**
   A. AXONS: from cell bodies located in the posterior gray horns (laminae VI and VII) decussate in the anterior white commissure (a few fibers may not cross and ascend on the ipsilateral side)
   B. FIBERS THEN ASCEND in the anterolateral system (in the anterior funiculus) as the anterior spinothalamic tract which ascends in the spinal cord to the medulla
   C. IN THE MEDULLA AND PONS: the exact course of this tract has not been definitively established
      1. It may become intimately associated with the medial lemniscus
      2. It may join the lateral lemniscus to form a spinothalamic lemniscus
   D. In the diencephalon, this fiber tract ends in the ventral posterolateral nucleus (VPL) of the thalamus

V. **Course of axons of third-order neurons (neuron III):** fibers leave the nerve cell bodies in the VPL nucleus to pass through the posterior limb of the internal capsule and corona radiata to end in the postcentral gyrus of the parietal lobe of the cerebral cortex (areas 3, 2, and 1)

Body

Leg

Arm

Postcentral gyrus: areas 3—1—2

L
A
H

VPL

Red nucleus

Section

Head

Arm

Leg

Putamen

INTERNAL CAPSULE

Post. part of post. limb

Ventral posterolateral nucleus of
thalamus (VPL)

MIDBRAIN

**LIGHT (CRUDE) TOUCH—
FROM BODY**

MEDULLA

Merkel's disks

Gracile
Cuneate } Nuclei

LOWER
MEDULLA

Lemniscal (sensory) decussation

Fasciculi { Gracile
Cuneate

Dorsal root gang.

Peritrichial
arborization

UPPER CORD

Anterior spinothalamic tract

Meissner's corpuscle

MIDDLE CORD

Dorsal funiculus

Pacinian corpuscle

LOWER CORD

Genital corpuscle

Panexy

82. GENERAL SOMATIC AFFERENT PATHWAYS: LIGHT (CRUDE) TOUCH—FROM FACE

I. Introduction: the facial area is especially sensitive to light touch

II. Receptors: the same type of receptors mentioned for the body are also found in the skin of the face and in the cornea, conjunctiva, and associated mucous membranes of the nasal and oral cavities

III. Cell bodies of first-order neurons (neuron I) are located in the semilunar (gasserian) ganglion

A. PERIPHERAL PROCESSES (DENDRITES) convey impulses from the receptors into the cell bodies via the three divisions of the trigeminal (V) nerve

B. CENTRAL PROCESSES (AXONS) make up part of the sensory root (portis major) of the trigeminal nerve which enters the lateral aspect of the pons

  1. In the pons, these fibers divide into short ascending and long descending branches

    a. Ascending division synapse in the main (chief) sensory nucleus of the trigeminal nerve

    b. Descending fibers enter the spinal tract of the trigeminal (V) nerve and descend for a short distance to end in the nucleus of V

IV. Cell bodies of second-order neurons (neuron II) are located in

A. SPINAL (DESCENDING) NUCLEUS OF V (ITS ROSTRAL PORTION)

  1. Their axons turn ventromedially toward the midline and from this point, their exact source is disputed

  2. Some may decussate and join the medial lemniscus

  3. Others may join the anterior spinothalamic tract

B. MAIN SENSORY NUCLEUS OF V (LOCATION OF MOST CELL BODIES OF NEURON II)

  1. Some of their axons may decussate in the pontine tegmentum (below the hypoglossal nucleus) and then ascend as the anterior trigeminal (trigeminothalamic) tract which ends in the ventral posteromedial (VPM) nucleus of the thalamus

  2. Other fibers ascend without crossing, as the posterior trigeminal (trigeminothalamic) tract which also ends in the VPM

V. Course of axons of third-order neurons (neuron III): axons from cell bodies in the VPM nucleus pass laterally through the posterior limb of the internal capsule and corona radiata to end in the postcentral gyrus (facial area)

Ventral posteromedial (VPM) nucleus

Parietal cortex: areas 3—1—2

Head area

VPM

Head area

Post. trigeminothalamic tract

Section

Post. part of
post. limb

Putamen

INTERNAL CAPSULE

MIDBRAIN

Ventral posteromedial nucleus
of thalamus (VPM)

Ant. trigeminothalamic tract

Chief sensory
nucleus V

Chief sensory nucleus of V

PONS

V

Medial lemniscus + ant. spinothalamic tracts

Ophthalmic V

Maxillary V

Mandibular V

Spinal tract V + nucleus

MEDULLA

Spinal tract V

LOWER MEDULLA

Spinal V nucleus

LIGHT (CRUDE) TOUCH— FROM FACE

Pansky

83. GENERAL SOMATIC AFFERENT PATHWAYS: CONSCIOUS AND UNCONSCIOUS PROPRIOCEPTION— FROM HEAD AND FACE

I. **Receptors** are associated with the peripheral ends of the three divisions of the trigeminal (V) nerve and are responsible for detecting conscious and unconscious proprioception. They include
   A. MUSCLE SPINDLES (p. 206): in muscles of mastication and perhaps in the tongue, extrinsic eye, and facial muscles to detect unconscious proprioception
   B. NEUROTENDON (GOLGI) ORGANS found dispersed in dense connective tissue of tendons convey unconscious proprioception
   C. FREE NERVE ENDINGS: in tendons, ligaments, and joint capsules are responsible for conscious proprioception
   D. PACINIAN CORPUSCLES: located deep in the dermis convey conscious proprioception

II. **Cell bodies of first-order neurons (neuron I)** are comparatively pseudounipolar and are found in the nucleus of the mesencephalic tract of V (located in the upper pons and mesencephalon)
   A. PERIPHERAL PROCESSES (FUNCTION AS DENDRITES) of the first-order neurons are heavily myelinated and take the following course
      1. They enter the three divisions of the trigeminal nerve and pass through the semilunar (gasserian) ganglion of the trigeminal nerve without synapsing
      2. They then enter the lateral aspect of the pons through both motor and sensory roots of the trigeminal nerve and ascend in the mesencephalic tract of V, on the same side, to reach the cell bodies in the mesencephalic nucleus
   B. THE CENTRAL PROCESSES of these same neurons function as axons
      1. Pass from the mesencephalic nucleus into the mesencephalic tract of V without crossing
      2. They descend in the mesencephalic tract to synapse in one of the following
         a. Main sensory nucleus of V which is concerned with conscious proprioception
         b. Motor nucleus of V which deals with motor reflexes (jaw jerk)
         c. Reticular formation
         d. Various other cranial nuclei in the area (facial nucleus)

III. **Course of second-order neurons (neuron II)**
   A. AXONS PASS FROM THE CELL BODIES in the main sensory nucleus of V and ascend through the brainstem in the dorsal trigeminal tract
   B. OTHER PATHWAYS are unclear and disputed
   C. FIBERS FOR CONSCIOUS PROPRIOCEPTION synapse in the VPM nucleus of the thalamus

IV. **Course of third-order neurons (neuron III):** the axons pass through the posterior limb of the internal capsule and corona radiata to end in the postcentral gyrus of the parietal cerebral cortex (areas 3, 2, and 1) in the area of the face

VPM of thalamus

VPM

Face area

Post. part of post. limb
of int. capsule

Postcentral gyrus: areas 3–1–2

H

Head area

Dorsal
trigeminothalamic tract

MIDBRAIN

Mesencephalic nucleus + tract of V

Muscle spindles

Flower-spray ending   Annulospiral ending

Reticular
formation

Main sensory nucleus of V

Motor nucleus of V

Main sensory nucleus V

Motor nucleus V

V   V₁

V₂

V₃

MIDPONS

Semilunar gang.

Muscles of
mastication

Golgi
tendon
organ

Free nerve
ending

VII nucleus

Pacinian corpuscle

LOWER PONS

VII nerve

VI

CONSCIOUS AND UNCONSCIOUS
PROPRIOCEPTION—FROM HEAD AND FACE

84. AFFERENT CEREBELLAR CONNECTIONS: SPINOCEREBELLAR TRACTS— GENERAL PROPRIOCEPTION

I. **Definition:** spinocerebellar fibers convey unconscious, proprioceptive information into the cerebellum concerning tension in muscle and the orientation of joints. Proprioception means the receiving of sensory information from muscles, tendons, and joints

II. **Functions:** anterior and posterior spinocerebellar tracts are primarily concerned with the lower extremities. The posterior spinocerebellar tract conveys information on individual extremity muscles, while the anterior spinocerebellar tract relays information concerning the status of the entire extremity

III. **Receptors for the first-order neurons in muscle, tendons, and joints**
   A. MUSCLE SPINDLES: adapted to register changes in muscle tension
   B. NEUROTENDINOUS (GOLGI TENDON) ORGANS: adapted to register stretch of a tendon due to muscle contraction
   C. PACINIAN CORPUSCLES: located in fascia, intermuscular septa, and joint capsules to record swelling and movements

IV. **Cell bodies of first-order neurons:** found in dorsal root ganglia

V. **Course of axons of first-order neurons (neuron I)**
   A. PROPRIOCEPTIVE AXONS (largest and most heavily myelinated fibers) collect in bundles to make up the medial division of the dorsal root
   B. AXONS ENTER THE POSTERIOR FUNICULUS of the spinal cord and bifurcate into long ascending and short descending branches which give off collaterals
   C. AXONAL TERMINALS AND COLLATERALS SYNAPSE with neurons in the base of the posterior gray horn and in the nucleus dorsalis (of Clarke)
   D. CLARKE'S NUCLEUS is a cell column found only from the eighth cervical through the third lumbar segments. Proprioceptive fibers entering cord segments below L3 ascend in the fasciculus gracilis to reach this nucleus; those entering above C8 ascend in the fasciculus cuneatus to reach the accessory (lateral) cuneate nucleus in the lower medulla

VI. **Course of axons of second-order neurons (neuron II)**
   A. NERVE CELL BODIES IN THE POSTERIOR FUNICULAR GRAY NUCLEUS (laminae V, VI, and VII) at lumbar and sacral levels send mostly crossed (some uncrossed) fibers through the anterior white commissure to ascend in the anterior or ventral spinocerebellar tract
      1. This is the most peripheral tract in the ventral part of lateral funiculus of cord
      2. In the medulla, the tract is positioned dorsal to the inferior olive and lateral to the lateral spinothalamic tract
      3. In the pons, this tract shifts even more laterally and then turns dorsally in the superior cerebellar peduncle to reach the anterior medullary velum
      4. By way of the anterior medullary velum, it enters the rostral portion of the contralateral (some remain uncrossed) cerebellar vermis. These fibers end as the mossy fibers of the paleocerebellum or anterior lobe
         a. Mossy fibers synapse with granule cells
         b. Granule cells synapse with Purkinje cells
   B. THE NERVE CELL BODIES IN CLARKE'S NUCLEUS (lamina VII, levels C8–L3) send ipsilateral fibers to the periphery of the dorsal portion of the lateral funiculus where they turn rostrally to form the dorsal or posterior spinocerebellar tract
      1. This tract runs in the superficial dorsolateral part of the lateral funiculus of cord
      2. In the medulla, it lies superficial to the spinal tract of V
      3. Fibers enter the vermis of the cerebellum through the inferior cerebellar peduncle and end as mossy fibers of the paleocerebellum

Granule cell

Purkinje cell

Ant. medullary velum

Sup. cerebellar peduncle

MIDPONS

Ventral spinocerebellar tract

Mossy fiber

LOWER PONS

TO PALEOCEREBELLUM

Ant. lobe

Declive

Inf. cerebellar peduncle

Uvula, tonsils, paraflocculus

UPPER MEDULLA

TO PALEOCEREBELLUM

Ventral spinocerebellar tract

Dorsal spinocerebellar tract

MIDDLE MEDULLA

**GENERAL
PROPRIOCEPTION—
SPINOCEREBELLAR
TRACTS**

Lateral cuneate nuc.

Annulospiral

Muscle spindles

Flower spray

LOWER MEDULLA

SPINOCEREBELLAR TRACTS

Ventral   Dorsal

Via gracile fasc.

All but cervical

All levels

Neurotendinous (Golgi) organ

All levels

UPPER CORD

Clarke's nucleus (column)

LOWER CORD

Pacinian corpuscle

Pansky

– 187 –

85. AFFERENT CEREBELLAR CONNECTIONS: FROM THE BRAINSTEM—PART I

I. **Cuneocerebellar tract** may be thought of as the upper extremity counterpart of the posterior spinocerebellar tract

A. REPORTS THE INSTANTANEOUS ACTIVITY from muscle spindles associated with the upper limb to the cerebellum, which then modifies muscle group activity for smooth, coordinated movements*

B. NERVE CELL BODIES in the accessory (lateral) cuneate nucleus of the lower dorsolateral portion of the medulla send ipsilateral fibers (dorsal arcuate fibers) to form the cuneocerebellar tract

   1. Impulses into the accessory cuneate nucleus are via A alpha afferents
   2. This tract enters the cerebellum through the inferior cerebellar peduncle
   3. Tract ends primarily in the forelimb region of the anterior lobe and pyramis

II. **Vestibulocerebellar tract**

A. VESTIBULAR (ARCHICEREBELLAR) FIBERS from the vestibular apparatus (see Vestibular pathways) to the cerebellum follow two routes

   1. Direct (primary) ascending branch of the vestibulocochlear nerve courses along the medial side of the inferior peduncle to enter the cerebellum and end as mossy fibers of the homolateral archicerebellum (or flocculonodular lobe)
   2. Indirect (secondary) fibers from the inferior and medial vestibular nuclei enter the cerebellum primarily through the juxtarestiform body (medial part of the inferior peduncle) to end bilaterally in the flocculonodular lobe and fastigial nuclei

B. THIS TRACT is essential for the maintenance of equilibrium, posture, and balance

III. **Trigeminocerebellar tract**

A. MESENCEPHALIC NUCLEUS OF V gives rise to secondary proprioceptive fibers which end in the dentate and emboliform nuclei

B. SENSORY NUCLEUS OF V gives rise to secondary fibers which end in the tonsil and anterior quadrangular lobule

C. SPINAL NUCLEUS OF V may give rise to crossed and uncrossed fibers which end in the tonsil and anterior quadrangular lobule

IV. **Reticulocerebellar tract**

A. FIBERS MAKING UP THIS TRACT originate in the medulla (from two regions) and in the pons

   1. From the medulla (lateral and paramedian reticular nuclei) fibers enter the inferior cerebellar peduncle to end primarily in the vermis of the anterior lobe (some fibers also pass to the pyramis and uvula of the posterior lobe)
      a. Most transmission within the reticular formation is modality nonspecific
      b. The lateral reticular nucleus, however, may receive tactile impulses from the spinal cord and convey these somatic impulses to numerous cerebellar cortical areas
   2. In the pons fibers from the reticular formation (specifically the reticulotegmental nucleus) enter the cerebellum through the middle peduncle to end in the vermis of the anterior and posterior lobes

B. THIS TRACT MAY SERVE AS AN IMPORTANT FEEDBACK CIRCUIT from the cerebrum to the cerebellum since many of the nuclei also receive input from sensorimotor cortex

---

*The rostral spinocerebellar tract has been described recently in the cat. It is considered the forelimb equivalent of the anterior spinocerebellar tract.

CEREBELLAR AFFERENTS
FROM THE BRAINSTEM

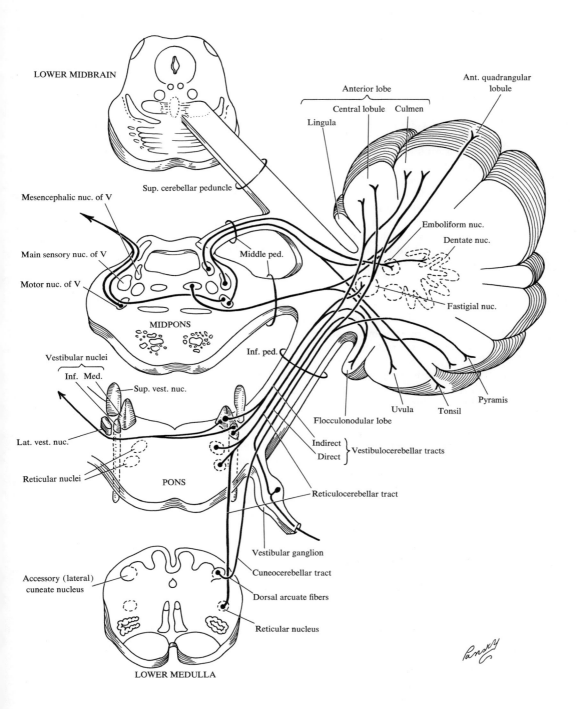

LOWER MIDBRAIN

Anterior lobe

Ant. quadrangular lobule

Central lobule   Culmen

Lingula

Sup. cerebellar peduncle

Mesencephalic nuc. of V

Emboliform nuc.

Dentate nuc.

Main sensory nuc. of V

Middle ped.

Motor nuc. of V

MIDPONS

Fastigial nuc.

Vestibular nuclei

Inf. ped.

Inf. Med.

Sup. vest. nuc.

Uvula   Tonsil   Pyramis

Flocculonodular lobe

Lat. vest. nuc.

Indirect
Direct } Vestibulocerebellar tracts

Reticular nuclei   PONS

Reticulocerebellar tract

Vestibular ganglion

Accessory (lateral)
cuneate nucleus

Cuneocerebellar tract

Dorsal arcuate fibers

Reticular nucleus

LOWER MEDULLA

86. AFFERENT CEREBELLAR CONNECTIONS: FROM THE BRAINSTEM—PART II

I. Olivocerebellar tract
A. FIBERS MAKING UP THIS TRACT arise from the inferior olivary complex
   1. Principal (chief) nucleus sends numerous fibers to hemispheres of the anterior and posterior lobes
   2. Medial accessory olivary nucleus sends fibers to the flocculonodular lobe, pyramis, uvula, declive, and fastigial nucleus
   3. Dorsal accessory nucleus sends fibers to the vermis of the anterior lobe and the declive and folium of the posterior lobe
B. COURSE OF THE OLIVOCEREBELLAR TRACT FROM THE NUCLEAR COMPLEX
   1. Fibers cross to the opposite side of the medulla, pass through the contralateral olive, and ascend into the above designated areas of the cerebellar cortex via the inferior peduncle
   2. This tract comprises the bulk of the inferior peduncle
C. MAJOR INCOMING FIBERS OF THE INFERIOR OLIVARY COMPLEX
   1. Fibers ascending in the spino-olivary tract arise from neurons in the dorsal gray horn of the cord on the contralateral side and transmit sensory impulses from the skin, joints, and some muscles
   2. Fibers descend from sensorimotor cortex and the red nucleus
D. INFERIOR OLIVARY COMPLEX serves as a relay for modulation and transmission of information from the cord, sensorimotor cortex, and red nucleus to the cerebellum

II. Tectocerebellar tract arises from the tectum of the mesencephalon (in the colliculi)
A. COURSE OF THE TECTOCEREBELLAR FIBERS
   1. Enter the anterior medullary velum from the inferior aspect of the inferior colliculus, undergo partial decussation in the velum, and end in the anterior lobe, posterior lobe (folium, tuber, pyramis), and adjacent areas of the hemispheres
B. THE TECTUM is the center for both visual and auditory reflexes
C. THE TECTOCEREBELLAR TRACT may be the most important connection between the auditory pathway and the cerebellum

III. Corticopontocerebellar tract
A. FIBERS FROM NERVE CELL BODIES IN THE CEREBRAL CORTEX (especially those concerned with somatic motor function) descend through the internal capsule and mesencephalon as pyramidal, frontopontine, and temporoparieto-occipitopontine tracts
B. IN THE PONS, these fibers terminate on or synapse with pontine nuclei on the same side (in its ventral aspect)
C. AXONS OF THE CELLS OF THE PONTINE NUCLEI form the pontocerebellar tract which crosses to the opposite side of the pons to enter the cerebellum through the middle cerebellar peduncle and end as mossy fibers of the neocerebellum (posterior lobe) of the vermis and both hemispheres

IV. Arcuocerebellar tract arises from the arcuate nuclei located on the ventral surface of the pyramids (medulla)
A. MAJOR ROUTES OF FIBERS LEAVING THE ARCUATE NUCLEI
   1. Ventral external arcuate fibers traverse the ventral and lateral surfaces of the medulla to enter the cerebellum via the inferior peduncle where most end in cortex with the pontocerebellar fibers
   2. Arcuatocerebellar fibers run dorsally near the midline to the floor of the 4th ventricle, turn laterally as the striae medullares, and enter the cerebellum via the inferior peduncle to end in cortex with pontocerebellar fibers
B. ARCUATE NUCLEI may be displaced pontine nuclei and serve to supplement the corticopontocerebellar tract

Parietal

Post. } Frontal
Ant.

Internal capsule

Occipital

Frontopontine fibers

Occipito-parieto-
temporo-pontine fibers

Tectum (sup. colliculus)

Red nucleus

UPPER MIDBRAIN

Cerebral peduncle

Tectocerebellar tract

Anterior lobe

LOWER MIDBRAIN

Sup. cerebellar
peduncle

C

CL

L

Corticopontine fibers

D

4th vent.

Middle
ped.

Rubro-olivary tract

F

T

MIDPONS

Inf. ped.

N

U

P

Olivocerebellar fibers
(int. arcuate)

4th vent.

Flocculonodular lobe

Posterior lobe

Arcuatocerebellar tracts

Dorsal accessory nucleus

Olivocerebellar tract

Medial accessory nucleus

Striae medullares

Inferior olive

Ventral ext. arcuate fibers

Spino-olivary tract

Pansky

87. EFFERENT CEREBELLAR CONNECTIONS: CORTICOFUGAL AND NUCLEOFUGAL

I. **Corticofugal pathways** are comprised of fibers that carry impulses away from cell bodies located in the cerebellar cortex

A. DIRECT CORTICOFUGAL FIBERS pass directly from the cerebellar cortex into the brainstem (relatively few in number) and include

1. Flocculovestibular tract: from the flocculus to the superior and lateral vestibular nuclei
2. Fibers from the anterior lobe (lingula, central lobule, and culmen) to vestibular nuclei
3. Fibers from the posterior lobe (nodulus and uvula) to vestibular nuclei
4. Fibers from the nodulus to the reticular formation of the medulla

B. INDIRECT OR CORTICONUCLEAR FIBERS pass from the cerebellar cortex to the intrinsic nuclei of the cerebellum and include

1. Fibers from the vermis to the fastigial and globose nuclei
2. Fibers from the medial part of the hemispheres next to the vermis pass to the globose and emboliform nuclei
3. Fibers from the lateral part of the hemispheres pass to the dentate nucleus

II. **Nucleofugal pathways** are comprised of fibers that carry impulses away from cell bodies located in the intrinsic nuclei out of the cerebellum proper

A. FASTIGIAL FIBERS

1. Fastigiobulbar tract (or cerebellovestibular tract) arises in the fastigial nucleus and its fibers may cross or remain on the same side
   a. Fibers enter the medial part of the inferior cerebellar peduncle to pass to the brainstem where they end in the vestibular nuclei
2. Uncinate bundle (or cerebellospinal tract) arises in the fastigial nucleus
   a. Fibers cross in the inferior cerebellar commissure and ascend a short distance with fibers of the superior cerebellar peduncle
   b. Fibers then curve around (hook = uncinate) the superior peduncle and leave the cerebellum via the inferior peduncle to enter the vestibular nuclei (both crossed and uncrossed fibers; contralateral fibers enter the reticular formation of the pons and medulla)

B. DENTATE FIBERS comprise the majority of efferent pathways from the cerebellum

1. Dentatorubral tract
   a. Arises primarily from cell bodies in the dentate nucleus (some fibers from the emboliform and globose nuclei)
   b. Exit the cerebellum via the superior peduncles which form the lateral wall of the fourth ventricle (superior portion)
   c. Enter the pontine tegmentum and pass into the tegmentum of the mesencephalon
   d. Most of the fibers cross to the contralateral side at the level of the caudal portion of the mesencephalon; a few uncrossed fibers probably end in pontine or in reticular nuclei
   e. After crossing, the tract breaks into fasciculi, the majority of which ascend to synapse with both small and large (primarily) cells of the red nucleus (descending fibers end in the reticulotegmental nucleus)
2. Dentatothalamic tract
   a. Course of fibers is identical to above tract except that the fascicles pass around the red nucleus without synapsing and continue to ascend into the diencephalon where they end in the ventral lateral (VL) nucleus of the thalamus
   b. Fibers from the VL nucleus ascend to the motor cortex through the internal capsule
3. Cerebellorubral tract arises mainly in the globose and emboliform nuclei and terminates in the red nucleus

Leg

Arm

Frontal cortex areas 4 and 6

Head

Section

Ventral lateral nucleus

Parvocellular
(small)

Red nucleus

Magnocellular
(large)

Ventral lateral thalamic nucleus

Dentatorubral + cerebellorubral tracts

Dentatothalamic tract

Reticulotegmental nucleus

Uncinate fasciculus

Ant. lobe

Dentate nucleus

Sup.
ped.

Pontine nuclei

Reticular nucleus

Middle ped.

MIDPONS

Inf. cerebellar peduncle

Post. lobe

Fastigial nucleus

Globose + emboliform
nuclei

Flocculonodular lobe

Cerebellovestibular tracts

Superior

Lateral

Medial

Inferior

MEDULLA

Reticular formation

Vestibular nuclei

EFFERENT CEREBELLAR CONNECTIONS

88. MOTOR SYSTEMS: PYRAMIDAL (CORTICOSPINAL) TRACT, SOMATIC EFFERENT—PART I

I. **Definition:** refers to that part of the motor system arising from specific areas of the cerebral cortex and descending through the pyramids to end on alpha motor neurons. The corticospinal tract is a direct link from cerebral cortex to spinal cord

II. **Functions:** related to precise and skilled movements of the extremities (hand and foot). Deals with the contraction of individual muscles and is the pathway for the selection of the "prime movers" for any muscular activity

III. **Cortical representation** of muscles (body parts) and movements as depicted in the motor homunculus remains unclear
   A. ELECTRICAL STIMULATION of specific cortical areas does elicit individual muscle contractions, but contractions are only fragments of a movement
   B. SYNTHESIS OF FRAGMENTS OF MOVEMENT into a useful movement occurs in the motor cortex (area 4)
   C. COMPLEX MUSCULAR ACTIVITY is controlled through areas of cortex other than the motor strip area
      1. Area 6 of the frontal lobe deals with the development of precise and skilled movements effected in area 4
      2. Areas 3, 2, and 1 of the parietal lobe also contribute efferent fibers
      3. Other cortical areas that also have a somatotopic organization are found in the upper top of the lateral fissure (secondary motor area) and on the medial aspect of the cerebral hemispheres (supplementary motor area)

IV. **Upper motor neuron cell bodies** are located in cortical gray areas 6, 4, 3, 2, and 1

V. **Course of upper motor neuron axons**
   A. EFFERENT (CORTICOFUGAL) FIBERS descend by way of corticospinal and corticobulbar pathways to the ventral gray columns of the cord and the cranial motor nuclei of the brainstem, respectively
   B. ALL CORTICOFUGAL FIBERS pass through the corona radiata
      1. Axons forming the corticospinal tract are long (a meter or longer in length)
      2. Pyramidal tract is comprised of about 1 million fibers, 60% of which originate from the frontal lobe cortex and 40% from parietal lobe cortex
      3. Majority (90%) of fibers in the pyramidal tract are small (1–4 nm diameter); 7–9% have diameters of 5–10 nm; and 1–3% have 10–20 nm diameters (Betz cells?)
   C. ALL FIBERS CONVERGE on and pass into the posterior limb of the internal capsule where they are compactly gathered, but are topographically localized
      1. Corticobulbar fibers are found most anteriorly in the posterior limb, next to the genu
      2. Fibers to the upper extremity are located in the rostral part; those to the lower extremity lie more caudally
      3. Sensory tracts from the thalamus to the parietal lobe lie posterior to the corticospinal fibers
   D. AXONS PASS THROUGH and form the cerebral peduncles (crus cerebri or basis pedunculi) along with other descending cortical fibers
      1. Pyramidal fibers remain in distinct groups as they pass into the peduncles at the rostral border of the midbrain and occupy the middle two thirds of the peduncles with the corticobulbar fibers more medial
      2. Corticobulbar fibers leave the corticospinal pathway at appropriate brainstem levels (some crossed and others uncrossed) to connect with motor nuclei of certain cranial nerves

INT. CAPSULE

Genu

Ant. limb

Corticobulbar tracts

Upper ext. + face
Trunk
Lower extremity
} Corticospinal tracts

Post. limb

MIDBRAIN

Pons

Upper medulla

Lower medulla

Corticospinal tract location

Trunk
Knee
Toes
Areas 4 and 6 frontal cortex
Fingers
Face
Tongue
Thalamus
Putamen
Globus pallidus
Internal capsule (post. limb + genu)

Pyramidal cells

Int. capsule

Cerebral peduncle

Pyramidal
(corticospinal) tract

Decussation of pyramidal tract

Lateral corticospinal tract

PYRAMIDAL
(CORTICOSPINAL) TRACT

III

IV

V

VI

VII

XII

XI

Corticobulbar tracts

N. ambiguus

Decussation

Ant.
Lat.
} Corticospinal tracts

Anterior (ventral)
corticospinal tract

Efferent segmental nerve

Ventral gray

Ant.

Lat.

89. MOTOR SYSTEMS: PYRAMIDAL (CORTICOSPINAL) TRACT, SOMATIC EFFERENT—PART II

I. Course of upper motor neuron axons (cont.)

A. IN THE PONS descending pyramidal tract fiber bundles split into smaller irregular bundles (running in the ventral pons) by scattered pontine nuclei; somatotopic localization is observed from medial to lateral as face, upper extremity, and lower extremity

B. IN THE MEDULLA fiber bundles recombine to form the pyramids on the ventral aspect of the medulla

C. IN THE LOWER MEDULLA fibers cross the ventral fissure to the opposite side
  1. 75–90% of the fibers decussate (pyramidal decussation) just above the medulla-cord junction
  2. Fibers to the upper extremity cross over higher in the pyramidal decussation than those to the lower extremity

D. IN THE SPINAL CORD the crossed pyramidal fibers form the lateral corticospinal tract in the lateral white column
  1. This tract gradually becomes smaller as it descends in the spinal cord
    a. 50% of the fibers end in the cervical spinal cord levels
    b. 20% end in the thoracic cord levels
    c. 30% of the fibers continue the full length of the spinal cord
  2. Some (about 10%) of these upper neurons end directly on alpha motor neurons of lamina 9 in the anterior gray horn; the others (as in the brainstem) synapse with connector neurons of laminae 4–7
  3. A few uncrossed fibers may run caudally in the lateral corticospinal tract of the same side (ventrolateral corticospinal tract or Barne's tract)

E. UNCROSSED FIBERS (10–25%) to the trunk muscles descend in the anterior white columns as the anterior corticospinal tracts
  1. These tracts are difficult to trace below the thoracic level, although some few fibers may continue throughout the cord
  2. Before ending, these fibers usually cross through the anterior white commissure to synapse either with anterior horn cells (alpha motor neurons) or connector neurons in laminae 7 and 8

II. Lower motor neurons

A. LOWER MOTOR NEURON CELL BODIES (ALPHA MOTOR NEURONS) are located in nuclei of the medial (for trunk muscles) and lateral (for extremity muscles) divisions of the anterior gray horns of the spinal cord

B. COURSE OF THE LOWER MOTOR NEURON AXONS
  1. Exit spinal cord through the ventral (anterior) roots and enter the spinal nerves at appropriate levels
  2. Are distributed with muscular branches of the dorsal and ventral rami to innervate skeletal muscles of the extremities and trunk

Apical dendrites

Pyramidal cell

Basilar dendrites

Axon

Corona radiata   Trunk   Knee

Toes

Thalamus

Int. capsule

Caudate

Ant. limb   Corticospinal tr.

Thalamus

Post. limb int. capsule

Putamen + globus pallidus

**PYRAMIDAL TRACT
(CORTICOSPINAL)**

Decussation

Anterior (ventral) corticospinal
tract (uncrossed fibers)

UPPER CORD

Terminal arborization

LOWER CORD

Parietal cortex   Frontal cortex
3–1–2   4 6

LATERAL VIEW OF HEMISPHERE

MIDBRAIN

Cerebral peduncle

Corticospinal (pyramidal) tract

PONS

JUNCTION OF PONS AND MEDULLA

Inferior olive

Pyramid of medulla oblongata

Pyramidal decussation

LOWER MEDULLA

Lateral (posterior) corticospinal
tract (crossed) fibers

Efferent segmental nerve

Panoky

– 197 –

90. MOTOR SYSTEMS: PYRAMIDAL (CORTICOBULBAR) TRACT—SPECIAL VISCERAL EFFERENT

I. **Definition and function:** the corticobulbar pathways (head and neck) are related to movements of muscles of facial expression and mastication, and tongue musculature

II. **Cortical representation and upper motor neuron location:** see Corticospinal Tracts

III. **Course of upper motor neuron axons**
   A. FIBERS CONVERGE as they pass through the corona radiata, with corticobulbar fibers lying in the most anterior portion of the posterior limb of the internal capsule
   B. CORTICOBULBAR FIBERS OCCUPY the medial portion of the middle two thirds of the cerebral peduncles of the mesencephalon
   C. FIBERS PASS INTO THE PONS along with the corticospinal fibers
   D. SOME CORTICOBULBAR FIBERS LEAVE the pyramidal bundle some distance above the nucleus for which they are destined and descend in region of the medial lemniscus
   E. MOST OF THE CORTICOBULBAR FIBERS MAY END in nuclei of the reticular formation (connector neurons) and use the reticular formation, medial longitudinal fasciculus (MLF), and the medial lemniscus to reach cranial motor nuclei
   F. IN THE MIDDLE PONS crossed (and some uncrossed) corticobulbar fibers pass to the motor nucleus of the trigeminal (V) nerve
   G. IN THE LOWER PONS crossed corticobulbar fibers pass to the anterior portion of the nucleus of the facial (VII) nerve which innervates the lower facial muscles; crossed and uncrossed fibers pass to the posterior portion of this same nucleus which innervates the upper facial muscles
   H. IN THE MEDULLA both crossed and uncrossed fibers go to the nucleus ambiguus

IV. **Lower motor neuron cell bodies** are located in
   A. MOTOR NUCLEUS OF THE TRIGEMINAL (V) NERVE
   B. NUCLEUS OF THE FACIAL (VII) NERVE
   C. NUCLEUS OF THE SPINAL ACCESSORY (XI) NERVE located in cervical spinal cord segments 1–5. Uncrossed upper motor neurons synapse with cell bodies of this nucleus
   D. NUCLEUS OF THE HYPOGLOSSAL (XII) NERVE

V. **Course of the lower motor neuron axons**
   A. DISTRIBUTED WITH THE MANDIBULAR DIVISION OF THE TRIGEMINAL NERVE
   B. DISTRIBUTED WITH THE FACIAL NERVE
   C. NUMEROUS FIBERS ALSO ARE DISTRIBUTED with the glossopharyngeal and vagus nerves and the cranial portion of the spinal accessory nerve
   D. DISTRIBUTED WITH THE SPINAL PORTION OF THE SPINAL ACCESSORY NERVE
   E. DISTRIBUTED WITH THE HYPOGLOSSAL NERVE

VI. **Structures innervated by the lower motor neurons**
   A. THE MANDIBULAR DIVISION OF THE TRIGEMINAL NERVE: muscles of mastication as well as the tensor veli palatini, tensor tympani, mylohyoid, and anterior belly of the digastric
   B. THE FACIAL NERVE: lower facial muscles (via fibers from anterior part of facial nucleus), upper facial muscles (via fibers from posterior part of facial nucleus), posterior belly of digastric, stapedius, and stylohyoid muscles
   C. THE GLOSSOPHARYNGEAL NERVE: the stylopharyngeus muscle
   D. THE VAGUS NERVE: muscles of the soft palate, pharynx, esophagus, and larynx
   E. SPINAL ACCESSORY NERVE
      1. Spinal portion to trapezius and sternocleidomastoid muscles
      2. Cranial portion to muscles of pharynx and larynx

VII. **Subthalamic and peduncular aberrants** supply the oculomotor (III), trochlear (IV), and abducens (VI) nuclei to help influence eye movements

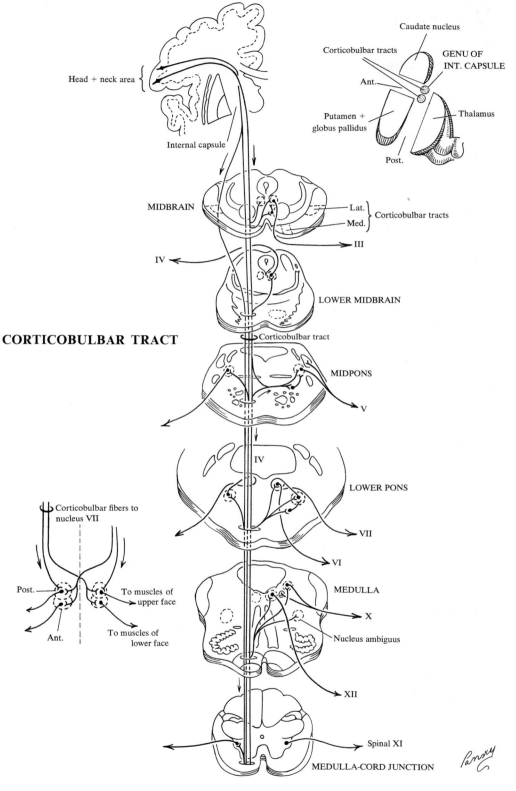

CORTICOBULBAR TRACT

Head + neck area

Internal capsule

Caudate nucleus

Corticobulbar tracts

GENU OF INT. CAPSULE

Ant.

Putamen + globus pallidus

Thalamus

Post.

MIDBRAIN

Lat.

Med.

Corticobulbar tracts

III

IV

LOWER MIDBRAIN

Corticobulbar tract

MIDPONS

V

IV

LOWER PONS

VII

VI

Corticobulbar fibers to nucleus VII

Post.

To muscles of upper face

Ant.

To muscles of lower face

MEDULLA

X

Nucleus ambiguus

XII

Spinal XI

MEDULLA-CORD JUNCTION

Pansky

91. MOTOR SYSTEMS: EXTRAPYRAMIDAL PATHWAYS

I. **Definition:** extrapyramidal system is comprised of all tracts, their nuclei, and feedback circuits that influence somatic motor activity of voluntary muscles except for the pyramidal pathways. It conveys its influence from the cerebral cortex, basal ganglia, and other subcortical nuclei indirectly to the spinal cord via multisynaptic connections*

II. **Functions:** this system deals with gross, more automatic, voluntary movements

III. **Structures related to and included in the "extrapyramidal" pathways**
   A. CEREBRAL CORTEX AND BASAL GANGLIA (CAUDATE, PUTAMEN, AND GLOBUS PALLIDUS)
   B. THALAMUS (CENTROMEDIAN NUCLEUS) AND SUBTHALAMIC NUCLEUS
   C. RED NUCLEUS AND SUBSTANTIA NIGRA (MESENCEPHALON)
   D. NUCLEI IN THE RETICULAR FORMATION (PONS AND MEDULLA)
   E. FEEDBACK CIRCUITS, TRACTS, AND PATHWAYS (CORTICORUBROSPINAL, CORTICORETICULOSPINAL, AND VESTIBULOSPINAL)

IV. **Main source of afferent fibers into the basal ganglia**
   A. CEREBRAL CORTEX (ALL CORTICAL AREAS) send nerve fibers into caudate nucleus and putamen (corticostriates), thalamus (ventrolateral), red nucleus and tectum, reticular formation, pons, and substantia nigra
   B. CENTROMEDIAN NUCLEUS OF THALAMUS sends fibers into caudate nucleus
   C. SUBSTANTIA NIGRA sends fibers to caudate nucleus and putamen (nigrostriates)
   D. CAUDATE NUCLEUS AND PUTAMEN send short connector neurons to the globus pallidus to synapse with upper motor neurons

V. **Main source of efferent fibers leaving the basal ganglia** is via the globus pallidus

VI. **Primary upper motor neuron cell bodies of the extrapyramidal system** are located in the cerebral cortex and globus pallidus

VII. **Course of the primary upper motor neuron axons**
   A. CEREBRAL CORTICAL FIBERS descend to the level of the superior colliculus of the midbrain and synapse in the red nucleus
   B. EFFERENT FIBERS FROM THE GLOBUS PALLIDUS form the ansa lenticularis and the lenticular fasciculus (some fibers may pass to the red nucleus and tegmentum)
   C. THE ANSA LENTICULARIS AND LENTICULAR FASCICULUS FIBERS turn medially to join the thalamic fasciculus
      1. Ventral anterior and ventral lateral nuclei of the thalamus receive fibers from the thalamic fasciculus
      2. Efferent fibers from these two thalamic nuclei pass to the motor and premotor cortex (areas 4 and 6) to form a feedback circuit for the control of motor activity

VIII. **Secondary upper motor neuron cell bodies** are located in the red nucleus and reticular formation of the mesencephalon and in other areas of the brainstem (vestibular nuclei and superior colliculus)

IX. **Course of secondary upper motor neuron axons**
   A. EFFERENT FIBERS FROM THE RED NUCLEUS decussate and form the rubrospinal tract which descends through brainstem in the lateral funiculus of the spinal cord
   B. FIBERS LEAVING THE RUBROSPINAL TRACT synapse in the gray matter of the spinal cord
   C. THESE CONNECTOR NEURONS, in turn, synapse either with alpha motor neurons (inhibitory) or with gamma motor neurons in the anterior gray horn

X. **Other descending tracts (vestibulospinal, reticulospinal, tectospinal, and MLF),** often considered part of the extrapyramidal system, are discussed separately

*The cerebellum and its related pathways are closely associated with but not usually included in the extrapyramidal system.

Cortical areas

2, 4, 5, 6, 8, 9, 19, 24

Corticocaudate tr.

Caudate nucleus

Corticothalamic tr.

Corticorubral tr.

Ansa lenticularis + fasciculus lenticularis

Putamen

Centromedian nucleus

Ventrolateral nuclei

Thalamus

Globus pallidus

Subthalamic nucleus

Corticonigral tr.

Superior colliculus

Inferior colliculus

Red nucleus

Rubroreticular tr.

MIDBRAIN

Rubrospinal tr.

Substantia nigra

Reticular formation

Deep tegmental nucleus

Reticulospinal tr.

PONS

Corticopontine + corticoreticular tracts

Rubroreticulospinal tr.

Tectospinal tr.

MEDULLA OBLONGATA

Inferior olive

Vestibular nucleus

Propriospinal tr.

SPINAL CORD

Reticulospinal tr.

Tegmentospinal tr.

Vesticulospinal tr.

Olivospinal tr.

Reticulospinal tr.

Pansky

92. MUSCLE REGULATION

I. Introduction: central nervous system influence on voluntary muscle is via the lower motoneurons and includes the anterior horn cells of the cord and branchial and somatic motor nuclei of the brainstem. The lower motoneuron and the defined group of muscle fibers it innervates are called a motor unit. There is normally no spontaneous firing of lower motoneurons. Maintenance of muscle tone and movement depends on an adequate stimulation from the CNS pathways, which include

A. PROPRIOCEPTIVE FEEDBACK: from muscle, tendons, and joint capsules over which they operate, via the spinal cord or brainstem

B. NONPROPRIOCEPTIVE SENSORY INPUT: from all areas, especially the skin, which are innervated at the same segmental level, via the spinal cord or brainstem

C. FEEDBACK COLLATERALS: from motor axons themselves (inhibitory)

D. SENSORY INPUT FROM OTHER SEGMENTS includes both proprioceptive and nonproprioceptive components via the spinal cord and brainstem

E. LABYRINTHINE PATHWAYS via the vestibulospinal and reticulospinal tracts

F. PATHWAYS IN THE BRAINSTEM RETICULAR FORMATION: originating in, or mediated by the rubrospinal and reticulospinal tracts

G. INDIRECT CEREBELLAR PATHWAYS: operating through the reticular formation and cerebral cortex, reticulospinal, and corticospinal tracts

H. CEREBRAL CORTICAL AREAS: pathways via the pyramidal tracts and cortically originating extrapyramidal system (COEP)—corticospinal and reticulospinal tracts

I. VIRTUALLY THE ENTIRE CNS is directly or closely involved in muscle control since it represents a complete range of environmental responses of which the organism is capable

II. Skeletal muscle functions

A. INVOLUNTARY (REFLEX) MOVEMENT: in response to environmental stimuli common to all forms of life with contractile tissue. Intrinsic to the segmental organization of the CNS (including the spinal cord and brainstem)

B. MAINTENANCE OF POSITION (POSTURE): including orientation related to gravity and maintenance of stance against gravity which includes responses related to man's upright position

C. EXECUTION OF VOLITIONAL MOVEMENTS: organized or set into motion by suprasegmental structures and not necessarily dependent on external stimuli

   1. Both B and C require integration, initiation, or both by suprasegmental parts of the brain

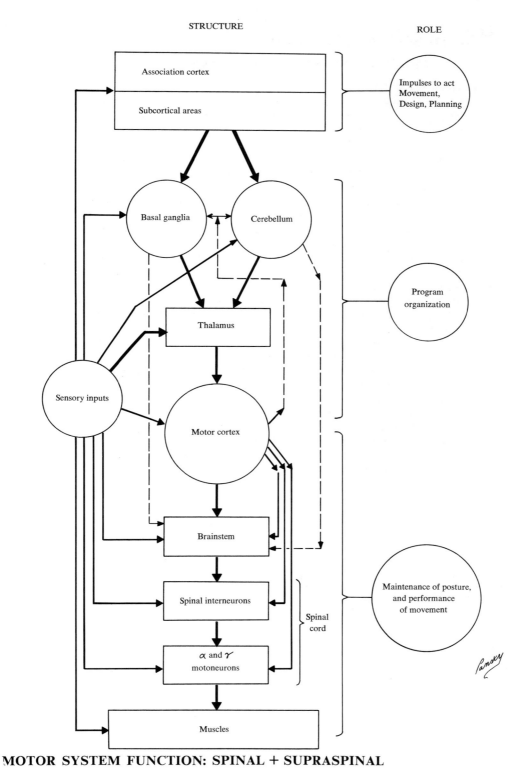

STRUCTURE
ROLE

Association cortex

Subcortical areas

Impulses to act
Movement,
Design, Planning

Basal ganglia

Cerebellum

Program
organization

Thalamus

Sensory inputs

Motor cortex

Brainstem

Spinal interneurons

Spinal
cord

α and γ
motoneurons

Maintenance of posture,
and performance
of movement

Muscles

**MOTOR SYSTEM FUNCTION: SPINAL + SUPRASPINAL
CENTERS AND CONNECTIONS**

93. MUSCLE REGULATION:
AFFERENT AND EFFERENT CONNECTIONS

I. Motor fibers

A. LOWER MOTONEURONS are cells whose axons leave the CNS. The axon branches or telodendria end as neuromuscular junctions or motor end-plates on skeletal muscle fibers, each with a single motor end-plate

B. THE CELL BODIES are large, polygonal with conspicuous Nissl bodies and are found in the branchial and somatic brainstem motor nuclei of cranial nerves III, IV, motor V (masticator), VI, VII, ambiguus, accessory (XI), and XII as well as in the anterior horns of the spinal cord
   1. In the motor nuclei, anatomically indistinct subgroupings correspond to individual muscles, muscle groups, or muscle actions
   2. In the spinal cord, flexor and adductor neurons lie deep to extensor and abductor neurons. Cells for distal muscles lie lateral to those for trunk structures

C. EACH MOTOR AXON, as it passes through the anterior white matter, gives off a recurrent collateral which passes back into the anterior horn to synapse with Renshaw cells (interneurons found in the margins of the anterior horn)

D. MOTOR AXONS are thick, heavily myelinated fibers, 10–20 μm in diameter and make up 40% of the large fibers in a muscle nerve. The other 60% are large afferent fibers, mostly from the spindles
   1. When the axon reaches the muscle, it branches extensively, supplying a group of muscle fibers with 1 motor end-plate each
   2. A single axon may supply only 3–10 muscle fibers of small muscles (eye muscles) or it may supply up to 200 fibers of large proximal muscles concerned with postural or coarse voluntary responses
   3. The group of fibers innervated by 1 axon (motor unit) does not correspond to a muscle bundle but extends over adjacent bundles, with much overlapping
   4. The range of tension developed by a single motor unit varies from about 0.1 g (in eye muscle) to about 50 g in a large limb muscle and is the result of fiber size and the number of fibers supplied

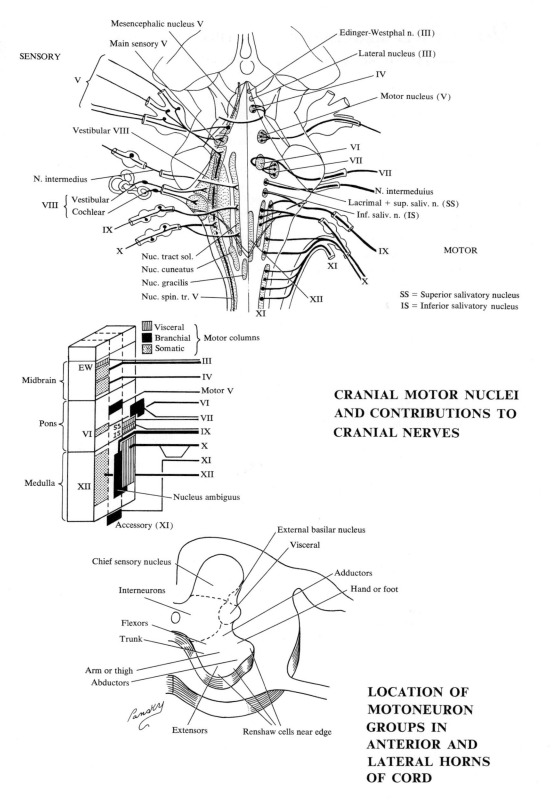

SENSORY

Mesencephalic nucleus V
Main sensory V
V
Vestibular VIII
N. intermedius
VIII { Vestibular
Cochlear
IX
X
Nuc. tract sol.
Nuc. cuneatus
Nuc. gracilis
Nuc. spin. tr. V
XI

Edinger-Westphal n. (III)
Lateral nucleus (III)
IV
Motor nucleius (V)
VI
VII
VII
N. intermeduius
Lacrimal + sup. saliv. n. (SS)
Inf. saliv. n. (IS)
IX
MOTOR
XI
X
XII

SS = Superior salivatory nucleus
IS = Inferior salivatory nucleus

Visceral
Branchial } Motor columns
Somatic
III
IV
Motor V
VI
VII
IX
X
XI
XII
EW
Midbrain
Pons
VI
SS
IS
Medulla
XII
Nucleus ambiguus
Accessory (XI)

CRANIAL MOTOR NUCLEI AND CONTRIBUTIONS TO CRANIAL NERVES

External basilar nucleus
Visceral
Chief sensory nucleus
Interneurons
Adductors
Hand or foot
Flexors
Trunk
Arm or thigh
Abductors
Pansky
Extensors
Renshaw cells near edge

LOCATION OF MOTONEURON GROUPS IN ANTERIOR AND LATERAL HORNS OF CORD

94. MUSCLE REGULATION: SENSORY FEEDBACK FROM MUSCLE AND TENDON—PART I

I. **Introduction:** naked nerve endings in musculoskeletal tissues convey pain, which may lead to reflex muscle spasm in response to inflammation or ischemia, but are probably not important in regulation of normal muscle activity. Endings dealing with proprioception are found in joint capsules (joint organ), near the attachments of the muscle to the periosteum or tendon (tendon organ), and in the muscle itself (muscle spindle)

A. JOINT ORGAN is associated with a subjective sense of limb position, and its major central connections are through the thalamus to the sensory cortex rather than to the cerebellum

B. TENDON OR GOLGI ORGAN is an encapsulated mechanoreceptor found at the ends of a muscle and is stretched either by active contraction or passive lengthening of the muscle

    1. Each is supplied by 1 to 2 large (12–18 μm) myelinated fibers (Aβ, group IB), which branch on any fasciculi of the tendon included in the capsule

    2. Exhibit little or no spontaneous firing without tension, but fire (without adaptation) if tension rises to 100–200 g. The firing rate is proportional to the logarithm of the tension

    3. Functions to protect against overstretching and rupture of the muscle

C. MUSCLE SPINDLE: a complex, encapsulated, elongated structure in the interstitial connective tissue of the muscle belly. Probably seen in all skeletal muscle, but less easily found in branchial than in somatic muscle

    1. Each contains 2 types of muscle fibers, receives 2 types of motor axons, and originates 2 types of sensory fibers. The muscle fibers are intrafusal as opposed to the extrafusal fibers that make up the bulk of the muscle

    2. Each contains 1 or 2 bag fibers, which are 7–8 mm long and 25 μm wide, centrally dilated, noncontractile "bags" containing several nuclei and show no cross-striations. The rest of the fiber has myofibrils and is striated

        a. Each bag fiber has 1–2 medium-sized myelinated gamma (γ) motor axons (2.5–4.5 μm) and 1–6 separate motor end-plates on the cross-striated regions

        b. Each bag is surrounded by spiral sensory fibers, annulospiral endings, from which originate large (10–18 μm) myelinated afferent fibers (Aβ, group IA)

        c. Small numbers of irregularly branching sensory, flower-spray endings, from which 4–10 μm myelinated afferent fibers originate (Aβ, group II)

        d. The second type of intrafusal fiber is the chain fiber, 4–5 mm long and 12 μm in diameter (4–8 per spindle)

            i. Their gamma motor axons are smaller (1–2 μm) and form irregular networks (trail endings) rather than discrete motor end-plates

            ii. Both annulospiral (sensory) and flower-spray (sensory) endings are present on chain fibers

    3. Stimulation of the large motor fibers presumably shortens the contractile ends of the bag fibers, increasing tension on the bag. Stimulation of the smaller motor fibers shortens the chain fibers and reduces tension on bag

        a. Differential functions of the annulospiral and flower-spray endings are unclear

Free nerve ending

Gamma (γ) motor fiber

III + IV fibers

II fibers

γ fiber

2° sensory ending (flower spray) II

Nuclear bag fiber

Nuclear chain fiber

γ fiber

1° sensory ending (annulospiral) 1A

Alpha (α) motor fiber (extrafusal)

Plate ending of β efferent fiber

Extrafusal muscle fibers

1B fibers

III + IV fibers

External capsule

Internal capsule

Nuclear bag fiber

2°

Nuclear chain fiber

Joint ending

Neurotendinous ending of Golgi

Extrafusal efferent (α) fiber

1° (annulospiral) ending of grp 1A afferent fiber

2° (flower spray) ending of group II afferent fiber

Trail ending of γ_2 efferent fiber

Subcapsular space

Plate ending of β efferent fiber

Plate ending of γ_1 efferent fiber

MAMMALIAN NEUROMUSCULAR SPINDLE

95. MUSCLE REGULATION: SENSORY FEEDBACK FROM MUSCLE AND TENDON—PART II

A. MUSCLE SPINDLE (CONT.)
1. Spindle function
   a. Spindle afferent fibers (IA) are excitatory mainly to motoneurons of the originating muscle, and this excitation involves only 1 synapse between the afferent fiber and the motor neuron (the only monosynaptic reflex known in man)
   b. Facilitation of synergists and inhibition of antagonists, acting across the same joint as the originating muscle and confined to the same side, involve interneurons since a single axon cannot be both inhibitory and excitatory
   c. The CNS origin of the spindle and motor fibers is in the same nuclear zone as the motor alpha fibers to the main bulk (extrafusal) muscles. Gamma motoneurons, like alpha, are in clusters associated with individual muscles or movements and resemble each other morphologically
      i. One function of gamma neurons is to maintain a constant spindle firing rate by causing the spindle fibers to shorten as the muscle shortens; i.e., alpha and gamma motoneurons fire at the same time
   d. The spindle is stretched when the muscle is stretched (like the tendon organ), but when the muscle actively contracts, the spindle is shortened but the tendon organ is stretched
   e. Both tendon organ and spindle are sensitive to stretching, but the spindle fires at lower levels of tension, 1–2 g
      i. The rate of firing is proportional to the logarithm of tension but is virtually nonadapting
   f. There is a servomechanism to maintain muscle length. If a muscle stretches, the spindle is stretched, the ratio of spindle firing increases, and motoneurons are stimulated and the muscle shortens
      i. Muscle length can be altered directly by stimulating alpha motoneurons (shortening the spindle increases the spindle firing and stimulates alpha motoneurons) or indirectly by stimulating gamma motoneurons
   g. Inhibitory feedback is provided via recurrent motor-axon collaterals that synapse with Renshaw cells which are directly inhibitory to the alpha motoneurons

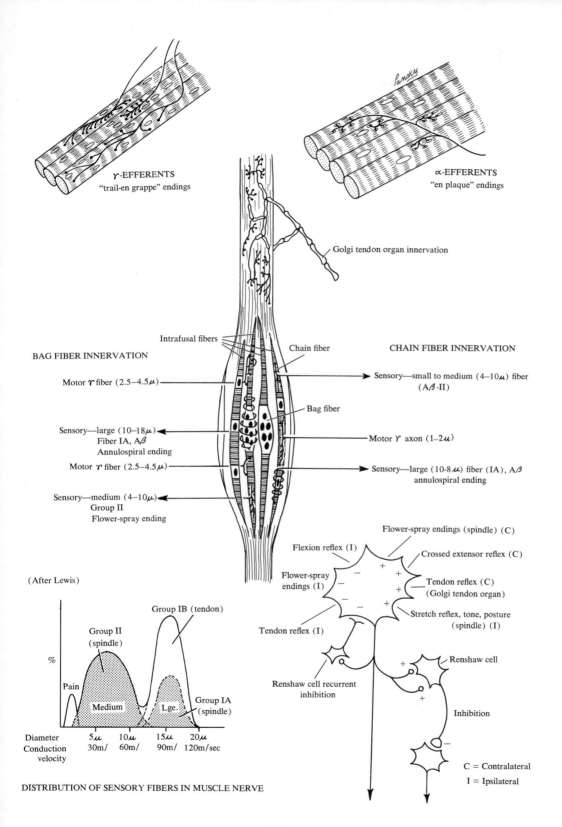

γ-EFFERENTS
"trail-en grappe" endings

α-EFFERENTS
"en plaque" endings

Pansky

Golgi tendon organ innervation

Intrafusal fibers

Chain fiber

BAG FIBER INNERVATION

CHAIN FIBER INNERVATION

Motor γ fiber (2.5–4.5μ)

Sensory—small to medium (4–10μ) fiber
(Aβ -II)

Bag fiber

Sensory—large (10–18μ)
Fiber IA, Aβ
Annulospiral ending

Motor γ axon (1–2μ)

Motor γ fiber (2.5–4.5μ)

Sensory—large (10-8μ) fiber (IA), Aβ
annulospiral ending

Sensory—medium (4–10μ)
Group II
Flower-spray ending

(After Lewis)

Flower-spray endings (spindle) (C)

Flexion reflex (I)

Crossed extensor reflex (C)

Flower-spray
endings (I)

Tendon reflex (C)
(Golgi tendon organ)

Group IB (tendon)

Group II
(spindle)

Stretch reflex, tone, posture
(spindle) (I)

Tendon reflex (I)

Renshaw cell

%

Pain

Renshaw cell recurrent
inhibition

Group IA
(spindle)

Medium

Lge.

Inhibition

Diameter   5μ   10μ   15μ   20μ
Conduction   30m/   60m/   90m/   120m/sec
velocity

C = Contralateral

I = Ipsilateral

DISTRIBUTION OF SENSORY FIBERS IN MUSCLE NERVE

96. MUSCLE REGULATION: SEGMENTAL MUSCLE REFLEXES

I. **Segmental muscle reflexes:** following spinal section, all muscles innervated from segments below the section are flaccid (have no tonus), and movement does not occur in the absence of local sensory input as from muscle spindles, tendon organs, and joint organs [proprioception; skin touch, pressure, temperature, or pain; or deep tissue sensation (pain)]

A. STRETCH EXTENSOR (MYOTATIC, DEEP TENDON) REFLEXES are also known as simple jerk or patellar reflexes. Their basis is that the muscle spindle responds to tension by increasing its firing rate, stimulating the alpha motoneurons of the stretched muscle(s) only

1. Characterized as 2 neuron, ipsilateral (restricted to one side), and intrasegmental (each receptor stimulates an afferent neuron which excites alpha neurons in the same segment)

2. If a flaccid muscle is stretched, it responds by contracting, the latter occurring only in the stretched muscle(s) and is physiologically induced by relaxation of the antagonist(s). There is no contralateral response; contraction is rapidly started, concluded, and abolished by dorsal root section

3. The reflex can be elicited without spinal section and is the basis of deep tendon reflexes and muscle clonus seen clinically

4. It is monosynaptic and not depressed by drugs that affect interneurons and cause reduced flexor response and reflex spasm

B. CLASP-KNIFE REFLEX: solely dependent on segmental input and cannot be shown in a flaccid spinal prep because it consists of a sudden relaxation of muscle when tension is very high

1. In decerebrate rigidity, if a limb is forcibly bent beyond a certain degree, rigidity abruptly disappears and the limb is freely movable. Seen also in patients with spasticity

2. The reflex originates in the tendon organs that respond to high tension by firing
   a. Is multisynaptic (involves interneurons) and causes inhibition of the stretched muscle and its synergists and facilitation of the antagonists
   b. Occasionally the corresponding contralateral muscle is also facilitated

C. FLEXOR RESPONSE: painful stimulation to a flaccid limb results in generalized flexion of the limb and is called withdrawal response

1. Accompanied by extension of the contralateral limb, and if the cord is intact between hindlimb and forelimb segments, flexion of the diagonally opposite limb also occurs

2. The phenomenon (local sign) implies topographic representation in segmental sensory and motor gray matter

3. Flexor response with crossed extension is seen in patients with spasticity but may be diffuse in the gray matter and appear as a "mass" reflex including incontinence of bowel and bladder

4. Stimuli reach the cord via Aδ or C fibers and connect with anterior horn cells only via interneurons, accounting for its long duration (compared to stretch reflex) and diffusion into the spinal gray over several segments on same and opposite side

5. All above are multisynaptic and may be intersegmental

STRETCH (MYOTATIC) REFLEX

CLASP-KNIFE REFLEX

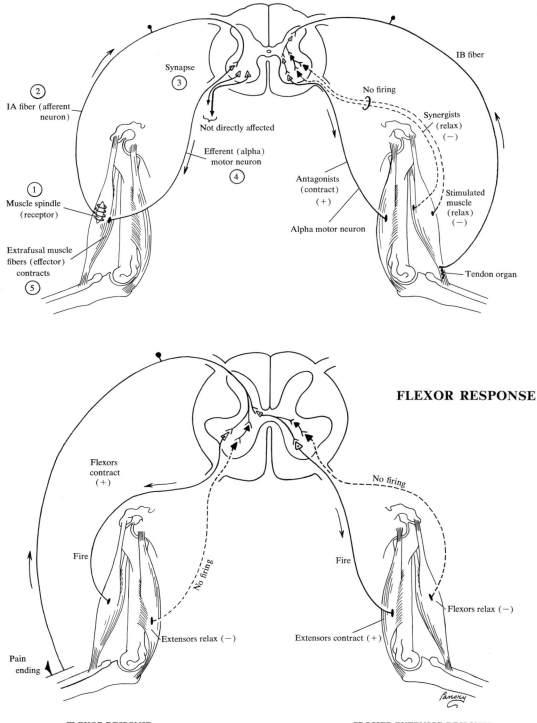

Synapse
③

② IA fiber (afferent neuron)

Not directly affected

① Muscle spindle (receptor)

Extrafusal muscle fibers (effector) contracts ⑤

Efferent (alpha) motor neuron ④

Antagonists (contract) (+)

Alpha motor neuron

IB fiber

No firing

Synergists (relax) (−)

Stimulated muscle (relax) (−)

Tendon organ

FLEXOR RESPONSE

Flexors contract (+)

Fire

No firing

No firing

Fire

Pain ending

Extensors relax (−)

Extensors contract (+)

Flexors relax (−)

Panosky

FLEXOR RESPONSE

CROSSED EXTENSOR RESPONSE

97. MUSCLE REGULATION:
OTHER REFLEX RESPONSES

I. **Gamma reflex loop** includes a gamma motoneuron, a neuromuscular spindle, an afferent neuron, an alpha motor neuron, and voluntary muscle
   A. CONTINUOUS MUSCLE CONTRACTION of a muscle mass results when muscle fibers are in various degrees of contraction in response to asynchronous volleys of alpha motor neurons. This continuous contraction depends on the gamma loop
   B. THE GAMMA CONTROL of spindle sensitivity to stretch by the contraction of the intrafusal muscle fibers is important because the gamma neurons influence the quality of sensory input from spindle to spinal cord and thus play a role in alpha motor neuron output
   C. THE GAMMA NEURONS are stimulated by IA fibers from the muscle spindle, from the periphery via cutaneous stimulation by direct connections through interneurons synapsing with the gamma neurons, and actively by upper motor neurons (originating in the brainstem reticular formation)
   D. TYPES OF GAMMA EFFERENT FIBERS
      1. Static fusimotor: more numerous; serve to adjust the sensitivity of the annulospiral and flower-spray endings to length alone
      2. Dynamic fusimotor serve to adjust the sensitivity of the annulospiral endings to the velocity of shortening

II. **Flexor (superficial) reflex** is a 3-neuron, disynaptic, ipsilateral, intersegmental flexor reflex (superficial, nociperceptive, or cutaneous)
   A. FLEXOR REFLEX includes skin receptors, afferent neurons, spinal intersegmental interneurons, and alpha motor neurons to flexor muscles of upper extremity
   B. REFLEX CAN BE FACILITATED by another 3-neuron reflex arc consisting of the flower-spray ending of the neuromuscular spindle, afferent neuron, spinal interneurons, and alpha motor neurons

III. **Polysynaptic extensor reflexes:** crossed extensor reflex and extensor thrust reflex
   A. CROSSED EXTENSOR REFLEX: stimuli from peripheral receptors that initiate flexion in an extremity are also involved with extension of the contralateral extremity
   B. EXTENSOR THRUST REFLEX: with pressure on the sole of the foot, the pressure receptors in the foot project influence that elicit extension of the stimulated extremity

IV. **Disynaptic flexor reflex (stretch flexor reflex or "mark time" reflex)** is involved in the stepping movements of a spinal animal. The flower-spray receptors are responsible. It is a 3-neuron, disynaptic, intersegmental flexor reflex (see Flexor Reflex)

V. **Monosynaptic flexor reflex (jaw jerk or pluck reflex):** an antigravity action resulting in teeth clenching
   A. REFLEX CAN BE EVOKED by a tap on the jaw, setting off volleys from the annulospiral endings of spindles in flexor muscles of mastication

VI. **Polysynaptic flexor reflex (flexor twitch):** reflex withdrawal of the hand from a stimulus (cutaneous pain or heat receptors), a primitive protective reflex which is polyneuronal, polysynaptic, ipsilateral, and intersegmental

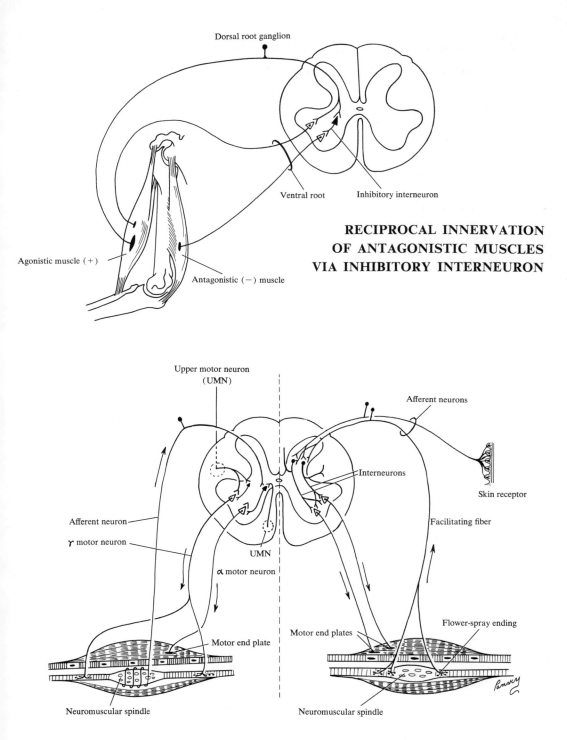

Dorsal root ganglion

Ventral root

Inhibitory interneuron

Agonistic muscle (+)

Antagonistic (−) muscle

**RECIPROCAL INNERVATION
OF ANTAGONISTIC MUSCLES
VIA INHIBITORY INTERNEURON**

Upper motor neuron
(UMN)

Afferent neurons

Interneurons

Skin receptor

Afferent neuron

Facilitating fiber

γ motor neuron

UMN

α motor neuron

Motor end plate

Motor end plates

Flower-spray ending

Neuromuscular spindle

Neuromuscular spindle

GAMMA REFLEX LOOP

FLEXOR REFLEX LOOP

98. MUSCLE REGULATION: ORGANIZATION OF SPINAL CORD GRAY MATTER

I. Stretch, clasp-knife, and flexor reflexes involve not only contraction of muscle groups but also relaxation of their antagonists

A. ALL REFLEXES AND ALL VOLUNTARY MOVEMENTS (RECIPROCAL INHIBITION) are thus involved, and there appears to be a built-in feature at the segmental level that helps us control and avoid useless concurrent contraction of opposing muscle groups.

B. INFLUENCES ACTING ON MOTONEURONS probably do so via two pools of interneurons, an extensor-positive, flexor-negative pool, and an extensor-negative, flexor-positive pool

   1. Reciprocal inhibition can be overridden voluntarily, as in pathologic rigidity of basal ganglion disease (mediated via pyramidal fibers)

C. AN EXTENSIVE INTRINSIC ORGANIZATION exists in the spinal gray matter, including some topographic representation in the posterior and anterior horns and interneuronal connections that have implications for posture and progressive movement

   1. Most animal progression is associated with rhythmic flexion of the limbs or trunk muscles on one side and then the other with corresponding contralateral extension

   2. In quadrupeds, diagonally opposite limbs assume similar positions (in crossed reciprocal innervation) for standing and moving

   3. Withdrawal of support of one limb (flexor response) implies transfer of support to another limb, as in the crossed extensor component of response

D. IN THE STEPPING RESPONSE, the flexion response or extensor-thrust response gives way to alternating extension and flexion on the two sides and appears to be independent of suprasegmental influence, implying the existence of successively higher orders of interneurons as well as a switching mechanism. Thus, there may be some circuitry in the spinal gray capable of alternating the expected responses instead of merging or suppressing them

POLYSYNAPTIC REFLEX PATHWAYS

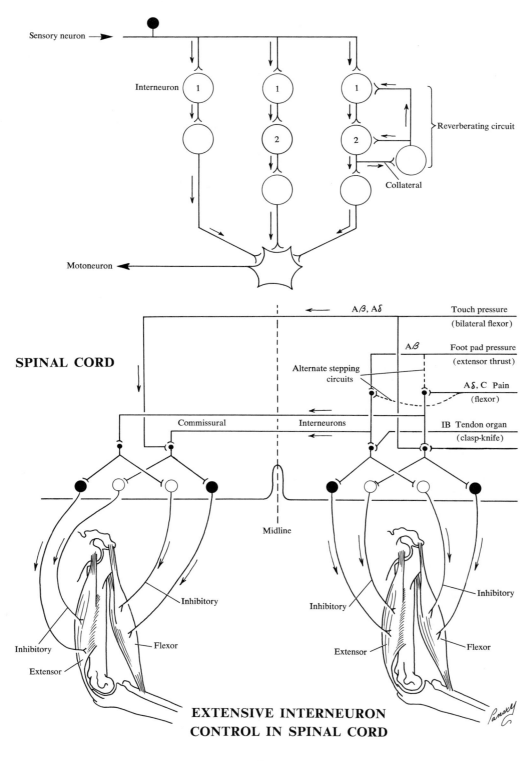

Sensory neuron →

Interneuron

1   1   1

Reverberating circuit

2   2

Collateral

Motoneuron ←

SPINAL CORD

A/β, Aδ — Touch pressure (bilateral flexor)

Aβ — Foot pad pressure (extensor thrust)

Alternate stepping circuits

Aδ, C Pain (flexor)

Commissural   Interneurons   IB Tendon organ (clasp-knife)

Midline

Inhibitory

Inhibitory

Flexor

Inhibitory

Flexor

Inhibitory

Extensor

Extensor

**EXTENSIVE INTERNEURON
CONTROL IN SPINAL CORD**

-215-

99. MUSCLE REGULATION: POSTURAL TONUS

I. **Origin of postural tonus:** muscles isolated from suprasegmental influences have minimal tonus but respond to segmental sensory input. Both extensor and flexor tonus are low. Animals do not normally collapse under gravity nor appear consciously preoccupied with posture maintenance. There is a continuous discharging suprasegmental center responsible for maintaining unbalanced muscle tone

A. IN QUADRIPEDS: the antigravity or postural muscles are the extensors in all limbs. In primates, postural muscles are extensors of the lower limbs and flexors of upper limbs. Arms are in semiflexed position for dealing with environment

B. SUGGESTION OF A CENTER(S) IN LOWER BRAINSTEM FOR POSTURAL TONUS whose activity can be modulated by descending influences and volitional movement

   1. If the brainstem is isolated from the cord (spinal animal), all muscles are flaccid; section of midpons or midmesencephalon results in continuous rigidity of postural muscles and animals are unable to initiate voluntary movement. Section at higher levels (upper midbrain or above hypothalamus) results in exaggerated postural tone, but these animals can perform purposeful movements (walk, run, chew, etc.) and can override the extensor rigidity

   2. Stimulation of reticular formation in medulla, pons, midbrain, and subthalamus facilitates extensor tonus; stimulation of medial reticular formation of medulla inhibits it. The latter requires activation by more rostral centers; the lateral is autonomous and originates continuous (tonic) discharges. Both centers act on motoneurons via the reticulospinal tract (probably on gamma fibers rather than alpha motoneurons)

     a. Reciprocal inhibition exists. Action is exerted indirectly via interneurons

     b. The lateral division of the vestibular nucleus and vestibulospinal tract also strongly facilitates postural tonus autonomously

II. **Modification of postural tonus for voluntary movement** by impulses in the proprioceptors of the neck muscles or in the labyrinth (reflects head position with respect to gravity and body position with respect to the head). Many pathways with respect to postural reflexes are still unknown

III. **Descending influences on postural tonus**

A. COEPS: cortically originating fibers of the extrapyramidal system project directly to the medial reticular area of the medulla and are inhibitory to postural tonus

B. MIDLINE (VERMAL) REGION OF CEREBELLUM (ESPECIALLY ANTERIOR LOBE) project via the fastigiobulbar and other tracts to the postmedullary reticular formation. The vermis is inhibitory to postural tonus. Projections also pass through the lateral ventrolateral thalamic nucleus to cortical areas 4 and 6

C. INTERMEDIATE AND LATERAL ZONES OF CEREBELLUM project via the red nucleus and thalamic nuclei to areas 4 and 6 of cortex. Appear to facilitate pyramidal (not extrapyramidal) fibers and cause generalized increased muscle tone

   1. There may also be a direct reduction of flexor tonus by decreased stimulation of large-celled red nucleus, thus, a reduction of rubrospinal tract activity

D. PYRAMIDAL TRACT: probably facilitatory to flexor and extensor stretch reflexes. Lack of this facilitation is more obvious in antigravity (postural) muscles

   1. It is of a monosynaptic nature with connections to motoneurons and may have combined or separate facilitatory effects on opposing muscle groups

   2. Unclear whether it is autonomous (intrinsic to cortex) or needs driving by the lateral and intermediate region of cerebellum

E. POSTURAL TONUS: maintained by the lateral reticular formation and lateral vestibular nucleus with variable effects originating in lateral and intermediate cerebellum and perhaps pyramidal tracts. It is inhibited by the medial reticular area of the medulla, but this is not autonomous and needs driving by midline cerebellar and COEPS fibers

POSTURAL TONUS

Lateral
Medial } Reticular formations

Midbrain

To medial
reticular formation

Autonomous discharge } Pons

Lateral
vestibular
nucleus } Medulla

Reticulospinal tract

Vestibulospinal tract

Inhibitory to postural tonus

Facilitatory to postural tonus

+ −   − +

Antagonists

Postural muscles

COEPS
(Cortical fibers of
extrapyramidal system)

+ Pyramidal cortex

− + Thalamus

Vermis

Cerebellum

Red nucleus

Intermediate

Lateral

Inhibitory to postural tonus

Pyramidal tract (facilitates
stretch reflex)

Lateral
Medial } Reticular formations

Lateral vestibular nucleus

**DESCENDING
INFLUENCES ON
BRAINSTEM POSTURAL
CENTERS**

−   +

Decrease postural tonus   Increase postural tonus

100. SPINAL CORD LESIONS—PART I

I. Tabes dorsalis (locomotor ataxia)
A. A RELATIVELY UNCOMMON SYNDROME today caused by syphilitic infection of the dorsal root ganglia and dorsal roots with secondary degenerative changes in the dorsal columns, particularly in the fasciculus gracilis bilaterally in the lower segments of the spinal cord
B. WHEN VIEWED GROSSLY the spinal cord, dorsal root ganglia, and dorsal roots commonly appear smaller and atrophic
C. MICROSCOPIC STUDY reveals the following changes
1. Number of normal nerve fibers in dorsal nerve roots (especially lumbrosacral) is decreased
2. Cellular infiltration of dorsal root ganglia
3. Progressive dorsal column degeneration
D. CLINICAL SIGNS AND SYMPTOMS
1. Paresthesia, hypesthesia, and intermittent attacks of sharp pain, which commonly occur early in neurosyphilis, are caused by irritation of dorsal nerve root fibers
2. Sensitivity to pain is diminished usually in the course of the disease (particularly about the Achilles tendon and in the testes)
3. Deep tendon reflexes (knee and ankle jerk) are diminished or absent because of interference with stretch reflex arc in the dorsal roots.
4. Muscle, joint, and vibratory sense impaired
5. Positive Romberg sign, hypotonic skeletal muscles, and ataxic gait
6. Patient walks with the legs apart, head bent forward, eyes look down (to try to compensate for the loss of kinesthetic sense), knees raised excessively high, and feet slap down on the ground

II. Syringomyelia
A. SYRINGOMEYELIA IS CHARACTERIZED by softening and cavitation around the central canal of the cervical enlargement of the spinal cord. Gliosis and cavitation may extend into other regions as the disease progresses
B. EARLY CLINICAL SIGNS reveal a loss of pain and temperature sensibility with segmental distribution in the upper limbs on both sides. This is due to the interruption of lateral spinothalamic fibers as they decussate ventral to the central canal
C. NO SENSORY IMPAIRMENT in lower limbs and body (spinothalamic tract intact)
D. SIMPLE TOUCH AND PRESSURE SENSES are preserved in the affected dermatomes of the upper limbs (sensory dissociation or dissociate anesthesia)
E. LATE MANIFESTATIONS of this neurologic disorder include
1. Lower motor neuron disturbances with atrophy of muscles when degeneration extends into the anterior gray horns
2. Pyramidal tract disturbances in lower limbs (cystic cavity compression)

III. Spinal cord transection (paraplegia)
A. THE FOLLOWING SIGNS AND SYMPTOMS appear immediately following complete transection of the spinal cord
1. Voluntary muscular activity and somatic and visceral reflex activity in the body below the lesion are absent
2. Sensibility from the body below the lesion is completely absent
3. Spinal shock (depressed motor and sensory activity) lasts 2 to 3 weeks
B. WITH TIME THE DISTAL PORTION OF THE TRANSECTION SPINAL CORD begins to function independently (autonomously) in the following sequence
1. Minimal reflex activity
2. Superficial flexor reflex activity noted
3. Alternation between flexor and extensor spasm activity
4. Deep extensor reflex activity predominates

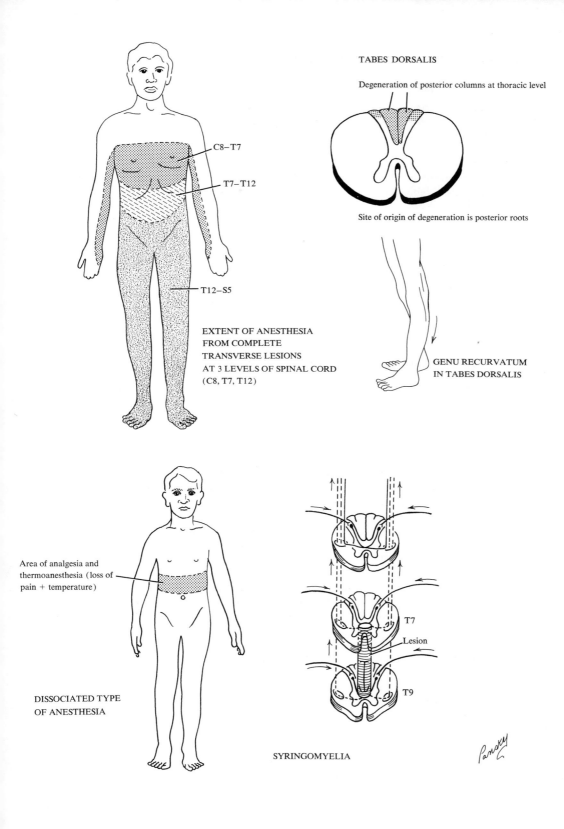

TABES DORSALIS

Degeneration of posterior columns at thoracic level

Site of origin of degeneration is posterior roots

GENU RECURVATUM
IN TABES DORSALIS

C8–T7

T7–T12

T12–S5

EXTENT OF ANESTHESIA
FROM COMPLETE
TRANSVERSE LESIONS
AT 3 LEVELS OF SPINAL CORD
(C8, T7, T12)

Area of analgesia and
thermoanesthesia (loss of
pain + temperature)

DISSOCIATED TYPE
OF ANESTHESIA

T7

Lesion

T9

SYRINGOMYELIA

101. SPINAL CORD LESIONS—PART II

I. Lesions of the anterior gray horns and ventral nerve roots (poliomyelitis)

A. ANTERIOR HORN LESIONS interrupt specific alpha and gamma motor neurons. Injury to these neurons results in either lower motor neuron paralysis or paresis (partial paralysis) of the muscle or muscle group innervated

B. LOWER MOTOR NEURON PARALYSIS is diagnosed by recognizing the following
   1. Voluntary muscle movements are completely absent and reflex activity cannot be elicited when all anterior horn cells and/or ventral root fibers to a particular muscle are involved
   2. Paralyzed muscles are flaccid (atonic) and deep tendon reflexes are absent
   3. Muscle fibrillations (single fiber contractions) and fasciculations (muscle twitching) may be recorded or observed respectively
   4. Muscle atrophy becomes apparent and skin becomes dry and cyanotic

C. Acute poliomyelitis or infantile paralysis is caused by a virus that damages anterior horn cells in the cervical and lumbar enlargements
   1. The extent of muscle paralysis depends on the number and distribution of lower motor neurons damaged beyond recovery with time
   2. Muscle atrophy begins several weeks after paralysis

II. Amyotrophic lateral sclerosis, a fatal disease of unknown origin

A. PATHOLOGICALLY, it is characterized by
   1. Degeneration of motor cells in the spinal cord (lower motor neurons), brainstem, and cerebral cortex (upper motor neurons)
   2. Secondary degeneration of axons in both the peripheral nerves and in the pyramidal tracts bilaterally

B. CLINICAL SIGNS AND SYMPTOMS associated with this disease include
   1. Characteristics of both upper and lower motor neuron lesions
      a. Most paralyzed muscles show signs of lower motor neuron degeneration which include paralysis, atrophy, fasciculations, and muscle weakness
      b. Some paralyzed muscles show signs of upper motor neuron degeneration which include spasticity, hyperreflexia, and possibly Babinski's sign
   2. This disease is characterized initially by weakness, atrophy, and fasciculations of muscles of the hand and arm, followed by spastic paralysis of legs

C. SENSORY CHANGES are usually not observed with amyotrophic lateral sclerosis

III. Spinal cord hemisection (Brown-Sequard syndrome)

A. A UNILATERAL TRANSVERSE LESION (HEMISECTION) of the spinal cord (resulting from a bullet or knife wound, spinal cord tumor, syringomyelia, etc.) will produce the Brown-Sequard syndrome

B. CLINICAL SIGNS AND SYMPTOMS associated with this syndrome include
   1. Upper motor neuron paralysis on the same side below the level of the lesion
   2. Lower motor neuron paralysis on the same side at the level of the lesion
   3. Cutaneous anesthesia on the same side at the segmental level of the lesion
   4. Hyperesthesia on the same side below the level of the lesion and at the segmental level of the lesion on the side opposite the lesion
   5. Loss of two-point discrimination, proprioception, and vibratory sensibility below the level of the lesion of the same side of the body
   6. Loss of pain and temperature below the level of the lesion on the side opposite the lesion

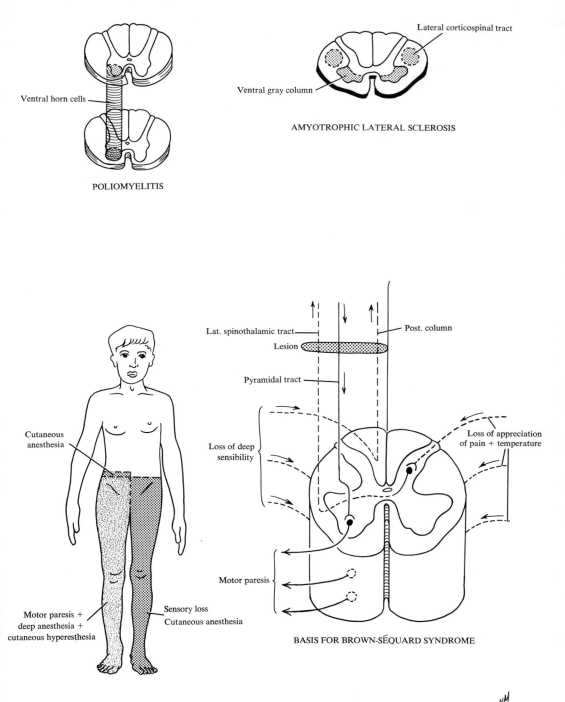

Ventral horn cells

POLIOMYELITIS

Lateral corticospinal tract

Ventral gray column

AMYOTROPHIC LATERAL SCLEROSIS

Lat. spinothalamic tract

Lesion

Pyramidal tract

Post. column

Cutaneous anesthesia

Loss of deep sensibility

Loss of appreciation of pain + temperature

Motor paresis

Motor paresis + deep anesthesia + cutaneous hyperesthesia

Sensory loss
Cutaneous anesthesia

BASIS FOR BROWN-SÉQUARD SYNDROME

LESION OF RT. HALF OF CORD (ABOUT T10)

Pansky

Unit Five

CRANIAL NERVES

102. OLFACTORY (I) AND OPTIC (II) NERVES

I. Olfactory nerve
A. FUNCTIONAL COMPONENTS: special visceral afferent (SVA)
   1. Receptors: neuroepithelial cells of the olfactory mucosa
   2. Location: nasal mucous membrane
   3. Distribution: upper $\frac{1}{3}$ of nasal cavity (septum and superior conchae)
   4. Cell bodies: in the nasal mucous membrane
   5. Termination: olfactory bulb
B. Symptoms of disease: anosmia or the loss of the sense of smell

II. Optic nerve
A. FUNCTIONAL COMPONENTS: special somatic afferent (SSA): refer to the organs of special sensation; e.g., visual, auditory, vestibular, etc.
   1. Review the anatomy of the eye (see Unit One)
   2. The gross and functional aspects of the visual pathways
      a. The primary neurons of the optic system are the receptor cells of the retina, the rods, and the cones
      b. The secondary neurons are the bipolar cells of the retina
      c. The tertiary neurons are the ganglionic cells of the retina and the axons of these neurons which form the optic nerve
         i. The axons run over the inner surface of the retina, in contact with the vitreous humor, and converge toward the optic papilla or optic disk
      d. The optic nerve passes through the orbit via the optic foramen, into the cranial cavity, and the 2 nerves converge to form the optic chiasma. The fibers from the nasal retina (temporal field of vision) cross in the chiasma and join the uncrossed fibers from the temporal retina (nasal field of vision) to form the optic tracts
      e. The optic tracts course posteriorly and end at the lateral geniculate bodies
         i. Most of the optic fibers terminate by synapsing with neurons in the lateral geniculate bodies
         ii. Some fibers bypass the bodies and course through the brachium of the superior colliculus or in the pretectal region to synapse in the colliculus and pretectal region, thus forming reflex pathways
      f. Axons of cells in the geniculate bodies project through the sublenticular and retrolenticular parts of the internal capsule to enter the geniculocalcarine tract (optic radiations) and terminate in the visual cortex around the calcarine fissure
      g. The fibers synapsing in the superior colliculus and the pretectal region are primarily concerned with visual reflexes
         i. Fibers from the superior colliculus enter the tectobulbar and tectospinal tracts, pass through the dorsal tegmental decussation, and terminate in the various motor nuclei of the cranial nerves and in the anterior gray column of the cervical segments of the spinal cord
         ii. Axons of the neurons in the pretectal region pass to the Edinger-Westphal nuclei (parasympathetic portion of the oculomotor nucleus) to establish a reflex that will constrict the pupil in response to strong light (loss of the light reflex without loss of vision indicates some pathologic process in the pretectal region)

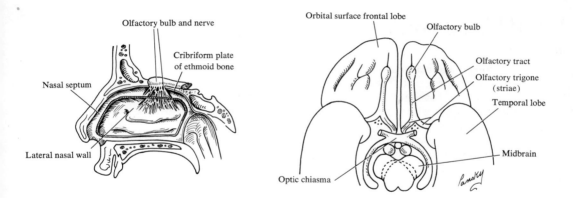

Olfactory bulb and nerve

Cribriform plate of ethmoid bone

Nasal septum

Lateral nasal wall

Orbital surface frontal lobe

Olfactory bulb

Olfactory tract

Olfactory trigone (striae)

Temporal lobe

Midbrain

Optic chiasma

OLFACTORY NERVE (I)

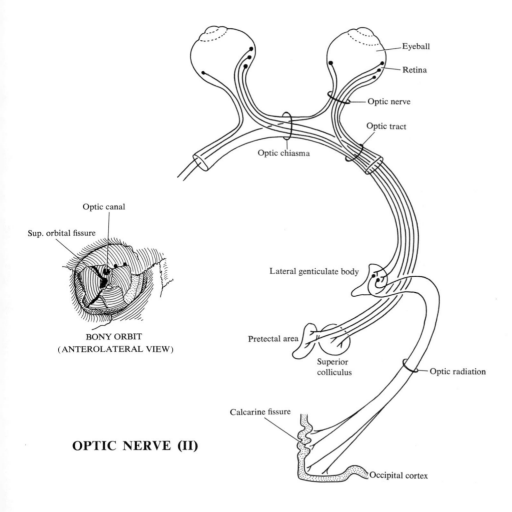

Eyeball

Retina

Optic nerve

Optic tract

Optic chiasma

Optic canal

Sup. orbital fissure

Lateral geniculate body

BONY ORBIT (ANTEROLATERAL VIEW)

Pretectal area

Superior colliculus

Optic radiation

Calcarine fissure

OPTIC NERVE (II)

Occipital cortex

103. OCULOMOTOR (III) NERVE

I. **Functional components** contain 2 functionally distinct types of motor fibers

A. GENERAL SOMATIC EFFERENT (GSE): to all the extraocular muscles except the lateral rectus and superior oblique

B. GENERAL VISCERAL EFFERENT (GVE): parasympathetic preganglionic fibers to the ciliary and episcleral ganglia
   1. Postganglionic fibers terminate in the sphincter pupillae and ciliary muscles and function to constrict the pupil by causing contraction of the sphincter muscle of the iris and relaxation of the suspensory ligament of the lens by contraction of the ciliary muscle

II. **Oculomotor nuclear complex** is located in the mesencephalon at the level of the superior colliculus, near the midline, ventral to the aqueduct, in a V-shaped trough formed by the medial longitudinal fasciculus (MLF)

A. THE NUCLEUS can be divided into
   1. A paired lateral nuclear group
   2. An unpaired paramedian caudal central nucleus
      a. Both 1 and 2 give rise to SE fibers, and both contain typical large to medium-sized somatic motor neurons
   3. The Edinger-Westphal nucleus, which gives rise to GVE fibers and is located dorsomedial to the rostral ⅔ of the lateral nuclei
      a. Consists of small neurons similar to those seen in the dorsal motor nucleus of the vagus nerve

B. SOME FIBER CONNECTIONS of the oculomotor complex
   1. The somatic efferent nuclei in the complex receive corticobulbar fibers, probably via relay by intercalated neurons in the reticular formation; fibers from the MLF, most of which originate in the vestibular nuclei and reflexly correlate eye position with that of the head (some fibers in the MLF, however, are intranuclear and associate the abducens and oculomotor nuclei for conjugate eye movement); fibers from the superior colliculus, probably via the intercalated neurons, placing eye movements under control of the optic reflex centers in the colliculus and thus under control of eye movement centers in the cerebrum that project to the colliculus; and fibers from the reticular formation
   2. The Edinger-Westphal nucleus receives
      a. Fibers from the pretectum that participate in the pupillary light reflex
      b. The visceral efferent components of the oculomotor nucleus also possess a convergence center responsible for the accommodation-convergence reaction
         i. This center receives fibers from the visual cortex directly or via the frontal eye centers

III. **The internal course of the oculomotor root fibers**

A. THE AXONS COURSE VENTRALLY from the oculomotor nucleus in bundles that pass medial to, lateral to, and through the red nucleus to emerge in the interpeduncular fossa medial to the basis pedunculi

IV. **Intracranial course of the oculomotor nerve**

A. THE NERVE COURSES FORWARD between the posterior cerebral and superior cerebellar arteries before piercing the dura and entering the cavernous sinus

B. IN THE SINUS, THE NERVE LIES SUPERIOR to cranial nerves IV and VI, and bends downward and laterally to enter the orbit through the medial aspect of the superior orbital fissure

V. **Symptoms of disease**

A. GSE: eye looks downward and outward with no convergence

B. GVE: pupils do not constrict to light or for accommodation

E.W. nucleus

IIIrd nucleus

DORSAL VIEW

IIIrd nerve
IV
V
VI
VII
VIII

VENTRAL VIEW

Int. carotid a.
III
IV
V
VI
V
Cavernous sinus

OCULOMOTOR DISTRIBUTION

Superior rectus m.

Ciliary ganglion

Levator palpebrae
superioris m.

Ciliary m

IIIrd nucleus

IIIrd nerve

Sphincter pupillae mm.

Inf. oblique m.

Inf. rectus m.

Short ciliary nn.

Frontal
IV
Lacrimal
Sup. division III
Inf. division III

Optic nerve

Ophthalmic artery

Nasociliary
VI
Superior orbital fissure

Pansky

Ant. med. nucleus (Perlia)

Lateral nucleus

Edinger-Westphal nucleus to
sphincter pupillae and ciliary mm.

Lev. palp. sup.
Sup. rect.

**III—NERVE
COMPLEX**

Med. rect.
Inf. obliq.
Inf. rect.

Oculomotor n. (III)

Caudal central nucleus
(Perlia)

Lateral nucleus

To inf. rectus

IV

Ocular convergence center
+ inf. rectus

To sup. oblique

104. TROCHLEAR (IV) AND ABDUCENS (VI) NERVES

I. Trochlear (IV) nerve
A. FUNCTIONAL COMPONENTS
   1. Somatic efferent fibers (GSE) fibers supply the superior oblique eye muscle
B. TROCHLEAR NUCLEUS is found in the mesencephalon at the level of the inferior colliculus in the ventromedian portion of the central gray
   1. The nucleus contains typical somatic motor neurons and appears to indent the MLF
C. INTERNAL COURSE OF THE TROCHLEAR ROOT FIBERS
   1. The trochlear nerve is the only cranial nerve that exits from the dorsal aspect of the brainstem
   2. Axons of motor neurons in the trochlear nucleus sweep dorsolaterally and caudally, in the outer margin of the central gray, toward the dorsal surface of the brainstem where they decussate completely
   3. The nerves exit the brainstem at the dorsal junction of the mesencephalon and pons, just lateral to the anterior medullary velum
D. INTRACRANIAL COURSE OF THE TROCHLEAR NERVE
   1. The thin IVth nerve swings around the lateral surface of the brainstem, passes between the posterior cerebral and superior cerebellar arteries, pierces the dura in the anterior attaching fold of the tentorium cerebelli
   2. The nerve then enters the cavernous sinus inferior to the IIIrd nerve and enters the orbit through the superior orbital fissure
E. SYMPTOMS OF DISEASE: eye tends to look slightly upward

II. Abducens (VI) nerve
A. FUNCTIONAL COMPONENTS
   1. Somatic efferent fibers (GSE): fibers supply the lateral rectus muscle of eyeball
B. ABDUCENS NUCLEUS is located in the lateral part of the median eminence in the floor of the 4th ventricle
   1. Fibers of the facial nerve loop (internal genu of the facial nerve) around the nucleus
   2. The facial colliculus marks the location of the genu of the facial nerve and the VIth nucleus in the floor of the 4th ventricle
C. CONNECTIONS OF THE ABDUCENS NUCLEUS
   1. Bilateral corticobulbar fibers via intercalated neurons
   2. Vestibular fibers via the MLF that may be concerned with control of eye movements
      a. Fibers in the MLF also serve to interconnect the abducens and oculomotor nuclei
D. INTERNAL COURSE OF THE ABDUCENS ROOT FIBERS
   1. Root fibers emerge from the medial aspect of the abducens nucleus and pass ventrally and somewhat caudally to exit from the brainstem at the caudal border of the pons
      a. In passing to its exit point, the abducens root fibers pass just lateral to the corticospinal tract
E. INTRACRANIAL COURSE OF THE ABDUCENS NERVE
   1. The nerve passes forward to pierce the dura and enters the lateral aspect of the cavernous sinus, lying inferior to the trochlear nerve
   2. The abducens nerve enters the orbit via the superior orbital fissure
F. SYMPTOMS OF DISEASE: the eye turns inward

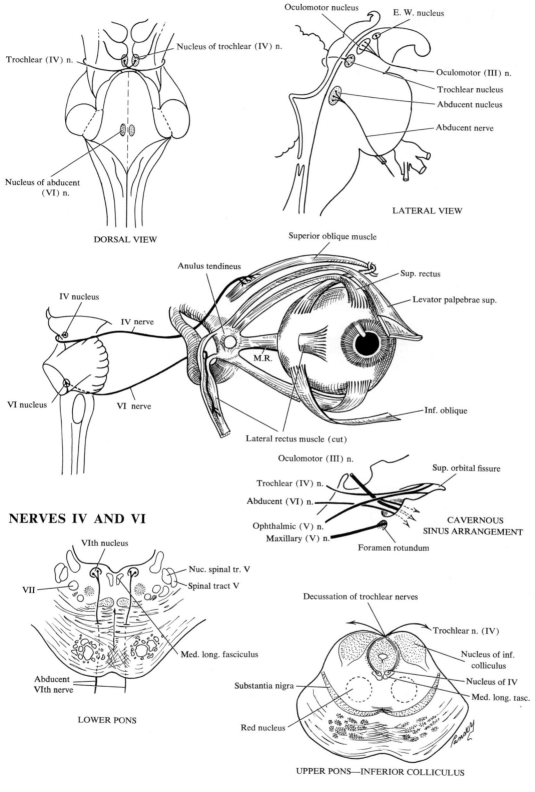

Nucleus of trochlear (IV) n.

Trochlear (IV) n.

Nucleus of abducent (VI) n.

DORSAL VIEW

Oculomotor nucleus

E. W. nucleus

Oculomotor (III) n.

Trochlear nucleus

Abducent nucleus

Abducent nerve

LATERAL VIEW

Superior oblique muscle

Anulus tendineus

Sup. rectus

Levator palpebrae sup.

IV nucleus

IV nerve

M.R.

VI nucleus

VI nerve

Inf. oblique

Lateral rectus muscle (cut)

NERVES IV AND VI

Oculomotor (III) n.

Sup. orbital fissure

Trochlear (IV) n.

Abducent (VI) n.

Ophthalmic (V) n.

Maxillary (V) n.

Foramen rotundum

**CAVERNOUS
SINUS ARRANGEMENT**

VIth nucleus

Nuc. spinal tr. V

Spinal tract V

VII

Med. long. fasciculus

Abducent
VIth nerve

LOWER PONS

Decussation of trochlear nerves

Trochlear n. (IV)

Nucleus of inf. colliculus

Substantia nigra

Nucleus of IV

Med. long. fasc.

Red nucleus

UPPER PONS—INFERIOR COLLICULUS

105. TRIGEMINAL (V) NERVE—PART I

I. Functional components: contains both sensory and motor components
A. GENERAL SOMATIC AFFERENT (GSA) FIBERS convey the impulses of touch, pain, and temperature from the head and face, from the teeth and mucous membrane of the oral cavity, and from the mucous membrane of the nasal cavity and sinuses as well as from the dura mater (portio major)
B. GENERAL PROPRIOCEPTION (GP) FIBERS: fibers of deep pressure and kinesthetic sense
   1. Convey impulses from the teeth, temporomandibular joint, hard palate, and from stretch receptors in the muscles of mastication
C. SENSORY PORTION OF THE TRIGEMINAL NERVE
   1. The somatic sensory column forms only the sensory nucleus of the Vth nerve and is a long column of gray matter in the dorsolateral areas extending from the upper midbrain to the upper cervical segments of the cord
      a. The sensory nucleus is often referred to as a series of contiguous nuclei (nucleus of the mesencephalic tract of the Vth nerve, principal sensory nucleus of the Vth nerve, and the nucleus of the spinal Vth tract)
      b. The cells of the nucleus receive, mainly via the Vth nerve, somatic sensory impulses from the head, except for the scalp in back of the head which is related to the 2nd and 3rd cervical dermatomes
         i. Some cutaneous branches of the VIIth, IXth, and Xth cranial nerves also send pain and temperature fibers to the spinal sensory nucleus of the Vth nerve
      c. Neurons that send afferent fibers to the trigeminal sensory nucleus have their cell bodies in the semilunar (gasserian) ganglion outside the CNS
         i. Proprioceptive fibers from the face, however, have their cell bodies within the mesencephalic nucleus of the Vth nerve
D. SPECIAL VISCERAL EFFERENT (SVE) FIBERS: motor portion of the trigeminal nerve
   1. Motor impulses pass to the muscles of mastication, the anterior belly of the digastric muscle, the tensor tympani muscle, the tensor veli palatini muscle, and the mylohyoid muscle
   2. Because the muscles of mastication are of branchial arch origin, the motor component is classified as SVE and the fibers are called branchiomotor

II. Trigeminal nuclear complex consists of 4 named nuclei (3 sensory, 1 motor)
A. THE 3 SENSORY NUCLEI (see above for details): the spinal nucleus, the main sensory nucleus, and the mesencephalic nucleus
B. THE MOTOR NUCLEUS: composed of typical multipolar motor neurons and located just medial to the incoming trigeminal root fibers and the main sensory nucleus
   1. Some connections of the motor nucleus
      a. The motor neurons of the Vth nerve are acted on by fibers from higher centers, as are most motor neurons; the most abundant of these being the corticobulbar fibers, many of which terminate on reticular neurons which in turn project to the motor neurons
      b. Bilateral secondary trigeminal fibers, sensory fibers from other cranial nerves, and fibers from the mesencephalic nucleus of the Vth nerve apparently activate the motor nucleus to provide reflex control of jaw muscles (especially to superficial stimuli from the tongue and oral cavity)

Secondary V pathways

Mesencephalic nucleus of V (proprioception)

Principal sensory nucleus of V

Sensory root of trigeminal n. (V)

Optic n.

Ant. clinoid process

Ophthalmic n.

Maxillary n.

III

IV

Trigeminal (gasserian) ganglion

VI

V

Mandibular n.

VII, VIII

IX, X, XI

Motor nucleus of V (masticatory nucleus)

Nucleus oralis

Nucleus interpolaris

Nucleus caudalis

Nucleus of spinal Vth tract

PASSAGE THROUGH
SUBARACHNOID SPACE
AND DURA
(MIDDLE CRANIAL FOSSA)

NUCLEI OF TRIGEMINAL NERVE (V)

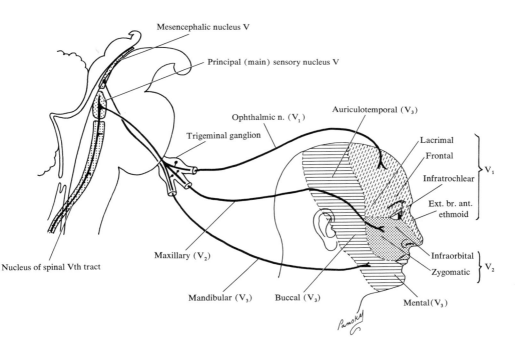

Mesencephalic nucleus V

Principal (main) sensory nucleus V

Ophthalmic n. (V_1)

Auriculotemporal (V_3)

Trigeminal ganglion

Lacrimal

Frontal

Infratrochlear

Ext. br. ant. ethmoid

V_1

Maxillary (V_2)

Infraorbital

Zygomatic

V_2

Nucleus of spinal Vth tract

Mandibular (V_3)

Buccal (V_3)

Mental (V_3)

Pansky

CUTANEOUS SENSORY AREAS

106. TRIGEMINAL (V) NERVE—PART II

I. Internal course of trigeminal root fibers
A. COURSE OF THE GSA COMPONENT: see below (portio major)
B. AXONS OF NEURONS IN THE MOTOR NUCLEUS pass on the medial side of the incoming sensory fibers, pass through the lateral side of the pontine tegmentum and the middle cerebellar peduncle at a midpontine level, and emerge as the portio minor, internal to the incoming sensory fibers

II. Intracranial course of the trigeminal nerve
A. THE PONS lies in the most anterior part of the posterior cranial fossa and only the Vth nerve attaches to it. This leaves laterally and has 2 roots: a very large sensory root and a small motor root. The two roots run anterolaterally from the posterior into the middle cranial fossa across a notch on the petrous bone which houses the trigeminal ganglion in the middle fossa
  1. The sensory nerves enter the ganglion via the ophthalmic division from the superior orbital fissure and the maxillary division via the foramen rotundum
  2. The motor root passes beneath the semilunar ganglion to join the mandibular division and exits the cranium through the foramen ovale where it is distributed to the face via its inferior alveolar and lingual branches
  3. The GSA fibers have their cell bodies in the ganglion and terminate in the spinal V nucleus (pain and temperature) and in the chief sensory nucleus (touch)

III. Symptoms of disease
A. GSA: anesthesia in the area of distribution
B. SVE: paralysis of the jaws as related to the muscles of mastication

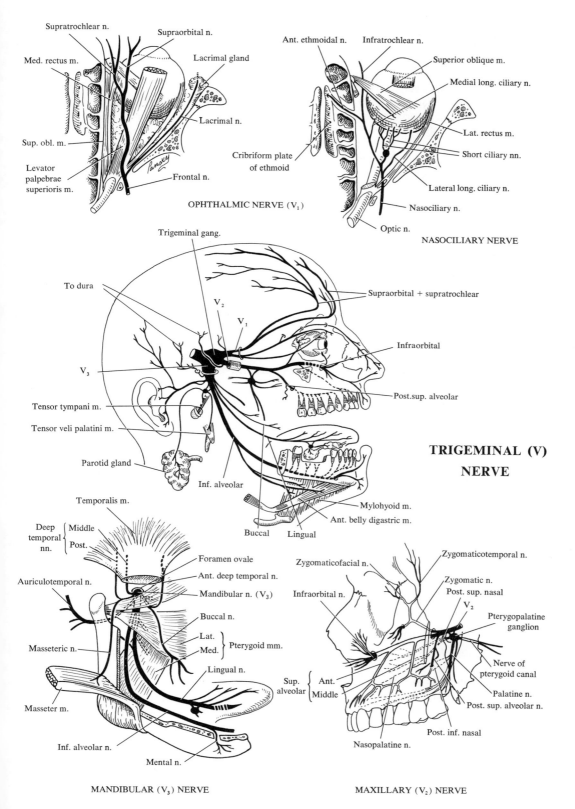

Supratrochlear n.
Supraorbital n.
Med. rectus m.
Lacrimal gland
Sup. obl. m.
Lacrimal n.
Levator
palpebrae
superioris m.
Frontal n.

OPHTHALMIC NERVE (V₁)

Ant. ethmoidal n.
Infratrochlear n.
Superior oblique m.
Medial long. ciliary n.
Lat. rectus m.
Short ciliary nn.
Cribriform plate
of ethmoid
Lateral long. ciliary n.
Nasociliary n.
Optic n.

NASOCILIARY NERVE

Trigeminal gang.
To dura
V₂
V₁
Supraorbital + supratrochlear
Infraorbital
V₃
Post.sup. alveolar
Tensor tympani m.
Tensor veli palatini m.
Parotid gland
Inf. alveolar
Mylohyoid m.
Ant. belly digastric m.
Buccal
Lingual

**TRIGEMINAL (V)
NERVE**

Deep
temporal
nn.
Temporalis m.
Middle
Post.
Auriculotemporal n.
Foramen ovale
Ant. deep temporal n.
Mandibular n. (V₃)
Buccal n.
Lat.
Med.
Pterygoid mm.
Masseteric n.
Lingual n.
Masseter m.
Inf. alveolar n.
Mental n.

MANDIBULAR (V₃) NERVE

Zygomaticofacial n.
Zygomaticotemporal n.
Infraorbital n.
Zygomatic n.
Post. sup. nasal
V₂
Pterygopalatine
ganglion
Nerve of
pterygoid canal
Sup.
alveolar
Ant.
Middle
Palatine n.
Post. sup. alveolar n.
Post. inf. nasal
Nasopalatine n.

MAXILLARY (V₂) NERVE

107. FACIAL (VII) NERVE—PART I

I. Functional components

A. GENERAL SOMATIC AFFERENT (GSA) FIBERS arise from neurons in the geniculate ganglion and innervate the external auditory meatus and skin behind the ear

1. Centrally, these sensory fibers probably synapse in the trigeminal sensory nuclei and are conveyed to the thalamus with secondary trigeminal fibers before they are relayed to the somesthetic cortex

B. SPECIAL VISCERAL AFFERENT (SVA) FIBERS have their cell bodies in the geniculate ganglion and convey taste from the anterior ⅔ of tongue via the chorda tympani

1. These fibers form a major portion of the nervus intermedius
2. Centrally, taste fibers enter the tractus solitarius to synapse in the upper portion of the nucleus of the tractus solitarius
    a. The central course of secondary gustatory fibers is not well known, but they are somehow transmitted to the thalamus and then to the cortical taste area located close to the area for somatic sensibility of the tongue
    b. Some gustatory secondary fibers also reach the hypothalamus and presumably influence autonomic function
    c. Taste fibers of the IXth and Xth cranial nerves are also distributed to the nucleus of the solitary tract

C. GENERAL VISCERAL AFFERENT (GVA) FIBERS have their cell bodies in the geniculate ganglion and convey impulses from the deep tissues of the face. Centrally, the fibers terminate in the nucleus of the tractus solitarius

D. GENERAL PROPRIOCEPTION (GP)

1. Receptors lie in the muscles of facial expression
2. Cell bodies are in the mesencephalic nucleus of the trigeminal
3. Termination is in the reticular formation of the brain stem by way of the mesencephalic nucleus

E. GENERAL VISCERAL EFFERENT (GVE) FIBERS

1. Supply parasympathetic preganglionic impulses to the submandibular and pterygopalatine ganglia. Postganglionic fibers from the ganglia innervate the submandibular, sublingual, and lacrimal glands along with glands in the mucous membranes of the nasal and oral cavities

F. SPECIAL VISCERAL EFFERENT (SVE) FIBERS: branchiomotor fibers

1. Innervate the superficial skeletal muscles of the face and scalp (muscles of facial expression), the platysma, the stylohyoid, the stapedius, and the posterior belly of the digastric muscles

II. The facial nuclei

A. THE SENSORY COMPONENTS OF THE VIITH NERVE utilize the trigeminal sensory nuclei (GSA, GP) centrally and the nucleus of the solitary tract (SVA)

B. THE SVE FIBERS originate in the facial motor nucleus located in the caudal pons just medial to the spinal nucleus of the Vth nerve. The facial nucleus is composed of several distinct cellular groups, but it is sufficient to know that

1. The dorsal portion of the nucleus supplies motor innervation to muscles of the upper face and forehead
2. The ventral portion supplies the lower facial muscles

C. THE GVE FIBERS originate in the visceral motor nucleus called the superior salivatory nucleus, which is difficult to discern, but is apparently continuous caudally with the inferior salivatory nucleus of the IXth cranial nerve

II

III

IV

Sensory V

VII

Nervus intermedius

VIII

VI

IX

X

XI

XII

BRAINSTEM (VENTRAL SURFACE) CRANIAL NERVES

Nervus intermedius (sensory)

Nucleus solitarius

VI nucleus

VII nucleus

VII nerve

Nervus intermedius (motor)

Superior salivatory nucleus

Genu of VIIth nerve

VIth nucleus

Sup. salivatory nucleus

Sup. gustatory nucleus

Nucleus solitarius

Trigeminal sensory nucleus

Medial longitudinal fasciculus

VIIth nucleus

GSA

GVA

SVA

GSA (to a lower level)

GVE

SVE

VIth nerve

Facial nerve

LOWER PONS AT LEVEL OF NUCLEI VI AND VII

Lat. semicircular canal

Int. acoustic meatus

Geniculate ganglion

Greater petrosal n.

VII n.

Chorda tympani n.

VIIth NERVE IN TEMPORAL BONE

Nervus intermedius

VII

Aperature of aqueductus vestibuli

VIII

Internal acoustic meatus

POSTERIOR PETROUS TEMPORAL BONE

-235-

108. FACIAL (VII) NERVE—PART II

I. **Connections of the facial motor nucleus:** many sources send fibers to the facial motor nucleus

A. SEVERAL IMPORTANT REFLEXES are initiated by optic, auditory, and other stimuli from around the face and mouth (some are clinically useful)

   1. Optic reflexes are conveyed to the 7th motor nucleus via the superior colliculus to mediate such reflexes as blinking or closing the eye in response to bright light
   2. Fibers from the superior olive mediate impulses of acoustic origin and produce reflexes such as blinking in response to a loud sound, or contraction or relaxation of the stapedius muscles in response to varying sound intensities
   3. Strong projections from sensory nuclei of nerve V pass to the facial nucleus
      a. The corneal reflex is an important reflex mediated by these projections
         i. When the cornea is touched, both eyes normally blink
         ii. Corneal reflex tests the integrity of the Vth sensory complex as well as the motor fibers and nuclei of the VIIth nerve
   4. Descending fibers from higher centers also influence the 7th motor nucleus
      a. Direct and indirect (impulses relayed by reticular neurons) corticobulbar fibers from the motor cortex presumably mediate voluntary control of face
         i. Clinical evidence indicates that corticobulbar fibers pass bilaterally to the dorsal part of the facial nucleus, but those to the ventral part of the nucleus arise in the contralateral motor cortex only; thus, unilateral damage to the corticobulbar fibers above the facial nucleus produces paralysis of the contralateral lower facial muscles only
      b. Other higher centers projecting to the 7th motor nucleus are red nucleus, mesencephalic reticular formation, globus pallidus, and perhaps the thalamus
         i. Many project indirectly to the facial nucleus via the reticular formation
         ii. Projections from the globus pallidus are probably involved in emotional (mimetic) facial movement. The ability of the lower facial muscles to respond to emotional stimuli even in the presence of a central-type facial paralysis (lesion above the 7th motor nucleus) is well documented

II. **Internal course of facial root fibers**

A. THE BRANCHIOMOTOR FIBERS have a complex course to their point of exit: emerge from the dorsal surface of the facial nucleus and pass dorsomedially to approach the floor of the 4th ventricle; the fibers then turn to ascend in a compact bundle and pass medial to the VIth nucleus; at the rostral pole of the latter, the fibers loop dorsolaterally over the nucleus and pass ventrolaterally between the spinal trigeminal complex and the motor nucleus of VII to exit from the brain stem near the pontomedullary junction. The loop around the VIth nucleus is the internal genu of the 7th nerve

III. **Intracranial course of the facial nerve**

A. TOGETHER WITH the nervus intermedius and the 8th cranial nerve (bound in a common meningeal sheath), the facial nerve passes laterally from its point of exit from the brainstem to enter the internal acoustic meatus

B. TO REACH THE FACE, fibers pass through the wall of the ear cavity (giving off the great petrosal nerve, nerve to stapedius muscle, and the chorda tympani) and exit via the stylomastoid foramen just behind the parotid gland and spread on face

VI. **Symptoms of disease**

A. SVE: facial paralysis (Bell's palsy)

B. GVE: poor secretion from gland of termination

C. SVA: ageusia on anterior $\frac{2}{3}$ of tongue

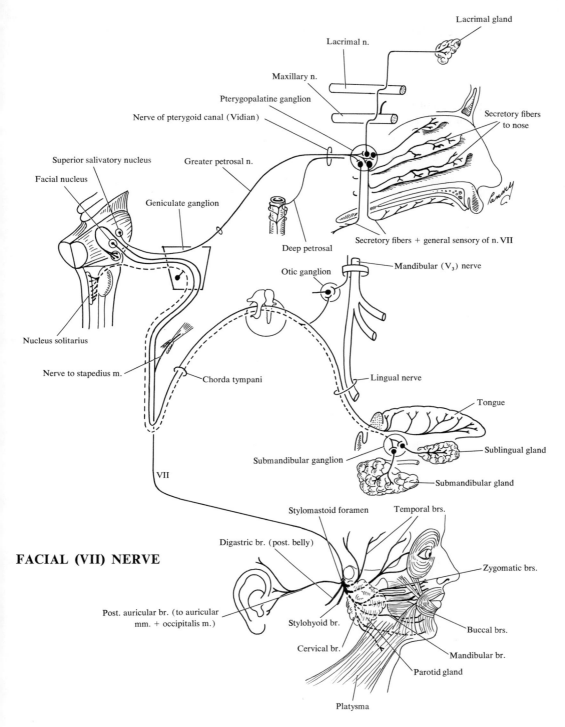

Lacrimal gland

Lacrimal n.

Maxillary n.

Pterygopalatine ganglion

Nerve of pterygoid canal (Vidian)

Secretory fibers to nose

Superior salivatory nucleus

Facial nucleus

Greater petrosal n.

Geniculate ganglion

Secretory fibers + general sensory of n. VII

Deep petrosal

Nucleus solitarius

Otic ganglion

Mandibular (V₃) nerve

Nerve to stapedius m.

Chorda tympani

Lingual nerve

Tongue

Sublingual gland

Submandibular ganglion

Submandibular gland

VII

Stylomastoid foramen

Temporal brs.

Digastric br. (post. belly)

FACIAL (VII) NERVE

Zygomatic brs.

Post. auricular br. (to auricular mm. + occipitalis m.)

Stylohyoid br.

Buccal brs.

Cervical br.

Mandibular br.

Parotid gland

Platysma

MUSCLES OF FACIAL EXPRESSION

109. COCHLEAR (VIII) NERVE: AUDITORY SYSTEM

I. Functional components

A. SPECIAL SOMATIC AFFERENT (SSA) FIBERS
1. Receptors: the hair cells located in the organ of Corti
2. Distribution: cochlear duct of the inner ear
3. Cell bodies: the spiral ganglion
4. Terminations: dorsal and ventral cochlear nuclei

II. Functional auditory system

A. THE AUDITORY PATHWAY TO THE CORTEX consists of at least a 4-neuron chain and, in addition, the auditory pathways are complicated by other small relay nuclei
1. The superior olivary complex and the nucleus of the lateral lemniscus are two such small relay nuclei (omitted here to simplify the initial study)
2. The primary neurons are bipolar cells located in the spiral ganglion within the bony spiral lamina
   a. The distal processes terminate in relation to the hair cells of the organ of Corti
   b. The central processes pass via the auditory portion of the VIIIth cranial nerve to terminate in the dorsal and ventral cochlear nuclei
3. The secondary neurons are located in the dorsal and ventral cochlear nuclei
   a. These axons pass rostrally to either decussate in and form the trapezoid body or synapse with neurons in the superior olivary nucleus
      i. The fibers that decussate enter the lateral lemniscus and ascend to the inferior colliculus
      ii. Some fibers in the lateral lemniscus pass through the commissure of the inferior colliculus of the opposite side
4. Axons of the third-order neurons in the inferior colliculus pass via the brachium of the inferior colliculus to the medial geniculate body
5. The fourth neuron of the chain begins in the medial geniculate body. Fibers pass from here through the sublenticular portion of the internal capsule to end in the auditory area of the cerebral cortex

III. Auditory reflexes

are affected by the inferior colliculus which gives fibers to the tectospinal and tectobulbar tracts which both cross to the opposite side via the dorsal tegmental decussation and run downward, giving off collateral and terminal branches to various cranial motor nuclei and to the anterior gray columns in the cervical segments of the spinal cord (fibers from these 2 tracts probably terminate in the reticular formation, as well)

IV. Symptoms of disease

A. THE SENSATION OF HEARING has bilateral cortical representation, thus, unilateral lesions in the CNS do not cause a complete loss of hearing in either ear
B. SYMPTOMS OF DISEASE in this system, however, could be deafness or a decrease in the ability to hear

BRAINSTEM (DORSAL VIEW)

Spinal V tract + nucleus

Vestibular nuclei

Inf. + lat.

Med.

MLF

Inf. cerebellar ped.

Dorsal } Cochlear
Ventral } nuclei

Med. lemniscus

N. VIII

Dorsal } Cochlear nuclei
Ventral }

Pyramid

Inf. olive

MEDULLA OBLONGATA AT LEVEL OF VIIIth NERVE

COCHLEAR NERVE (VIII)

Transverse temporal
(auditory)
gyrus of Heschle

Geniculocortical fibers

Tectorial membrane

Inner hair cells

Outer hair cells

Medial geniculate body

Brachium of inf.
colliculus

Inferior colliculus

MIDBRAIN

Lateral lemniscus

Basilar membrane

Cochlear n.

Spiral ganglion

ORGAN OF CORTI

Nuclei of lateral lemniscus

ISTHMUS

Superior canal

Lateral canal

Lateral lemniscus

Post. canal

Restiform body

Dorsal } Cochlear nuclei
Ventral }

Int. acoustic meatus

Superior olivary nucleus

U

S

Cochlear (VIII) nerve

Cochlear duct

Trapezoid body   MEDULLA

110. VESTIBULAR (VIII) NERVE: VESTIBULAR SYSTEM—PART I

I. Functional components

A. SPECIAL SOMATIC AFFERENT (SSA) FIBERS
1. Receptors: hair cells of the cristae and maculae
2. Distribution: ampullae of the semicircular canals of the inner ear, utriculus of the inner ear, and the sacculus of the inner ear
3. Cell bodies: in the vestibular ganglion
4. Terminations: vestibular nuclei and the cerebellum

II. Functional vestibular system consists primarily of a 3-neuron chain

A. THE PRIMARY NEURON is a bipolar cell in the vestibular ganglion in the internal auditory meatus
1. The distal processes pass to the crista ampullaris of the semicircular canals and to the maculae of the utricle and saccule
2. The central processes pass in the vestibular portion of the VIIIth nerve to enter the brainstem at the junction of the pons and medulla
3. Majority of vestibular fibers terminate in the vestibular nuclei, but some pass directly into the cerebellum via the inferior cerebellar peduncle

B. THE SECONDARY NEURONS are located in 1 of 4 vestibular nuclei: superior nucleus of Bechterew, lateral nucleus of Deiters, medial nucleus of Schwalbe, and the inferior nucleus (spinal or descending nucleus)
1. Secondary connections are numerous
   a. Fibers from the superior nucleus enter the medial longitudinal fasciculus (MLF) of the same side and turn rostrally to end among the cells of cranial nerves III, IV, and VI
   b. Fibers from the lateral nucleus (the majority) descend, on the same side, through the brainstem, then down the cord in the anterior funiculus as the vestibulospinal tract, and terminate among cells of the ventral gray column
   c. Fibers from the inferior nucleus pass medially, cross the midline, and descend in the MLF, to terminate in the nucleus of the accessory nerve and in the ventral columns of the cervical cord
   d. Fibers from the medial nucleus cross the midline, enter the opposite MLF, and bifurcate into fibers that ascend and descend in that fascicle to terminate just as other fibers do in it
      i. Some fibers from the medial nucleus enter the reticular formation of both sides to be distributed to motor nuclei of the autonomic system in the brainstem, then go to the intermediolateral column of the cord via the reticulospinal tract

C. THE THIRD-ORDER NEURONS (EFFERENT NEURONS)
1. Axons from the ventral columns of the cord, from lower cervical to lower sacral levels, are distributed to skeletal muscles of trunk and extremities
   a. Completes a typical somatic reflex and is important in relating body movements with those of the head; significant in maintaining balance
2. Fibers from the ventral columns of cord in the cervical region and from nucleus of the spinal portion of accessory nerve are distributed to muscles of neck (sternocleidomastoid and trapezius); important in maintaining the normal position of the head
3. Axons from somatic efferent columns of the brainstem, particularly nuclei of the IIIrd, IVth, and VIth nerves, supply extrinsic muscles of eyeball and help coordinate eye movements with those of head
4. Fibers from autonomic centers, especially the dorsal motor nucleus of X and intermediolateral cell columns of the cord, account for nausea, vomiting, and pallor manifested after overstimulation of the vestibular system

Superior (Bechterew)

Lateral (Deiters)

Medial (Schwalbe)

Inferior (Spinal)

VESTIBULAR NUCLEI (DORSAL VIEW)

IV

I

II

III

VI

IV

Motor
Sensory } Roots of V

Pons

VII

VIII

IX
X
XI
XII

Medulla

VESTIBULAR NERVE
(VIII)

MLF

UPPER MEDULLA

Sup. vest. nucl.

Med. vest. nucl.

Lat. vest. nucl.

MLF

Inf. vest. nucl.

Reticulospinal tract

Vestibular ganglion

Vestibular nerve

Vestibulospinal
tract

Vestibulospinal
tract

Spinal accessory
nucleus

Medial longitudinal facsiculus

Cervical cord

Motor neuron

Gelatinous substance

Hair tufts

Hair cells

Supporting cells

Nerve fibers

CRISTA

Sup.

Lat.

Post.

Ampulla

Utricle

Saccule

Cochlea

Otoconia

Gelatinous layer

Hair tufts

Hair cells

Supporting cells

Nerve fiber

MACULA

111. VESTIBULAR (VIII) NERVE: VESTIBULAR SYSTEM—PART II

I. **Cerebellar influence on the vestibular system:** cerebellum does exert control over the activities initiated by the stimulation of the vestibular receptors

A. DIRECT PATHWAYS

  1. Afferent neurons travel through ascending fibers to terminate in the nodulus and flocculus of the cerebellum

  2. From cells in the cortex of the nodulus and flocculus, fibers descend to terminate in the fastigial nucleus of the cerebellum

    a. From the fastigial nucleus, axons descend via the medial segment of the inferior cerebellar peduncle to end in the vestibular nuclei

    b. Other axons from the fastigial nucleus cross the midline and descend to the vestibular nuclei of the opposite side via the uncinate bundle (of Russell)

    c. Additional fibers from the vestibular nuclei are carried by the pathways already described (vestibulospinal tract, MLF, and reticular formation) to the same terminations

B. INDIRECT PATHWAYS

  1. The afferent neurons terminate in the superior and lateral vestibular nuclei

  2. From cells mainly in the superior nucleus (but also lateral), fibers ascend in the medial segment of the inferior cerebellar peduncle to terminate both in the flocculonodular lobe and in the lingula and uvula of the cerebellum

  3. From cells in the cerebellar cortex of the above regions, fibers descend both to the fastigial and to the globose nuclei of the cerebellum

    a. Axons from the above descend by the same pathways and have the same terminations as described above for the direct route

II. **Clinical considerations**

A. DISEASE IN THE VESTIBULAR MECHANISM expresses itself in disturbances of equilibrium, balance, and coordination

  1. Simple observation of the patient standing or walking becomes an integral part of the examination

  2. Since the system is so closely tied to reflex movements of the eye, such reflexes are used to localize a lesion

B. TESTS

  1. Rotation (Barany)

  2. Caloric

C. SPECIAL PATHOLOGY

  1. Vertigo: most characteristic symptom of vestibular disease

  2. Inflammation of the labyrinth

  3. Ménière's disease: paroxysmal attacks of vertigo associated with unilateral tinnitus and deafness

  4. Basal or pontine tumors or hemorrhage can involve the extra- or intramedullary course of the afferent fibers of the vestibular nerve

Sup. oblique m.

Sup. rectus m.

Inf. rectus m.

Inf.
oblique
m.

IIIrd nucleus

Trochlear nucleus

Sup. vest.
nuc.

Lat. vest.
nuc.

Med. vest.
nuc.

Inf. vest.
nuc.

**VERTICAL PLANE EYE
MOVEMENTS**

Medial
recti mm.

Lateral rectus muscle

Oculomotor (III) n.

IIIrd nucleus

Abducent (VI) nerve

Parabducent nucleus

Abducent nucleus

Medial

Superior

Lateral

Inferior

Vestibular nuclei

Medial longitudinal
fasciculus

Reticulospinal tract

Vestibulospinal tract

Vestibular nerve

**HORIZONTAL PLANE
EYE MOVEMENTS**

Fastigial nuclei

Lingula

Superior vest. nuc.

Juxtarestiform body

Cerebellum

Uvula

Nodulus

Flocculus

Inferior

Lateral   Medial

Vestibular nuclei

Nodulus

Flocculus

Reticulospinal tract

Vestibulospinal tract

Vestibular nerve

INDIRECT PATHWAYS

MLF

Vestibular n.

DIRECT PATHWAYS

VESTIBULOCEREBELLAR + CEREBELLOVESTIBULAR CONTROL

112. GLOSSOPHARYNGEAL (IX) AND VAGUS (X) NERVES—PART I

I. **Functional components:** the nerves are discussed together because of their close relationship. Both nerves are mixed and contain the same functional components

A. GENERAL SOMATIC AFFERENT (GSA) FIBERS: for vagus (X), only
   1. Receptors: external auditory meatus
   2. Cell bodies: jugular ganglion
   3. Termination: spinal V nucleus
   4. Symptoms of disease: anesthesia in the area of distribution

B. GENERAL VISCERAL AFFERENT (GVA) FIBERS
   1. In the IXth and Xth nerves, arise from cell bodies in the petrosal and inferior nodose ganglion, respectively
      a. Those of the IXth nerve convey touch, pain, and temperature from the posterior portion of the tongue, tonsil, and eustachian tube
      b. Those from the Xth convey sensory sensations from the pharynx, larynx, trachea, and the viscera of the thorax and abdomen

C. SPECIAL VISCERAL AFFERENT (SVA) FIBERS: fibers of taste
   1. The IXth nerve from the posterior $\frac{1}{3}$ of tongue; Xth fibers from the epiglottis and tongue adjacent to it
   2. Their cell bodies are in the petrosal and nodose ganglia, respectively
   3. Centrally, all visceral afferent fibers of IX, X (and also taste fibers of VII) will enter the solitary nucleus
      a. Secondary pathways to the thalamus and cerebral cortex are not well known, but apparently do exist
      b. There is a definitive gustatory receptive cortex
      c. Other visceral sensations are not definitively localized
      d. The solitary nucleus does send numerous projections to the centers in the reticular formation
      e. The respiratory center, i.e., is located dorsal to the olivary complex and responds to vagal and other neural impulses as well as to changes in CO_2 concentration in the blood
      f. There are also cardiovascular and vomiting centers in the reticular formation that receive secondary visceral afferent fibers
      g. Numerous fibers also pass from the solitary nucleus to cranial nerve nuclei, especially to the dorsal motor nucleus of X
   4. Of special interest is the SVA component of the IXth nerve from the carotid sinus to the CNS
      a. Afferent impulses are initiated by increased carotid arterial pressure and conveyed centrally by the sinus nerve
         i. Centrally, collaterals are given off to the dorsal motor nucleus of X
         ii. The impulses are transmitted to preganglionic neurons of the vagus nerve which send their axons to postganglionic neurons in the atria of the heart
         iii. Stimulation of the sinus causes a decrease in heart rate and blood pressure

Nucleus solitarius
and fasciculus

Inferior salivatory nucleus

Nucleus ambiguus

Spinal tract and
nucleus V

Superior ganglion

Inferior ganglion

IX

MEDULLA

SVE — GSA
GVE — SVA
GVA

Inf. salivatory
nucleus

IX

X

Nucleus of spinal V tract

IX

X

Dorsal motor nucleus X

Nucleus ambiguus

BRAINSTEM NUCLEI (DORSAL VIEW)

**GLOSSOPHARYNGEAL (IX)
NERVE COMPONENTS**

Inf. salivatory nu.

Inf. (petrous) ganglion + tympanic n.

Sup. (jugular) ganglion

Nerves to fenestra cochleae + vestibuli

Small superf. petrosal n.

Otic ganglion

Parotid gland

Nu. solitarius

IX

Pharyngeal wall

Nucleus ambiguus

Communication with
auricular (X)

Tympanic
plexus

Tonsil + ant. and
post. pillars

Nodose (X) ganglion

Sup. cervical gang.

To eustachian tube

Tongue

Cut styloglossus

Hyoglossus

Stylopharyngeus m.

**DISTRIBUTION OF
GLOSSOPHARYNGEAL
(IX) NERVE**

Carotid sinus

Carotid body

Sup. + middle pharyngeal constrictors

113. GLOSSOPHARYNGEAL (IX) AND VAGUS (X) NERVES—PART II

I. Functional components (cont.)

A. GENERAL VISCERAL EFFERENT (GVE) FIBERS

    1. Of the IXth nerve arise from neurons in the inferior salivatory nucleus and pass as parasympathetic fibers to the otic ganglion via the inferior petrosal nerve. Postganglionic fibers innervate the parotid gland

    2. Those in the Xth nerve arise in the dorsal motor nucleus of X and pass as preganglionic parasympathetic fibers to terminate in parasympathetic ganglia innervating the thoracic and abdominal viscera

B. SPECIAL VISCERAL EFFERENT (SVE) FIBERS

    1. Of the IXth nerve arise in the rostral pole of the nucleus ambiguus and innervate the stylopharyngeus muscle

    2. Those in the Xth nerve arise from the middle portion of the nucleus ambiguus and innervate the voluntary striated muscles of the larynx and pharynx

    3. The nucleus ambiguus is a column of motor neurons found in the medulla about halfway between the spinal V nucleus and the inferior olive

        a. Axons in its rostral portion contribute fibers to the IXth nerve

        b. Axons from its middle portion pass obliquely dorsally and medially to join the other exiting fibers of the Xth nerve

        c. Neurons in its caudal portion contribute to the XIth nerve

II. Symptoms of disease

A. THE IXTH NERVE

    1. GVA: interference with the afferent limb of the swallowing reflex

    2. SVA: ageusia on the posterior $\frac{1}{3}$ of the tongue

    3. GVE: decreased secretion of the parotid gland

    4. SVE: some difficulty in swallowing

B. THE XTH NERVE

    1. GSA: anesthesia in the area of distribution (external auditory meatus)

    2. GVA

        a. Loss of reflex control of circulation (heart rate, etc.)

        b. Disturbance in the depth of respiration

    3. SVA: anesthesia in the area of distribution (epiglottis and adjacent tongue)

    4. GVE

        a. Loss of reflex control of circulatory system (heart rate, blood pressure)

        b. Poor digestion due to decreased secretion of digestive enzymes and a slowing of peristalsis

    5. SVE: difficulty in swallowing and speaking

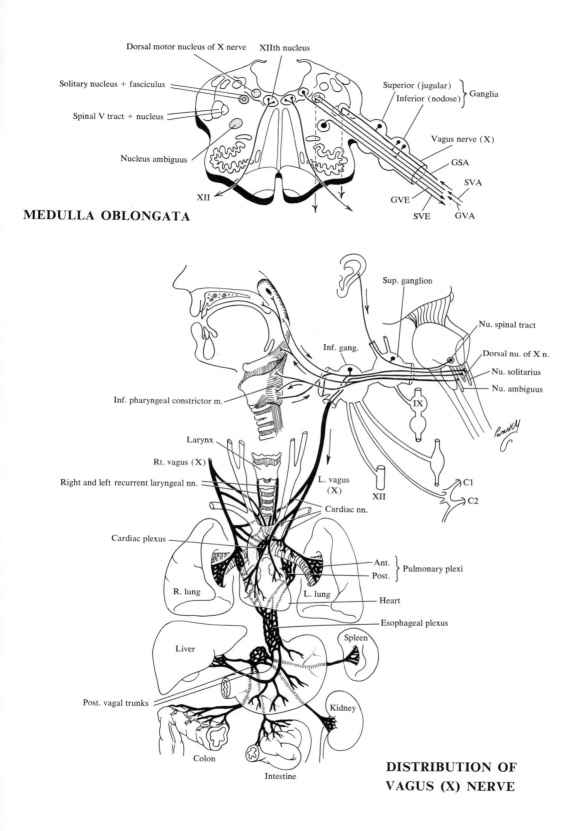

Dorsal motor nucleus of X nerve   XIIth nucleus

Solitary nucleus + fasciculus

Superior (jugular)
Inferior (nodose) } Ganglia

Spinal V tract + nucleus

Vagus nerve (X)

Nucleus ambiguus

GSA

SVA

GVE

XII

SVE   GVA

MEDULLA OBLONGATA

Sup. ganglion

Nu. spinal tract

Inf. gang.

Dorsal nu. of X n.

Nu. solitarius

Nu. ambiguus

Inf. pharyngeal constrictor m.

IX

Larynx

Rt. vagus (X)

L. vagus (X)

Right and left recurrent laryngeal nn.

XII

C1

C2

Cardiac nn.

Cardiac plexus

Ant.
Post. } Pulmonary plexi

R. lung

L. lung

Heart

Esophageal plexus

Spleen

Liver

Post. vagal trunks

Kidney

Colon

Intestine

**DISTRIBUTION OF
VAGUS (X) NERVE**

-247-

114. ACCESSORY (XI) NERVE

I. Functional components
A. THE XITH NERVE is generally considered to be an efferent nerve, although some sensory fibers (GP) probably exist in muscles supplied by its SVE component
B. SPECIAL VISCERAL EFFERENT (SVE) FIBERS: to the intrinsic muscles of the larynx and the sternocleidomastoid and trapezius muscles

II. Nuclei of the XIth nerve
A. THE NERVE HAS TWO DIVISIONS, a spinal and cranial
   1. The spinal portion arises from the spinal nucleus of XI in the anterior horn, extending from the 5th or 6th cervical vertebra to the middle of the pyramidal decussation
   2. The cranial portion arises from the caudal pole of the nucleus ambiguus to supply the intrinsic muscles of the larynx

III. Connections of the nuclei
A. BILATERAL CORTICOBULBAR
B. COLLATERAL from other cranial nuclei and incoming sensory systems
C. RETICULAR FORMATIONS, etc.

IV. Internal and intracranial course of the XIth nerve root fibers
A. FIBERS OF THE SPINAL PORTION
   1. Exit from the lateral aspect of the spinal cord between the dorsal and ventral nerve roots
   2. The rootlets of this portion of the XIth nerve unite to form the external branch of XI which ascends in the spinal canal just posterior to the dentate ligaments and enters the skull through the foramen magnum to eventually exit from the skull via the jugular foramen with nerves IX and X
B. FIBERS OF THE CRANIAL PORTION pass from the caudal pole of the nucleus ambiguus to join the fibers of the vagus nerve to exit with that nerve through the jugular foramen
   1. They later form the motor portion of the recurrent laryngeal nerves which supply the intrinsic muscles of the larynx

V. Symptoms of disease: difficulty in turning the head and possibly in swallowing

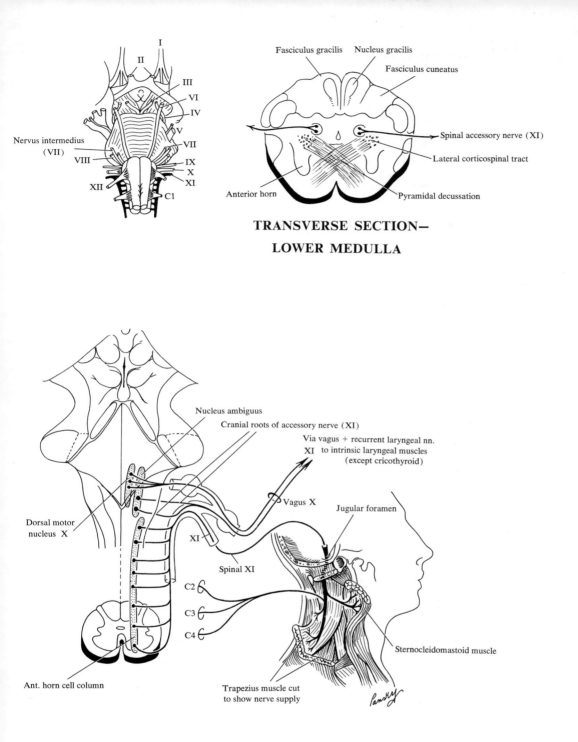

Nervus intermedius (VII)

Fasciculus gracilis Nucleus gracilis Fasciculus cuneatus

Spinal accessory nerve (XI)

Lateral corticospinal tract

Anterior horn

Pyramidal decussation

TRANSVERSE SECTION— LOWER MEDULLA

Nucleus ambiguus

Cranial roots of accessory nerve (XI)

Via vagus + recurrent laryngeal nn.
XI to intrinsic laryngeal muscles
    (except cricothyroid)

Vagus X

Jugular foramen

Dorsal motor nucleus X

XI

Spinal XI

C2

C3

C4

Sternocleidomastoid muscle

Ant. horn cell column

Trapezius muscle cut to show nerve supply

DISTRIBUTION OF ACCESSORY NERVE (XI)

115. HYPOGLOSSAL (XII) NERVE

I. Functional components
A. THE XIITH NERVE is classified primarily as a GSE nerve to the skeletal musculature (extrinsic and intrinsic) of the tongue, but it probably carries some impulses (GP) originating from muscle spindles
B. GENERAL SOMATIC EFFERENT (GSE) FIBERS
  1. Origin: hypoglossal nucleus
  2. Termination: all extrinsic and intrinsic muscles of the tongue

II. Nucleus of the XIIth nerve
A. THE NUCLEUS IS LOCATED near the floor of the 4th ventricle just beneath the hypoglossal trigone
  1. It is a longitudinal cell column of typical somatic motor neurons extending from the caudalmost portion of the medulla to the level of the stria medullaris
  2. The nucleus consists of several distinct cell groups, each group supplying certain muscles of the tongue
    a. The medially placed groups, i.e., probably innervate the intrinsic muscles of the tongue

III. Connections of the XIIth nucleus
A. CORTICOBULBAR FIBERS are involved where voluntary movement of the tongue is involved
  1. Some of these apparently end directly on neurons in the nucleus
  2. Many of the corticobulbar impulses are apparently relayed by crossed and uncrossed reticular neurons
    a. Some uncrossed corticobulbar fibers end predominantly in the medially placed cell groups that control the intrinsic tongue muscles, which work symmetrically
B. OTHER FIBERS, primarily from the sensory nuclei of V and the nucleus solitarius, enter the hypoglossal nucleus to activate the reflex actions of sucking, swallowing, chewing, etc.

IV. Internal course of the hypoglossal root fibers
A. THE AXONS of the hypoglossal neurons gather in bundles and leave the ventral surface of the nucleus
B. THE ROOTLETS course in a ventral and slightly lateral direction, traverse the medullary reticular formation just lateral to the medial lemniscus, pierce the most medial part of the inferior olive, and exit the brainstem in the ventrolateral sulcus (preolivary) between the pyramid and the inferior olive
C. THE CLOSE RELATIONSHIP of the emerging fibers of XII and the pyramidal tract is the anatomic basis for the syndrome called alternating hypoglossal hemiplegia

V. Intracranial course of the XIIth nerve: the large to medium-sized, myelinated fibers of the XIIth nerve exit as 10–15 rootlets that soon join to leave this skull via the hypoglossal canal in the occipital bone

VI. Symptoms of disease: paralysis of the tongue

Nuc. gracilis

Nuc. cuneatus

Internal arcuate fibers

Inf. olive

Pyramid

XII

Hypoglossal (XII) nerve

LOWER MEDULLA

Hypoglossal nucleus

DORSAL VIEW

Dorsal nuc. X

Restiform body

MLF

Nuc. ambiguus

Medial lemniscus

Inferior olive

XII

Hypoglossal (XII) nerve

UPPER MEDULLA

(Xth NERVE OMITTED)

Pyramids

Medulla oblongata

Hypoglossal (XII) nerve

Hypoglossal (XII) nerve

Hypoglossal canal

Nodose ganglion (X)

Sup. cervical ganglion

Styloglossus m. (XII)

Intrinsic tongue mm. (XII)

Hypoglossal nucleus

C1

C2

C3

Genioglossus m. (XII)

Hyoglossus m. (XII)

Geniohyoid m.

Descendens hypoglossi

Thyrohyoid m.

Descendens cervicalis

Sternohyoid m.

Ansa hypoglossi

Omohyoid m.

Sternothyroid m.

DISTRIBUTION OF

HYPOGLOSSAL (XII) NERVE

116. SUMMARY OF CRANIAL NERVES

Nerve	Attachment to Brain	Cranial Exit	Cells of Origin	Chief Components	Chief Functions
I. Olfactory (sensory)	Olfactory bulb	Cribriform plate	Nasal mucosa (epithelial hair cells)	Special visceral or somatic afferent	Smell
II. Optic (sensory)	Optic chiasma	Optic canal	Retina (ganglion cells)	Special somatic afferent	Vision
III. Oculo-motor (motor)	Midbrain, at medial border of cerebral peduncle	Superior orbital fissure	Midbrain (IIIrd nucleus)	Somatic efferent	Movements of eyes (lid elevation)
			Midbrain (Edinger-Westphal nucleus)	General visceral efferent (para-sympathetic)	Miosis and accom-modation
IV. Trochlear (motor)	Midbrain, below inferior colliculus	Superior orbital fissure	Midbrain	Somatic efferent	Movements of eyes (superior oblique muscle)
V. Trigeminal (mixed)	Side of pons	Superior orbital fissure, foramen rotundum, and foramen ovale	Pons (motor V nucleus)	Special visceral efferent	Chiefly movements of mandible
			Trigeminal ganglion	General somatic afferent	Sensation in head
VI. Abducent (motor)	Lower border of pons	Superior orbital fissure	Pons	Somatic efferent	Movements of eyes (lateral rectus muscle)
VII. Facial (mixed)	Lower border of pons	Stylomastoid foramen	Pons	Special visceral efferent	Facial expression
			Pons	General visceral efferent (para-sympathetic)	Secretion of tears and saliva
			Geniculate ganglion	Special visceral afferent	Taste

Nerve	Attachment to Brain	Cranial Exit	Cells of Origin	Chief Components	Chief Functions
VIII. Vestibulo-cochlear (sensory)	Lower border of pons	Does not leave skull	Vestibular ganglion	Special somatic afferent	Equilibration
			Spiral ganglion	Special somatic afferent	Hearing
IX. Glosso-pharyngeal (mixed)	Medulla, lateral to olive	Jugular foramen	Medulla (nucleus ambiguus)	Special visceral efferent	Elevation of pharynx
			Medulla (dorsal nucleus)	General visceral efferent (para-sympathetic)	Secretion of saliva
			Inferior ganglion	General visceral afferent	Sensation in tongue and pharynx; visceral reflexes
			Inferior ganglion	Special visceral afferent	Taste
X. Vagus (mixed)	Medulla, lateral to olive	Jugular foramen	Medulla (nucleus ambiguus)	Special visceral efferent	Movements of larynx
			Medulla (dorsal nucleus)	General visceral efferent (para-sympathetic)	Movements and secretion of thoracic and abdomi-nal viscera
			Inferior ganglion	General visceral afferent	Sensation in pharynx, larynx, and thoracic and abdominal viscera. Also visceral reflexes
			Inferior ganglion	Special visceral afferent	Taste
			Superior ganglion	General somatic afferent	Sensation in external ear

Nerve	Attachment to Brain	Cranial Exit	Cells of Origin	Chief Components	Chief Functions
XI. Accessory (2 parts: cranial, spinal)	Medulla, lateral to olive	Jugular foramen	Medulla (nucleus ambiguus)	Special visceral (?) efferent	Movements of pharynx and larynx
			Spinal cord (cervical)	Special visceral efferent	Movements of head and shoulder
XII. Hypoglossal (motor)	Medulla, between pyramid and olive	Hypoglossal canal	Medulla	Somatic efferent	Movements of tongue

117. PARASYMPATHETIC GANGLIA ASSOCIATED WITH CRANIAL NERVES

Ganglion	Location	Parasympathetic Root	Sympathetic Root	Chief Distribution
Ciliary	Lateral to optic nerve	Oculomotor nerve	Internal carotid plexus	Ciliary m. and sphincter pupillae Dilator pupillae and tarsal muscles
Pterygopalatine	In pterygopalatine fossa	Greater petrosal (VII) and n. of pterygoid canal	Internal carotid plexus	Lacrimal gland
Otic	Below foramen ovale	Lesser petrosal (IX)	Plexus on middle meningeal artery	Parotid gland
Submandibular	On hyoglossus muscle	Chorda tympani (VII) by way of lingual n.	Plexus on facial artery	Submandibular and sublingual glands

THE DIENCEPHALON

118. DIENCEPHALON: EPITHALAMUS

I. **Introduction:** the diencephalon lies rostral to the brainstem, begins at the level of the posterior commissure and extends to the interventricular foramina, includes the structures surrounding the third ventricle, and is subdivided into four major parts: epithalamus, thalamus, hypothalamus, and subthalamus

A. EPITHALAMUS is located dorsomedial to the thalamus in the dorsal wall of the posterior portion of the third ventricle, inferior to the splenium of the corpus callosum. This division consists of the pineal body, posterior commissure, habenular commissure, habenula (habenular nuclei), tela choroidea, and the stria medullaris thalami

1. The membranous tela choroidea stretches between the stria medullaris thalami of the opposite sides to form most of the roof of the third ventricle. Remaining structures form only a small posterior part of the roof

2. Habenular trigone is the small, triangular, depressed area just anterior to the superior colliculi and consists of
   a. Habenular nuclei receive nerve fibers from the stria medullaris thalami
   b. Habenular commissure joins the habenular nuclei
   c. Habenulopeduncular tract (retroflex tract of Meynert) is comprised of nerve fibers from the habenular nucleus to the interpeduncular ganglion of the mesencephalon

3. Posterior commissure is a bundle of white fibers that crosses the midsagittal plane immediately rostral to the superior colliculi at the point where the third ventricle becomes continuous with the cerebral aqueduct. It contains fibers
   a. Connecting the two superior colliculi
   b. From the pretectal nuclei
   c. From nuclei of the posterior commissure
   d. From the olivary nuclei

4. Pineal body is the main structure of the epithalamus
   a. Located in the depression between the superior colliculi
   b. Formed by an evagination of the roof of the diencephalon
   c. Attached by a stalk that is continuous with both the habenular and posterior commissures
   d. Its recess (pineal recess) is the cavity of the evagination formed by the separation of the dorsal and ventral laminae of the stalk
   e. In man, it secretes melatonin, which aids in the regulation of the secretion of gonadotropins
   f. This gland also secretes melanocyte-stimulating hormones which contain endorphin molecules, important in pain control
   g. In older persons, the pineal body tends to accumulate granules of calcium salts
   h. Clinical significance: the pineal body is located in the midline, and upon calcification (after 20 years) it serves as a useful marker of midline structures on a roentgenogram. A shift of this structure to one side may represent a shift of brain structures due to a mass lesion on the opposite side or from an atropic lesion on the same side

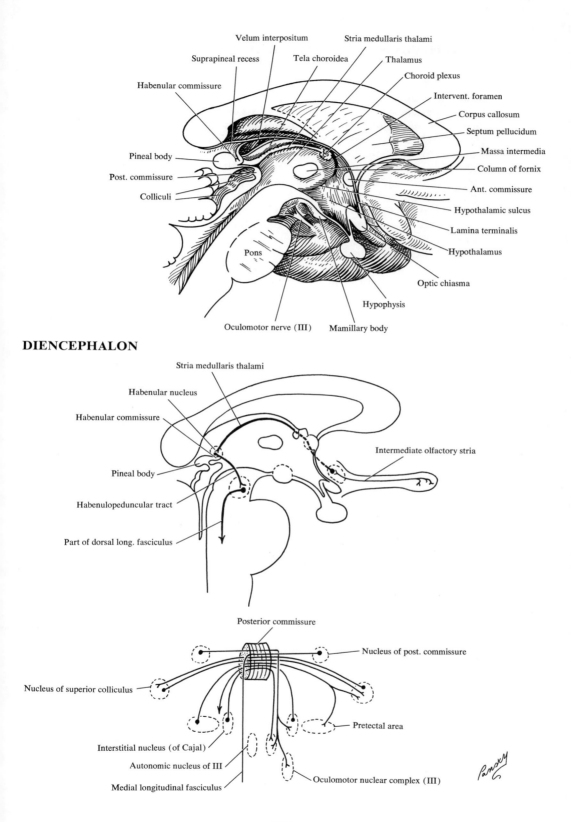

Velum interpositum

Suprapineal recess

Stria medullaris thalami

Tela choroidea

Thalamus

Habenular commissure

Choroid plexus

Intervent. foramen

Corpus callosum

Septum pellucidum

Massa intermedia

Pineal body

Column of fornix

Post. commissure

Ant. commissure

Colliculi

Hypothalamic sulcus

Lamina terminalis

Hypothalamus

Pons

Optic chiasma

Hypophysis

Oculomotor nerve (III)

Mamillary body

DIENCEPHALON

Stria medullaris thalami

Habenular nucleus

Habenular commissure

Intermediate olfactory stria

Pineal body

Habenulopeduncular tract

Part of dorsal long. fasciculus

Posterior commissure

Nucleus of post. commissure

Nucleus of superior colliculus

Pretectal area

Interstitial nucleus (of Cajal)

Autonomic nucleus of III

Oculomotor nuclear complex (III)

Medial longitudinal fasciculus

Pansky

119. THALAMUS: ANATOMY AND NUCLEI

I. **Introduction:** the thalamus is the most dorsal of the four subdivisions of the diencephalon and forms the dorsal half of each lateral wall of the third ventricle. Major ascending pathways of the auditory, visual, and somatosensory systems have their final subcortical relays here. The large, ovoid (egg-shaped) gray thalami of both sides often fuse (70%) in the midline of the third ventricle to form the interthalamic adhesion or massa intermedia

   A. THE THALAMIC RADIATION: comprised of tracts emerging from the lateral thalamic surface, entering the internal capsule and terminating in the cerebral cortex. The thalamus is the major source of afferent fibers to the cortex

   B. THE THALAMUS IS ALSO THE RELAY CENTER for both cerebellar projections to the motor cortex and major efferent fibers from the basal ganglia

   C. THALAMIC ANATOMY is best seen by dividing its more than 30 nuclei into three gray areas: anterior, medial, and nuclear, formed by a vertical sheet of white matter, the internal medullary lamina, which divides each of the thalami

   1. Intralaminar nuclei are made up of several small groups of nuclei which are enclosed within the medullary lamina

   2. The reticular nucleus is separated from the main nuclear mass by the external medullary lamina, a layer of myelinated fibers on the lateral surface of the two thalami adjacent to the internal capsule

   D. FOUR EXTERNAL PROMINENCES are present on the surface of each of the thalami: the anterior tubercle, pulvinar, medial, and lateral geniculate bodies

II. **Four major groups of the more than 30 thalamic nuclei**

   A. ANTERIOR NUCLEAR GROUP corresponds to and is visualized as the anterior tubercle. Afferent fibers from the mamillary bodies run in the mamillothalamic tract; efferent fibers project to the cingulate gyrus (cortex)

   B. MEDIAL NUCLEAR GROUP includes the gray substance medial to the internal medullary lamina (intralaminar nuclei); the dorsal medial nucleus, which projects to the frontal cortex lying anterior to the precentral gyrus (motor cortex); and the centromedian nucleus which connects with the corpus striatum

   C. MIDLINE NUCLEAR GROUP includes groups of cells lying just beneath the ependymal lining of the third ventricle and in interthalamic adhesion. These nuclei connect with nuclei in the hypothalamus and with periaqueductal gray matter

   D. LATERAL NUCLEAR GROUP: located anterior to and includes the pulvinar. It lies between the internal and external medullary laminae. It is further subdivided into a dorsal and ventral tier of nuclei

   1. Pulvinar is the largest nucleus of the dorsal tier and connects with cortices of the parietal and temporal lobes

   2. Other nuclei of the dorsal tier

      a. Lateral dorsal nucleus: caudal to the anterior nucleus. Has fiber connections with the cingulate gyrus, precuneus, and mamillary bodies

      b. Lateral posterior nucleus receives fibers from medial and lateral geniculate bodies, from the ventral posterior nuclear complex and has reciprocal connections with cortices of precuneus and superior parietal lobule

   3. Lateral and medial geniculate bodies: nuclei are part of the ventral tier

      a. Lateral geniculate body: an oval elevation on lateral part of posterior end of thalamus. It receives fibers from the optic tract and projects fibers to visual cortex as the geniculocalcarine (optic) radiations

      b. Medial geniculate body: located under the pulvinar, lateral to midbrain. It receives auditory fibers from the lateral lemniscus and inferior colliculus, and projects fibers to Heschl's gyri (temporal cortex) as the auditory radiations

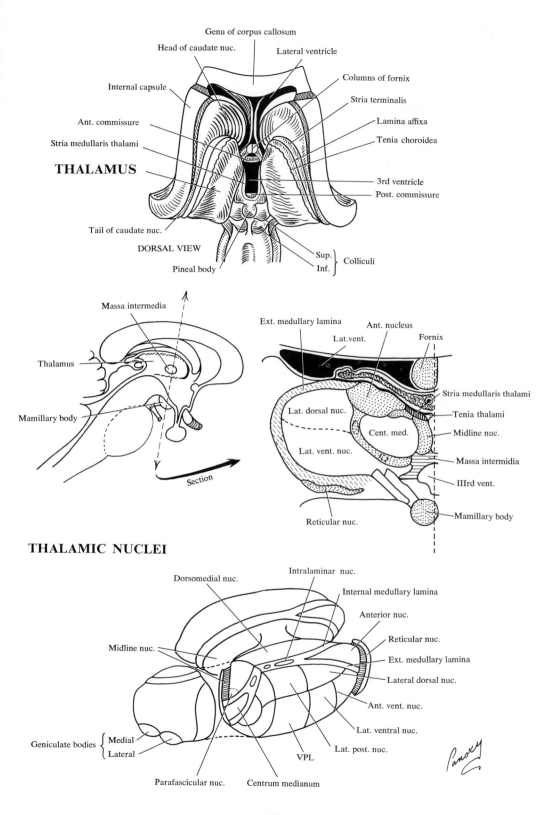

Genu of corpus callosum

Head of caudate nuc.

Lateral ventricle

Columns of fornix

Stria terminalis

Internal capsule

Lamina affixa

Ant. commissure

Tenia choroidea

Stria medullaris thalami

THALAMUS

3rd ventricle

Post. commissure

Tail of caudate nuc.

DORSAL VIEW

Sup.
Inf. } Colliculi

Pineal body

Massa intermedia

Ext. medullary lamina

Ant. nucleus

Lat.vent.

Fornix

Thalamus

Stria medullaris thalami

Lat. dorsal nuc.

Tenia thalami

Cent. med.

Midline nuc.

Lat. vent. nuc.

Massa intermidia

Mamillary body

IIIrd vent.

Reticular nuc.

Mamillary body

THALAMIC NUCLEI

Dorsomedial nuc.

Intralaminar nuc.

Internal medullary lamina

Anterior nuc.

Reticular nuc.

Midline nuc.

Ext. medullary lamina

Lateral dorsal nuc.

Ant. vent. nuc.

Lat. ventral nuc.

Geniculate bodies { Medial
Lateral

Lat. post. nuc.

VPL

Parafascicular nuc.

Centrum medianum

120. THALAMUS: VENTRAL TIER NUCLEI

I. Four major groups of thalamic nuclei (cont.)
A. LATERAL NUCLEAR GROUP (CONT.)
1. Other nuclei of the ventral tier (from anterior to posterior)
   a. Ventral anterior nucleus receives input from substantia nigra, globus pallidus, other thalamic nuclei (midline and intralaminar), reticular formation, and cerebral cortex. Its output goes to the caudate nucleus and premotor cortex (area 6)
   b. Ventral lateral nucleus receives input from the globus pallidus, substantia nigra, other thalamic nuclei, reticular formation, cerebral cortex, cerebellum (dentatothalamic tract), and the red nucleus (rubrothalamic tract). It has reciprocal connections with the precentral gyrus (area 4)
   c. Both of the above nuclei are specific motor nuclei for the integration and transmission of information from the motor control pathways in the motor system to the motor cortex of the cerebrum. Both send efferent fibers to widespread areas in the frontal cortex
2. The ventral posterior nuclear complex consists of a ventral posteromedial nucleus (VPM), which is also called the semilunar or arcuate nucleus, and a ventral posterolateral nucleus (VPL)
   a. These nuclei relay somatic sensory system information to the cortex
   b. The VPL nucleus is where second-order neurons of the spinothalamic and medial lemniscus pathways from the trunk and limbs terminate
   c. The VPM nucleus is the site of termination of neurons of the trigeminal pathways from the head
   d. Axons from the VPL and VPM nuclei project to the postcentral gyrus (primary somatosensory cortex) via the middle thalamic radiation in the posterior limb of the internal capsule

II. Summary of nuclear areas and thalamic nuclei

Anterior nucleus ——————————————————————— ANTERIOR AREA
Midline nucleus ————————————
Dorsomedial (medial nucleus)   } MEDIAL AREA
Centromedian (intralaminar) nucleus ——————

Lateral dorsal nucleus ————————
Lateral posterior nucleus   } DORSAL TIER
Pulvinar ———————————

Ventral anterior nucleus ————————
Ventral lateral nucleus
Ventral posterolateral nucleus   } LATERAL AREA
Ventral posteromedial nucleus   } VENTRAL TIER
Lateral geniculate body
Medial geniculate body
Reticular nuclei ———————

III. Clinical considerations
A. SPECIFIC NEUROSURGICAL LESIONS in the ventral anterior thalamic nuclei interrupt connections between the basal ganglia and the cerebral cortex and serve to decrease rigidity and tremor in patients with Parkinson's disease
B. LESIONS THAT DESTROY THE VENTRAL POSTERIOR NUCLEUS OF THE THALAMUS produce a contralateral hemianesthesia with loss of sensory modalities of the face, trunk, and limbs. This is often caused by an infarction of a branch of the posterior cerebral artery

THALAMIC NUCLEI

Midline nuc. (M)
Dorsomedial nuc. (MD)
Intralaminar nuc.
Dorsolateral nuc. (LD)
Ant. nuc. (A)
Reticular nuc. (R)
Ant. vent. nuc. (VA)
Lat. vent. nuc. (VL)
Posterolateral nuc. (LP)
Posterolat. vent. nuc. (VPL)
Centromedian nuc. (CM)
Int. medullary lamina
Interthalamic adhesion
Pulvinar
Geniculate bodies { Med. / Lat. }

X-SECTIONS OF THALAMUS
(ANT. TO POST.)

A
VA
LD
A
MD
VL
MD
LP
VPL
MD
LP
VPL
M
CM
MD
LP
VPI
CM
P
MG
LG

CORTICAL PROJECTIONS

Caudate
Putamen
Cingulate
Globus pallidus
Frontal
A
VA
MD
VL
M
CM
VP
Pulvinar
MG
LG
Premotor frontal
Motor frontal
Sensory frontal
Parietotemporal occipital
Visual occipital
Auditory temporal

THALAMUS + CORTICAL CONNECTION AREAS

Midline nuc. (14)
MD (12)
LP (6)
LD (1)
CM (11)
Pulvinar (3)
A (10)
VA (9)
VL (8)
VPL (5)
Medial (1) / Lateral (2) } Geniculate bodies

LATERAL VIEW

8
5
6+7
9
11
2
3
1
12
14

MEDIAL VIEW

8
5
9
10
6+7
3
14
12
14
2
3
14

Panosky

–261–

121. FUNCTIONAL THALAMUS AND SUBTHALAMUS

I. Functions of the thalamus

A. INTEGRATION, CORRELATION, AND RELAY of information for the sensory, motor, consciousness, limbic, and visual systems

B. FOCUSING ATTENTION: by making certain cortical sensory areas more receptive and other areas less receptive

C. CONSCIOUS PERCEPTION AND INTERPRETATION OF PAIN: the only peripheral sensory stimulus to be interpreted at the level of the thalamus

II. Functional classification of thalamic nuclei

A. SPECIFIC NUCLEI possess reciprocal connections with specific areas of the cerebral cortex that have specific motor or sensory functions: these include most of the ventral tier nuclei of the lateral area, namely
   1. Anterior nucleus (some include these with the association nuclei)
   2. Ventral anterior (in part)
   3. Ventral lateral and ventral posterior nuclei
   4. Lateral and medial geniculates

B. SUBCORTICAL NUCLEI have no direct fiber connections with cerebral cortices and include thalamic reticular nucleus, intrathalamic nucleus, dorsomedial nucleus (in part), and the ventral anterior nucleus (in part)

C. ASSOCIATION NUCLEI have reciprocal fiber connections with association areas of the cerebral cortex and include the lateral dorsal nucleus, the lateral posterior nucleus, the pulvinar, and the dorsomedial nucleus (in part)

III. Subthalamus (ventral thalamus) lies between the tegmentum of the midbrain and the dorsal thalamus

A. THE SUBTHALAMUS IS LOCATED ventral to the thalamus and posterolateral to the hypothalamus. Lateral to it lies the internal capsule

B. THE SUBTHALAMUS CONTAINS nuclei and pathways associated with the control circuits of the basal ganglia

C. THE RED NUCLEUS AND SUBSTANTIA NIGRA extend into the subthalamus from the midbrain

D. THE SUBTHALAMUS NUCLEUS (BODY OF LUYS) is the major structure within the ventral thalamus. It is a lens-shaped mass of gray matter dorsolateral to the upper end of the substantia nigra

E. THE ANSA LENTICULARIS is the major pathway through this region and conveys fibers from the globus pallidus to the ventral anterior thalamic nucleus

F. THE FIELDS OF FOREL lie anterior to the red nucleus at this level and are formed by fibers from the globus pallidus
   1. Field H is the ventromedial portion
   2. Field H_1 is the dorsomedial portion
   3. Field H_2 (fasciculus lenticularis) is the ventrolateral portion which runs medially
   4. The ansa lenticularis joins field H_2 as it bends acutely in field H
   5. The thalamic fasciculus (Forel's field H_1) extends through field H_1 to reach the anterior ventral nucleus of the thalamus
   6. The zona inserta is a thin gray zone above the fasciculus lenticularis

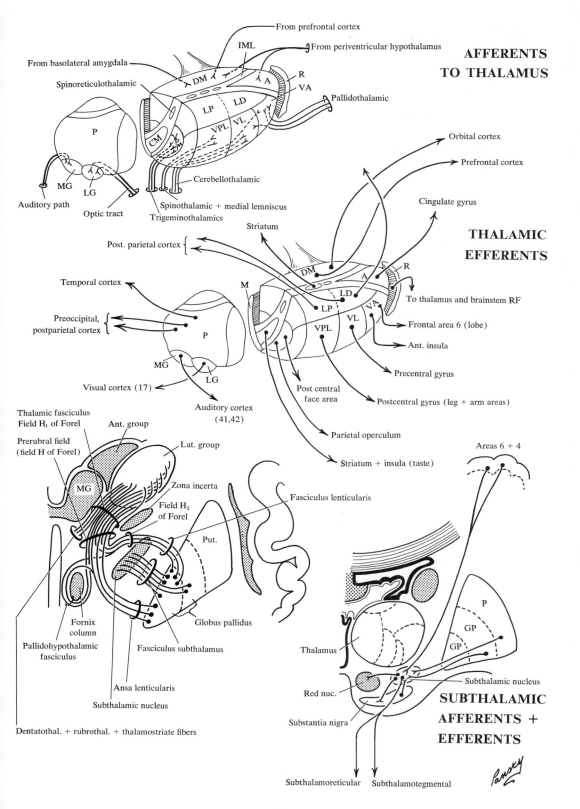

From prefrontal cortex

IML

From periventricular hypothalamus

**AFFERENTS
TO THALAMUS**

From basolateral amygdala

DM

R

A

Spinoreticulothalamic

VA

Pallidothalamic

LP

LD

P

VPL

VL

Orbital cortex

CM

Prefrontal cortex

MG

LG

Cingulate gyrus

Auditory path

Cerebellothalamic

Optic tract

Spinothalamic + medial lemniscus

Trigeminothalamics

Striatum

Post. parietal cortex

**THALAMIC
EFFERENTS**

DM

R

A

Temporal cortex

M

LD

LP

VA

To thalamus and brainstem RF

Preoccipital,
postparietal cortex

P

VPL

VL

Frontal area 6 (lobe)

Ant. insula

MG

LG

Precentral gyrus

Visual cortex (17)

Post central
face area

Postcentral gyrus (leg + arm areas)

Auditory cortex
(41,42)

Parietal operculum

Thalamic fasciculus
Field H_1 of Forel

Ant. group

Striatum + insula (taste)

Prerubral field
(field H of Forel)

Lat. group

Areas 6 + 4

MG

Zona incerta

Field H_2
of Forel

Fasciculus lenticularis

Put.

P

GP

GP

Fornix
column

Globus pallidus

Subthalamic nucleus

Pallidohypothalamic
fasciculus

Fasciculus subthalamus

Thalamus

Red nuc.

**SUBTHALAMIC
AFFERENTS +
EFFERENTS**

Ansa lenticularis

Subthalamic nucleus

Substantia nigra

Dentatothal. + rubrothal. + thalamostriate fibers

Subthalamoreticular   Subthalamotegmental

Pansky

122. HYPOTHALAMUS

I. **Introduction:** hypothalamus is located on the inferior aspect of the diencephalon and bordered by lamina terminalis, anteriorly; midbrain, posteriorly; hypothalamic sulcus, dorsally; third ventricle, medially; and the subthalamic nucleus, laterally

II. **Hypothalamic structures seen on the ventral aspect of the diencephalon**
   A. MAMILLARY BODIES: paired, spherical (pea-sized), white masses inferior to the third ventricle, caudal to the tuber cinereum, and rostral to the posterior perforated substance
   B. TUBER CINEREUM: that part of the hypothalamic floor between the mamillary bodies and the optic chiasm
   C. MEDIAN EMINENCE: funnel-like extension of the tuber cinereum which is a hypothalamic derivative of the pituitary (hypophysis) gland
   D. INFUNDIBULUM: a hollow, ventral extension of the median eminence and includes the infundibular stalk to the posterior lobe (pars nervosa) of the hypophysis
   E. NEUROHYPOPHYSIS consists of the median eminence, the infundibular stalk and the neural lobe (posterior lobe) of the hypophysis
   F. OPTIC CHIASM AND OPTIC TRACTS delineate the floor of the hypothalamus
   G. TWO HYPOTHALAMIC AREAS are seen on each side of the 3rd ventricle on either side of a sagittal plane through the fornix and mamillothalamic tract
      1. Lateral nuclear area: positioned in the lateral part of the tuber cinereum and includes the following nuclei
         a. Lateral preoptic nucleus: between anterior commissure and supraoptic nucleus
         b. Supraoptic nucleus straddles the lateral part of the optic chiasm
         c. Lateral nucleus: in the floor of hypothalamus lateral to arcuate nucleus
         d. Tuberal nucleus: in a position dorsal to the tuber cinereum
      2. Medial nuclear area is subdivided into three regions based on external anatomic features, as indicated
         a. Supraoptic (anterior) region: located dorsal to optic chiasm. Includes
            i. Medial preoptic nucleus: medial extension of the preoptic nucleus; medial and lateral preoptic nuclei are part of the telencephalon developmentally, but usually are included with the hypothalamic nuclei because it is hypothalamic functionally
            ii. Anterior nucleus: in the plane with and between supraoptic and paraventricular nuclei
            iii. Paraventricular nucleus: a group of cells dorsal in position to the anterior nucleus and lying close to the lining of the 3rd ventricle
            iv. Supraoptic nucleus closely parallels the optic chiasm. Occupies both lateral nuclear area and anterior region of medial nuclear area
         b. Middle (tuberal) region: located dorsal to the tuber cinereum and includes
            i. Dorsomedial nucleus: a mass lying above the ventromedial nucleus
            ii. Ventromedial nucleus: an ovoid mass of cells lying anterior to the mamillary bodies and posterior to the supraoptic nucleus
            iii. Arcuate (infundibular) nucleus: located in the floor of the hypothalamus ventral to the ventromedial nucleus
         c. Posterior (mamillary) region: positioned dorsal and immediately anterior to the mamillary bodies and includes
            i. Mamillary nuclei: found in mamillary bodies. Consist of a medial nucleus, larger of the two and responsible for protuberance of mamillary body, and a lateral nucleus, located between lateral border of the medial mamillary nucleus and base of brain
            ii. Posterior nucleus: dorsal to the mamillary nuclei and located in the lateral wall of the hypothalamus

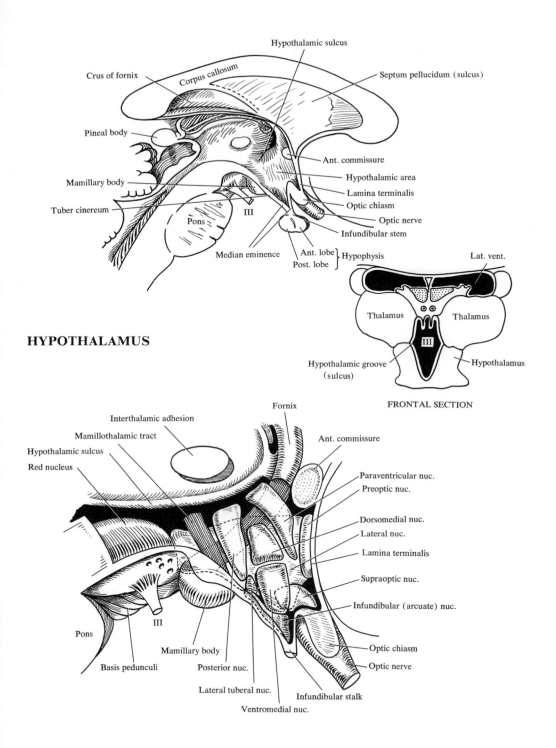

Hypothalamic sulcus

Crus of fornix

Corpus callosum

Septum pellucidum (sulcus)

Pineal body

Ant. commissure

Hypothalamic area

Mamillary body

Lamina terminalis

Optic chiasm

Tuber cinereum

III

Optic nerve

Pons

Infundibular stem

Median eminence

Ant. lobe
Post. lobe
} Hypophysis

Lat. vent.

Thalamus

Thalamus

III

Hypothalamic groove
(sulcus)

Hypothalamus

HYPOTHALAMUS

FRONTAL SECTION

Fornix

Interthalamic adhesion

Mamillothalamic tract

Ant. commissure

Hypothalamic sulcus

Red nucleus

Paraventricular nuc.

Preoptic nuc.

Dorsomedial nuc.

Lateral nuc.

Lamina terminalis

Supraoptic nuc.

Infundibular (arcuate) nuc.

III

Pons

Optic chiasm

Mamillary body

Optic nerve

Basis pedunculi

Posterior nuc.

Lateral tuberal nuc.

Infundibular stalk

Ventromedial nuc.

HYPOTHALAMIC NUCLEI— SCHEMATIC

123. HYPOTHALAMIC FIBER CONNECTIONS AND NUCLEAR SUMMARY

I. Summary of hypothalamic nuclei

A. THE NUCLEI ARE LOOSELY ORGANIZED GROUPS OF CELLS in the walls of the third ventricle. They are not very well defined or distinct but appear as more or less well-circumscribed cell groups in a more diffuse conglomerate of cells

B. THE HYPOTHALAMUS MAY BE DIVIDED into medial and lateral areas
   1. The medial hypothalamus has many short axonal connections among cells which allow excitation to spread in any given direction from any focus to establish many self-exciting networks
   2. Large neurons of the lateral hypothalamus have more specific axonal connections. The many neuronal connections of the medial forebrain bundle traverse this lateral hypothalamic area

II. Fiber connections of the hypothalamus include those to the spinal cord, brainstem, cerebrum, as well as intrahypothalamic nuclear fibers

A. AFFERENT FIBER CONNECTIONS
   1. Stria terminalis: a small fiber bundle running in the terminal sulcus (between thalamus and caudate nucleus) and carrying fibers from the amygdaloid nucleus to the anterior, ventromedial, and preoptic nuclei of the hypothalamus
   2. Pallidohypothalamic fibers pass from the lentiform nucleus to the ventromedial hypothalamic nucleus
   3. Thalamohypothalamic fibers originate in the medial and midline thalamic nuclei
   4. Fornix (corticomamillary fibers) conveys fibers from the hippocampus to the mamillary bodies and possibly the preoptic nuclei
   5. Medial forebrain bundle (olfactohypothalamic fibers) is a phylogenetically ancient fiber system comprised of many neuronal connections interposed in the bundle
      a. Extends throughout the entire lateral hypothalamus
      b. Continues rostrally through the preoptic area to the olfactory region and caudally to the midbrain
      c. Carries fibers from nuclei in the paraolfactory area and corpus striatum to the hypothalamus
      d. Composed of short relay and long fibers
      e. Conduction is from the lateral to the more medial areas with short axon neurons traversing the medial forebrain bundle or collaterals arising from fibers in the bundle

B. EFFERENT FIBER CONNECTIONS begin as a conspicuous bundle of fibers, the fasciculus mamillaris princeps, in the mamillary nuclei and divide to form
   1. Mamillothalamic fasciculus (hypothalamic-thalamic tract) which ends in the anterior thalamic nucleus
   2. Mamillotegmental fasciculus ends in the dorsal tegmental and interpeduncular nuclei of the midbrain
   3. Dorsal longitudinal fasciculus (periventricular system) is located in the periventricular gray of the brainstem and connects the hypothalamus with the reticular formation of the mesencephalon and parasympathetic nuclei
   4. Hypothalamicohypophyseal tract is made up of fibers that arise in the supraoptic and paraventricular nuclei and carry neurosecretory products to the neurohypophysis (pars nervosa of the pituitary)
   5. Medial forebrain bundle receives fibers from the hypothalamus and continues to the paramedian cell groups of midbrain tegmentum and in the mesencephalon reticular formation and raphe nuclei

HYPOTHALAMIC AFFERENTS (INPUT)

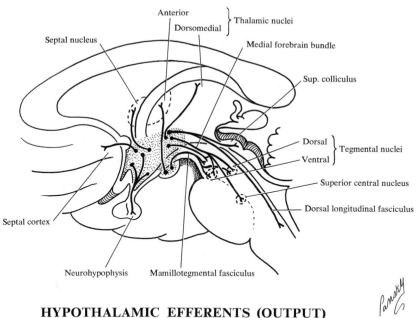

HYPOTHALAMIC EFFERENTS (OUTPUT)

124. HYPOTHALAMIC FUNCTIONAL CONSIDERATIONS

I. **The hypothalamus** functions as the primary control center for the visceral system and serves to integrate activity for consciousness, visceral, limbic, and endocrine systems. It is considered to be diversified in activities and includes regulatory influence on the autonomic nervous system. Stimulation of the anterior area nuclei is excitatory to the restorative (parasympathetic) activities, while stimulation of the posterior nuclei is preparative (sympathetic). Except for the supraoptic and paraventricular nuclei, specific functions have not been determined for individual nuclei. The following functions are controlled by the hypothalamus

A. THE SUPRAOPTIC AND PARAVENTRICULAR NUCLEI contain neurons whose axons convey neurosecretory droplets in the hypothalamohypophyseal tract to the pars nervosa (posterior lobe of pituitary)
  1. Supraoptic nucleus is associated with the release of ADH (antidiuretic hormone)
  2. Paraventricular nucleus is associated with the release of oxytocin
B. THE HYPOTHALAMUS ALSO INFLUENCES THE SECRETORY ACTIVITY of the pars distalis (anterior lobe of the pituitary)
C. THE HYPOTHALAMUS HAS CERTAIN NEURONAL ORGANIZATIONS referred to as "centers" which subserve other "centers" in the brainstem via complex synaptic pathways. These centers in the brainstem are concerned with the autonomic regulation of respiration, vomiting, micturition, and cardiovascular activity
D. TEMPERATURE IS CONTROLLED in part by the hypothalamus. Thermoreceptor-sensitive neurons (center) in the hypothalamus initiate the loss or conservation of heat by vasodilatation, vasoconstriction, sweating, shivering, etc.
  1. The anterior hypothalamic center regulates loss of heat
  2. The posterior hypothalamic center regulates the conservation of heat
E. WATER METABOLISM AND "DRINKING" are influenced by the stimulation of the lateral hypothalamus
F. FOOD INTAKE AND FEEDING BEHAVIOR are controlled in the lateral hypothalamic areas, while the "satiety" of feeding is controlled in the medial (ventromedial) areas. Lesions in the ventromedial area usually lead to hyperphagia and obesity; lateral hypothalamic lesions cause aphagia and ultimate emaciation
G. SLEEP is regulated through the consciousness system which has connections with the posterior hypothalamus. Lesions here may cause excessive sleep (hypersomnia)
H. SEXUAL FUNCTIONS are influenced by the hypothalamus and its complex interconnections with other parts of the brain
I. EMOTIONS are considered in the realm of hypothalamic activities since it is part of the limbic system

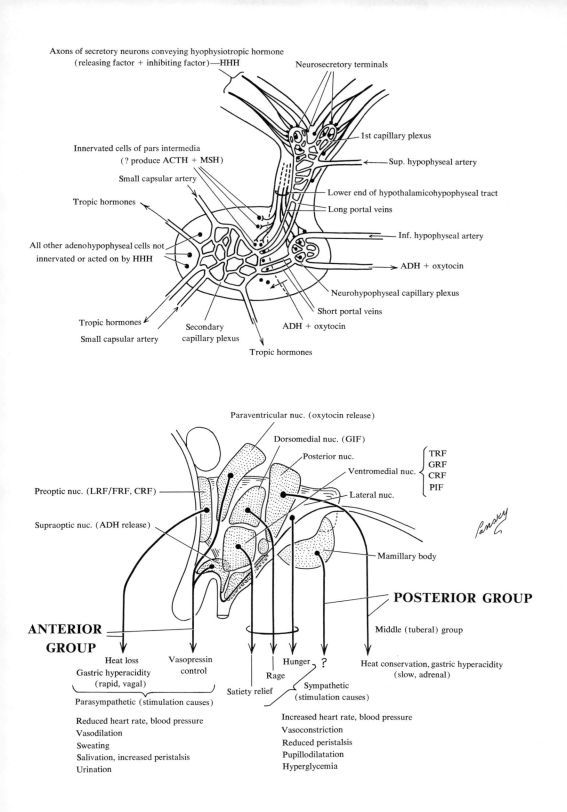

Unit Seven

THE BASAL GANGLIA

125. BASAL GANGLIA: GENERAL CONSIDERATIONS AND DEFINITIONS

I. Definitions: the *basal ganglia* are the three large subcortical gray (nuclear) masses derived from the telencephalon

A. THESE SUBCORTICAL COLLECTIONS are comprised of neurons and intermixed finely myelinated axons

B. THE GANGLIA ARE PHYSICALLY SEPARATED from the diencephalon by the internal capsule, but intimately related to the cerebral cortex, thalamus, subthalamus, and substantia nigra

C. THE TERM "BASAL GANGLIA" was originally used to include all gray matter in the diencephalon and telencephalon, but the term is currently used to designate only the following structures

    1. Caudate nucleus: made up of a head which forms lateral wall of anterior horn of the lateral ventricle, a body overlying the lateral part of the dorsal thalamus, and a tail located above temporal horn of lateral ventricle

    2. Lenticular (lentiform) nucleus: divided into an outer larger portion, the putamen, and an inner portion, the globus pallidus, by the lateral medullary lamina (a vertical plate of white matter). The whole nucleus has the size and shape of a Brazil nut. Unlike the caudate nucleus, it has no ventricular surface, lies deeply buried in the white matter of the cerebral hemisphere. It is closely applied to the lateral surface of the internal capsule which separates it from the caudate nucleus rostrally and thalamus caudally. In current usage, the individual terms (putamen and globus pallidus) are preferred over lenticular nucleus

        a. Putamen (shell): the largest and most lateral part of the basal ganglia. It underlies the insula and is continuous anteriorly with the head of the caudate nucleus. It is composed primarily of small medium-sized nerve cells

        b. Globus pallidus (pale globe): the nucleus that comprises the medial portion of the lenticular nucleus located between internal capsule and putamen. Fibers from this nucleus enter ansa lenticularis, lenticular fasciculus, and other tracts associated with the extrapyramidal system

D. STRIATUM is a collective term for caudate nucleus and putamen because of the striated appearance, while corpus striatum refers to the combination of caudate nucleus, putamen, and globus pallidus

    1. Corpus striatum is a single gray mass during early development, but becomes separated by the fibers of the internal capsule into two distinct cellular masses, the caudate and the lenticular nuclei

    2. The corpus striatum is the part of the basal ganglia considered to be concerned with somatic motor functions. Neurologists often use the term "extrapyramidal motor system" to group together the corpus striatum and certain related brainstem nuclei considered to subserve these somatic motor functions

E. THE AMYGDALA (AMYGDALOID NUCLEAR COMPLEX) AND CLAUSTRUM, included by some authorities with the basal ganglia, but for clinical purposes they are not considered to be part of the basal ganglia

F. THE NEOSTRIATUM is comprised of caudate and putamen

G. THE PALEOSTRIATUM consists of globus pallidus (lateral and medial portions)

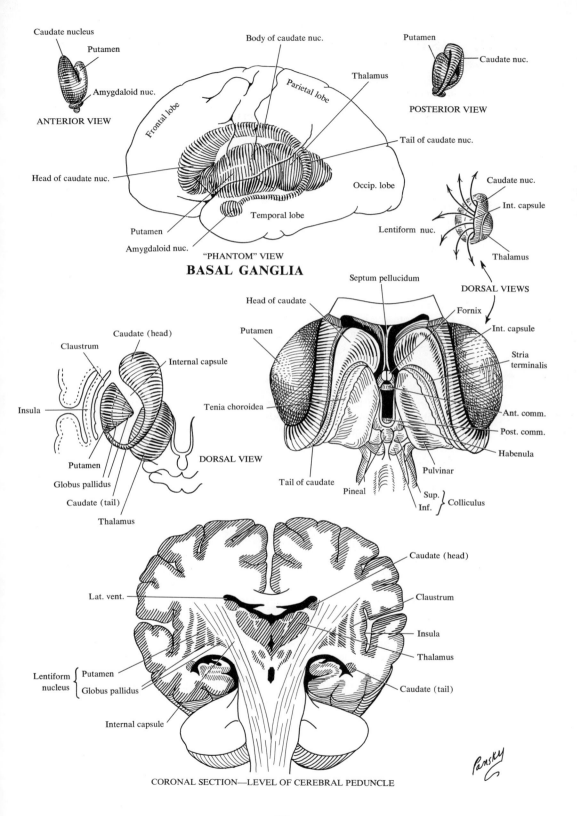

Caudate nucleus
Putamen
Amygdaloid nuc.
ANTERIOR VIEW

Head of caudate nuc.

Putamen
Amygdaloid nuc.

Body of caudate nuc.
Thalamus
Parietal lobe
Frontal lobe
Tail of caudate nuc.
Occip. lobe
Temporal lobe
"PHANTOM" VIEW

Putamen
Caudate nuc.
POSTERIOR VIEW

Caudate nuc.
Int. capsule
Lentiform nuc.
Thalamus
DORSAL VIEWS

BASAL GANGLIA

Caudate (head)
Claustrum
Internal capsule
Insula
Putamen
Globus pallidus
Caudate (tail)
Thalamus
DORSAL VIEW

Head of caudate
Putamen
Tenia choroidea
Tail of caudate
Pineal

Septum pellucidum
Fornix
Int. capsule
Stria terminalis
Ant. comm.
Post. comm.
Habenula
Pulvinar
Sup.
Inf. } Colliculus

Lat. vent.

Lentiform { Putamen
nucleus { Globus pallidus

Internal capsule

Caudate (head)
Claustrum
Insula
Thalamus
Caudate (tail)

CORONAL SECTION—LEVEL OF CEREBRAL PEDUNCLE

126. BASAL GANGLIA: ANATOMIC RELATIONSHIPS

I. Anatomic relationships

A. THE CORPUS STRIATUM is shown to be interrupted by massive fiber bundles of the internal capsule when viewed in horizontal and coronal sections of the brain

1. In a rostral coronal section, it is revealed that the heads of the caudate and putamen are "fused" along the ventromedial surface of the caudate, thus indicating their common origin. There is a topographic relationship, with rostral cortex projecting to the head of the caudate

2. In a slightly more posterior coronal section, the caudate and lenticular nuclei are completely separated by the fiber bundles of the internal capsule

B. THE INTERNAL CAPSULE in horizontal section is V-shaped with the apex directed medially with the following spatial relationships

1. The anterior limb or arm of the internal capsule separates caudate and lenticular nuclei. The anterior limb is comprised of
   a. Thalamocortical and corticothalamic fibers which are reciprocal connections between the lateral thalamic nucleus and the frontal lobe of the cerebral cortex
   b. Frontopontine fibers from the frontal lobe to the pontine nuclei
   c. Fibers from the caudate nucleus to the putamen. (*Note:* the existence of these fibers is currently in question.)

2. The posterior limb of the internal capsule separates the thalamus and the lenticular nucleus and is subdivided into three portions which contain the following designated fibers or fiber tracts
   a. Lenticulothalamic portion
      i. Corticobulbar tract and corticospinal tract (nerve fibers to superior limb lie anterior to those to inferior limb)
      ii. Corticorubral tract: comprised of nerve fibers from cortex of the frontal lobe to red nucleus which run closely with those in the corticospinal tract
   b. Retrolenticular part
      i. Fibers from lateral nucleus of thalamus to postcentral gyrus
   c. Sublenticular (below lenticular nucleus) part
      i. Parietotemporopontine fibers from parietal and temporal lobe cortex to pontine nuclei
      ii. Auditory radiations from medial geniculate body to transverse temporal gyrus or Heschl's gyrus (areas 41, 42)
      iii. Optic fibers from lateral geniculate body to area 17

C. THE CAUDATE NUCLEUS is an elongated subcortical nuclear mass

1. Head portion is contiguous with anterior perforated substance
2. Lies adjacent to inferior border of anterior horn of lateral ventricle
3. Tapers as it continues posteriorly and inferiorly as the tail which enters the roof of the inferior (temporal) horn of lateral ventricle to end at level of the amygdala

D. THE LENTICULAR NUCLEUS is wedge-shaped and located within the limbs of internal capsule between insula (laterally), caudate nucleus, and thalamus (medially). It is subdivided into the putamen, the larger convex portion lying medial to the insular cortex, and the globus pallidus, the smaller medial, triangular-shaped portion which appears lighter in color because of its many myelinated fibers

Thalami

Cerebral cortex

Corona radiata

Caudate nuc.

Int. capsule

Lentiform nucleus

II

Colliculi

Frontopontile tract

Anterior limb

Anterior thalamic radiations

Caudate

Genu

Putamen

Claustrum

Corticobulbar + sensory
radiations to area 8

Corticorubral + corticotegmental
from area 6 + sensory radiations

GP   GP

Corticobulbar + sensory
rad. to area 4

Globus pallidus

T

Head of caudate

Ant. limb

Genu

Post. limb

Thal.

Cortico-
spinal

{ Upper
Trunk
Lower

Post. limb—sublenticular part

Auditory radiations

Occipitotemporo-(parieto)
pontine tracts

Sensory radiations
(areas 3,1,2,5,7)

Post. limb
postlenticular part

Visual radiations

Post. thalamic radiations,
corticotectal + corticotegmental tracts

Int.capsule (ant. limb)

Int. capsule (post. limb)

Caudate head

Ext. capsule

Putamen

Insula

Globus pallidus

Claustrum

Thalamus

INTERNAL CAPSULE

Panky

127. BASAL GANGLIA: FIBER CONNECTIONS AND PATHWAYS

I. Principal afferent connections

A. CEREBRAL CORTEX contributes majority of afferent projections to caudate and putamen (neostriatum)
   1. Fibers from the rostral cortex project ipsilaterally to the head of the caudate
   2. Sensorimotor cortex (areas 3, 2, 1) sends fibers mainly to the tail of the caudate
B. THALAMUS is the most important source of subcortical afferents to the caudate nucleus and putamen
   1. Centromedian nucleus sends afferent fibers to the putamen
   2. Intralaminar nuclei projects to the caudate nucleus
C. SUBSTANTIA NIGRA sends afferents to the caudate nucleus and globus pallidus
D. NEOSTRIATUM (CAUDATE AND PUTAMEN) sends afferents to the globus pallidus
E. SUBTHALAMIC NUCLEUS sends fibers to the medial portion of the globus pallidus

II. Principal efferent connections

A. CAUDATE AND PUTAMEN (NEOSTRIATUM) project efferents to both medial and lateral portions of globus pallidus and the substantia nigra
B. GLOBUS PALLIDUS (PALEOSTRIATUM) provides the major outflow system for the corpus striatum
   1. Lateral portion gives rise to majority of pallido-subthalamic fibers
   2. Medial portion gives rise to four fiber bundles
      a. Lenticular fasciculus originates from the dorsal aspect of the medial portion of the globus pallidus and runs rostral to the subthalamic nucleus and ventral to the zona incerta as Forel's field H_2
      b. Ansa lenticularis fibers merge with the fibers of the lenticular fasciculus in the prerubral field (field H of Forel). These two bundles then project dorsolaterally as a component of the thalamic fasciculus
      c. Thalamic fasciculus (Forel's field H_1) is a complex bundle which contains pallidothalamic, rubrothalamic, and dentatothalamic fibers
         i. It projects fibers to VA, VL, centromedian, and intralaminar thalamic nuclei
         ii. The region where thalamic fasciculus enters the thalamus is perhaps the most strategic single location in the brain
            (a) Fibers from the corpus striatum and cerebellum converge at this point to enter the specific thalamic nuclei
            (b) Fibers exit the thalamus at this point to project to motor and premotor cerebral cortex
C. SUBSTANTIA NIGRA sends efferent fiber projections to the medial portion of VA and VL nuclei of the thalamus and the neostriatum

PRINCIPAL AFFERENT CONNECTIONS— BASAL GANGLIA

Caudate nucleus

Cerebral cortex

Thalamus {
Intralaminar nuclei

Centromedian nucleus

Red nucleus

Subthalamic nucleus

Substantia nigra

Putamen

Globus pallidus

Tail of caudate nucleus

PRINCIPAL EFFERENT CONNECTIONS

Lenticular fasciculus

Caudate nucleus

Putamen

Globus pallidus

Ant. + lat. ventral nuclei of thalamus

Dentatothalamic fibers

Red nucleus

Substantia nigra

Ansa lenticularis

Pallidothalamic fibers

To midbrain tegmentum

Subthalamic nucleus

Zona inserta

Putamen

TA   TL

TM

Thalamic fasciculus (field H₁ of Forel)

Dentatothalamic, rubrothalamic + thalamostriate fibers

Prerubral field (field H of Forel)

Fasciculus lenticularis

Claustrum

Field H₂ of Forel

Globus pallidus

Pallidohypothalamic fasciculus

Ansa lenticularis

Fasciculus subthalamus

Subthalamic nucleus

128. BASAL GANGLIA: FUNCTIONAL AND CLINICAL CONSIDERATIONS—PART I

I. **Introduction:** the basal ganglia control system includes putamen, globus pallidus, caudate nucleus, ventral anterior nucleus of the thalamus, subthalamic nucleus, substantia nigra, and their connections in the cerebral hemispheres and diencephalon

A. THIS CONTROL SYSTEM OR CIRCUIT exerts a profound influence on motor activities, but the precise mechanism of action for specific areas has not been demonstrated

B. THE BASAL GANGLIA CONTROL CIRCUIT is a modulating system that functions to either facilitate or inhibit motor activities
  1. Excessive discharge (electrical stimulation) of the caudate nucleus slows or suppresses motor function
  2. Electrical stimulation of the globus pallidus causes contraversive turning and postural changes
  3. Lesions in the subthalamic nucleus usually suppress motor activity
  4. Lesions in the striatum (caudate nucleus and putamen) cause deficits in the more complex motor functions

II. **Pathologic conditions:** when the delicate balance in this modulating system (facilitation vs. suppression) is disrupted, the following pathologic symptoms or signs of abnormalities in motor function are observed

A. LOSS OR REDUCTION OF MOVEMENT (HYPOKINESIA): the term "hypokinesia" refers to the reduction in initiation, implementation, and facility of execution of movement and is characterized by
  1. Tone: there is usually an increase in tone (hypertonia) which occurs as a direct effect on alpha motor neurons and is known as rigidity. Depending upon the degree of hypertonia, some or all of the following clinical signs may be observed
    a. Movements of limbs or the entire body are initiated slowly and stopped with difficulty
    b. Conscious and emotional movements may be suppressed
    c. Face is masklike (fixed facial expression)
    d. Abnormal postures may be assumed
    e. Patient may appear paralyzed due to exaggerated tone
    f. Arm swing during walking is absent
    g. Tendon reflexes usually appear normal
  2. Rigidity: when hypertonicity is widespread over most of the body, the condition is called rigidity and manifests itself in one of the following
    a. Cogwheel rigidity: increased resistance to passive movement, usually with "cogwheel jerks" but without the clasp-knife phenomenon of spasticity
    b. Plastic rigidity: increased resistance to passive movement is constant, continuous, and smooth (not jerky)

B. INVOLUNTARY MOVEMENTS (HYPERKINESIA): greatly increased activity in the dopaminergic pathway may result in the following types of involuntary movements
  1. Tremor may occur when the body is at rest (postural, static, or alternating) or during voluntary muscular activity (intention or action)
    a. Alternating tremor is the result of the alternating contraction of opposing muscle groups such as adductors-abductors and flexors-extensors. The "pill-rolling" phenomenon is an example of tremor in the hand. All alternating tremors are usually suppressed or absent during voluntary activity and may be completely absent during sleep
    b. Intention tremor occurs during voluntary muscular activity and becomes more exaggerated near the end of the voluntary act

TREMOR

RIGIDITY

TREMOR WITH RIGIDITY—
EXAGGERATED WITH
VOLUNTARY ACT

MASKLIKE "FIXED"
FACIAL EXPRESSION

RIGIDITY

UNCONTROLLED
CONTINUED
MOVEMENT

SLOW TO BEGIN
MOVEMENT

RIGIDITY

Flexed

Flexed

Flexed

Flexed

PARKINSON STATE

129. BASAL GANGLIA: FUNCTIONAL AND CLINICAL CONSIDERATIONS—PART II

I. Pathologic conditions (cont.)

A. INVOLUNTARY MOVEMENTS (CONT.)

1. Athetosis is the disorder in which there are slow, twisting, involuntary movements of the extremities, particularly of the fingers and hands. It may occur at rest or during voluntary muscle movements and is referred to as "serpentine movement"
2. Dystonia is the disorder that involves abnormal posturing of the trunk or an extremity. Movements are slow, involuntary, and accompanied by bizarre, grotesque twisting of the shoulder, pelvic girdle, and trunk
3. Chorea is characterized by sudden, jerky, involuntary movements, along with grimacing, twitching of facial muscles, and faulty vocalization. These movements occur at rest or during voluntary acts
4. Ballism is characterized by involuntary movements of an entire limb which begins in the proximal portion and passes in wavelike movements distally. There may be gross, rapid, flinging movements when the subthalamic nucleus is involved. *Hemiballism* is the condition when both limbs on the same side are involved. *Monoballism* denotes only one limb is involved. This disorder is the only one involving the basal ganglia whose onset may occur suddenly

II. Specific hypokinetic and hyperkinetic disorders

A. PARALYSIS AGITANS (PARKINSON'S DISEASE), the most common type of basal ganglia disease, is characterized by the following symptoms, clinical signs, and traits

1. Massive degeneration of the cerebral cortex, globus pallidus, and substantia nigra (usually accompanied by lesions in the substantia nigra)
2. Alternating tremor occurs at the rate of 3–5 per second
3. Decreased striatal dopamine level which results in rapid firing of cells of the striatum; normally dopamine of the substantia nigra inhibits this rapid firing of cells of the striatum
4. Loss of the normal modulating influence of the basal ganglia on motor activities of the pyramidal and extrapyramidal systems
5. Cogwheel and plastic rigidity may be observed along with akinesia
6. Flexed posture throughout the body may result
7. Cause is unknown (idiopathic), but often follows encephalitis, carbon monoxide poisoning, manganese poisoning, or other drug toxicity

B. HUNTINGTON'S CHOREA is a progressive, dominantly inherited disorder of adults with onset usually between the ages of 35 and 40 with the following signs and symptoms

1. Moderate to severe choreiform motor disability; chorea is random with continuous occurrence of brief, abrupt, jerky movements of body parts
2. Diffuse degeneration with widespread loss of neurons in the cerebral cortex and basal ganglia; atrophy of caudate nucleus and putamen allows dilatation of the lateral ventricles
3. General and progressive mental deterioration (dementia)
4. Decreased muscle tone
5. Reflexes and sensations are usually normal

C. SYDENHAM'S CHOREA is a disorder in children or adolescents often associated with rheumatic fever with the following symptoms and signs

1. Sudden, irregular, involuntary, and purposeless movements
2. Petechial (small hemorrhagic spots) lesions may be present in corpus striatum

CONGENITAL
CHOREOATHETOSIS

SPASMOTIC DYSTONIA

Flexion

Smoothed out,
anxious facies
(little mobility)

Flexion

Flexion

PARALYSIS AGITANS
(PARKINSON'S DISEASE)

POSITION OF HAND AND
FINGERS DURING ATHETOTIC
MOVEMENT

ATHETOSIS

Unit Eight

FUNCTIONAL
CEREBRAL CORTEX

130. CEREBRAL CORTEX: GENERAL CONSIDERATIONS

I. **Definition:** the cerebral cortex or pallium is the term applied to the thin layer of gray matter which covers the white matter of the convoluted cerebral hemispheres. This gray matter or cortex is comprised of cell bodies of neurons, neuroglia, nerve fibers, and blood vessels

A. GRAY APPEARANCE: characteristic of the cerebral cortex imparted by its rich vascular supply and relative absence of myelinated nerve fibers

B. NEARLY 14 BILLION NEURONS have been estimated to be in the 2200 sq cm of cortex within both cerebral hemispheres

C. THICKNESS OF THE CORTEX: averages 2.5 mm. It varies considerably in different areas: 4.5 mm in the motor area of the precentral gyrus; 1.5 mm in the calcarine fissure

II. **Summary of the development of the cerebral cortex** (also see p. 30)

A. DEVELOPS FROM THE TELENCEPHALON, the most rostral of the three original brain vesicles

B. THE EARLY TELENCEPHALIC VESICLE reveals three zones or concentric layers
1. Matrix (germinal or ependyma) layer surrounds the lateral ventricles
2. Mantle (intermediate) layer gives rise to the white matter of the cerebrum
3. Marginal (outer) layer gives rise to the cortical layer of the cerebrum

C. COLUMNAR EPITHELIAL CELLS, originally present in the most primitive neural tube, extend through the above three layers

D. CELLS MIGRATE FROM THE MANTLE LAYER into the marginal layer to form the cerebral cortex by the end of the second month
1. The cortex increases in thickness as cells migrate into and become differentiated in the marginal zone

E. CELLS IN THE CORTICAL ZONE become organized into layers (horizontally arranged) between 6–8 months
1. Cells formed at the same time tend to migrate to the same layer and remain together there
2. Newly formed cells migrate through these previously formed layers to form more superficial layers—"inside-out" neuronal migration

III. **Summary of cerebral function:** based upon known information on cortical organization in terms of cells and fibers, the following general statements can be made about cortical functions

A. CORTICAL LAYERS I–III (MOLECULAR, EXTERNAL GRANULAR, AND EXTERNAL PYRAMIDAL) have numerous stellate cells which indicates that these three layers are important for association and higher functions such as memory, interpretation of sensory input, and certain discriminative functions

B. INTERNAL GRANULAR LAYER (IV) is mainly a receptive layer (thalamocortical fibers end here)

C. LAYERS V AND VI (GANGLIONIC AND MULTIFORM) are primarily efferent layers that contain nerve cell bodies whose axons enter the corticospinal tract

IV. **Interrelation of cortical nerve cells:** histologic information concerning these neurons, their processes, and intracortical relationships has been obtained by using special silver impregnation methods such as the Golgi method. The latter is especially useful when studying neuronal relationships and their axonal and dendritic arborizations. In Nissl-stained sections, cell bodies can be studied; and with the Weigert method, the course and distribution of myelin fibers can be investigated

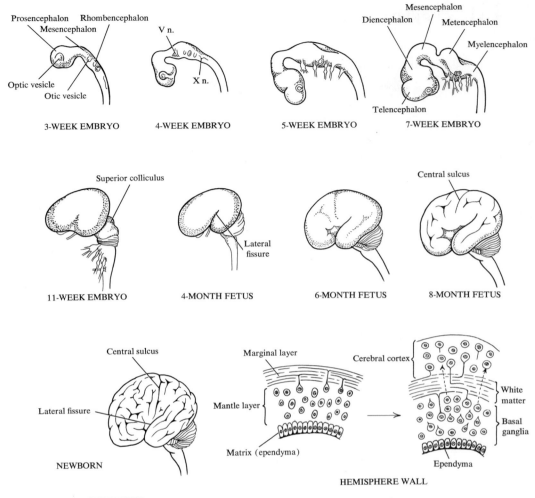

3-WEEK EMBRYO   4-WEEK EMBRYO   5-WEEK EMBRYO   7-WEEK EMBRYO

11-WEEK EMBRYO   4-MONTH FETUS   6-MONTH FETUS   8-MONTH FETUS

NEWBORN   HEMISPHERE WALL

CEREBRAL CORTEX

LAYERS OF
CEREBRAL CORTEX
(DIAGRAMMATIC)

I
II
III
IVa
IVb
V
VI
Central white matter

131. CEREBRAL CORTEX: CORTICAL CELLS

I. **Cortical cells:** five types of neurons are found in the cerebral cortex. These cells have been demonstrated by several neurohistologic stains (Golgi, Nissl, and Weigert)

A. PYRAMIDAL CELLS: most characteristic and numerous cells of the cortex
1. Cell bodies are pyramidal or triangular in shape and possess prominent Nissl granules in their cytoplasm
2. They vary in size from 10 to more than 100 μm in height
   a. Small pyramidal cells measure 10–20 μm; medium-sized cells 20–30 μm; large cells 30–70 μm; and giant cells 70–100 μm (and larger)
3. One or more types (sizes) of pyramidal cells may be found in all but the outermost cortical (molecular) layer
4. Giant pyramidal cells (Betz cells) are characteristic of the precentral gyrus (motor area 4)
5. All pyramidal cells are usually oriented with their apices directed toward the cortical surface
6. Apical dendrites branch profusely within the cortical layers that lie superficial to the location of the cell body
7. Other dendrites sprout from the base and body of these cells and usually extend horizontally within the same layer of the cortex
8. The axon typically arises from the base of the cell body and enters the underlying white matter as one of the efferent cortical fibers

B. GRANULE CELLS (STELLATE) are small, star- or polygon-shaped neurons with scant cytoplasm (second most numerous type)
1. Cell bodies vary in size from 4–8 μm
2. Dendrites are numerous, short, and highly branched in all directions
3. The axon is short and branches near the parent cell body
4. These cells are present in all cortical layers, but most numerous in layer IV
5. They are classified as Golgi type II neurons and are correlative in function

C. FUSIFORM (POLYMORPHIC MULTIFORM) CELLS are usually spindle-shaped, but they may show wide variation in morphology (such as pyramidal or ovoid)
1. Limited to the deepest layer of the gray cortex
2. Their long axis is perpendicular to the cortical surface
3. Dendrites sprout from the two poles of the cell with more superficial dendrites ascending toward the surface and deeper ones branching in the same cortical layer. Their axons leave the cell body to enter the white matter

D. HORIZONTAL CELLS (OF CAJAL) are small neurons with fusiform cell bodies (usually spear- or spindle-shaped) with large nuclei and scant cytoplasm
1. Found only in the outermost layer (molecular) of the cortex; their axis is parallel to the surface
2. Dendrites spread out from each end of the cell body in a plane that is also parallel to the surface of the cortex
3. Axon leaves the cell body and spreads out parallel to the surface in the outermost layer of the cortex
4. These cells are correlative in function

E. CELLS OF MARTINOTTI (ASCENDING AXON CELLS) are small and usually polygonal in shape but may be round or triangular
1. Cell bodies are found in all cortical layers with possible exception of the outer (molecular) layer. These cells are associative in function
2. Dendrites are short and spread out in all directions from the cell body
3. Axon ascends from cell body toward the cortical surface and branches in the outermost layer of cortex

HUMAN CEREBRAL CORTEX

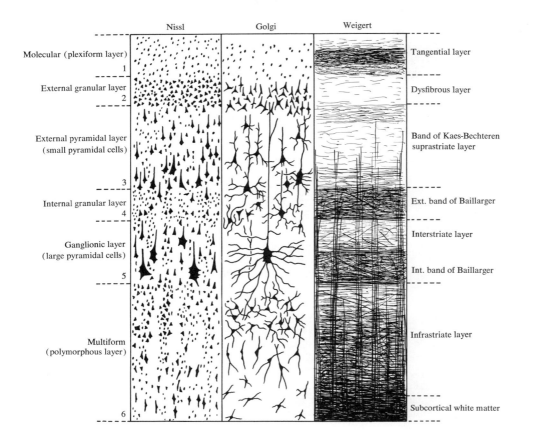

	Nissl	Golgi	Weigert	
Molecular (plexiform layer) 1				Tangential layer
External granular layer 2				Dysfibrous layer
External pyramidal layer (small pyramidal cells) 3				Band of Kaes-Bechteren suprastriate layer
Internal granular layer 4				Ext. band of Baillarger
Ganglionic layer (large pyramidal cells) 5				Interstriate layer
				Int. band of Baillarger
Multiform (polymorphous layer)				Infrastriate layer
6				Subcortical white matter

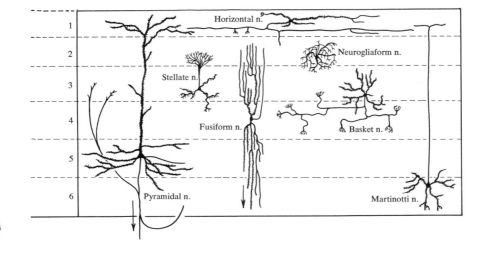

- Horizontal n.
- Neurogliaform n.
- Stellate n.
- Fusiform n.
- Basket n.
- Pyramidal n.
- Martinotti n.

132. CEREBRAL CORTEX: CORTICAL LAYERS

I. **Cortical layers:** the major portion (90%) of the cerebral cortex in man is characterized by having six layers of cells in the adult. This six-layer cellular arrangement is found in the neopallium (new cortex) which is also termed neocortex, isocortex, and homogenetic cortex. Olfactory cortex or paleopallium (old cortex) and the hippocampal formation (archipallium) usually have three basic cellular layers and are termed the allocortex or heterogenetic cortex. The six basic layers seen in the neocortex are, from the pial surface

A. MOLECULAR LAYER I (200 μm): often called the plexiform layer because of the fairly dense fiber plexus of axons of horizontal cells and dendrites of deeper lying pyramidal and granule cells
   1. Horizontal cells of Cajal are distinguishing cells (also see granule cells)

B. EXTERNAL GRANULAR LAYER II (200 μm) is made up of some small pyramidal cells and numerous small stellate cells
   1. Round cell bodies of these stellate cells, seen in Nissl preparations, have led to the designation of these cells as granule cells
   2. Apical dendrites enter the molecular layer
   3. Axons of granule cells descend to either terminate within the cortex or extend into the white matter

C. EXTERNAL PYRAMIDAL LAYER III (450 μm) is made up mainly of pyramidal cells, but it also contains some small granule and Martinotti cells
   1. Apical dendrites of pyramidal cells ascend into the molecular layer
   2. Axons from pyramidal cells descend into the multiform layer

D. INTERNAL GRANULAR LAYER IV (200 μm): homogenetic cortex has a preponderance of small stellate or granular cells. The latter are more numerous in granular heterogenetic cortex of the postcentral sensory cortex. In contrast, granule cells are obscured by large pyramidal cells in the agranular heterogenetic cortex of the precentral motor area so that the motor cortex is called agranular
   1. Axons of granule cells are short and usually remain within layer IV (some may descend into deeper layers)
   2. Layer IV has a large number of horizontal, myelinated nerve fibers (external band of Baillarger) from the thalamus which branch and end here

E. INTERNAL PYRAMIDAL (GANGLIONIC) LAYER V (250 μm): made up mainly of cell bodies of medium- and large-sized pyramidal neurons
   1. Stellate and Martinotti cells are present but not in abundance
   2. Apical dendrites of large pyramidal cells extend into the molecular layer; those of smaller pyramidal cells penetrate into layer IV and branch
   3. Axons of pyramidal cells descend into the white matter mainly as projection fibers (some association fibers)
   4. Internal band of Baillarger is formed by horizontal myelinated fibers in the deeper part of this layer
   5. Precentral motor cortex contains giant pyramidal cells of Betz in this layer to make its identification easy

F. FUSIFORM (MULTIFORM) LAYER VI (250 μm): predominant cell is the fusiform cell
   1. Spindle-shaped cells oriented with their long axis vertical to surface
   2. Some Martinotti, stellate, and small pyramidal cells may be present here
   3. Dendrites from large fusiform cells ascend into the molecular layer; those from small cells pass into layer IV
   4. Fusiform cell axons descend as projection and association fibers
   5. All fibers entering or leaving the cortex must pervade this layer
   6. Motor cortex may contain a few giant pyramidal cells of Betz in this layer, making it difficult to distinguish from the overlying ganglionic layer

Agranular   Frontal   Parietal   Granular   Polar

**MAJOR TYPES OF
CEREBRAL CORTEX**
(According to von Economo and Koskinas)

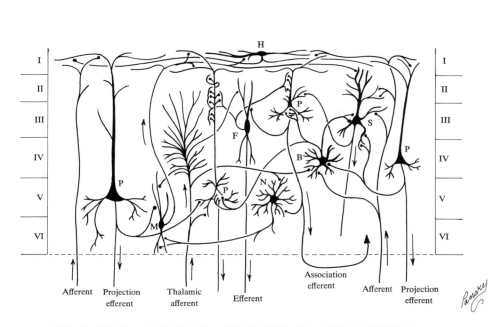

H

I
II
III
IV
V
VI

P   F   P   B   N   S   P   M

Afferent   Projection efferent   Thalamic afferent   Efferent   Association efferent   Afferent   Projection efferent

SOME INTRACORTICAL CIRCUITS OF NEOCORTEX

133. CEREBRAL CORTICAL AREAS: PREMOTOR AREAS

I. Introduction: the distribution pattern of cerebral cortical cells varies in different regions of the brain. Based on morphologic characteristics (some obvious, others very subtle) several cytoarchitectural maps have been drawn, dividing the cerebrum into areas. Some areas have known specialized functions, while for others no clear function has been demonstrated. These stated functions have been arrived at primarily by electrically stimulating the area and observing the response elicited. Histologic surveys based on differences in the arrangement and types of cells, and in the pattern of myelinated fibers, have resulted in several fundamental cortical maps (Campell, 1905, recognized some 20 cortical areas; Brodmann, 1909, extended the number of fields to 52; Economo, 1929, increased the number to 109; and Vogts, 1919, subdivided the human cerebral cortex into more than 200 fields). Brodmann's cortical maps with 47–52 areas are still used for descriptive purposes

II. Motor functions of specific cortical areas
A. PRIMARY MOTOR AREA (BRODMANN'S AREA 4)
   1. Located in precentral gyrus, along the anterior wall of central sulcus, and in the most anterior part of the paracentral lobule
   2. Cortex is unusually thick (4.5 mm)
   3. This area is classified as heterogenetic, agranular in structure (layer VI is obscured by an increase in pyramidal cells)
   4. Ganglionic layer (V) contains giant pyramidal cells of Betz
   5. Corticospinal (pyramidal) tract arises in part from area 4
      a. This tract carries impulses for highly skilled voluntary movements to lower motor neurons
      b. Largest corticospinal fibers may be axons of Betz cells found in layer V
      c. Hemispherectomy or complete decortication results in degeneration of corticospinal fibers
   6. Area 4 controls fine, highly skilled, voluntary movements of the body, especially the facial muscles and flexor musculature of the extremities
   7. Though the point of location of cortical centers for specific movements of body parts varies in individuals, the sequence of these centers seems to be constant. In Penfield's homunculus, the center for the pharynx (swallowing) is in the most inferior portion of the precentral gyrus followed upward (superiorly) by the tongue, jaw, lips, face, fingers, hand, etc.
B. PREMOTOR AREA (AREA 6)
   1. Located in front of area 4 and continues on the medial surface (superior frontal gyrus) to the cingulate gyrus
   2. Histologically, area 6 resembles area 4
      a. Giant pyramidal cells of Betz are not present, however, in area 6
      b. Internal granular layer (layer IV) is obscured by pyramidal cells (agranular heterogenetic cortex)
   3. This area is concerned with the development of motor skills
   4. Ablation of this specific area alone does not result in paresis, hypertonia, or grasp reflexes

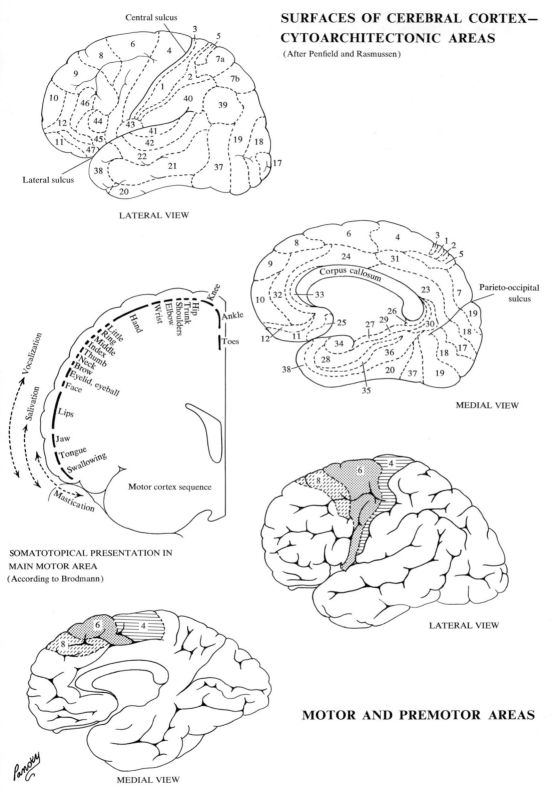

SURFACES OF CEREBRAL CORTEX— CYTOARCHITECTONIC AREAS
(After Penfield and Rasmussen)

Central sulcus

Lateral sulcus

LATERAL VIEW

Corpus callosum

Parieto-occipital sulcus

MEDIAL VIEW

Knee
Hip
Trunk
Shoulders
Elbow
Wrist
Hand
Little
Ring
Middle
Index
Thumb
Neck
Brow
Eyelid, eyeball
Face
Lips
Jaw
Tongue
Swallowing

Ankle
Toes

Vocalization
Salivation
Mastication

Motor cortex sequence

SOMATOTOPICAL PRESENTATION IN
MAIN MOTOR AREA
(According to Brodmann)

LATERAL VIEW

MOTOR AND PREMOTOR AREAS

MEDIAL VIEW

134. CEREBRAL CORTICAL AREAS: SUPPLEMENTARY MOTOR, FRONTAL, AND OCCIPITAL EYE FIELDS AND PREFRONTAL AREAS

I. Motor functions (cont.)

A. SUPPLEMENTARY MOTOR AREAS: no assigned number
   1. Located in the superior frontal gyrus on its medial surface just rostral to primary motor area 4
   2. Somatotopic pattern (sequence for representation of body parts) is reversed from that of motor area 4
   3. This area is concerned with the following movements as determined by electrically stimulating the area
      a. Assumption of posture
      b. Complex patterned muscle movements
      c. Rapid, infrequent, incoordinate muscle activities
   4. This area deals with bilateral, synergistic muscle activities

B. FRONTAL EYE FIELD (AREA 8)
   1. Located primarily in the caudal aspect of the middle frontal gyrus
   2. Histology within area 8 varies (more than one cytoarchitectural plan)
   3. This area is concerned with voluntary eye movements; these movements are independent of visual stimuli
   4. Stimulation of the area results in conjugate deviation of the eyes to the opposite side. These same movements can be made by asking the patient to look right/left
   5. The exact pathway for eliciting these responses is unknown, but the superior colliculus is probably involved

C. OCCIPITAL EYE FIELD: no assigned number
   1. Located in the occipital lobe, but not well localized
   2. Stimulation of the occipital cortex results in conjugate deviation of the eyes to the opposite side
   3. This area is concerned with involuntary eye movements; these movements are induced by visual stimuli
   4. These eye fields in the two cerebral hemispheres are interconnected by commissural fibers through the splenium, but the pathway is unknown

D. PREFRONTAL CORTEX (AREAS 9–12)
   1. Located rostral to motor and premotor cortical areas and comprises the remaining part of the frontal lobe
   2. Receives fibers from the dorsomedial nucleus of the thalamus
   3. Receives long association fibers from all lobes of the cerebrum
   4. This region (areas 9–12) is concerned with abstract thinking, mature judgment, foresight, tactfulness, and self-control
   5. Patients with large lesions (tumors or cortical atrophy from syphilis) of the frontal lobe often reveal the following traits
      a. Poor personal habits
      b. Vulgar speech
      c. General euphoria (clownish behavior)
      d. Poor self-control

LATERAL VIEW

Frontal motor eye field

Occipital eye field

Prefrontal
cortex

LATERAL VIEW

Corpus collosum

MEDIAL VIEW

Supplementary motor
areas

MEDIAL VIEW

Head, eyes, + trunk turn toward opposite side

Sensory aura in contralateral leg + synergistic movements
+ vague disagreeable sensations

Eyes turn to opposite side of body

Eyes turn to contralateral side
No visual aura

Complex optical hallucinations
(lights, colors, flames)

Primary motor

Primary sensory

Head, eyes, + trunk rotate to contralateral side

Chewing, licking, swallowing movements

Auditory hallucinations

FUNCTIONAL AREAS OF CORTEX BY ELECTRIC STIMULATIONS

135. CEREBRAL CORTEX: SENSORY FUNCTIONS OF CERTAIN CORTICAL AREAS

I. Primary sensory or somesthetic cortex (areas 3, 1, 2)

A. THE PRIMARY SENSORY CORTEX is necessary to recognize the source, quality, and quantity (severity) of pain and temperature; appreciate simple touch, light pressure, and vibrations from bony prominences; and discern fine discriminating touch and position and movement of the body parts (proprioception and kinesthetic sense)

  1. Located in the postcentral gyrus, including its medial extension into the paracentral lobule (posterior portion)
  2. Histologically the postcentral gyrus consists of three narrow strips of cortex (areas 3, 1, 2) that differ in structure: area 3 is granular heterogenetic cortex, while areas 1 and 2 are homogenetic and thicker
  3. The postcentral gyrus receives fibers from VPM and VPL nuclei of the thalamus
     a. These nuclei serve as relays for fibers from the medial lemniscus, spinothalamic tracts, and trigeminothalamic tracts
     b. Sensory modalities are detected and recognized at the level of the thalamus
  4. The various regions of the contralateral half of the body are represented as inverted in the postcentral gyrus, in a pattern corresponding to that of the motor cortex (area 4)

B. SECONDARY (SOMATIC) SENSORY CORTEX

  1. Located around the most ventral aspect of the pre- and postcentral gyri and continues posteriorly along the superior lip of the lateral fissure for a limited distance
  2. Receives impulses from several sensory modalities but pain appears to predominate
  3. Sequence for the location of body parts is in reverse to that given for the primary sensory cortex (no cortical areas identified for face, mouth, and swallowing mechanism)
  4. Representation of most body parts (extremities) is mainly contralateral

CEREBRAL HEMISPHERE—
CYTOARCHITECTONIC AREAS

LATERAL VIEW

MEDIAL VIEW

PRIMARY SENSORY CORTEX

SENSORY
SEQUENCE

Foot
Toes
Trunk
Neck
Head
Shoulder
Arm
Elbow
Forearm
Wrist
Hand
Little
Ring
Middle
Index
Thumb
Eyes
Nose
Face
Upper lip
Lips
Lower lip
Teeth, gums, jaw
Tongue
Pharynx
Intra-abdominal

Third somatosensory area

Primary (1st) somatosensory
area (3,1,2)

Second somatosensory area

LATERAL VIEW

Third somatosensory area

Primary somatosensory area
(3,1,2)

SOMATOSENSORY AREAS

MEDIAL VIEW

136. CEREBRAL CORTEX: VISUAL AND AUDITORY CORTEX

I. Primary visual cortex (area 17)
A. AREA 17 surrounds the calcarine fissure and is known as *the striate area*
  1. This cortical area is very thin (1.5 mm)
  2. Classified as granular, heterogenetic cortex
    a. Layer IV is thick and packed with stellate (granule) cells
  3. Outer band of Baillarger (Gennari) appears in layer IV of area 17
  4. This area gives rise to projection fibers to
    a. The pons (corticopontine)
    b. The tectum of the midbrain (corticotegmental)
    c. To the oculomotor complex and abducens nuclei
      i. Conjugate deviation of the eyes (involuntary) is initiated through these centers
      ii. Fixation on objects (both stationary and moving) is possible through these fiber connections
      iii. Accommodation reflexes necessary for the adjustment to near/far vision may be initiated through cortical connections with these nuclei
  5. Primary visual cortex receives the geniculocalcarine tract from the lateral geniculate body

II. Secondary visual cortex (areas 18 and 19)
A. THESE VISUAL PERCEPTION AREAS (18 AND 19) deal with the interpretation of the printed word and the spatial localization of an object
  1. Located in cortical regions that surround the primary visual area on its medial and lateral surfaces and account for visual perception
  2. Area 18 receives fibers from area 17 and sends fibers to area 19; both areas 18 and 19 send fibers into area 17 (reverberating circuits)

III. Primary auditory cortex (areas 41 and 42)
A. AREAS 41 AND 42 are concerned with the perception of simple auditory impulses
  1. Located in the floor of the lateral fissure on the transverse temporal gyri (Heschl's convolutions or gyri)
  2. Auditory input into the medial geniculate nucleus is mainly from the contralateral organ of Corti (some input from the ear of the same side) and it is the medial geniculate nucleus that is the main source of fibers (auditory radiations) into these auditory cortical areas
  3. A lesion in the primary auditory cortex on one side results in diminished auditory acuity in both ears, with the greater loss in the opposite ear
  4. A unilateral lesion in the auditory areas makes it more difficult to detect the direction from which sound comes

IV. Auditory association cortex (area 22)
A. AREA 22 is essential for the interpretation and appreciation of intricate sounds in the dominant hemisphere
  1. It is located on the lateral surface of the superior temporal gyrus (Wernicke's area)
  2. It has reciprocal fiber connections with areas 41 and 42
  3. Fibers from areas 41 and 42 into this area are responsible for the turning of the head and eyes to follow sounds
  4. Area 22 receives and synthesizes language data from sensory areas into a comprehensive sensory language pattern

VISUAL AREAS IN THE OCCIPITAL LOBE (AREAS 17, 18, 19)

LATERAL VIEW

19

18

17

MEDIAL VIEW

Area of Broca—motor speech

45 44

40

39

22

37

Auditory

Visuoauditory

Wernicke's area—motor speech

Posterolateral ventral sulcus nucleus

Lateral ventral nucleus

Lateral posterior nucleus

Medial dorsal nucleus

Lateral geniculate body

Pulvinar

Medial geniculate body

CORTEX AND SPECIFIC THALAMIC CONNECTIONS

Frontal association area

Higher intellectual + psychic functions

Motor area

Sensory area

Hand skills

Speech 2

Auditory association area

Speech 3

Memory patterns

Motor speech

Principal visual cortex

Visual association areas

GENERALIZED FUNCTIONAL AREAS OF LEFT CORTEX

– 297 –

Unit Nine

AUTONOMIC NERVOUS SYSTEM

137. AUTONOMIC NERVOUS SYSTEM: BASIC PLAN

I. **Introduction:** the peripheral nervous system consists of 2 main divisions, the autonomic nervous system (ANS) and the craniospinal nervous system, which are highly integrated both anatomically and functionally

A. BY DEFINITION, the ANS or "involuntary system" is entirely a motor system and comprises all efferent (motor) nerves not otherwise classified as somatic motor

B. AUTONOMIC RESPONSES usually occur without conscious control or awareness

C. AUTONOMIC NEURONS are distributed to smooth muscle and cardiac muscle, arrector pili muscle (hair follicle), and myoepithelial cells (glands)

D. THE ANS IS ANATOMICALLY SUBDIVIDED according to location of its preganglionic cell bodies, into sympathetic (thoracolumbar) and parasympathetic (craniosacral) components.

II. **Structural organization of the ANS** consists of numerous paravertebral, prevertebral (collateral), cranial, and terminal ganglia associated with either cranial or spinal nerves

A. PARAVERTEBRAL AND PREVERTEBRAL GANGLIA belong to the sympathetic division
1. The paravertebral ganglia form 2 sympathetic chains extending from the base of the skull to the coccyx
2. The prevertebral ganglia are located in perivascular plexuses, the abdominal aorta being the most prominent site for these plexuses
3. Paravertebral and prevertebral ganglia are connected to spinal cord segments T1 through L2 or 3 by means of spinal nerves from these same segments

B. CRANIAL AND TERMINAL GANGLIA are part of the parasympathetic division. Cranial ganglia are associated with cranial nerves 3, 7, and 9; terminal ganglia are found within or near the various internal organs and are associated with cranial nerve 10 and spinal cord segments S2, 3, 4

C. THE ANS IS A 2-NEURON CHAIN SYSTEM
1. The preganglionic neuron has its cell body within the CNS, and its axon carries the impulse from the CNS to a ganglion where it synapses
2. The postganglionic neuron has its cell body in the ganglion, and its axon leaves the ganglion to innervate smooth muscle, cardiac muscle, and glands. Postganglionic neurons outnumber preganglionics by a ratio of about 32:1, thus, a single preganglionic may serve to stimulate a number of postganglionic fibers

III. **Morphologic and functional distinction between autonomic and somatic motor outflow**

A. SENSORY AND INTERNUNCIAL NEURONS to both autonomic and somatic efferent neurons are structurally and functionally similar (same afferent pathway feeds both)

B. THE SOMATIC SYSTEM has its cell bodies located exclusively within the CNS anterior gray horn; the ANS has its first-order preganglionic neuron cell bodies within the CNS, but the second-order postganglionic neuron cell bodies are located outside the CNS in a ganglion

C. IN SOMATIC REFLEX ARCS, inhibition is exerted by one neuron upon another, but never by a nerve cell upon an effector (muscle cell—central inhibition)

D. IN AUTONOMIC REFLEX ARCS, presynaptic fibers may inhibit preganglionic neurons. In addition, some autonomic postganglionic fibers inhibit the act of the effector organs they innervate

E. UNLIKE SOMATIC CONTROL, STRUCTURES INNERVATED BY THE ANS can contract, secrete, and function normally in the absence of the nervous system, e.g., the small intestine completely isolated from the CNS continues to show peristalsis and responds to stretch and other stimuli—autoregulation

F. UNLIKE THE WELL-DEVELOPED END-PLATE REGION OF THE SKELETAL MUSCLE FIBERS, THE ANS lacks a well-defined neuromuscular or neuroglandular junction. Instead, many varicosities (swellings) are found along the entire length of the terminal nerves

Smooth muscle

Postganglionic

Preganglionic

Sweat glands

Dorsal root

Dorsal root ganglion

Pain receptors

Peripheral n.

Sensory neuron

Gray ramus

Postganglionic

White ramus

Motoneuron

Ventral root

Intermediolateral column

Skeletal muscle

Heart

Postganglionic

Preganglionic

Sensory

Arrector pili m.

AUTONOMIC NERVOUS SYSTEM

Nonskeletal m. (viscus)

Prevertebral (collateral) ganglion

Sympathetic ganglion (chain)

Sphincter pupillae + ciliary m. — Ciliary ganglion

Lacrimal gland

Pterygopalatine gang.

Sublingual + submandibular glands

Submand. g.

Parotid g.

Otic ganglion

III

VII
IX
X

Dilator of pupil

Head sweat glands + blood vessels

Hypophysis

Superior
Middle
Inferior } Cervical ganglia

Thyroid gland

Bronchi + lungs {

Heart

Esophagus

Stomach

Small intestine

Proximal colon

Liver

Pancreas

Kidneys

T1

5

10

L1

Upper limb sweat glands + blood vessels

} Heart

Stomach

Small intestine

Proximal colon

Celiac g.

Suprarenal gland

Kidney

Urinary bladder + genitalia

Distal colon + rectum

S
2
3
4

Inf. mes. gang.

Distal colon

Rectum
Urinary bladder
Genitalia

Lower limb sweat glands + blood vessels

– 301 –

138. SYMPATHETIC DIVISION OF THE AUTONOMIC NERVOUS SYSTEM—PART I

I. **Introduction:** most sympathetic ganglia are linked together in chainlike structures called the sympathetic chain or trunk (bilateral) extending from the cranium to the coccyx

A. THE GANGLIA OF EACH CHAIN are arranged segmentally except in the cervical region where they fuse into 3 large ganglia: superior, middle, and inferior cervical

II. **Preganglionic sympathetic axons** originate from cell bodies in the lateral gray column of the spinal cord at levels T1 through L2 or L3

A. THESE FIBERS exit the spinal cord in the ventral roots (along with somatic motor axons), but separate from the spinal nerves to form white rami communicantes

   1. These fibers enter the sympathetic chain, synapse with postganglionic cells in the chain ganglia, or pass through to synapse in prevertebral ganglia

   2. Preganglionic fibers do not necessarily synapse with a ganglion cell in the adjacent area, but may enter the chain and ascend or descend for several segments before synapsing. Connecting strands between ganglia of these chains are partly explained by these ascending-descending preganglionic fibers

B. PREGANGLIONIC WHITE RAMI are myelinated (B-type fibers), have a glistening white appearance in fresh specimens (not evident in cadaver), and their cell bodies are located at levels T1 to L2 or L3

III. **Postganglionic sympathetic axons** originate from cell bodies in the chain ganglia and return to the spinal nerves via the gray rami communicantes or go directly to various viscera

A. POSTGANGLIONIC NEURONS arise from all levels of the sympathetic chains and run in close association with blood vessels. Although its origin is limited, it ultimately fans out to cover the entire body, including the head, neck, and pelvic contents

   1. Postganglionic axons that emerge from chain ganglia as gray rami are mostly non-myelinated (C-type fibers)

B. EACH SPINAL NERVE receives a gray ramus which is distributed with the nerve to smooth muscle of blood vessels, sweat glands, and arrector pili muscles

   1. Gray rami, associated with all pairs of spinal nerves, outnumber white rami

C. COLLATERAL OR PREVERTEBRAL GANGLIA are other collections of postganglionic cell bodies outside the sympathetic trunks belonging to the sympathetic division

   1. Chief among the named prevertebral/collateral ganglia are the celiac, superior mesenteric, and inferior mesenteric ganglia

      a. These ganglia are located at origins of major arteries having the same names

      b. Collections of preganglionic axons called thoracic splanchnic nerves pass through the sympathetic chains without synapsing and pass through the diaphragm to synapse on postganglionic cells of these named ganglia

      c. These splanchnic nerves are examples of an exception to the rule that the sympathetic division has very short preganglionic axons

   2. Other less well-defined collateral ganglia (ovarian, hypogastric, etc.) are present in association with extensive autonomic nerve plexuses

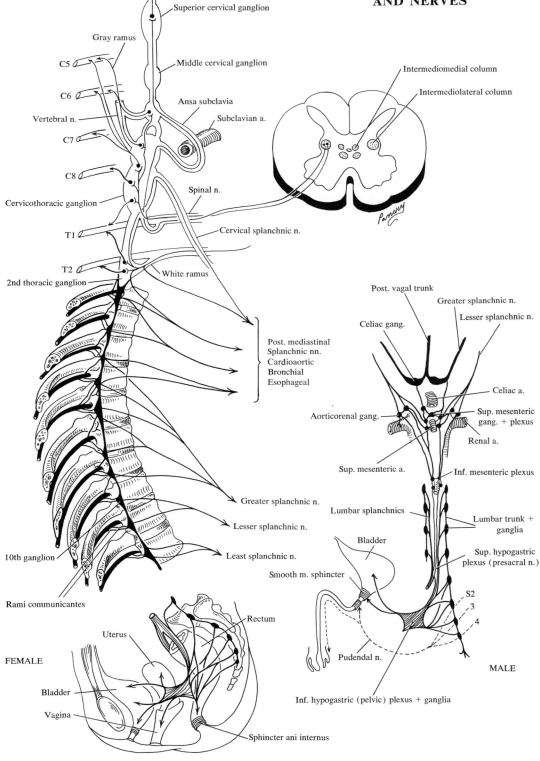

Superior cervical ganglion

Gray ramus

C5

Middle cervical ganglion

C6

Ansa subclavia

Vertebral n.

Subclavian a.

C7

C8

Spinal n.

Cervicothoracic ganglion

T1

Cervical splanchnic n.

T2

White ramus

2nd thoracic ganglion

Intermediomedial column

Intermediolateral column

Post. mediastinal
Splanchnic nn.
Cardioaortic
Bronchial
Esophageal

Greater splanchnic n.

Lesser splanchnic n.

Least splanchnic n.

10th ganglion

Rami communicantes

Post. vagal trunk

Greater splanchnic n.

Lesser splanchnic n.

Celiac gang.

Celiac a.

Aorticorenal gang.

Sup. mesenteric
gang. + plexus

Renal a.

Sup. mesenteric a.

Inf. mesenteric plexus

Lumbar splanchnics

Lumbar trunk +
ganglia

Bladder

Sup. hypogastric
plexus (presacral n.)

Smooth m. sphincter

S2

Rectum

3

Uterus

4

FEMALE

Pudendal n.

MALE

Bladder

Vagina

Sphincter ani internus

Inf. hypogastric (pelvic) plexus + ganglia

139. SYMPATHETIC DIVISION OF THE AUTONOMIC NERVOUS SYSTEM—PART II

I. Summary of the general pattern for the sympathetic division

A. ALL PREGANGLIONICS originate from spinal levels T1 through L2 or L3. Thus, innervation of the viscera, head, and extremities must originate from the thoracolumbar outflow regardless of the point at which they finally emerge from the chain

B. CERTAIN DISSECTABLE NERVES emerging from the chain contain mostly preganglionic fibers (thoracic splanchnics), while others contain mostly postganglionic fibers (cardiac sympathetics and most sympathetic fibers to the head, neck, and extremities)

C. GRAY RAMI COMMUNICANTES emerge from the chain and join all spinal nerves, being distributed in a segmental pattern to the body wall and musculature in accordance with the segmental distribution of spinal nerves

D. SYMPATHETIC FIBERS emerging from the superior cervical ganglion are distributed to the head, following the arterial supply of that region

E. THORACIC VISCERA receive their sympathetic innervation from fibers emerging from the chain at levels C1 through T5

F. MOST OF THE ABDOMINAL VISCERA are supplied by the thoracic splanchnics which arise from the sympathetic chain at levels T5 through T12; the remainder of the abdominal and pelvic viscera are supplied by fibers emerging from the sympathetic chain at levels L1 through S3.

Postganglionic

Ciliary g. → Eye
→ Lacrimal gland
Pterygopalatine g. → Mucosa of nose + palate
→ Sublingual gland
Submandibular g. → Submandibular gland
→ Mucosa of mouth
Otic g. → Parotid gland

Cervical ganglia

Superior
Middle
Inferior

White rami

T 1
2
3
4
5
6
7
8
9
10
11
12
L 1
2
3

Greater splanchnic n.

Lesser splanchnic n.

Preganglionic

Gray rami

Sup. mesenteric g.

Inf. mesenteric g.

Sympathetic trunk
(paravertebral ganglia)

Celiac g.

Heart
Larynx
Trachea
Bronchi
Lungs
Blood v.

Esophagus
Stomach
Liver + ducts
Pancreas
Spleen
Adrenal medulla
Small intestine

Large intestine

Rectum
Kidney

Bladder

Postganglionic

Sex organs

Ext. genitalia

SYMPATHETIC
NERVOUS SYSTEM

Inf. mesenteric ganglion

Sympathetic { L1
2

Somatic { S2
3
4

Inf. hypogastric plexus

Pudendal n.

Detrusor m. of bladder wall
(relaxes)

Bladder

Smooth m. of vesicle neck
(contraction)

Ext. sphincter of urethra
(contraction)

Dorsal n. of penis   Skeletal mm. of perineum

Pansky

SYMPATHETIC—RETENTION MECHANISM

140. PARASYMPATHETIC DIVISION OF THE AUTONOMIC NERVOUS SYSTEM

I. **Introduction:** preganglionic fibers of the craniosacral outflow originate from the brainstem and from sacral lateral gray horns of levels S2, S3, and S4 of the spinal cord. Fibers from the cranial portion are distributed through cranial nerves III (oculomotor), VII (facial), IX (glossopharyngeal), and X (vagus)

A. THE GANGLIA OF THE PARASYMPATHETIC DIVISION are widely scattered and irregularly spaced when compared with the sympathetic division
   1. Parasympathetic ganglia lie close to the organ innervated or, in some cases, within the wall of the organ supplied, accounting for the characteristically short postganglionic fibers of this division
   2. Most preganglionic neurons run without interruption from their CNS origin either to the wall of the viscus they supply or where they synapse with terminal ganglion cells associated with the pelvic plexuses

B. PARASYMPATHETIC FIBERS OF CRANIAL NERVES III, VII, AND IX supply ganglia of the head, whereas cranial nerve X (vagus, wanderer) supplies fibers to the heart, lungs, and most of the abdominal viscera via the prevertebral plexuses

C. THE SACRAL PORTION supplies fibers to the remainder of the abdominal contents not supplied by the vagus and to most of the pelvic viscera. The pelvic nerve (nervus erigens) supplies parasympathetic fibers to most of the large intestine and to the pelvic viscera and genitalia via the hypogastric plexus

II. **The craniosacral subdivision of the autonomic nervous system:** preganglionic fibers of the cranial and sacral subdivisions form synaptic relations with postganglionic neurons in cranial autonomic or terminal ganglia respectively

A. CRANIAL PARASYMPATHETICS: in the cranial region, 4 parasympathetic ganglia are related topographically to the major divisions of the trigeminal (V) nerve
   1. Ciliary ganglion: located along the lateral surface of the optic nerve
   2. Pterygopalatine ganglion: lies in the pterygopalatine fossa near the maxillary nerve
   3. Submandibular ganglion: positioned over the submandibular gland near the lingual nerve
   4. Otic ganglion: located medial to the mandibular nerve as it exits the foramen ovale

B. SACRAL PARAMPATHETICS
   1. Preganglionic fibers exit the spinal cord at levels S2, S3, S4 to form the pelvic nerve (nervus erigens) and pass to the terminal ganglia of the pelvic plexuses, myenteric and submucosal plexuses of the descending colon and rectum
   2. Postganglionic fibers: from terminal ganglia supply the urinary bladder, the descending colon, the rectum, and accessory reproductive organs
   3. Sacral autonomic fibers innervate all viscera not supplied by the vagus nerve

C. LARGEST SOURCE OF PREGANGLIONIC PARASYMPATHETICS is the dorsal motor nucleus of the vagus nerve which supplies most thoracic and abdominal viscera except those in the pelvis
   1. In the thorax: preganglionic vagal fibers enter the pulmonary, esophageal, and cardiac plexuses to be distributed to the terminal (intrinsic) ganglia of the heart and bronchial musculature
   2. In the abdomen: preganglionic vagal fibers go to the stomach and pass through the celiac and subsidiary plexuses, to end in the terminal ganglia of the intestine, liver, pancreas, and kidneys
      a. These terminal ganglia form the extensive ganglionated plexuses of Auerbach (myenteric) and of Meissner (submucosal) which extend the entire length of the digestive tube and are composed of many small aggregates of ganglion cells interconnected by delicate fiber bundles

Oculomot. (III) n.   Ciliary g.   Dilator pupillae m.

Ciliary m.

Lacrimal gland

Pterygopalatine g.   Mucosa of nose + palate

Midbrain

Facial (VII) n.

Medulla   Submandibular g.   Sublingual gl.

C1   Vagus (X) n.   Mucosa of mouth

Submandibular gl.

Otic g.

T1   Glossopharyngeal (IX) n.   Parotid gl.

Larynx
Trachea
Bronchi

Heart

Lung

L1   Liver, gallbladder ducts   Pancreas   Spleen

Esophagus   Kidney

S2   Stomach   Abdominal blood vessels
3
4   Sml. intestine
Pelvic n. (n. erigens)   Proximal colon   Distal colon

Cecum

PARASYMPATHETIC NERVOUS SYSTEM

Ureter

Bladder

Prostate

Ext. genitalia   Sex organs   Rectum

C3
Somatic 4   Phrenic n. to diaphragm   Diaphragm contracts: indirect pressure on bladder
5

T5   Thoracic nn.: abdominal mm contract—downward pressure on bladder

Somatic

T12

Intrinsic plexus   Detrusor muscle (contracts)

Smooth m. of vesicle neck (relaxes)

S2
Parasympathetic 3   Pelvic nerves (nervi erigentes)
4   Ext. sphincter (relaxes)

Urination

PARASYMPATHETIC–EXPULSION MECHANISM

Pansky

– 307 –

141. AUTONOMICS OF THE HEAD

I. **Preganglionic fibers from white rami of the upper thoracic nerves** go mainly to the superior cervical ganglion

II. **Postganglionic fibers from the superior cervical ganglion** are distributed to the lower 4 cranial nerves, the upper 3 or 4 cervical nerves, the pharynx, the superior cervical cardiac nerve, and the external and internal carotid arteries

III. **Nerve plexuses:** formed about the carotid arteries by postganglionic fibers from the superior cervical ganglion. From these, postganglionics pass through cranial autonomic ganglia (without synapsing) to join cranial nerves for distribution to face and scalp

IV. **Dual, sympathetic-parasympathetic innervation** is also provided to the salivary glands, intrinsic muscles of the eye, and mucous membranes of the pharynx and nose
   A. FOUR PAIRS OF CRANIAL AUTONOMIC GANGLIA, each of which receives a sympathetic, parasympathetic, and sensory root, mediate this dual autonomic supply (only parasympathetic nerves synapse here; others pass through without synapse)
      1. The ciliary ganglion: located between the optic nerve and the lateral rectus muscle and is comprised of
         a. A parasympathetic root from cells of the Edinger-Westphal nucleus and the medial portion of the IIIrd nucleus
         b. A sympathetic root of postsynaptic fibers from the superior cervical ganglion via the carotid plexus
         c. A sensory root from the nasociliary branch of the ophthalmic (V) nerve
         d. Short ciliary nerves leave ganglion with postganglionic (III) fibers to sphincter pupillae and ciliary muscles, postganglionic sympathetics to the dilator pupillae muscle, and sensory fibers from the eyeball
      2. The pterygopalatine ganglion is located in the pterygopalatine fossa, closely associated with the maxillary (V) nerve, and is composed of
         a. A parasympathetic preganglionic root from cells of the superior salivatory nucleus via the glossopalatine nerve, great petrosal nerve, and Vidian nerve (VII)
         b. A sympathetic postganglionic root from the internal carotid plexus via the deep petrosal nerve which joins the great petrosal nerve to form the vidian nerve
         c. A sensory root from the maxillary (V) nerve, but some fibers arise in cranial nerves VII and IX via the tympanic plexus and vidian nerve
         d. Postganglionic parasympathetic fibers carry VII secretory fibers to lacrimal gland
      3. The otic ganglion is located medial to the mandibular nerve just below the foramen ovale in the infratemporal fossa and is composed of
         a. A parasympathetic preganglionic root from the inferior salivatory nucleus (medulla) which is distributed via the glossopharyngeal (IX) nerve (tympanic branch), tympanic plexus, and lesser petrosal nerve to the ganglion
         b. A sympathetic postganglionic root from the superior cervical ganglion follows the plexus on the middle meningeal artery to the ganglion
         c. A sensory root receives fibers from the IXth nerve and from the geniculate ganglion of the VIIth nerve via the tympanic plexus and the lesser petrosal nerve
         d. Postganglionics of IX follow the auriculotemporal (V) nerve to the parotid gland
      4. The submandibular ganglion is located on medial side of mandible between lingual nerve and mandibular duct, lying over submandibular gland; composed of
         a. A parasympathetic preganglionic root from the superior salivatory nucleus via the glossopalatine, chorda tympani, and lingual nerves
         b. A sympathetic postganglionic root from the plexus along the maxillary artery
         c. A sensory root from the geniculate ganglion which is distributed via the glossopalatine, chorda tympani, and lingual nerves
         d. Postganglionic (VII) fibers to submandibular and sublingual glands (secretory)

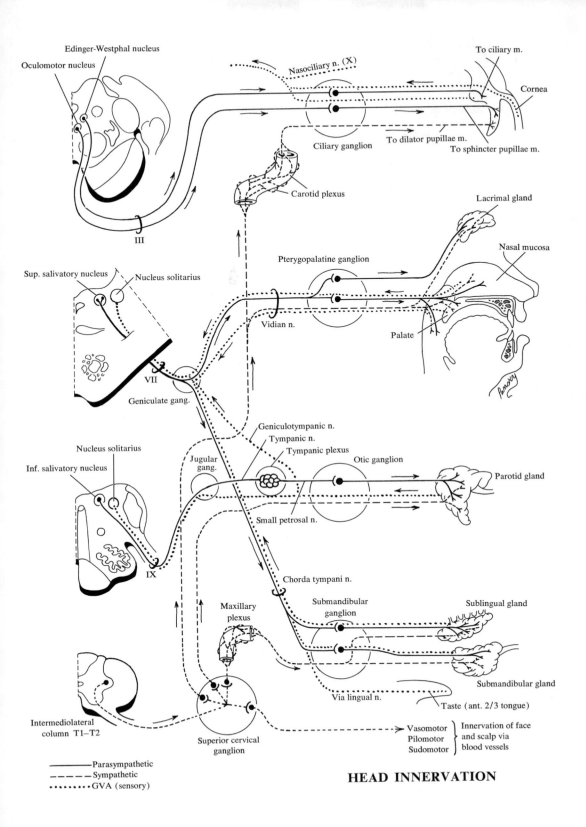

Edinger-Westphal nucleus

Oculomotor nucleus

To ciliary m.

Nasociliary n. (X)

Cornea

Ciliary ganglion

To dilator pupillae m.

To sphincter pupillae m.

Carotid plexus

Lacrimal gland

Nasal mucosa

Pterygopalatine ganglion

Sup. salivatory nucleus

Nucleus solitarius

Vidian n.

Palate

VII

Geniculate gang.

Geniculotympanic n.

Tympanic n.

Tympanic plexus

Otic ganglion

Nucleus solitarius

Inf. salivatory nucleus

Jugular gang.

Parotid gland

Small petrosal n.

IX

Chorda tympani n.

Maxillary plexus

Submandibular ganglion

Sublingual gland

Submandibular gland

Via lingual n.

Taste (ant. 2/3 tongue)

Intermediolateral column T1–T2

Vasomotor
Pilomotor
Sudomotor

Innervation of face and scalp via blood vessels

Superior cervical ganglion

——— Parasympathetic
– – – – Sympathetic
········· GVA (sensory)

HEAD INNERVATION

142. FUNCTIONAL ANATOMY OF THE AUTONOMIC NERVOUS SYSTEM

I. Transmitter chemicals of the ANS

A. IN BOTH DIVISIONS OF THE ANS, the chemical transmitter for the ganglionic synapse between pre- and postganglionic fibers is acetylcholine

   1. The transmitter at the junction between parasympathetic postganglionic fibers and effector organs is also acetylcholine

   2. The transmitter between the sympathetic postganglionic fiber and its effector organ is norepinephrine (a member of the catecholamine family)

      a. An exception is sympathetic postganglionic fibers to sweat glands which release acetylcholine

      b. One ganglion, the adrenal medulla, in the sympathetic system never develops long postganglionic fibers but on activation of its preganglionic nerves, the cells discharge their transmitter into the bloodstream

         i. Releases a mixture of 80% epinephrine and 20% norepinephrine

         ii. Are really hormones rather than neurotransmitters and are transmitted via blood and intestinal fluid to receptor sites on effector cells sensitive to them

B. FIBERS THAT RELEASE ACETYLCHOLINE are called cholinergic fibers; those that release norepinephrine are called adrenergic fibers

C. MANY DRUGS STIMULATE OR INHIBIT the synaptic functions of the ANS

   1. Those whose actions "mimic" the actions of the sympathetic system are called sympathomimetic drugs (amphetamine and phenylephrine)

   2. Several choline compounds and the mushroom poison muscarine mimic parasympathetic actions and are called parasympathomimetic drugs

D. DRUGS THAT BLOCK THE ACTIONS of the autonomic system are sympatholytic and parasympatholytic. The plant belladonna (deadly nightshade) yields atropine which is a parasympatholytic drug

D. DRUGS THAT BLOCK THE ACTIONS of the autonomic system are sympatholytic and parasympatholytic. The plant belladonna (deadly nightshade) yields atropine which is a parasympatholytic drug

II. Actions and dual innervation of the ANS: the actions of the ANS depend not only on the chemical released by the postganglionic cell but also on the effector cell's receptor sites and intracellular machinery

A. AN IMPORTANT CHARACTERISTIC of autonomic nerves is dual innervation of most effector organs, often with opposite effects. Skeletal muscle cells, in contrast, do not receive dual innervation since somatic nerves are excitatory

   1. Action potentials over the sympathetic nerves to the heart increase heart rate; over the parasympathetic fibers decrease heart rate

   2. In the intestine, activation of sympathetic fibers reduces smooth muscle contraction in the intestinal wall; parasympathetics increase contraction

B. TO PREVENT THE TWO SYSTEMS FROM CONFLICTING, they are usually activated reciprocally; i.e., as activity of one system is enhanced, the other is depressed

C. BECAUSE SMOOTH MUSCLES, HEART MUSCLE, AND GLANDS PARTICIPATE AS EFFECTORS in all homeostatic control of our internal environment, there is a wide array of effects

   1. Sympathetic system helps the body to cope with challenges from the outside environment and involves stress and emotions (fear and rage)

   2. Parasympathetic system is more responsible for internal maintenance, such as digestion, defecation, and urination, and is more active during recovery or rest

D. AUTONOMIC RESPONSES usually occur without conscious control or awareness, but

   1. Discrete visceral and glandular responses can be learned; animals have learned to control their own heart rates or manipulate vessel dilatation or contraction

   2. Small segments of the autonomic response can be regulated independently

   3. Autonomic responses, made up of many components, are variable but rather than being gross, undiscerning discharges are finely tailored to specific bodily demands

CHEMICAL TRANSMITTERS

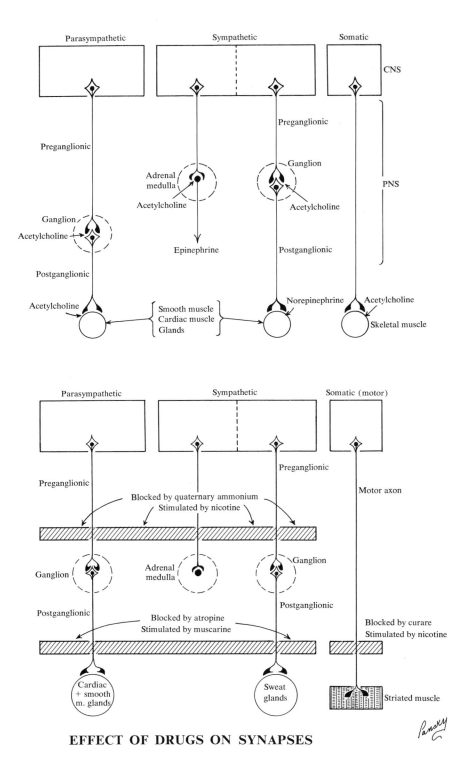

EFFECT OF DRUGS ON SYNAPSES

Unit Ten

THE CEREBELLUM

143. CEREBELLUM: GROSS ANATOMY

I. **General information:** the cerebellum occupies most of the posterior cranial fossa, lies dorsal to the brainstem, and is attached to the pons, medulla, mesencephalon by three pairs of thick fiber bundles, the cerebellar peduncles, found on the ventral aspect of the cerebellum. They include

A. SUPERIOR CEREBELLAR PEDUNCLE (BRACHIUM CONJUNCTIVUM): the bridge between the midbrain and the cerebellum, with fibers largely projecting from the cerebellum toward the midbrain and thalamus

B. MIDDLE CEREBELLAR PEDUNCLE (BRACHIUM PONTIS): the bridge between the pons and the cerebellum, with fibers projecting to the cerebellum

C. INFERIOR CEREBELLAR PEDUNCLE (RESTIFORM BODY): between medulla and cerebellum, with fibers projecting both to and from the cerebellum

II. **Other relationships**

A. THE FOURTH VENTRICLE separates the most medial portion of the ventral surface of the cerebellum from the pons and medulla

B. THE TENTORIUM CEREBELLI (TRANSVERSE FOLD OF DURA) separates the cerebellum from the occipital lobe of the cerebrum

III. **The cerebellum** is composed of two large lateral masses, the cerebellar hemispheres and a midline portion known as the vermis (wormlike)

A. ITS SURFACE is comprised of numerous parallel folia with each primary folium being subdivided under the surface into secondary and tertiary folia

B. IN MIDSAGITTAL SECTION (THROUGH THE VERMIS), the cerebellum has the appearance of a highly branched tree (arbor vitae)

   1. Cerebellar cortex is a layer of gray matter which covers the surface
   2. The medullary center is an internal core of white matter consisting of fibers from the cortex to cerebellar nuclei, incoming fibers which terminate on neurons in the gray matter and fibers connecting different cortical areas

C. EMBEDDED IN THE MEDULLARY CORE are gray substance, the cerebellar nuclei: the dentate nucleus (largest, most lateral, irregular), the emboliform nucleus (vertical plate of cells), globose nuclei (one or more rounded gray areas), and the fastigial nuclei (near the midline, in the core of the vermis). These nuclei are seen in each of the cerebellar hemispheres

IV. **Lobes of cerebellum**

A. ANTERIOR LOBE (PALEOCEREBELLUM) lies anterior to the primary fissure and consists of most of the vermis and of the anterior (superior) aspect of the cerebellar hemispheres. This phylogenetically old lobe is primarily associated with proprioceptive (spinocerebellar) and exteroceptive input from the head and body, including some from the vestibular system. It has a significant role in the regulation of muscle tone

B. POSTERIOR LOBE (NEOCEREBELLUM): the largest part of the cerebellum positioned between the other two lobes. It consists of the main bulk of the cerebellar hemispheres and part of the vermis. This phylogenetically new lobe (essentially a mammalian structure) receives connections of the cerebrum and plays an essential role in the muscular coordination of phasic movements

C. FLOCCULONODULAR LOBE (ARCHICEREBELLUM) consists of the paired flocculi of the hemispheres (small appendages in the posterior inferior region) and the unpaired nodulus, which is the inferior part of the vermis. The archicerebellum represents the cerebellar portion of the vestibular system and phylogenetically is its oldest part. It is actually a specialized portion of the somatic afferent column that is dominated by direct, indirect, and feedback connections of the vestibular system. It serves a significant role in muscle tone, equilibrium, and posture through its influences on the trunk muscles

CEREBELLAR PEDUNCLES

Superior peduncle
Ant. medullary velum
Middle peduncle
Rhomboid fossa
Inferior peduncle
F. Luschka
Choroid plexus
Tela choroidea
F. Magendie

ROOF OF FOURTH VENTRICLE

To thalamus   Corticopontine t.
Superior peduncle
Rubro-olivary t.
Middle peduncle
Cerebellovestibular +
reticular tracts
Inferior peduncle
Ventral spinocerebellar t.
Reticulo-olivovestibulocerebellar
tracts
Post. spinocerebellar t.

Anterior lobe
Posterior lobe
Posterolateral fissure

MEDIAL VIEWS OF RIGHT SIDE

CEREBELLUM

Central lobule
Lingula
Culmen
Primary fissure
Declive
Folium
Tuber
Pyramis
Nodule   Uvula
Posterolateral fissure

Central lobule
Lingula
Primary fissure
Ant.
Medullary
velum
Post.
Med. velum
Ant. (sup.)
Culmen
Declive
Folium
Tuber
Post. (inf.)
Pyramis
Nodule
Uvula

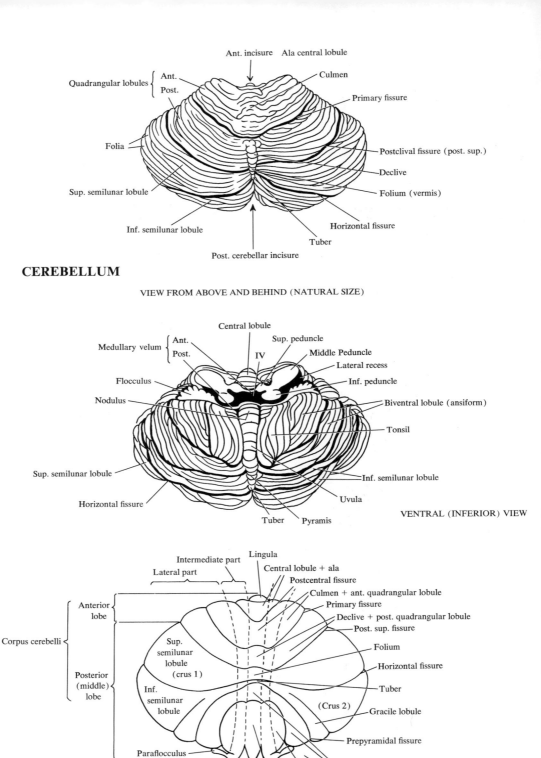

CEREBELLUM

VIEW FROM ABOVE AND BEHIND (NATURAL SIZE)

VENTRAL (INFERIOR) VIEW

FIGURE 9.

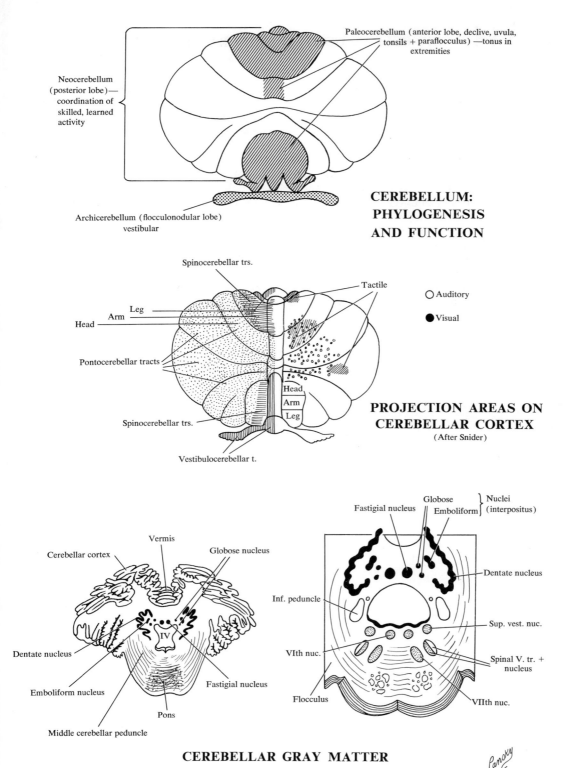

Paleocerebellum (anterior lobe, declive, uvula, tonsils + paraflocculus) —tonus in extremities

Neocerebellum (posterior lobe)— coordination of skilled, learned activity

Archicerebellum (flocculonodular lobe) vestibular

CEREBELLUM: PHYLOGENESIS AND FUNCTION

Spinocerebellar trs.

Tactile

Leg
Arm
Head

○ Auditory
● Visual

Pontocerebellar tracts

Head
Arm
Leg

Spinocerebellar trs.

PROJECTION AREAS ON CEREBELLAR CORTEX
(After Snider)

Vestibulocerebellar t.

Vermis

Cerebellar cortex

Globose nucleus

Fastigial nucleus

Globose
Emboliform

} Nuclei (interpositus)

Inf. peduncle

Dentate nucleus

IV

Sup. vest. nuc.

Dentate nucleus

VIth nuc.

Spinal V. tr. + nucleus

Emboliform nucleus

Fastigial nucleus

Flocculus

VIIth nuc.

Pons

Middle cerebellar peduncle

Panosky

CEREBELLAR GRAY MATTER

FIGURE 10.

144. CEREBELLAR CORTEX: GENERAL LAYERS

I. **General information:** cortical surface area is greatly increased by numerous long gyri known as folia. Its total surface area is 75% that of the cerebral cortex. Some 85% of its free surface is hidden from surface view and lies deep within the sulci of the folia. All cerebellar folia have the same microscopic structure and consist of three layers of neurons

II. **Internal structure of the cerebellum** consists of a highly convoluted cortex which is superficial to the central core of white matter

A. MOLECULAR LAYER: the outer layer which is the synaptic integrative layer. It receives its input mainly from the granular layer and deep cerebellar nuclei and partly from outside the cerebellum. It consists mainly of nerve fibers (axons and dendrites) and some basket cells

B. PURKINJE CELL LAYER: the middle cell layer. It is the output layer of the cerebellar cortex which discharges to the intrinsic cerebellar nuclei

C. GRANULAR LAYER: the innermost layer immediately adjacent to the white medullary core and the main receptive layer which receives input from outside the cerebellum

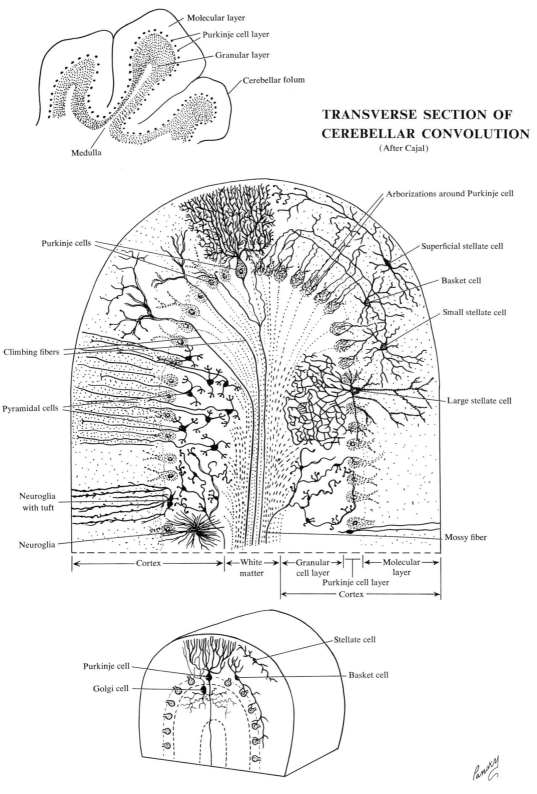

Molecular layer
Purkinje cell layer
Granular layer
Cerebellar folum
Medulla

**TRANSVERSE SECTION OF
CEREBELLAR CONVOLUTION**
(After Cajal)

Arborizations around Purkinje cell
Purkinje cells
Superficial stellate cell
Basket cell
Small stellate cell
Climbing fibers
Pyramidal cells
Large stellate cell
Neuroglia with tuft
Neuroglia
Mossy fiber

Cortex
White matter
Granular cell layer
Purkinje cell layer
Molecular layer
Cortex

Stellate cell
Purkinje cell
Basket cell
Golgi cell

145. CEREBELLAR CORTEX: CELL AND NERVE TYPES AND THEIR ROLES

I. Granule cells are small multipolar neurons whose nerve cell bodies are densely packed in the granular layer

A. AXONS ascend into molecular layer and bifurcate as a T to form parallel fibers that run parallel to the long axis of the folium

B. PARALLEL FIBERS synapse with dendritic branches of Purkinje cells which also ascend into the molecular layer. As many as 40,000 parallel fibers may synapse (all excitatory) on each Purkinje cell

C. DENDRITES are synaptically related to mossy fibers which are terminal branches of axons of cells that lie outside the cerebellum

II. Purkinje cells are arranged side by side in a single row between the molecular layer and the inner granular layer

A. DENDRITES ascend into the molecular layer where they branch profusely

B. AXONS run through the granular layer into the white matter where recurrent collateral branches are given off to other Purkinje cells. The axons end in cerebellar nuclei (or Deiter's cells)

III. Basket cells: their cell bodies are in the lower part of the molecular layer

A. DENDRITES are also in the molecular layer and are excited by parallel fibers

B. AXONS run perpendicular to parallel fibers and give collaterals to many Purkinje cells. They end in a rete of terminals (a basket appearance) about the cell bodies of the Purkinje cells

C. FUNCTION: inhibit the Purkinje cells

IV. Stellate cells are similar to basket cells and are often found in a small clump of mossy fibers. A volley of impulses in the stellate cells produces a line of excited Purkinje cells (with adjacent parallel bands of Purkinje cells being inhibited by the basket cells)

V. Golgi cells: cell bodies are found in the upper granular cell layer

A. DENDRITES branch extensively within the molecular layer and make contact with parallel fibers of the Purkinje cells

B. AXONS arborize extensively in the granular layer to eventually synapse on dendrites of the granule cells

C. FUNCTION: inhibit the mossy fiber, granule cell synapse

D. EXCITED BY parallel fibers and climbing fibers collaterals; inhibited by Purkinje cell axon collaterals

VI. Mossy fibers are axons that provide the primary cerebellar inputs from pontine nuclei and spinocerebellar pathways

A. CELL BODIES lie outside the cerebellum

B. EACH MOSSY FIBER ENDS in 3–5 mosslike rosettes that synapse with several granule cells. Each granule cell may receive input from as many as six mossy fibers

C. FUNCTION: to excite the granule cell

VII. Climbing fibers are axons of neurons whose cell bodies are located in the inferior olivary nucleus

A. A CLIMBING FIBER MAKES CONTACT with a single Purkinje cell and branches progressively as it "climbs" its dendrites

B. A CLIMBING FIBER SYNAPSES with primary and secondary dendritic branches of Purkinje cells at many sites (100–300)

C. EACH ACTION POTENTIAL in a climbing fiber causes a giant EPSP which, in turn, produces a burst of repetitive Purkinje cell firing

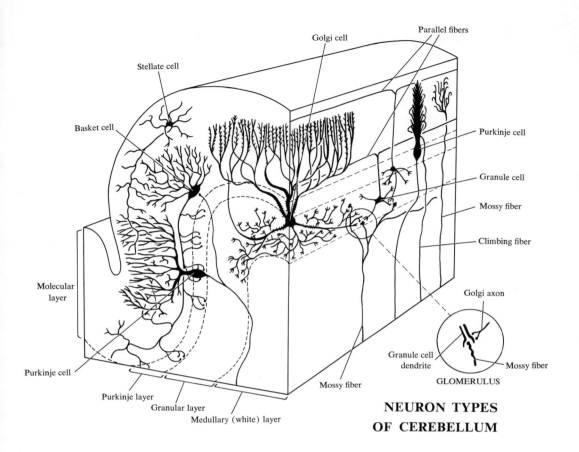

Stellate cell

Basket cell

Golgi cell

Parallel fibers

Purkinje cell

Granule cell

Mossy fiber

Climbing fiber

Molecular layer

Purkinje cell

Purkinje layer

Granular layer

Medullary (white) layer

Mossy fiber

Golgi axon

Granule cell dendrite

Mossy fiber

GLOMERULUS

NEURON TYPES OF CEREBELLUM

CEREBELLAR "WIRING DIAGRAM"

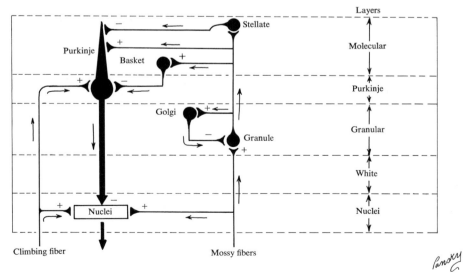

Layers

Stellate

Molecular

Purkinje

Basket

Golgi

Purkinje

Granule

Granular

White

Nuclei

Nuclei

Climbing fiber

Mossy fibers

Panofsky

146. SUBCORTICAL CEREBELLAR NUCLEI

I. **General location:** four pairs of nuclei are buried deep within the white medullary core of the cerebellum and are positioned dorsal and dorsolateral to the fourth ventricle

II. **Name and relationship**
A. DENTATE NUCLEI are the largest and most lateral of these deep nuclei. Each resembles a crumbled paper bag with its opening directed medially when cut in the horizontal plane
B. EMBOLIFORM NUCLEI* lie just medial to the dentate and usually appear somewhat plug(embolus)-shaped in their position, near to what appears to represent the opening or hilus of the dentate nucleus
C. GLOBOSE NUCLEI* are usually represented as two or more small ovoid nuclear masses just medial to the emboliform
D. FASTIGIAL NUCLEI are the most medial and are located just lateral to the fastigium (roof) of the fourth ventricle

III. **Relationship of cerebellar cortex and subcortical nuclei:** in general, the cerebellar nuclei receive mainly the axons from ipsilateral Purkinje cells
A. LATERAL AREAS OF THE CORTEX project to laterally positioned dentate nuclei
B. PARAMEDIAN AREAS OF CORTEX project to the emboliform and globose nuclei
C. MEDIAN AREA (VERMIS) OF CORTEX project to the globose and fastigial nuclei. The latter nucleus also receives some fibers from the vestibular nuclei

*The emboliform and globose nuclei are collectively called the nucleus interpositus.

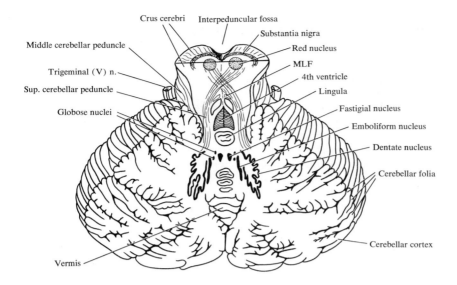

Crus cerebri — Interpeduncular fossa

Middle cerebellar peduncle

Substantia nigra

Red nucleus

MLF

Trigeminal (V) n.

4th ventricle

Sup. cerebellar peduncle

Lingula

Globose nuclei

Fastigial nucleus

Emboliform nucleus

Dentate nucleus

Cerebellar folia

Cerebellar cortex

Vermis

**SECTION THROUGH CEREBELLUM AND MESENCEPHALON—
LEVEL OF SUPERIOR CEREBELLAR PEDUNCLE**

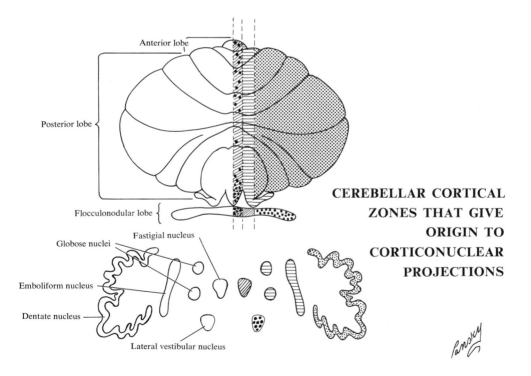

Anterior lobe

Posterior lobe

Flocculonodular lobe

**CEREBELLAR CORTICAL
ZONES THAT GIVE
ORIGIN TO
CORTICONUCLEAR
PROJECTIONS**

Globose nuclei

Fastigial nucleus

Emboliform nucleus

Dentate nucleus

Lateral vestibular nucleus

147. SUMMARY OF CEREBELLAR CONNECTIONS

I. General: all fibers carrying information to and from the cerebellum do so by way of the cerebellar peduncles

A. INCOMING (AFFERENT) FIBERS

   1. Superior cerebellar peduncle (brachium conjunctivum) is the main efferent pathway to the cerebellum. Its afferent connections are the ventral (anterior) spinocerebellar tract to the paleocerebellum; the tectocerebellar tract from the midbrain tectum to the ipsilateral neocerebellum which represents the optic and auditory input, and the trigeminocerebellar tract which conveys general afferent impulses from the head

   2. Middle cerebellar peduncle (brachium pontis) is entirely afferent. It conveys fibers from contralateral pontine nuclei which receive innervation from the cerebral cortex (corticopontocerebellar tract). Each cerebellar hemisphere monitors the activity of the opposite cerebral hemisphere

   3. Inferior cerebellar peduncle (restiform body) is predominantly an efferent pathway. Conveys "unconscious" exteroceptive and proprioceptive fibers from the spinal cord and vestibular system

      a. Posterior (dorsal) spinocerebellar tract projects directly to paleocerebellum

      b. Accessory (lateral) reticular nuclei, gracile and cuneate nuclei, and other medullary reticular nuclei which receive input from the spinal cord, project to the ipsilateral paleo- and neocerebellum

      c. Accessory inferior olivary nuclei (paleo-olivary) receive cord input and project to the paleocerebellum

      d. Main inferior olivary nucleus (neo-olivary) receives major input from the cerebral cortex, caudate nucleus, globus pallidus, red nucleus, and brainstem reticular formation and project crossed fibers to all parts of the neocerebellum, dentate nucleus, and nucleus interpositus

      e. Direct vestibulocerebellar fibers project to the archicerebellum of the same and opposite sides

B. OUTGOING (EFFERENT) FIBERS: no direct cerebellospinal pathways exist

   1. Superior cerebellar peduncle consists primarily of the efferent fibers from the dentate, emboliform, and globose nuclei, and the entire outflow completely crosses over in the lower midbrain as the decussation of the brachium conjunctivum

      a. The dentate nucleus is used to designate efferent tracts via the dentatorubral, dentatothalamic, and dentatoreticular fibers

      b. Most fibers from the dentate nucleus project rostrally to the thalamus (ventral lateral and thalamic reticular nuclei) with some fibers also to the red nucleus, inferior olive, and brainstem reticular formation

      c. While the dentate projects to the thalamus, the globose and emboliform nuclei project mainly to the red nucleus, inferior olive, and reticular formation

      d. This results in two additional loops which permit the cerebellum to monitor other areas of the nervous system.

         i. Cortico/ponto/cerebello/dentato/rubro/thalamo/cortical pathway

         ii. Cerebello/dentato/rubro/olivo/cerebellar pathway

   2. Middle cerebellar peduncle has no efferent fibers

   3. Inferior cerebellar peduncle: outflow is through the juxtarestiform body (on medial aspect of the peduncle) and includes the following

      a. Fastigiobulbar tract: from the fastigial nuclei to the vestibular nuclei (primarily the lateral nucleus) and reticular nuclei of the pons and medulla

         i. The uncinate fasciculus of Russell (hooked bundle) formed by fibers from the nuclei fastigii of both sides, hooks around the rostral border of the superior cerebellar peduncle before passing through the restiform body

      b. Some direct fibers from the flocculonodular cortex pass to the vestibular nuclei

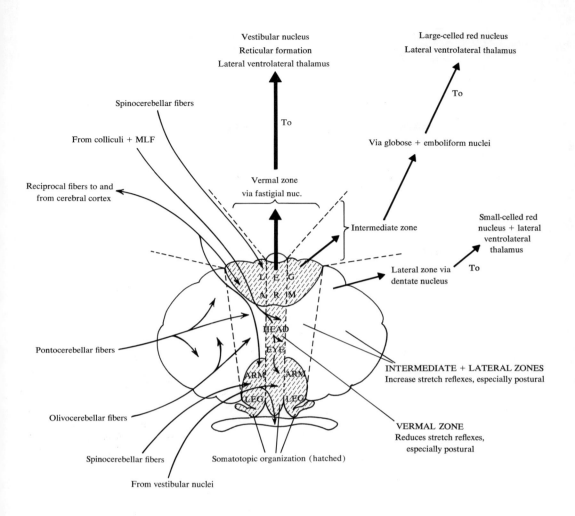

Vestibular nucleus
Reticular formation
Lateral ventrolateral thalamus

Large-celled red nucleus
Lateral ventrolateral thalamus

To

To

Spinocerebellar fibers

From colliculi + MLF

Via globose + emboliform nuclei

Reciprocal fibers to and
from cerebral cortex

Vermal zone
via fastigial nuc.

Intermediate zone

Small-celled red
nucleus + lateral
ventrolateral
thalamus

Lateral zone via
dentate nucleus

To

L E G

A R M

HEAD

EYE

ARM   ARM

LEG   LEG

Pontocerebellar fibers

INTERMEDIATE + LATERAL ZONES
Increase stretch reflexes, especially postural

Olivocerebellar fibers

VERMAL ZONE
Reduces stretch reflexes,
especially postural

Spinocerebellar fibers

Somatotopic organization (hatched)

From vestibular nuclei

SCHEMATIC OF FLATTENED CEREBELLAR CORTEX
PARASAGITTAL ZONES + MAJOR CONNECTIONS

(After Lewis)

CEREBELLAR AFFERENT FIBERS

FIGURE 11.

Areas 4 and 6 (frontal cortex)

Area 22 (temporal cortex)

Area 6

Area 4

Internal medullary lamina

Ventrolateral nucleus
of thalamus

Area 22

Dentatothalamic tract

Red nucleus

MIDBRAIN

Paleocerebellum

Sup. ped.

Crossed + uncrossed
cerebelloreticular fibers

Fastigial nucleus

Reticular formation

Middle ped.

PONS

Uncinate fasciculus (of Russell)
or fastigobulbar tract

Inf. ped.

Reticular formation

Dentate nuc.

MEDULLA

Emboliform +
globose nuclei

Paleocerebellum

Archicerebellum

Vestibular nuclei

Crossed vestibulospinal tract

Flocculovestibular fibers

Motor nerve

Uncrossed vestibulospinal tract

SPINAL CORD

CEREBELLAR EFFERENT FIBERS

FIGURE 12.

148. SUMMARY OF FUNCTIONAL CEREBELLAR ANATOMY

I. Normal cerebellar function

A. THE CEREBELLUM IS RESPONSIBLE FOR MUSCLE SYNERGY (coordination) throughout the body

B. THE CEREBELLUM COORDINATES the action of muscle groups and times their contractions so that movements are performed smoothly and accurately

C. THE CEREBELLUM ACTS AS A MONITOR of other centers of the brain. It is not the initiator of muscle contraction or movement in general, but monitors centers that do initiate such motion

D. VOLUNTARY MOVEMENTS can proceed without its assistance, but such movements are clumsy and disorganized. Lack of motor skills as a result of cerebellar dysfunction is called asynergia or cerebellar ataxia

E. THE CEREBELLUM IS THE MODULATOR (REGULATOR) of motor activities and is integrated with the vestibular system for the maintenance of muscle tone and equilibrium; with general proprioceptive and exteroceptive receptors, and with the auditory and visual systems

F. THE CEREBELLUM IS BASICALLY A SOMATIC AFFERENT ORGAN (head ganglion of proprioception)
   1. Although the cerebellum receives large numbers of afferent fibers, conscious perception does not occur in it, nor do its efferent fibers give rise to conscious sensations elsewhere in the brain
   2. The cerebellum receives and processes unconscious afferent stimuli from various afferent (sensory) centers and passes the information on to the motor systems to be utilized by them to make the necessary corrections in specific muscles or muscle groups (via the cerebellar-cerebral cortical connections or interconnections)
   3. As far as can be determined, the cerebellum plays no role in the appreciation of conscious sensations or in intelligence

II. Cerebellar dysfunction

A. ATAXIA is caused by cerebellar damage and is manifested in the following ways
   1. Disturbances of posture and gait may be very pronounced. Lesions of the midline region of the cerebellum cause difficulty in maintaining an upright stance. The gait is staggering like that seen in drunkenness
      a. The loss of equilibrium is due to the lack of muscle synergy and not to a defect in the pathway of conscious proprioception
   2. Decomposition of movement: an action that requires the cooperative movement of several joints is not properly coordinated, but is broken down into its component parts
   3. Dysmetria is shown by the inability to stop a movement at the desired point. In reaching the hand toward an object, the patient either overshoots the goal or stops before it is reached
   4. Dysdiadochokinesis is the inability to stop one movement and follow it immediately by the directly opposite action, as in the case of rapid alternating movements of pronation-supination of the hands
   5. Scanning speech is due to asynergy of the muscles used in speaking. The spacing of sounds is irregular with pauses in the wrong places

B. HYPOTONIA: muscle tone is decreased and may be ascertained by palpation

C. ASTHENIA: muscles affected by cerebellar lesions are weaker and tire more easily than normal muscles

D. TREMOR: of cerebellar dysfunction is usually an *intention tremor.* It is evident during purposeful movements, but absent or diminished with rest

E. NYSTAGMUS is present with cerebellar lesions being accounted for by irritation of vestibular fibers in the cerebellum. It may also be due to the effect of pressure on the vestibular nuclei of the brainstem ventral to the cerebellum

CEREBELLAR CONNECTIONS—
SUMMARY

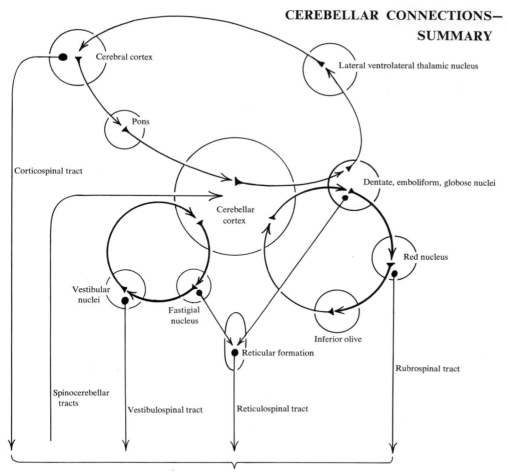

Cerebral cortex

Lateral ventrolateral thalamic nucleus

Pons

Corticospinal tract

Dentate, emboliform, globose nuclei

Cerebellar cortex

Red nucleus

Vestibular nuclei

Fastigial nucleus

Inferior olive

Reticular formation

Rubrospinal tract

Spinocerebellar tracts

Vestibulospinal tract

Reticulospinal tract

Connecting to motor systems

POSITION AS RESULT OF
CEREBELLAR ATAXIA

CONTORTIONIST MOVEMENT
AS RESULT OF CEREBELLAR
HYPOTONIA

Unit Eleven

THE RETICULAR
FORMATION, LIMBIC
AND OLFACTORY SYSTEMS

149. RETICULAR FORMATION: GENERAL CONSIDERATIONS

I. **Definition:** an area in the tegmentum of the midbrain and pons and its continuation into the medulla and upper spinal cord, consisting of small groups of nerve cells interspersed among horizontally and vertically running nerve fibers, and excluding the cranial nerve nuclei and roots of the long fiber tracts

II. **Anatomy of the reticular formation:** general considerations
   A. THE RETICULAR FORMATION (RF) BEGINS JUST ROSTRAL TO THE PYRAMIDAL DECUSSATION and extends rostrally through the mesencephalon
   B. THE RF IS FOUND ONLY IN THE MEDULLA, PONS, AND MIDBRAIN and is a phylogenetically old part of the brain (constitutes prominent part of brainstem in lower forms)
   C. THE RF IS COMPRISED OF THOSE AREAS OF THE BRAINSTEM that are characterized by being made up of diffuse aggregations of cells of different types and sizes, separated by fibers extending in all directions
   D. NERVE ROOTS AND BUNDLES OF MYELINATED AND UNMYELINATED FIBERS derived from phylogenetically more recent structures intersect the RF
   E. CIRCUMSCRIBED CELL GROUPS such as the red nucleus, olivary nucleus, and cranial nerve nuclei are not included in the RF
   F. THE RF IS SEPARATED from the ependyma of the 4th ventricle and aqueduct by a zone of subependymal (periaqueductal) central gray matter
   G. IN THE LOWER MEDULLA, the RF lies ventral to the central canal, appears to be continuous with the gracilis and cuneate nuclei dorsally, and extends laterally almost to the spinal nuclei of the 5th (trigeminal) cranial nerve
   H. IN THE UPPER MEDULLA, the RF lies between the central gray matter (dorsally) and the inferior and accessory olives (ventrally) and extends laterally to the minor tract area and trigeminal spinal nucleus
   I. IN THE PONTINE TEGMENTUM, the RF lies between the subependymal gray matter and the lemnisci
   J. IN THE MIDBRAIN, the RF reaches its largest extent and extends from the periaqueductal gray matter to reach the substantia nigra and completely surrounds the red nucleus

III. **Specific reticular nuclei**
   A. IN THE MEDULLA, the five major nuclei will be considered
      1. Lateral reticular nucleus is located near the surface of the medulla and extends from the level of the caudal border of the inferior nucleus to the midolivary level; it projects to the cerebellum
      2. Paramedian nucleus lies adjacent to the midline of the medulla and also projects to the cerebellum
      3. Ventral reticular nucleus is comprised of cells scattered in the medial two thirds of the area bounded by the inferior olive (ventrally), gracilis and cuneate nuclei (dorsally), and the nucleus of the trigeminal spinal tract (laterally)
      4. Magnocellular nucleus lies just rostral to the ventral reticular nucleus; comprised of large neurons intermingled with some smaller ones
      5. Parvicellular nucleus is located medial to the spinal nucleus of V and ventral to the vestibular nuclei. It extends from midolivary level to rostral limit of the medulla. Considered "sensory" because collaterals from sensory system terminate here
   B. IN THE TEGMENTUM OF THE PONS, three separate nuclei are mentioned, but two of these are rostral extensions of the medullary reticular nuclei
      1. Caudal pontine reticular nucleus is a forward extension of the magnocellular nucleus of the medulla and is comprised of large cells
      2. Rostral pontine reticular nucleus is located in the medial portion of the tegmentum and is comprised only of small nerve cell bodies
      3. Parvicellular reticular nucleus is located in the lateral tegmentum of the pons; it is a forward extension of the parvicellular nucleus

Medial geniculate body

Cerebral aqueduct

Sup. colliculus

Mesencephalic reticular nucleus

Nucleus of Darkschweitsch

Interstitial nucleus of Cajal

Substantia nigra

Oculomotor nucleus

Medial lemniscus

Red nucleus

Cerebral peduncle

UPPER MIDBRAIN

Spinal V tract + nucleus of V

Parvicellular reticular nucleus

VII nuc.

Caudal pontine (magnocellular) reticular nucleus

VIII

VII

VI

LOWER PONS

**RETICULAR NUCLEI
OF BRAINSTEM**

Fastigial n.

Globose n.

Emboliform n.

Dentate n.

Magnocellular reticular nucleus

IVth vent.

Spinal tract V + nucleus of V

Parvicellular reticular nucleus

Paramedian reticular nucleus

IX

VIII

Inf. olive

Med. lemniscus

UPPER MEDULLA

Nucleus and fasciculus gracilis

Nucleus and fasciculus cuneatus

Nucleus of trigeminal spinal tract

Ventral reticular nucleus

X

X

Lateral reticular nucleus

Paramedian reticular nucleus

Pyramid

XII

LOWER MEDULLA

-333-

150. RETICULAR FORMATION: SPECIFIC RETICULAR NUCLEI AND RELATED INFORMATION

I. Specific reticular nuclei (cont.)

A. IN THE MIDBRAIN, the reticular formation (RF) is a forward extension of the pontine reticular formation, and the mesencephalic reticular nucleus is the major nucleus found in this part of the brainstem
1. Mesencephalic reticular nucleus is surrounded by the red nucleus, tectum, and ascending lemnisci
2. From the midbrain, the RF extends into the diencephalon as the zona incerta of the subthalamus
3. Pallidotegmental and corticofugal fibers from the precentral gyrus end in this part of the RF which suggests a "motor" function
4. Mesencephalic RF does not send fibers as far caudally as the spinal cord

II. Some points of importance relative to reticular nuclei and formation

A. LARGE CELLS OF THE RF are restricted to its medial part (medial two thirds) where they are intermingled with both small and large cells
B. THE LATERAL ONE THIRD OF THE RF has only small cells
C. THE LARGEST CELLS are found in more caudal locations of the RF (i.e., caudal pons and medulla)
D. NO TYPICAL GOLGI TYPE II CELLS (cells with short axons) in the RF
E. ALL AXONS OF THE RF CELLS appear to project at least some distance caudal and/or rostral
F. DENDRITES OF RF CELLS are long and radiating. Most of the dendrites are spread out in a plane perpendicular to the long axis of the brainstem
G. EACH HALF OF THE RF BORDERS (peripherally) on the long ascending and descending fiber bundles traversing the brainstem (MLF, medial lemniscus, spinothalamic tract, etc.) and on particular nuclei, which in certain places intrude somewhat into its territory
H. INDISTINCT BOUNDARIES exist between different reticular nuclei. However, rather obvious architectonic differences between relatively small areas of the RF do exist. There are also differences between areas with regard to fiber connections and functions

III. Reticular activating system (ascending activating system or activating system)

A. WHEN ONE GOES FROM A DROWSY OR RELAXED STATE to one of attention or alertness, the EEG shows low voltage fast activity. This is referred to as desynchronization, activation, or arousal reaction
1. This reaction can be elicited by stimulating a great variety of structures within the CNS
2. Lesions placed in the central parts of the brainstem result in the manifestation of behavioral somnolence and electrocortical synchrony
B. THE RF DOES NOT CORRESPOND to the reticular activating system. The RF is too restrictive and is a morphologic designation
C. THE RETICULAR ACTIVATING SYSTEM does include part of the RF as well as the "nonspecific" thalamic nuclei (midline, intralaminar, and reticular)
1. These nonspecific nuclei project to hypothalamic and other "specific" thalamic nuclei
2. These nonspecific nuclei receive fibers from the RF and the spinal cord
3. The thalamic nuclei project to the cerebral cortex to influence its level of activity
D. RAPHE NUCLEI receive fibers from and give off fibers to the same part of the nervous system as does the RF with only minor differences
E. NEURONS OF THE RAPHE NUCLEI are rich in serotonin, whereas some components of reticular nuclei of the tegmentum of the pontomesencephalic brainstem are rich in norepinephrine and dopamine

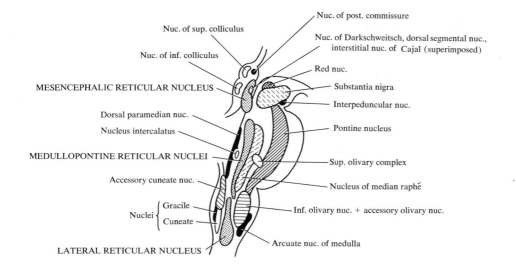

Nuc. of post. commissure

Nuc. of sup. colliculus

Nuc. of Darkschweitsch, dorsal segmental nuc., interstitial nuc. of Cajal (superimposed)

Nuc. of inf. colliculus

Red nuc.

MESENCEPHALIC RETICULAR NUCLEUS

Substantia nigra

Interpeduncular nuc.

Dorsal paramedian nuc.

Pontine nucleus

Nucleus intercalatus

MEDULLOPONTINE RETICULAR NUCLEI

Sup. olivary complex

Accessory cuneate nuc.

Nucleus of median raphé

Nuclei { Gracile / Cuneate }

Inf. olivary nuc. + accessory olivary nuc.

Arcuate nuc. of medulla

LATERAL RETICULAR NUCLEUS

BRAINSTEM FORMATIONS (UNSEGMENTED)

LOCALIZATION OF FUNCTION
IN RETICULAR FORMATION

Ascending reticular fibers to thalamic reticular nuclei

Secondary pain area

Alerting center

RETICULAR FORMATION

Postural facilitatory zone

Sleep center

Postural inhibitory zone

Level of respiratory centers

Reticulospinal fibers

TRANSVERSE SECTION OF
BRAINSTEM—EXTENT OF RETICULAR FORMATION

151. RETICULAR FIBER CONNECTIONS

I. Afferent fiber connections to the RF come from the following sources:
A. SPINORETICULAR FIBERS
    1. A massive influx of these direct fibers ascend in the spinal cord in the anterolateral funiculus. They appear to reach all parts of the RF bilaterally, but most are distributed to the medial two thirds of the RF and to the magnocellular nucleus of the medulla and pontine reticular nuclei
    2. RF located in pons-mesencephalon transitional zone of brainstem also receives a large influx of spinoreticular fibers
B. SPINOTHALAMIC TRACT FIBERS and collaterals terminate on lateral reticular nucleus which in turn projects to the cerebellum
C. CRANIAL NERVE FIBERS listed below send many sensory collaterals to RF
    1. Trigeminal sensory nucleus fibers
    2. Vestibular pathway fibers
    3. Ascending auditory pathway fibers
    4. Optic fibers from superior colliculus enter RF via tectoreticular fibers
D. CEREBELLUM
    1. Fastigial nuclei via uncinate fasciculus are distributed throughout the medial two thirds of the medullary reticular formation
    2. Dentate nuclei via the superior cerebellar peduncle
E. HYPOTHALAMUS sends fibers from the lateral nuclear group to the mesencephalic reticular formation
F. GLOBUS PALLIDUS (medial portion) provides small input to the mesencephalic tegmentum RF
G. CEREBRAL CORTEX fibers to the RF are prominent (mainly from sensorimotor area).
    1. Corticoreticular fibers supply the region of the magnocellular and pontine tegmental nuclei
    2. Fibers from cerebral cortex also project directly to mesencephalic RF

II. Efferent fibers from the RF pass mainly to spinal cord, thalamus, and other nuclei within and above the brainstem
A. RETICULOSPINAL FIBERS (DESCENDING TO THE SPINAL CORD)
    1. From medial two thirds of pontine RF to end in laminae VII and VIII
    2. From medial two thirds of medullary RF fibers descend (crossed and uncrossed) to end in laminae VII and VIII of spinal cord
    3. The reticulospinal fibers terminate on interneurons of the spinal cord
    4. No somatotopic pattern is noted in the origin of fibers
    5. Many RF fibers contain biogenic amines and serotonin

III. Ascending fibers
A. GIANT CELLS in caudal portion of RF do not have ascending projections
B. ASCENDING FIBERS originate mainly from pontine and medullary nuclei
C. ASCENDING RF FIBERS are mostly uncrossed, but some do decussate or cross
D. MANY RF CELLS FROM PONTINE AND MEDULLARY REGIONS have axons that project beyond the mesencephalon
E. AN IMPORTANT GROUP OF EFFERENT FIBERS of the mesencephalic reticular formation passes rostrally via mamillary peduncle, dorsal longitudinal fasciculus, and medial forebrain bundle to mamillary bodies, lateral hypothalamic areas, and preoptic and septal areas, respectively

Limbic lobe

Cerebral cortex

Caudate nucleus

Basal ganglia

From hypothalamus

Red nucleus

Thalamus

Spinal V (somatic)

VIII vestibular (balance)

VIII cochlear (auditory)

VII, IX, X solitary (taste + visceral)

Cerebellum

Spinotectal + spinothalamic }

Spinoreticular }

Somatic + visceral

INPUT TO BRAINSTEM RETICULAR FORMATION

To limbic lobe (via median forebrain bundle)

Cerebral cortex

Reticulothalamic + intralaminar projections

Intralaminar thalamic nuc.

To hypothalamus
(via dorsal long. fasc.)

Basal ganglia

Nigropallidal + reticulosubthalamic
tracts

Tectocerebellar tract

Red nucleus

Tectospinal tract

Reticuloerebellar tract

Reticulospinal tract { Ventral

Dorsal

Rubrospinal tract

OUTPUT FROM BRAINSTEM RETICULAR FORMATION

152. LIMBIC AND OLFACTORY SYSTEMS

I. **Introduction:** the term "limbic system" was derived from the concept of a "limbic lobe" presented in 1787 by the French anatomist Broca. The word "limbic" means to border, fringe, or ring, and the term "limbic system" was first used to describe a series of structures that form a border or ring about the medial wall of the cerebrum. Various approaches have been used to classify limbic structures and to relate them to each other based on a common principle such as (1) evolutionary development, (2) role in olfaction, and (3) fiber connections and interconnections

II. **Limbic lobe** designates brain tissue that surrounds the rostral brainstem and interhemispheric commissures, and it underlies the neocortical mantle. It appears early phylogenetically and demonstrates a certain consistent morphology both grossly and microscopically. Grossly it includes the following structures
   A. SUBCALLOSAL AND CINGULATE GYRI
   B. CINGULUM, a long association pathway, which connects the orbital frontal cortex and hippocampal cortex
   C. PARAHIPPOCAMPAL GYRUS AND UNDERLYING HIPPOCAMPAL FORMATION
   D. ISTHMUS AND VARIOUS GYRI SURROUNDING THE OLFACTORY FIBERS

III. **Structures in the limbic system** form a ring or loop positioned deep on the medial surface of the cerebrum and include the following gray areas and fiber bundles
   A. LIMBIC LOBE: ring-shaped convolution on the medial cerebrum
   B. AMYGDALA: amygdaloid nuclear complex
   C. HYPOTHALAMUS: specifically the mamillary bodies
   D. CERTAIN MIDBRAIN NUCLEI that have reciprocal relations with the hypothalamus
   E. THALAMUS: anterior and dorsomedial nuclei
   F. FORNIX, INDUSIUM GRISEUM (SUPRACALLOSAL GYRUS), AND MAMILLOTHALAMIC TRACT
   G. STRIA TERMINALIS, STRIA MEDULLARIS, AND HABENULA NUCLEI
   H. SEPTAL AREA AND NUCLEI
   I. ORBITOFRONTAL CORTEX, PREPYRIFORM CORTEX, AND OLFACTORY TUBERCLE

IV. **Primary roles of the limbic system:** numerous neuroanatomic substrates underlying man's behavioral and emotional expression reside in the limbic system. This system is functionally associated with
   A. EMOTIONAL ACTIVITIES essential for the self-preservation of the individual (e.g., feeding, fight, and flight)
   B. ACTIVITIES ESSENTIAL FOR THE PRESERVATION OF THE SPECIES (e.g., mating, procreation, and care of the young)
   C. VISCERAL ACTIVITIES associated with both of the above
   D. MECHANISMS FOR MEMORY
   E. NUMEROUS OTHER ACTIVITIES OF THE HYPOTHALAMUS (e.g., modulation and regulation)

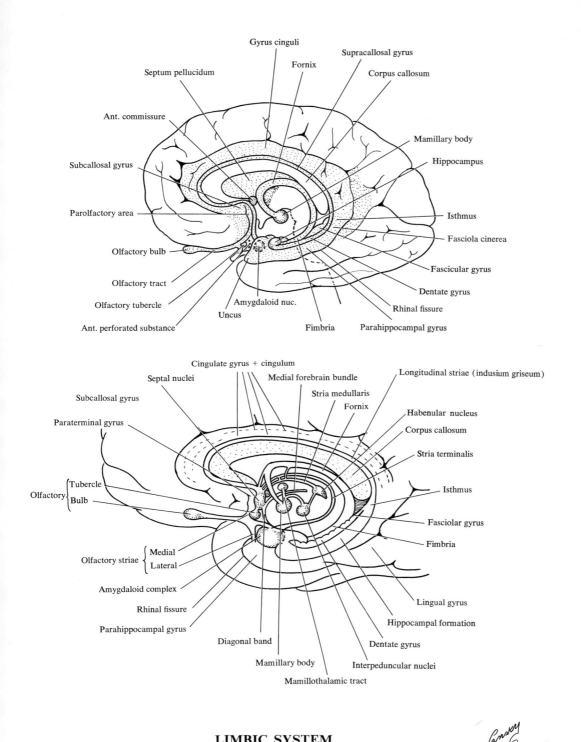

Gyrus cinguli

Septum pellucidum

Fornix

Supracallosal gyrus

Corpus callosum

Ant. commissure

Mamillary body

Hippocampus

Subcallosal gyrus

Parolfactory area

Isthmus

Fasciola cinerea

Olfactory bulb

Fascicular gyrus

Olfactory tract

Dentate gyrus

Olfactory tubercle

Rhinal fissure

Amygdaloid nuc.

Ant. perforated substance

Uncus

Fimbria

Parahippocampal gyrus

Cingulate gyrus + cingulum

Septal nuclei

Medial forebrain bundle

Longitudinal striae (indusium griseum)

Subcallosal gyrus

Stria medullaris

Fornix

Habenular nucleus

Paraterminal gyrus

Corpus callosum

Stria terminalis

Olfactory { Tubercle / Bulb

Isthmus

Fasciolar gyrus

Fimbria

Olfactory striae { Medial / Lateral

Amygdaloid complex

Rhinal fissure

Lingual gyrus

Parahippocampal gyrus

Hippocampal formation

Diagonal band

Dentate gyrus

Mamillary body

Interpeduncular nuclei

Mamillothalamic tract

LIMBIC SYSTEM

153. HIPPOCAMPAL FORMATION

I. **The hippocampal formation** consists of the hippocampus (Ammon's horn), dentate gyrus, and subicular complex (part of the parahippocampal gyrus which continues with the hippocampus)

A. IT DEVELOPS EMBRYOLOGICALLY on the medial wall of the temporal lobe along the hippocampal fissure
1. The hippocampal fissure parallels the choroid fissure, the point of invagination of the choroid plexus into the temporal horn of the lateral ventricle
2. Both fissures arch anteriorly and inferiorly from the interventricular foramina to the tip of the temporal horn of the lateral ventricle
3. The hippocampal fissure deepens, and the invaginated part extends into the temporal horn as the hippocampus or ram's horn as it is often called
4. The dentate and parahippocampal gyri form from the lips of this fissure

B. THE ALVEUS: a thin layer of white matter comprised of axons from the cells of the hippocampus, line the whole ventricular surface of the hippocampal formation, and constitutes the main efferent pathway
1. Pyramidal cell is the principal cell type of the hippocampus
2. Basket cells, a second cell type, lie proximal to the pyramidal cells

C. THE HIPPOCAMPUS has the following three major cellular layers
1. Plexiform layer is the external layer located adjacent to the ventricular surface (alveus) which contains axons of pyramidal cells and afferent neurons
2. Pyramidal layer is comprised of pyramidal cells
3. Polymorphic cell layer
4. Secondary lamina are formed (from alvear surface outward): stratum oriens (comprised of basal dendrites and basket cells), pyramidal layer, stratum radiatum, stratum lacunosum, and stratum moleculare. The last three are the molecular layer

D. THE FIMBRIA is formed by axons in the alveus which course over the hippocampus to its medial border. These fibers make up the beginning of the fornix system. Its thin free border is continuous and contiguous with the epithelium of the choroid fissure

E. PARAHIPPOCAMPAL GYRUS (ENTORHINAL CORTEX, AREA 28) extends from the hippocampal fissure to the collateral sulcus laterally. Its superior portion is called the subiculum

F. MOST OF THE CORTEX of the parahippocampal gyrus has a neocortex (isocortex) structure of six layers. The subiculum, hippocampal formation, and the dentate gyrus have an allocortex structure of three layers

G. THE INDUSIUM GRISEUM (LONGITUDINAL STRIAE OR SUPRACALLOSAL GYRUS) is a thin gray vestigial convolution of the hippocampal formation
1. It remains positioned above the developing fibers of the corpus callosum
2. It contains two thin bands of myelinated fibers (white matter) which appear as the medial and lateral longitudinal ridges or striae on the superior surface of the corpus callosum in each cerebral hemisphere

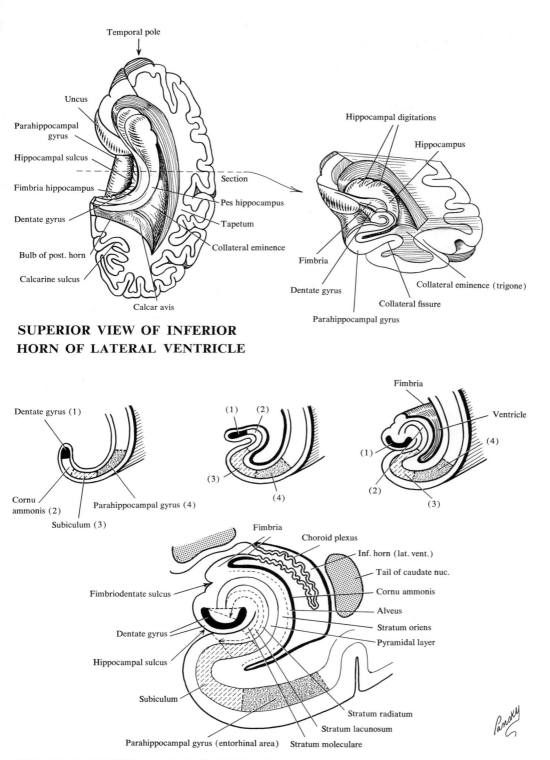

SUPERIOR VIEW OF INFERIOR
HORN OF LATERAL VENTRICLE

CORONAL SECTION OF INFERIOR HORN OF LATERAL VENTRICLE

154. FIBER CONNECTIONS OF THE HIPPOCAMPAL FORMATION: AFFERENTS

I. Afferent fiber connections

A. ENTORHINAL CORTEX (AREA 28) is one primary site where afferent fibers arise
  1. This cortical area is a part of the pyriform lobe (olfactory association cortex)
  2. No direct nerve fibers connect this lobe with the olfactory area
  3. Fibers from area 28 give rise to two major pathways
    a. Medial portion sends fibers via the "alvear path" to the hippocampus
    b. Lateral portion sends fibers via the "perforant path" which traverses the subiculum on its way to the hippocampus and dentate gyrus
  4. Axons making up these two pathways establish synaptic contacts with dendrites of pyramidal cells in the hippocampus. They do not make direct contact with granule cells in the dentate gyrus

B. DIAGONAL BAND FIBER COMPONENTS from the medial septal nucleus enter the hippocampal formation via the fimbria of the fornix and are distributed in a diffuse fashion to cells in the hippocampal formation

C. THE CINGULUM projects fibers to the entorhinal cortex which in turn relays impulses on to the hippocampus

D. THE INDUSIUM GRISEUM may send some fibers to the hippocampal formation

E. CORTICAL AREAS such as the piriform and prefrontal cortices contribute some fibers to the hippocampus

F. ANTERIOR THALAMIC NUCLEUS projects to the presubicular portion of the hippocampal formation

G. OLFACTORY, VISUAL, AUDITORY, GUSTATORY, AND SOMATOSENSORY CORTICAL AREAS indirectly supply the hippocampal formation via synaptic contacts with area 28 (entorhinal cortex). None of the direct afferent pathways to the hippocampal formation appears to transmit olfactory impulses

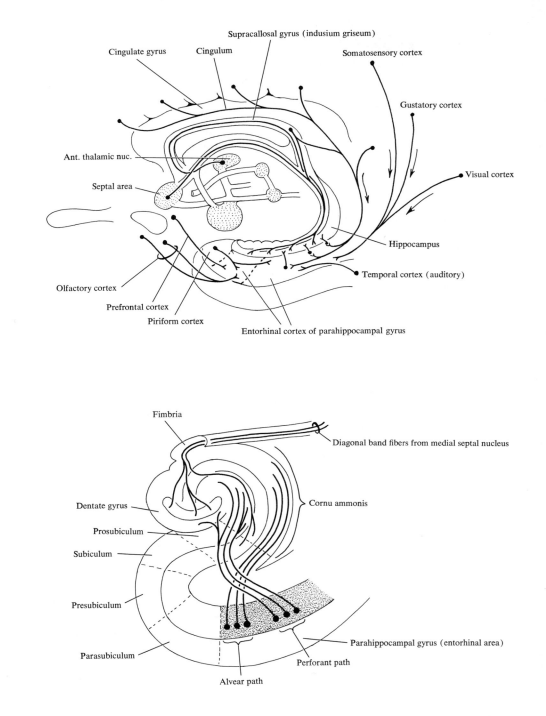

FIBER CONNECTIONS OF HIPPOCAMPAL FORMATION— AFFERENTS

155. FIBER CONNECTIONS OF THE HIPPOCAMPAL FORMATION: EFFERENTS

I. Efferent fiber connections

A. THE FORNIX comprises the main efferent fiber system of the hippocampal formation

1. Myelinated axons of the large pyramidal cells of the hippocampus form the thin surface lining of the lateral ventricle (temporal horn) known as the alveus. These fibers converge to form the fimbria of the fornix

2. The crura of the fornix are terms applied to the fimbriae at the level of the posterior end of the hippocampus as they bend backward and arch under the splenium of the corpus callosum

3. Hippocampal or fornical commissure is made up of a thin sheet of crossing fibers from the ventral aspect of the fornix that serve to connect the two hippocampi. This commissural connection is usually poorly developed in man

4. Body of the fornix is formed as the two crura join and run forward inferior to the corpus callosum to the level of the rostral margin of the thalamus

5. Anterior columns of the fornix are formed as the body splits into two separate fiber bundles that bend and continue inferiorly between the anterior commissure (ventrally) and the interventricular foramina (dorsally)

   a. The columns of the fornix extend into the hypothalamus where they divide into medial and lateral parts

   b. Most of the fibers in the columns end in the mamillary bodies

   c. Some fibers exit the columns to enter the anterior thalamic nuclei

6. Precommissural and postcommissural fornix are formed by fibers within the columns from the fimbria which regroup and separate off from the columns as components rostral and caudal to the anterior commissure

   a. Precommissural fibers pass rostral to the anterior commissure and (1) supply the septal area, (2) fibers located near the anterior pole of the hippocampal formation extend to the most lateral region of the dorsal septal area, and (3) fibers located more caudally in the hippocampal formation extend to the more medial parts of the dorsal septal area

   b. Postcommissural fibers pass just caudal to the anterior commissure and

      i. Descend through the hypothalamus to supply the mamillary bodies, adjacent parts of the hypothalamus, anterior thalamic nucleus, and perhaps areas of the mesencephalon

      ii. Fibers take origin from subicular complex cells instead of the pyramidal cells of the hippocampus

B. MAMILLOTHALAMIC TRACT is comprised of fibers that connect the mamillary body and the anterior thalamic nucleus of the same side

1. The anterior thalamic nucleus has reciprocal connections with the cingulate gyrus (cortex) via fibers that run through the medullary core

2. These and other connections show the complex pathways through which impulses from the hippocampal formation may be projected to different parts of the neuroaxis

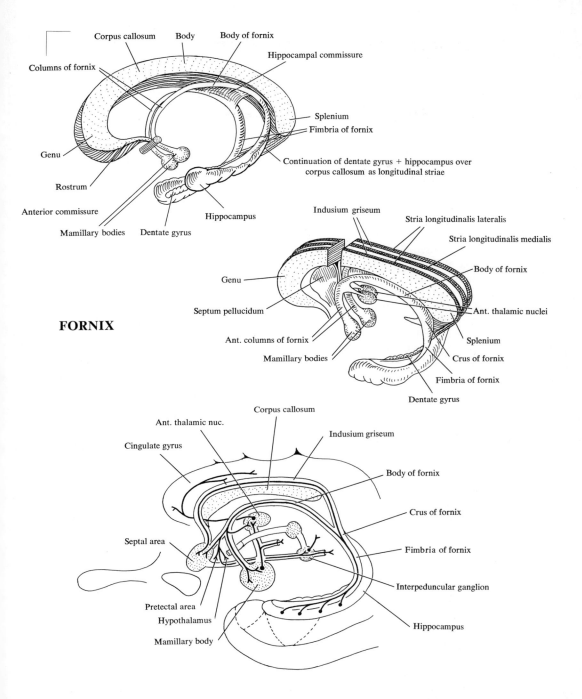

FORNIX

Labels, top figure:
Corpus callosum Body Body of fornix
Hippocampal commissure
Columns of fornix
Splenium
Fimbria of fornix
Genu
Continuation of dentate gyrus + hippocampus over
corpus callosum as longitudinal striae
Rostrum
Anterior commissure
Hippocampus
Mamillary bodies Dentate gyrus

Labels, middle figure:
Indusium griseum
Stria longitudinalis lateralis
Stria longitudinalis medialis
Body of fornix
Genu
Septum pellucidum
Ant. thalamic nuclei
Ant. columns of fornix
Mamillary bodies
Splenium
Crus of fornix
Fimbria of fornix
Dentate gyrus

Labels, bottom figure:
Corpus callosum
Ant. thalamic nuc.
Indusium griseum
Cingulate gyrus
Body of fornix
Crus of fornix
Septal area
Fimbria of fornix
Interpeduncular ganglion
Pretectal area
Hypothalamus
Mamillary body
Hippocampus

FIBER CONNECTIONS OF HIPPOCAMPAL FORMATION— EFFERENTS

156. FUNCTIONS OF THE HIPPOCAMPAL FORMATION

I. **Introduction:** both morphologic and functional evidence supports the thesis that the hippocampal formation takes part in a wide variety of activities such as endocrine, learning, recent memory, expression of emotion, hypothalamic, and autonomic function

A. ENDOCRINE FUNCTIONS are regulated or modulated by the hippocampal formation

  1. In the ventral region of this structure, estradiol-concentrating neurons are numerous and closely packed. Corticosterone is also localized in heavy concentrations in the ventral aspect of this formation

  2. Corticosterone injections cause a general inhibitory effect on the neurons in the hippocampus

  3. Hippocampal lesions (or section of the fornix) usually result in the disruption of the diurnal rhythm of ACTH release

  4. Stimulation of the hippocampal formation caused the inhibition of ovulation in spontaneously ovulating rats

B. LEARNING AND RECENT MEMORY ACTIVITIES, as revealed in experimental animal studies and human clinical investigations, strongly implicate a significant role for the hippocampal formation. EEG and ablation experiments support the thesis that the formation is involved in learning processes

  1. In humans with lesions of the hippocampus, learning and memory disabilities are common. The Korsakoff syndrome is such a disturbance which demonstrates both retrograde and antegrade amnesia. The patient has difficulty in remembering newly acquired information and in recalling events in the recent past

    a. Results from toxic effects of alcohol or from vitamin B deficiency resulting in damage to either the hippocampus or its target neurons

  2. Patients subjected to temporal lobectomies (for the treatment of psychomotor epilepsy) exhibit a disorder referred to as short-term memory defect which is characterized by not being able to learn and remember new simple facts, such as names, but are able to recall events that occurred in the distant past and those just prior to surgery with little or no difficulty

C. EXPRESSION OF EMOTION, HYPOTHALAMIC ACTIVITIES, AND AUTONOMIC FUNCTIONS
The hippocampal formation plays a role in controlling aggressive behavior, but its effects upon aggression are not uniform as observed when

    a. Stimulation of areas in the vicinity of the amygdala appears to facilitate the onset of aggressive behavior

    b. Stimulation of areas close to the septal area suppresses such reactions

    c. Hippocampal formation may exert its effects on hypothalamic aggression via its differential synaptic connections in the septal area since both the amygdala and areas closer to the septal area project their axons to different areas of the septum. The septal area may serve as a relay nucleus for impulses from the hippocampal formation to the hypothalamus

  2. In humans, aggressive behavior has been associated with lesions, tumors, and seizure activity involving the hippocampal formation and temporal lobe

    a. Seizures of the temporal lobe involving the hippocampus (partial or temporal lobe seizures) may result in confusion and cause hallucinations involving gustatory and auditory sensations

    b. An attempt to direct or move a patient with temporal lobe seizures may initiate aggressive behavior

  3. Hippocampectomized animals appear more active, more exploratory, and exhibit increased "startle reaction" in response to sudden loud noises

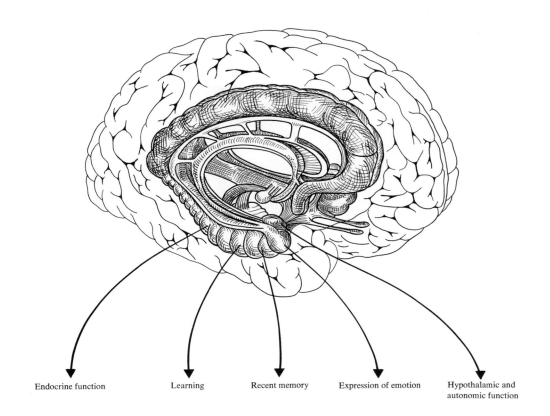

Endocrine function   Learning   Recent memory   Expression of emotion   Hypothalamic and autonomic function

FUNCTIONS OF THE HIPPOCAMPAL FORMATION

157. AMYGDALA (AMYGDALOID BODY OR NUCLEAR COMPLEX): ANATOMY AND CONNECTIONS

I. Introduction: the term "amygdala" means almond-shaped and is used to describe this series of nuclear masses (nerve cell bodies) and associated nerve fibers that are located in the dorsomedial portion of the temporal lobe of the cerebrum and form the medial, anterior, and superior portions of the ventral tip of the temporal horn of the lateral ventricle. It is continuous with the uncus and closely related to the hippocampus. It fuses with the tail of the caudate nucleus, and the stria terminalis originates from its more caudal portion. In addition to subnuclear groups, it also consists of a cortical mantle which includes the piriform lobe. It is also intimately related to the adjacent cortex known as the prepiriform area (rostral levels) and the periamygdaloid cortex (caudal levels)

II. The amygdaloid nuclear complex is divided into two main groups of nuclei
A. CORTICOMEDIAL AMYGDALOID is continuous with the anterior perforated substance, diagonal band, putamen, caudate nucleus, and cortical areas surrounding the uncus
   1. This portion of the complex is small in man and includes the anterior area (poorly differentiated), lateral olfactory tract nucleus (poorly developed), medial amygdaloid nucleus, and the cortical amygdaloid nucleus
B. Basolateral amygdaloid is the largest and best differentiated part of the nuclear complex. It is continuous with the claustrum and cortex of the parahippocampal gyrus and includes the lateral, basal, and accessory basal nuclei

III. Fiber connections of the amygdaloid complex are incompletely established in the human brain. Information on nerve fiber connections has been obtained from electrophysiologic and nerve degeneration experiments in cats and dogs
A. AFFERENT CONNECTIONS are closely linked with cortical tissue and include
   1. Olfactory bulb and anterior olfactory nucleus fibers enter the corticomedial nuclear group via the lateral olfactory stria
   2. Prepiriform and piriform cortices send indirect olfactory connections to the basolateral group of nuclei
   3. Orbitofrontal cortex has reciprocal connections with the basolateral group
   4. Diagonal band nuclei send direct fiber connections to the basolateral group
   5. Inferior temporal gyrus sends direct connections to the lateral nucleus
   6. Medial thalamus sends fibers to the anterior amygdala
   7. Ventromedial hypothalamus: to the amygdala (diffusely) via stria terminalis
   8. Brainstem reticular formation (dopaminergic neurons) to central amygdala
B. EFFERENT CONNECTIONS
   1. Stria terminalis is the most completely established efferent pathway
      a. It arises mainly from corticomedial group of nuclei and adjacent piriform cortex
      b. It conveys fibers to the septal areas, medial preoptic area, anterior medial hypothalamus, ventromedial hypothalamic nucleus, habenular nucleus (via stria medullaris), and the anterior perforated substance
   2. Ventral amygdalofugal pathways take origin in the basolateral amygdala, piriform and periamygdala cortices
      a. They extend to the anterior perforated substance, preoptic area, septal area, dorsomedial nucleus of the thalamus, hypothalamic nuclei, parahippocampal gyrus, and brainstem reticular formation via the medial forebrain bundle

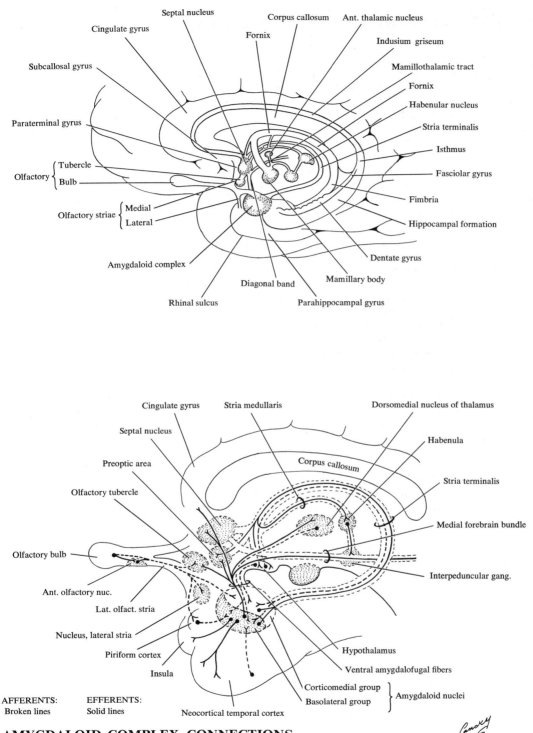

AMYGDALOID COMPLEX CONNECTIONS

AFFERENTS: Broken lines

EFFERENTS: Solid lines

158. FUNCTIONS OF THE AMYGDALA

I. Olfaction does not seem to be closely related to the amygdaloid complex since bilateral lesions or destruction does not impair olfactory discrimination

II. Emotional behavior is regulated by the amygdala through its effects upon the hypothalamus. These effects upon the hypothalamic rage mechanism may be inhibitory or facilitatory, though the former is often more pronounced
  A. LESIONS of this complex in rats, cats, and monkeys result in "taming effects"
  B. HYPEREMOTIONAL STATE may also result from lesions in this same area
  C. ELECTRICAL STIMULATION OF THE CORTICOMEDIAL GROUP OF NUCLEI results in the inhibition of aggressive behavior; and stimulation of the lateral nucleus facilitates aggressive activities
  D. STIMULATION OF THE AMYGDALOID REGION in humans causes a feeling of confusion, fear, and amnesia of events that occurred during the stimulation

III. Endocrine function is influenced by the amygdaloid complex as indicated
  A. NEURONS IN THE AMYGDALA have the ability to concentrate estrogen
  B. OVULATION IS INDUCED by the stimulation of the corticomedial group of nuclei
  C. OVULATION IS PREVENTED by transection of the stria terminalis
  D. ACTH IS RELEASED when the nasolateral nucleus is stimulated or when a lesion is made in the corticomedial group
  E. GROWTH HORMONE IS RELEASED when the basolateral nucleus is stimulated, whereas the corticomedial group serves to inhibit its release

IV. Somatic responses, observed when the amygdala is stimulated, include
  A. TURNING OF THE HEAD AND EYES to the side opposite that stimulated
  B. EXHIBITING COMPLEX RHYTHMIC MOVEMENTS related to chewing and swallowing

V. Autonomic responses, observed when the amygdala is stimulated, include
  A. CHANGES IN THE RESPIRATORY RATE, RHYTHM, AND AMPLITUDE as well as inhibition of respiration
  B. INCREASES AND DECREASES IN ARTERIAL BLOOD PRESSURE and changes in heart rate
  C. INCREASED SECRETION OF DIGESTIVE JUICES which may cause peptic ulcers

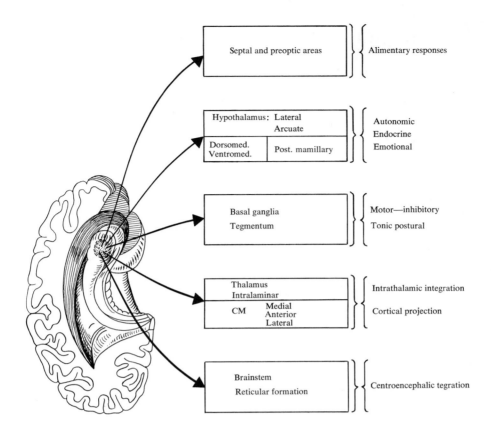

| Septal and preoptic areas | }} { | Alimentary responses |

| Hypothalamus: Lateral / Arcuate / Dorsomed. Ventromed. / Post. mamillary | }} { | Autonomic / Endocrine / Emotional |

| Basal ganglia / Tegmentum | }} { | Motor—inhibitory / Tonic postural |

| Thalamus / Intralaminar / CM / Medial / Anterior / Lateral | }} { | Intrathalamic integration / Cortical projection |

| Brainstem / Reticular formation | }} { | Centroencephalic tegration |

GENERAL FUNCTIONS OF THE AMYGDALA OLFACTORY SYSTEM

159. OLFACTORY SYSTEM: FUNCTIONAL ANATOMY

I. **The rhinencephalon (nose-brain)** consists of the olfactory nerves, bulbs, tracts, striae, paraolfactory (subcallosal) areas, anterior perforated substance, prepiriform cortex, amygdaloid complex, and the anterior olfactory nuclei. Vertebrates more dependent on olfactory senses (macrosomatic) than on vision (microsomatic) have more prominent rhinencephalic structures. Man is an example of the latter

A. PERIPHERAL OLFACTORY RECEPTORS are located in small specialized areas of the nasal mucosa referred to as the olfactory epithelium
   1. It is usually pseudostratified columnar epithelium in type
   2. It is located specifically on the superior concha, roof of the nasal cavity, and on the superior portion off the nasal septum. The mucosa lies above the path of the major air currents entering the nose so that odorous molecules must diffuse up to the cells or be drawn up by respiratory changes
   3. The receptor cells are modified neurons originally from the CNS itself. They have two processes (bipolar): one passes to the brain to form the olfactory nerve; the other has fine cilia and extends out to the surface of the olfactory mucosa
      a. There are about 100 million of these cells in the olfactory epithelium interspersed among sustentacular cells
      b. The mucosal end of the olfactory cell forms a knob called the olfactory vesicle from which large numbers of olfactory hairs or cilia (0.3 μm in length) project into the mucus that coats the inner surface of the nasal cavities. These projecting olfactory hairs are believed to react to odors in the air and then to stimulate the olfactory cells
      c. Spaced among the olfactory cells in the olfactory membrane are many small glands of Bowman that secrete mucus onto the surface of the membrane
   4. To be detected, an odorous substance must release molecules that diffuse into the air and pass into the nose to the region of the olfactory mucosa. Molecules then dissolve in the layer of mucus over the receptors and form a relation with the receptor that depolarizes the membrane to initiate an action potential in the afferent nerve fiber

II. **Physiologic basis for discrimination of the thousands of different odor qualities** is speculative

A. THERE ARE NO APPARENT MICROSCOPIC DIFFERENCES in the receptor cells, yet, meaningful differences are thought to exist at the molecular level

B. ODOR DISCRIMINATION may depend on a large number of different types of interactions between the odor-substance molecules and receptor sites. Assuming the receptor cells have 20–30 different types of receptor sites, each may be able to interact with many different odor molecules but responds best to a specific molecule

C. OLFACTORY DISCRIMINATION depends not only on action-potential patterns generated in the different afferent neurons but varies also with attentiveness. The state of the olfactory mucosa is affected by congestion, hunger (sensitivity is greater when hungry), sex (women have a keener sensitivity) and smoking (decreases sensitivity)

D. KNOWLEDGE OF THE ODOR of a substance is aided by the stimulation of other receptors accounting for odors of pungence, acridity, coolness, and irritation

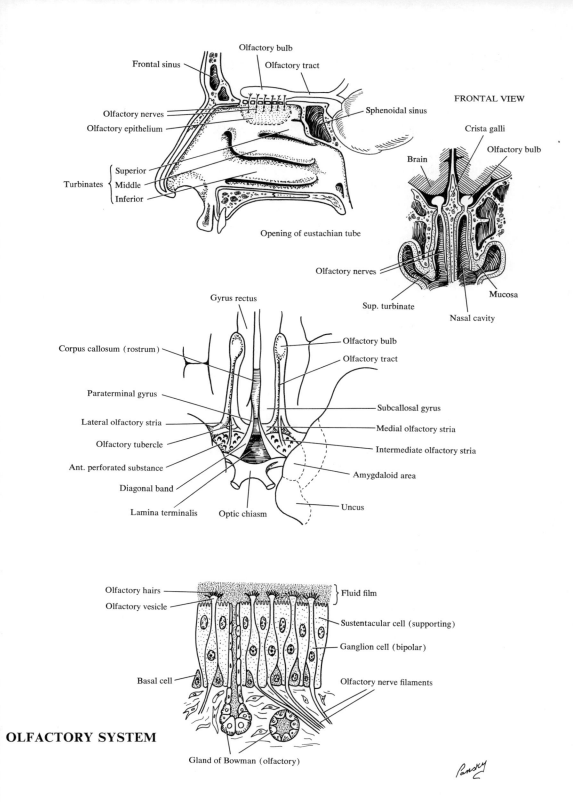

Frontal sinus

Olfactory bulb

Olfactory tract

Olfactory nerves

Olfactory epithelium

Sphenoidal sinus

FRONTAL VIEW

Crista galli

Olfactory bulb

Brain

Turbinates { Superior / Middle / Inferior }

Opening of eustachian tube

Olfactory nerves

Sup. turbinate

Mucosa

Nasal cavity

Gyrus rectus

Corpus callosum (rostrum)

Olfactory bulb

Olfactory tract

Paraterminal gyrus

Subcallosal gyrus

Lateral olfactory stria

Medial olfactory stria

Olfactory tubercle

Intermediate olfactory stria

Ant. perforated substance

Amygdaloid area

Diagonal band

Lamina terminalis

Optic chiasm

Uncus

Olfactory hairs

Fluid film

Olfactory vesicle

Sustentacular cell (supporting)

Ganglion cell (bipolar)

Basal cell

Olfactory nerve filaments

OLFACTORY SYSTEM

Gland of Bowman (olfactory)

-353-

160. OLFACTORY NERVES, BULBS, AND STRIAE

I. **The olfactory nerves, bulbs, and striae** comprise the most peripheral portion of the olfactory apparatus and possess the following relationships
   A. RECEPTOR CELLS possess elongated proximal processes (grouped as filaments) which pass through the openings of the cribriform plate of the ethmoid bone as the olfactory nerves, which then pierce the surface of the olfactory bulb to synapse with second-order neurons in the rostral part of the bulb
   1. Mitral cells are large, triangular-shaped second-order neurons that synapse with olfactory fila, tufted cells, or granule cells
   2. Tufted cells are more peripheral in position and smaller than mitral cells. They synapse with olfactory fila in a bushy synapse known as the olfactory glomerulus
   3. Granule cells are the smallest second-order neurons of the bulb and serve an inhibitory function
   B. THE ANTERIOR OLFACTORY NUCLEUS is located in the caudal portion of the bulb and consists of clusters of neurons. The cells of origin of the olfactory part of the anterior commissure are located here
   C. THE PAIRED OLFACTORY TRACTS are narrow caudal projections of the bulbs which lie in the olfactory sulcus of the cerebrum. Each tract bifurcates into lateral, medial, and intermediate striae which take the following routes
   1. Most of the fibers of the medial olfactory stria enter the subcallosal area, then the rostral portion of the anterior commissure, and finally return to the contralateral olfactory bulb
   2. Fibers that end in the olfactory trigone within the anterior perforated substance appear to be mitral cell processes
   3. Fibers of the lateral stria are mainly mitral cell processes, and they run along the lateral margin of the anterior perforated substance to end in the prepiriform cortex in the vicinity of the uncus. This area (prepiriform cortex) is designated as the primary olfactory cortex. Area 28, the parahippocampal gyrus, is considered the secondary olfactory cortex. Take special note that the olfactory system, unlike the other sensory systems, does not utilize the thalamus as a relay center

II. **Clinical significance**
   A. ANOSMIA: total or partial loss of smell may be caused by the following factors
   1. The common cold is perhaps the most frequent cause of the loss of smell
   2. Other causes include nasal irritants, ethmoid bone fractures, meningitis, drug intoxications, psychoses, and uncinate fits
   B. HYPEROSMIA has been associated with hysterias and drug addiction
   C. THE SUBARACHNOID SPACE, which is in close physical proximity to the olfactory mucosa, may also become involved in certain infections involving the olfactory system

OLFACTORY BULB

Nerve fibers
Olfactory tract

Internal
granular layer

Mitral layer

Molecular + ext.
granular layer

Synaptic
glomeruli

Olfactory nerves

Ant. olf. nucleus

Int. granule neurons

Mitral + tufted neurons

Glomeruli

Olfactory nerve cell

Internal granule cell

Mitral cell

Glomerulus

Olfactory bulb

Int. granule cell

Olfactory nerve cell

Tufted cell

Ant. olfactory nucleus

Pyramidal cell

Olfactory tract

Olfactory efferent fiber

Gyrus
recti

Medial olfactory striae

Septal area

Intermediate
Lateral

Olfactory striae

Nucleus, lateral olfactory stria

Anterior commissure

Olfactory tubercle

Ant. perforated substance

Amygdaloid nucleus

Primary olfactory cortex

OLFACTORY PATHWAYS

Unit Twelve

ORGANS OF
SPECIAL SENSATION

161. ANATOMY OF TASTE

I. Introduction

A. THE RECEPTORS for the sense of taste are chemoreceptors that respond to certain chemicals in the external environment near the receptors

B. IN MAN, the sense of taste is less important than the receptors for vision or hearing (not true for most other animals)

C. TASTE AND SMELL affect a person's appetite, initiate digestion, and aid in the avoidance of harmful substances, but do not exert strong or essential influences

D. IT IS NORMALLY IMPOSSIBLE TO PERCEIVE PURE TASTE without sensing an overlay of smell, because true taste is a much cruder and less sensitive sensation than smell, and taste thresholds are significantly higher than thresholds of smell

II. The anatomy of taste

A. THE SPECIALIZED RECEPTOR ORGANS for the sense of taste consist of about 10,000 taste buds located on the tongue, roof of the mouth (palate), tonsillar pillars, pharynx, and larynx (the latter receptors are not activated by substances in the mouth but fire during swallowing)

B. THE TASTE BUDS ON THE ANTERIOR $\frac{2}{3}$ OF THE TONGUE are located on the free surface of the fungiform papillae, with each papilla containing from 1–8 taste buds. These papillae, which are scattered singly, are especially numerous near the tip and margin of the tongue

   1. In the region of the V-shaped sulcus terminalis is a linear series of 3–14 large circumvallate papillae which cover a portion of the posterior $\frac{1}{3}$ of the tongue. The numerous taste buds on these papillae are found along the sides of the papillae, facing the troughs that surround the circumvallate papillae. The taste buds of the vallate papillae atrophy in old age, but apparently there is little to no decrease in taste sensitivity with age

   2. Moderate numbers of taste buds are found on the foliate papillae which are located along the posterolateral surface of the tongue

C. THE TASTE BUDS

   1. No structural difference in the taste buds or taste cells of the young adult has been shown to exist

   2. Each taste bud contains upward of 25–40 neuroepithelial taste cells and other less differentiated cells, the sustentacular cells, which act as reserve cells to replenish the taste cells when they die out

     a. The receptor taste cells are continuously turning over and each mature taste cell is replaced every 200–300 hours

   3. The outer tips of the taste cells are arranged around a minute taste pore

   4. From the tips of the taste cells, several microvilli (taste hairs), about 2–3 μm long and 0.2 μm wide, protrude outward through the taste pore to approach the cavity of the mouth, are bathed by mouth fluids, and provide the receptor surface for taste

     a. The receptor sites are probably located on the surface of the microvilli

     b. Not only does each taste cell have many different types of receptor cells, but the proportion of different types varies from taste cell to taste cell, thus, each taste cell can respond to several stimuli, and there is no evidence that a specific taste cell responds to but one group of stimuli

     c. Another regulatory factor for the types of stimuli is that the proteins on the cell membrane may act to open or close "gates" of the membrane to different types of potential taste stimuli

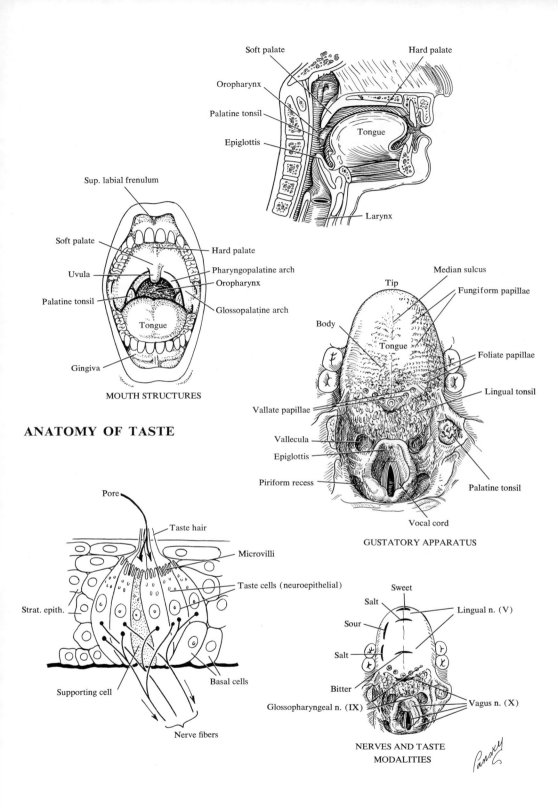

Soft palate

Oropharynx

Palatine tonsil

Epiglottis

Hard palate

Tongue

Larynx

Sup. labial frenulum

Soft palate

Uvula

Palatine tonsil

Tongue

Gingiva

Hard palate

Pharyngopalatine arch

Oropharynx

Glossopalatine arch

MOUTH STRUCTURES

ANATOMY OF TASTE

Median sulcus

Tip

Body

Tongue

Fungiform papillae

Foliate papillae

Lingual tonsil

Vallate papillae

Vallecula

Epiglottis

Piriform recess

Palatine tonsil

Vocal cord

GUSTATORY APPARATUS

Pore

Taste hair

Microvilli

Taste cells (neuroepithelial)

Strat. epith.

Supporting cell

Basal cells

Nerve fibers

Sweet

Salt

Sour

Salt

Bitter

Lingual n. (V)

Glossopharyngeal n. (IX)

Vagus n. (X)

**NERVES AND TASTE
MODALITIES**

Pansky

– 359 –

162. SENSATION OF TASTE

I. Taste nerve fibers: interwoven among the taste cells is a branching terminal network of several taste nerve fibers that are stimulated by the taste cells. These fibers invaginate deeply into folds of the taste cell membranes so that there is an extremely intimate contact between the taste cell and the nerves

A. ONE NERVE FIBER may innervate several receptor cells and one neuroepithelial receptor cell may be innervated by the nerve fibers of several nerve cells

B. THERE IS NO ONE-TO-ONE RELATIONSHIP between nerve and cell, thus the receptor cells lack specificity in the kind of chemical they respond to as well as the way they connect to the brain (each cell and each neuron respond to several taste qualities)

C. THE AFFERENT FIBERS, however, show different firing patterns in response to different substances; e.g., a fiber may fire faster when stimulated by salt, but only sporadically when by sugar (others may respond just the opposite). Thus, quality is achieved, and awareness depends on the number of neurons stimulated rather than the specific ones
   1. The variations in intensity of taste are transmitted by the differences in the frequency of firing of nerve impulses
   2. Adaptation is rapid (in seconds or minutes), and frequency signals quantity as well
   3. In addition, information is increased by temperature, texture, and odor

II. Mechanism of stimulation: the mechanism by which taste receptors are stimulated and action potentials generated is not known, but there are many theories

A. IT HAS BEEN SUGGESTED THAT STEP 1 IS A LOOSE BINDING of ions or individual dissolved molecules that enter the taste pores with specific sites on the receptor cell membrane (substances must be dissolved to be tasted)

B. THE CHEMICAL-SUBSTANCE-RECEPTOR SITE COMBINATION alters the cell membrane, forming pores through which ions move to change the membrane potential of the cell
   1. It is known that the membrane of the receptor cells depolarizes when the cell is chemically stimulated
   2. The fact that a single receptor will respond to more than one quality of taste is explained by the single receptor cell having different sites, each able to bind with different types of molecules

III. Taste sensations are divided into 4 basic modalities or groups: sweet, sour, salt, and bitter

A. DIFFERENT TYPES OF TASTE BUDS OR RECEPTOR CELLS, to support the specificity, have not been identified. In fact, a single receptor cell can respond to varying degrees to many different substances

B. SALT AND SWEET, however, are perceived more acutely on the tongue; bitter and sour are perceived most acutely on the palate

MODALITIES OF TASTE
(MOST SENSITIVE AREAS)

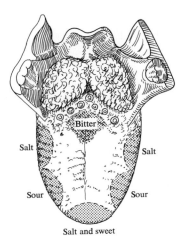

Bitter

Salt   Salt

Sour   Sour

Salt and sweet

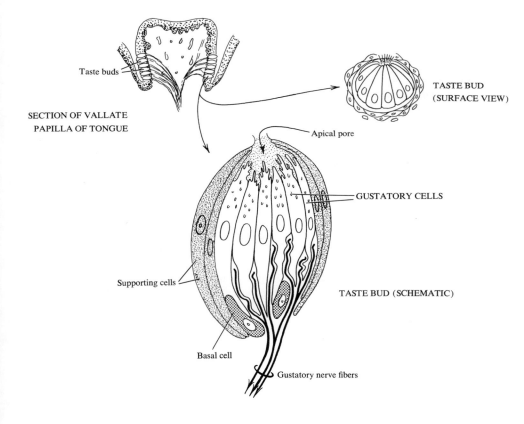

Taste buds

SECTION OF VALLATE
PAPILLA OF TONGUE

TASTE BUD
(SURFACE VIEW)

Apical pore

GUSTATORY CELLS

Supporting cells

TASTE BUD (SCHEMATIC)

Basal cell

Gustatory nerve fibers

163. PATHWAYS OF TASTE

I. **First-order fibers:** taste fibers of the facial (via the chorda tympani), glossopharyngeal, and vagus nerves pass into the medulla and end in the rostral portions of the nucleus solitarius (this portion being called the nucleus gustatorius)

A. TASTE IMPULSES are conveyed from the anterior $\frac{2}{3}$ of the tongue by the fibers of the chorda tympani (VII) and from the posterior $\frac{1}{3}$ of the tongue, palate, and pharynx by fibers in the branches of the glossopharyngeal (IX) and vagus (X) nerves

B. THE SENSORY NERVES terminate either in the taste buds or as free nerve endings in the connective tissues and epithelium other than the taste buds. Many of the free nerve endings are of terminals of general sensory fibers of the trigeminal (V) nerve and must not be confused with those for taste

II. **Second-order fibers** have axons which probably ascend as uncrossed fibers in association with the medial lemniscus and reticular formation to their termination in the most medial part of the ventral posteromedial (VPM) nucleus of the thalamus

A. SOME FIBERS OF THE GUSTATORY PATHWAYS from the nucleus solitarius to the thalamus may cross (not proven) and ascend in association with the contralateral medial lemniscus

III. **Third-order fibers** from the VPM nucleus pass through the posterior limb of the internal capsule and terminate in the lower end of the postcentral gyrus (opercular region of the parietal lobe) and possibly in the cortex of the insula and superior temporal gyrus

IV. **The nuances of true taste** are the result of transmission via many patterns of afferent impulses in the various nerve fibers and of processing in the nuclei of the taste pathways (comparable to the transmission of sound qualities)

A. A DECREASE IN TASTE SENSITIVITY WITH AGE is paralleled by a decrease in total number of taste buds after the age of 40. The center of the tongue with only a few taste buds is, in any event, relatively taste blind

B. THE ABILITY TO TASTE CERTAIN CHEMICALS has an hereditary basis. Some people have a taste "blindness" to certain substances that others can taste

C. SOME TASTE ABNORMALITIES are known to be associated with alterations of body metabolism as a consequence of the action of certain chemical substances

D. TASTE MAY HAVE A ROLE IN THE PRESERVATION OF LIFE ITSELF in some animals in that it enables some animals to select foods that contain nutrients or ions that they need to survive; i.e., the rat needs calcium salts to survive and selects foods high in this substance

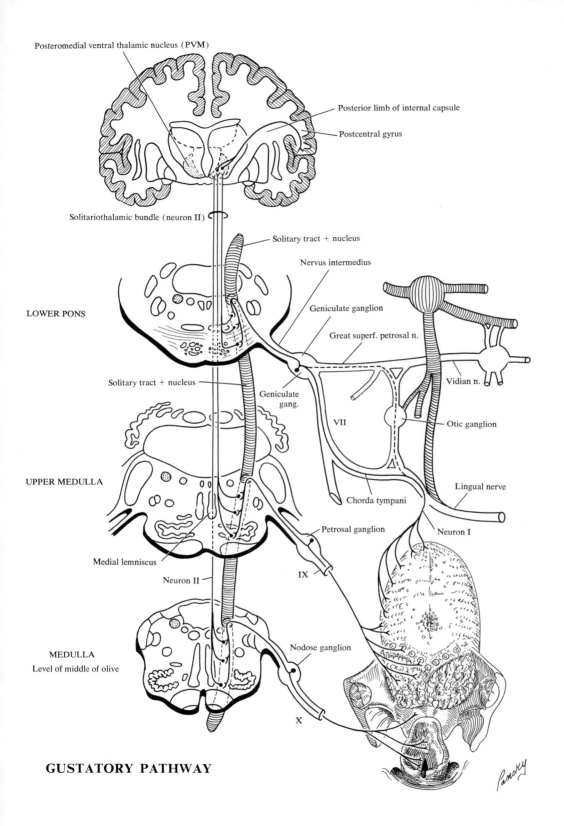

Posteromedial ventral thalamic nucleus (PVM)

Posterior limb of internal capsule

Postcentral gyrus

Solitariothalamic bundle (neuron II)

Solitary tract + nucleus

Nervus intermedius

Geniculate ganglion

Great superf. petrosal n.

LOWER PONS

Vidian n.

Solitary tract + nucleus

Geniculate gang.

VII

Otic ganglion

UPPER MEDULLA

Lingual nerve

Chorda tympani

Petrosal ganglion

Neuron I

Medial lemniscus

IX

Neuron II

MEDULLA
Level of middle of olive

Nodose ganglion

X

GUSTATORY PATHWAY

– 363 –

164. ANATOMY OF THE EAR

I. The ear is usually considered to be composed of 3 portions: outer, middle, and inner ears

A. THE OUTER EAR tends to gather and direct sound waves toward the structures of hearing
1. Pinna: fleshy, cartilaginous flap attached to the skull
2. External auditory meatus: tube lined with thin skin and lubricated by the brownish waxy cerumen, produced by ceruminous glands in the wall

B. THE MIDDLE EAR: air-filled cavity which contains the eardrum or tympanum, three small ear bones or ossicles, the tendons of 2 small muscles (tensor tympani and stapedius), and communicates posteriorly with the mastoid air cells and anteriorly with the nasopharynx via the auditory or eustachian tube
1. The eardrum (tympanic membrane) forms the lateral wall of the middle ear cavity and is composed of skin (laterally), mucous membrane (medial), and is set into vibration by sound waves
2. The ear ossicles: malleus (lateral), incus (intermediate), and stapes (medial)
    a. Act as a system of levers to transmit eardrum vibrations to the fluid-filled cavities of the inner ear (stapes to vestibule via the foramen ovale)
    b. The bones decrease the amplitude of the eardrum oscillation by a factor of 10 but increase the pressure at the inner ear by a factor of 13
3. Tensor tympani tendon attaches to the malleus; stapedius tendon to the stapes
    a. Contraction limits movement of ossicles, prevents damage with loud sounds
4. The auditory (eustachian) tube allows equalization of air pressure on the 2 sides of the eardrum, preventing rupture

C. THE INNER EAR consists of a series of cavities and channels (osseous labyrinth) hollowed out in the petrous temporal bone which contains a closed membranous labyrinth of living tissues. The osseous labyrinth is filled with perilymph; the membranous labyrinth with endolymph
1. Bony labyrinth consists of semicircular canals, a vestibule, and a cochlea and opens into the subarachnoid space via the perilymphatic duct
2. Membranous labyrinth
    a. The utriculus and sacculus lie in the bony vestibule and contain the maculae, which are structures for sensing static posture. The essential mechanism of operation of the maculae is the force of gravity acting on the otoliths and the bending of hair cells of the maculae. Nerve fibers from the latter pass to the cerebellum, over cranial nerve VIII, where appropriate muscular responses are initiated to correct head positions
    b. The membranous semicircular canals lie in their bony counterparts, are 3 in number, and are placed in 3 mutually perpendicular planes (superior, lateral, and posterior). Head movement, or centrifugal force, causes the fluid in the canals to move toward or away from the sense organs located in the ampullae (the cristae) near the terminal ends of the canals
        i. The cristae bend and impulses pass over the VIIIth nerve to the cerebellum where muscular responses correct body position. Thus, the maculae signal a held (static) position; the cristae are triggered by motion
    c. The cochlear duct is the membranous structure of the cochlea and divides the latter into 3 compartments or scalae: scala tympani and vestibuli, filled with perilymph, and the duct itself (scala media) filled with endolymph. The duct contains the spiral organ of Corti (of hearing), which rests on the basilar membrane, a tapered structure composed of 25,000 strands of connective tissue. The organ of Corti is an organized complex of supporting cells and hair cells (neuro-epithelial sensory end organs: 3500 inner and 20,000 outer hair cells). The specialized cells also contain microvilli or stereocilia

External auditory meatus

Semicircular canals

Malleus Incus

Cochlea

Acoustic nerve (VIII)

Auricle

Tympanic membrane

Stapes

Middle ear chamber

Eustachian tube

Int. carotid a.

Pharynx

Ligaments

Malleus Incus

Stapedius m. (VII)

Stapes

Tensor tympani m. (V)

Tympanic membrane

Sup. semicircular canal

Post. semicircular canal

Ampulla

Cochlear duct

Utricle

Bony labyrinth

Lat. semicirc. canal

Saccule

Ductus reuniens

Utriculosaccular duct

Endolymphatic sac

Vestibular aqueduct (endolymph)

Semicircular canal

Saccule

Perilymph

Cochlear canal

Cochlea

Cochlear aqueduct

2nd tympanic membrane (note inner membranous labyrinth)

VIII

Scala vestibuli

Scali tympani

Cochlear duct

Scala vestibuli

Cochlear duct

Spiral ganglion

Cochlea

Scala tympani

EAR

Vestibular lip

Tectorial membrane

Outer hair cells

Inner hair cell

Outer tunnel

Internal spiral sulcus

Cells of Hensen

Border cell

Cochlear n. (VIII)

Spiral lig.

Basilar membrane

Outer Inner

Tympanic lip

Outer phalangeal cells

Pillars

Inner phalangeal cell

ORGAN OF CORTI

Pansky

165. SENSATION OF HEARING: GENERAL INFORMATION

I. Hearing
A. GENERAL INFORMATION
1. Sound energy is transmitted through air as a disturbance of air molecules; since there are no air molecules in a vacuum, there is no sound in a vacuum
2. The disturbances of air molecules that make up a sound wave consist of
   a. Regions of compression, in which air molecules are close together and pressure is high, alternating with areas of rarefaction, where the molecules are farther apart and the pressure is lower
3. The individual molecules travel only short distances, but the disturbances passed from one molecule to another can travel many miles. It is in these disturbances or sound waves that sound energy is transmitted
4. The sound dies out only when enough of the original sound energy has been dissipated so that a sound wave can no longer disturb the air molecules around it
5. When the waves of rarefaction and compression are regularly spaced, the tone is said to be pure (tuning fork)
   a. Waves of speech and other sounds are not regularly spaced but are complex waves of many frequencies of vibration
6. Sounds keenly heard by human ears are from sources vibrating at frequencies of 1000 to 4000 Hz, but the entire range of frequencies that are audible extends from 20 to 20,000 Hz
7. The frequency of vibration of the sound source (wave) is related to the pitch we hear (analyzed in the cochlea, due to selective vibration of different areas of the organ of Corti for different pitches)
   a. The faster the vibration, the higher is the pitch
8. We also detect loudness (intensity), analyzed partially by the cochlea, due to the amplitude of the bounce of the basilar membrane; and tonal quality or timbre of sound, which appears to be a function of the brain
   a. The difference between the packing or pressure of air molecules in a zone of compression and a zone of rarefaction gives us amplitude and is related to the loudness of the sound
9. The degree of purity of the sound wave (number of sound frequencies in addition to the fundamental tone) is related to the quality or timbre of the sound
10. We can distinguish about 400,000 different sounds

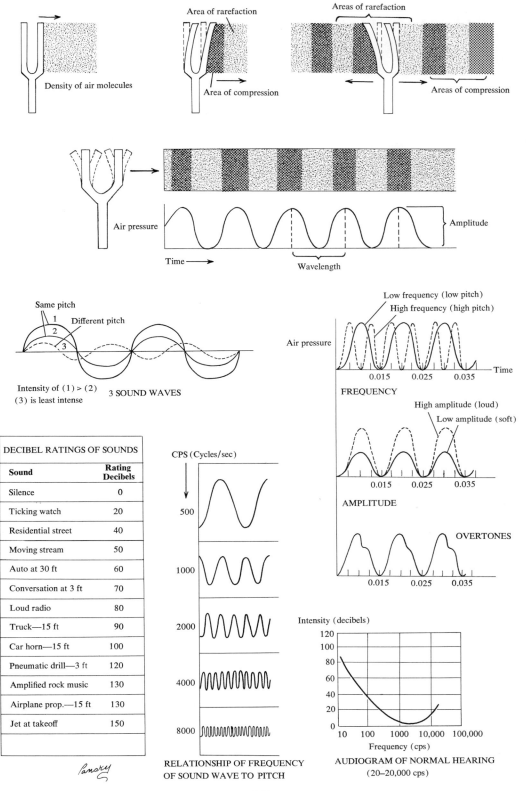

Density of air molecules

Area of rarefaction

Area of compression

Areas of rarefaction

Areas of compression

Air pressure

Amplitude

Time

Wavelength

Same pitch

1

Different pitch

2

3

Intensity of (1) > (2)
(3) is least intense

3 SOUND WAVES

Low frequency (low pitch)

High frequency (high pitch)

Air pressure

0.015   0.025   0.035

Time

FREQUENCY

High amplitude (loud)

Low amplitude (soft)

0.015   0.025   0.035

AMPLITUDE

OVERTONES

0.015   0.025   0.035

DECIBEL RATINGS OF SOUNDS	
Sound	**Rating Decibels**
Silence	0
Ticking watch	20
Residential street	40
Moving stream	50
Auto at 30 ft	60
Conversation at 3 ft	70
Loud radio	80
Truck—15 ft	90
Car horn—15 ft	100
Pneumatic drill—3 ft	120
Amplified rock music	130
Airplane prop.—15 ft	130
Jet at takeoff	150

CPS (Cycles/sec)

500

1000

2000

4000

8000

RELATIONSHIP OF FREQUENCY
OF SOUND WAVE TO PITCH

Intensity (decibels)

120
100
80
60
40
20
0

10   100   1000   10,000   100,000

Frequency (cps)

AUDIOGRAM OF NORMAL HEARING
(20–20,000 cps)

166. ANATOMY OF HEARING: SOUND TRANSMISSION

I. **Anatomy of hearing: the transmission of sound**

A. THE FIRST STEP IN HEARING is usually the entrance of the pressure wave into the external auditory meatus
   1. The waves reverberate from the side and end of the canal to fill it with the continuous vibrations of pressure waves
   2. The air molecules, under slightly higher pressure during a wave of compression, push against the tympanic membrane, causing it to bow inward
      a. The distance the membrane moves, although small, is a function of the force and velocity with which the air molecules hit it and is related to the loudness of the sound
   3. During the following wave of rarefaction, the membrane moves back to its original position
   4. The tympanic membrane is very highly sensitive and responds to all varying pressures of the sound waves, vibrating slowly for slow-frequency sounds and rapidly for high tones
      a. The membrane separates the external ear canal from the middle ear cavity and the pressures in the two air-filled chambers are usually equal, at atmospheric pressure. However, differences are possible as a result of atmospheric changes and this may distort the membrane and cause pain
      b. The eustachian tube has a slitlike ending and is normally closed, but can be opened by yawning, swallowing, or sneezing to equalize the pressure

B. THE SECOND STEP IN HEARING consists of transmission of the sound energy from the tympanic membrane, through the middle ear, to the receptor cells in the inner ear, which are surrounded by fluid
   1. The middle ear serves to transfer movements of air in the outer ear to the air-filled chambers of the inner ear
   2. The tympanic membrane is coupled by a chain of 3 small ossicles, the malleus, incus, and stapes, to a membrane-covered opening, the oval window (fenestra ovale)
      a. The total force on the oval window is the same as on the membrane, but due to its smaller size, the force per unit area (pressure) is multiplied 15–20 times
      b. Additional advantage is reached through the lever action of the 3 ossicles
      c. Thus, the tiny amounts of energy involved are transferred to the inner ear with little loss
   3. In addition, the amount of energy transmitted to the inner ear is further modified by contraction of 2 small muscles in the middle ear that alter tension on the tympanic membrane (tensor tympani muscle) and the position of the stapes (stapedius muscle) on the oval window
      a. The muscles protect the receptor apparatus from the intense sound stimuli and aid intent listening over certain frequencies

EXTERNAL EAR

Middle ear

Ossicles of middle ear

Inner ear

Internal auditory meatus

Auricle

Cochlea

External auditory meatus

Eustachian tube

Tympanic membrane

RIGHT TYMPANIC MEMBRANE

Malleolar prominence

Pars flaccida

Long crus of incus

Malleus

Umbo

Pars tensa

Cone of light

Malleus

Ext. auditory meatus

Incus

Stapes

Tympanic membrane

Umbo

Middle ear cavity

Aditus to antrum

Tympanic antrum

Epitympanic recess

Tegmen tympani

Post. ligament of incus

Incus

Chorda tympani n.

Head of malleus

Ant. process of malleus

Ant. lig. of malleus

Tensor tympani muscle

Stapes

Mastoid air cells

Cochleariform process

Malleus

Incus

Stapes

Facial (VII) n.

Tympanic membrane

Stapedius muscle

Tensor tympani m.

Handle of malleus

Auditory tube

Ant. process of malleus

Carotid a.

Styloid process

Pyramid of tympanic cavity

Lateral process of malleus
(Handle)

TYMPANIC CAVITY SEEN
FROM MIDDLE EAR

Pansky

167. ANATOMY OF HEARING: INNER EAR RECEPTORS

I. **The pressure wave** pushes in on the tympanic membrane, the chain of ossicles rocks the footplate of the stapes against the oval window, and it bows into the vestibule of the inner ear, creating a pressure wave in the inner ear scala vestibuli, which takes two pathways to dissipate

A. THE WAVE MAY PASS TO THE END OF THE SCALA VESTIBULI, pass around the end of the basilar membrane via the helicotrema, and return via a second chamber, the scala tympani, where it finally ends at still another membrane-covered window, the round window (fenestra cochleae) which bows out into the middle ear cavity

B. MOST OF THE PRESSURE WAVE, HOWEVER, IS TRANSMITTED TO THE BASILAR MEMBRANE, which is deflected into the scala tympani

1. The basilar membrane is narrow and relatively stiff in the cochlea closest to the middle ear cavity, but becomes more elastic as it extends the length of the cochlear spiral. The stiff proximal part vibrates immediately in response to pressure changes transmitted in the scala vestibuli, but the distal portion responds more slowly

2. With each change in pressure in the inner ear, the wave of vibrations is made to travel down the basilar membrane and frequencies of the incoming wave are "sorted out" by the membrane. The region of maximal displacement varies with frequency of vibration of the sound source

   a. Membrane properties are such that the cochlea nearest the oval window and middle ear resonates best with high-frequency tones and has its greatest amplitude of vibration when high-pitched tones and short waves are heard. The traveling wave dies out once it is past this region

   b. The basilar membrane acts as a filter and low-frequency waves progress farther along the membrane. Thus, lower tones (low pitch or frequency and long waves) travel out along the membrane for a greater distance and are heard at the apex

3. Where the basilar membrane is maximally displaced, the stimulation of the hair cells (receptors), that ride on it, is the greatest

   a. Fine hairs on the receptor cells are in contact with the overhanging tectorial membrane, an acellular protein structure similar to epidermal keratin, with specialized parts such as Hardesty's membrane and Hensen's stripe, into which the hair cells are embedded

   b. As the basilar membrane is displaced by pressure waves in the scala vestibuli, the hair cells move in relation to the tectorial membrane and are displaced. Incoming sound energy is transformed from vibrating molecules of pressure waves to electric events in the hair cells. Hair cell movement causes depolarization of the hair cells. However, the cells are easily damaged by high-intensity waves

C. THE GENERATOR POTENTIALS formed by the activated hair cells lead to action potentials in the peripheral endings of the afferent nerves and are transmitted to the CNS

1. The greater the loudness (energy of the sound wave), the greater is the basilar membrane movement. Intensity discrimination depends on length of the membrane set into motion

2. The greater the amplitude of the generator potential, the greater the frequency of action potentials in the afferent nerve

3. Musical sounds, chords, and harmonics are the result of several frequencies vibrating in simple periodic oscillations (rhythms). Noises are the result of frequencies not in periodic oscillation

D. SOUND IS LOCALIZED to the side where it is loudest. Comparison of its onset and intensities at each ear are clues in localizing its source

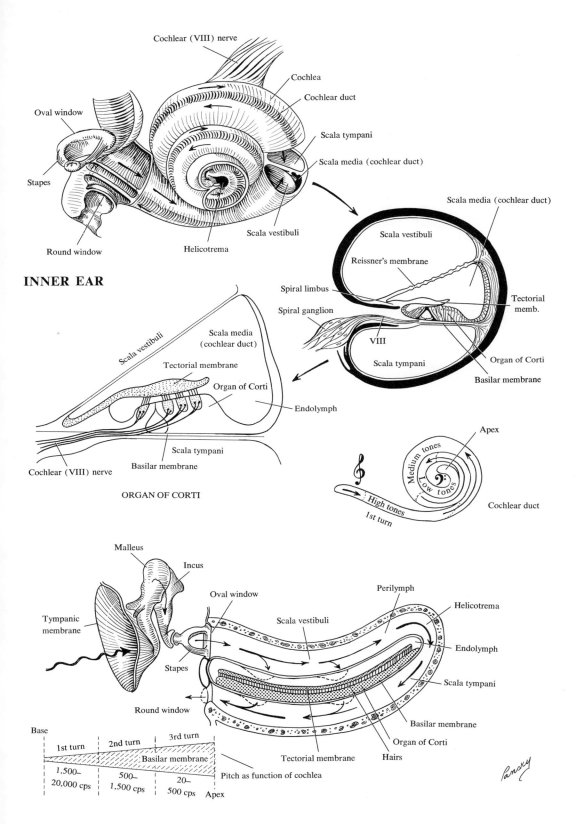

Cochlear (VIII) nerve

Cochlea

Cochlear duct

Scala tympani

Scala media (cochlear duct)

Oval window

Stapes

Round window

Helicotrema

Scala vestibuli

INNER EAR

Scala media (cochlear duct)

Scala vestibuli

Reissner's membrane

Spiral limbus

Spiral ganglion

VIII

Scala tympani

Tectorial memb.

Organ of Corti

Basilar membrane

Scala vestibuli

Scala media (cochlear duct)

Tectorial membrane

Organ of Corti

Endolymph

Scala tympani

Cochlear (VIII) nerve

Basilar membrane

ORGAN OF CORTI

Apex

Medium tones

Low tones

High tones

1st turn

Cochlear duct

Malleus

Incus

Oval window

Perilymph

Helicotrema

Tympanic membrane

Scala vestibuli

Endolymph

Stapes

Scala tympani

Round window

Basilar membrane

Organ of Corti

Hairs

Tectorial membrane

Base

1st turn

2nd turn

3rd turn

Basilar membrane

Pitch as function of cochlea

1,500–20,000 cps

500–1,500 cps

20–500 cps

Apex

Pansky

168. AUDITORY PATHWAYS

I. **The ascending pathway of the auditory system** is composed of sequences of neurons, arranged in series and in parallel. A precise designation of 2nd-, 3rd-, and 4th-order neurons is difficult. The basic sequence is

A. NEURONS OF THE 1ST ORDER: from cell bodies in the spiral ganglion of the cochlear nerve, extending from the spiral organ of Corti, terminate centrally in the dorsal and ventral cochlear nuclei

B. NEURONS OF THE 2ND ORDER: from cell bodies in the cochlear nuclei, axons ascend as crossed fibers in the lateral lemnisci to the inferior colliculus

1. Fibers from the dorsal cochlear nucleus decussate through the posterior tegmentum as the dorsal acoustic stria, ascend in the contralateral lemniscus, and end in the nucleus of the lateral lemniscus and inferior colliculus

2. Fibers from the dorsal part of the ventral cochlear nucleus pass dorsal to the inferior cerebellar peduncle, decussate through the intermediate tegmentum as the intermediate acoustic stria, ascend in the contralateral lemniscus, and end in the nucleus of the lateral lemniscus and inferior colliculus

3. Fibers from the ventral cochlear nucleus pass through the anterior tegmentum as the large ventral acoustic stria (part of the trapezoid body). These fibers (1) end in the ipsilateral and contralateral reticular formation, superior olivary nuclei, and nuclei of the trapezoid body, and (2) ascend in the contralateral lateral lemniscus and end in the contralateral nucleus of the lateral lemniscus and inferior colliculus

a. Decussating fibers of the ventral acoustic stria and those from the superior olivary and trapezoid nuclei are collectively called the trapezoid body

C. NEURONS OF THE 3RD ORDER: from cell bodies in the inferior colliculus, axons pass through the brachium of the inferior colliculus to the medial geniculate body (parvocellular or dorsal portion) of the thalamus

D. NEURONS OF THE 4TH ORDER: from cell bodies in the medial geniculate body, axons pass via the auditory (geniculocortical) radiations to the primary auditory cortex (transverse temporal gyri of Heschl, areas 41 and 42)

E. ALTHOUGH MEANINGFUL CONSCIOUS HEARING requires the neural activity of the cerebral cortex, the latter is not essential for auditory reflexes. "Listening" and reacting to cochlear input can occur with only the lower brain stations intact. The nucleus of the inferior colliculus and brainstem reticular formation may be the integrative centers for this reflex

II. **Descending (centrifugal) efferent fibers in the auditory pathways** run parallel to the ascending fibers and are integrated in the feedback control of auditory input and are involved with processing and sharpening the ascending auditory influences by channeling the essential neural information (signal) and inhibiting and suppressing unwanted neural activity or noise

A. CORTICOGENICULATE FIBERS: temporal cortex to medial geniculate body

B. CORTICOCOLLICULAR FIBERS: temporal cortex to inferior colliculus

C. GENICULOCOLLICULAR FIBERS: from medial geniculate body to inferior colliculus

D. COLLICULAR EFFERENTS: from inferior colliculus to nuclei of the superior olivary complex, lateral lemniscus, and dorsal and ventral cochlear nuclei

E. THE EFFERENT COCHLEAR BUNDLE OF FIBERS: from the superior olivary complex to the hair cells of the spiral organ of Corti. Consists of many crossing fibers together termed the olivocochlear bundle and of fewer uncrossed fibers, the peduncle of the olive. Conveys inhibitory influences directly to the hair cells

III. **Deafness—three types:** transmission, a defect of the eardrum or ossicle activity (e.g., osteosclerosis); cochlear, a defect in the organ of Corti; or central, injury to the auditory cortex or auditory nerve

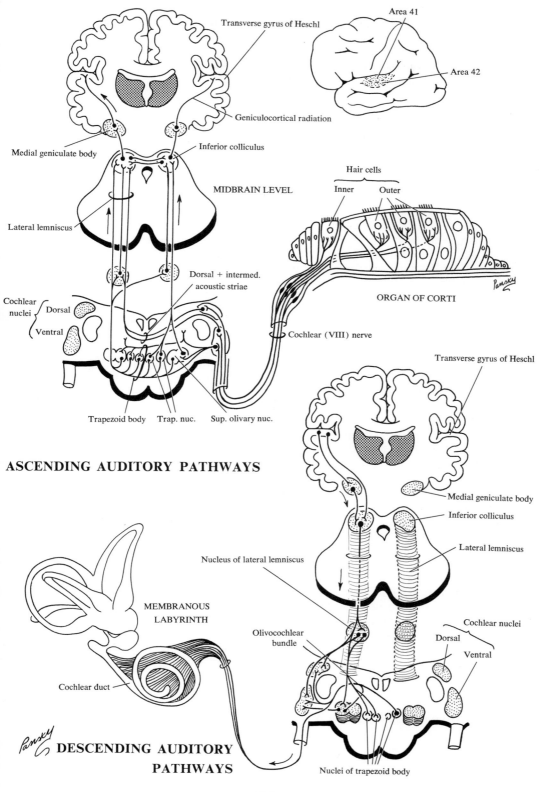

Transverse gyrus of Heschl

Area 41

Area 42

Geniculocortical radiation

Medial geniculate body

Inferior colliculus

MIDBRAIN LEVEL

Hair cells

Inner   Outer

Lateral lemniscus

Dorsal + intermed.
acoustic striae

ORGAN OF CORTI

Cochlear
nuclei { Dorsal

Ventral

Cochlear (VIII) nerve

Trapezoid body   Trap. nuc.   Sup. olivary nuc.

ASCENDING AUDITORY PATHWAYS

Transverse gyrus of Heschl

Medial geniculate body

Inferior colliculus

Nucleus of lateral lemniscus

Lateral lemniscus

MEMBRANOUS
LABYRINTH

Olivocochlear
bundle

Cochlear nuclei

Dorsal

Ventral

Cochlear duct

**DESCENDING AUDITORY
PATHWAYS**

Nuclei of trapezoid body

-373-

169. ANATOMY OF THE VESTIBULAR SYSTEM—
PART I

I. **The vestibular system** contains mechanoreceptors specialized to detect changes in both motion and position of the head
A. THE RECEPTORS are part of the vestibular apparatus which is housed in the bony channels or labyrinths of the inner ear, bilaterally
    1. The vestibular apparatus is a membranous labyrinth which forms 3 semicircular canals with their enlarged ends or ampullae. The latter contain the cristae ampullaris, which are the sensory receptors composed of hair cells and supporting cells. In addition, there are the utricle and saccule which lie in the bony vestibule and contain a ridge or macula composed of hair cells and supporting cells. Endolymph fills the membranous labyrinth
        a. The 3 canals, on each side, are arranged at right angles to each other
        b. The actual receptors of the canals are hair cells that lie at the ends of the afferent neurons
        c. The sensory hairs are ensheathed by a gelatinous mass which blocks the channel of the semicircular canals at that point
        d. The hair cells are 2ndary receptor cells of the vestibular sense organs and act as transducers. Two types of sensory hairs are seen in the hair cells
            i. Stereocilia (modified microvilli) and kinocilia (modified cilia)
            ii. Each hair cell has about 70 stereocilia (sensory hairs) and 1 kinocilium
        e. The vestibular sensory receptors consist of 2 types of sensory cells and supporting cells
            i. Sensory cell type I is flask-shaped with a round base and constricted neck enclosed by a chalice-shaped afferent nerve ending (associated with vestibular afferent fiber synapses)
            ii. Sensory cell type II is a cylindric columnar cell with a basically located synaptic region. Several afferent and efferent nerve terminals synapse with these hair cells

II. **Receptor system of canals**
A. WHEN THE HEAD IS MOVED, the bony labyrinth, membranous labyrinth, and the attached bodies of the hair cells turn with them, and the canals thus signal the rate of change of motion of the head
B. THE ENDOLYMPH of the membranous labyrinth is neither attached to bone nor is it automatically pulled with it, but due to inertia, the fluid tends to keep its original position
C. THE HAIRS are pulled against the relatively stationary column of endolymph and are bent, as the bodies of the hair cells move with the skull
D. SPEED AND MAGNITUDE OF HEAD MOVEMENT determine the degree to which the hairs bend and the hair cells are stimulated
E. THE HAIRS SLOWLY RETURN TO NORMAL POSITION as inertia is overcome
    1. The hair cells are stimulated during changes in rate of motion (during acceleration of the head) but, during motion at a constant speed, stimulation of the hair cells ceases

Superior semicircular canal

Endolymph

Ampulla

Utricle

Perilymph

Posterior semicircular canal

Lateral semicircular canal

Ampulla

Ampulla

Cochlear duct

Perilymph

Saccule

Round window

Endolymphatic sac

Posterior canal

Lateral canal

Superior canal

VESTIBULAR SYSTEM

Lateral ampulla

Superior ampulla

Macula utriculi

Macula sacculi

Organ of Corti

Cochlear duct

Post. ampulla

Neuroepithelial areas
(black)

Vestibular ganglia

Superior

Inferior

Vestibular
nerve

Macula of sacculus

Sensory
epithelium

(Section)

Ampulla of semicircular canal
(sectioned)

Sensory epithelium

Axis of sensitivity

Stereocilium

Kinocilium

Stereocilium

Type I

Kinocilium

Type II

Kinocilium

Stereocilium

Vestibular system
hair cell

Nerve chalice

Synaptic bar

Afferent ending

Supporting cell

Efferent fiber

BM

Myelin sheath

170. ANATOMY OF THE VESTIBULAR SYSTEM— PART II

I. Receptor system of canals (cont.)

A. THE MECHANISM OF SYNAPTIC TRANSMISSION or the nature of the transmitter substance is not known even though the junction between the hair cells and their afferent nerve fibers has the features of a chemically mediated synapse

B. THE AFFERENT NERVE FIBERS are activated at a relatively slow resting frequency even when the head is motionless

C. THE SHEARING FORCE that bends the hairs on the receptor cells is related to the frequency of action potentials in the afferent nerve. When the hairs are bent one way, the rate of firing speeds up; when they are bent in the opposite direction, the firing frequency slows down

D. IF THE HEAD IS TURNED LEFT, the hairs of the left canal move in that direction causing a higher rate of firing action potentials in the left vestibular nerve

    1. In the right canal, the hair movement is in the opposite direction, and the rate of firing in the right nerve drops below resting level

    2. Thus, there is an imbalance of input to the CNS that is very significant

II. The utricle and saccule contain the receptors of the vestibular system that provide information about the position of the head relative to the direction of the forces of gravity. The macula utriculi is maximally stimulated when the head is bent forward or backward and minimally when the head is erect. The macula sacculi is maximally stimulated when the head is bent to the side

A. THE RECEPTOR CELLS OF THE UTRICLE AND SACCULE are mechanoreceptors sensitive to the movement of their projecting cilia or hairs

    1. The hair cells of the utricle and saccule are collected into groups from which the hairs protrude into a gelatinous substance

    2. Tiny calcium carbonate stones or otoliths are embedded in the gelatinous covering of the hair cells making the substance heavier than the endolymph fluid that surrounds it

    3. When the head is bent, the gelatinous-otolith material changes position, pulled by the gravitational forces, to the lowest point in the utricle and saccule

        a. The shearing forces of the gelatinous-otolith substance against the hair cells bend the hairs and stimulate the receptor cells

B. INFORMATION FROM THE VESTIBULAR APPARATUS serves two purposes

    1. It controls the muscles that move the eyes so that in spite of changes in head position, the eyes remain fixed on the same point

    2. The information is used in reflex mechanism to maintain the upright posture

        a. In cats, dogs, etc., the vestibular apparatus plays an important role in postural fixation of the head, orientation in space, and reflexes accompanying locomotion

        b. In man, very few postural reflexes are known to depend primarily on vestibular input, even though the vestibular apparatus is referred to as the sense organs of balance

STRUCTURE OF CRISTA OF AMPULLA OF SEMICIRCULAR CANAL

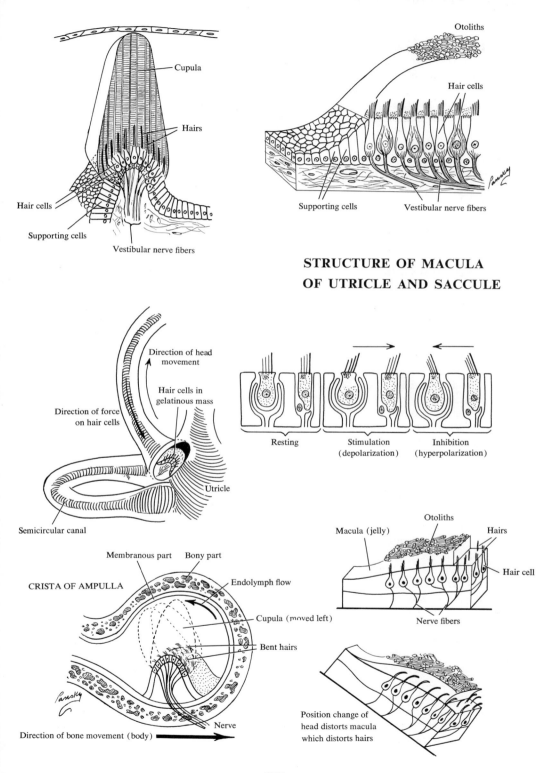

Cupula

Hairs

Hair cells

Supporting cells

Vestibular nerve fibers

Otoliths

Hair cells

Supporting cells

Vestibular nerve fibers

STRUCTURE OF MACULA
OF UTRICLE AND SACCULE

Direction of head movement

Hair cells in gelatinous mass

Direction of force on hair cells

Utricle

Semicircular canal

Resting

Stimulation (depolarization)

Inhibition (hyperpolarization)

Membranous part   Bony part

CRISTA OF AMPULLA

Endolymph flow

Cupula (moved left)

Bent hairs

Direction of bone movement (body)

Nerve

Macula (jelly)

Otoliths

Hairs

Hair cell

Nerve fibers

Position change of head distorts macula which distorts hairs

-377-

171. VESTIBULAR PATHWAYS

I. **The vestibular pathways** consist of suprasegmental pathways distributed to the brainstem, spinal cord, and cerebellum
A. THE CELL BODIES of the 19,000 nerve fibers of each vestibular nerve lie in the vestibular (Scarpa's) ganglion which is divided into a pars superior and pars inferior
    1. Nerves from receptors of the anterior and lateral semicircular canals and the utricle have their cell bodies in the pars superior
    2. Nerves from the posterior semicircular canal and saccule have their cell bodies in the pars inferior
B. THE PRIMARY VESTIBULAR FIBERS pass into the upper medulla deep to the inferior cerebellar peduncle, bifurcate into short ascending and descending branches, and terminate in specific regions in each of the 4 vestibular nuclei (lateral, medial, superior, and inferior) in the upper lateral medulla, and in the nodule of the archicerebellar cortex (flocculonodular lobe)
    1. The primary vestibular fibers from the cristae (semicircular canals) are distributed mainly to parts of the superior and medial nuclei; those from the maculae (utricle and saccule) terminate in portions of the medial and inferior vestibular nuclei
    2. Vestibular nuclei receive input primarily from the labyrinth and archicerebellum
    3. Influence from the fastigial nuclei to the vestibular nuclei are facilitatory; those from the cerebellar cortex (vermis and flocculonodular lobe) are inhibitory
C. THE VESTIBULAR NUCLEI project secondary vestibular fibers to
    1. Upper brainstem (pons and midbrain) via the medial longitudinal fasciculus (MLF)
    2. Cervical spinal segments via the medial vestibulospinal tract
    3. The full length of the spinal cord via the lateral vestibulospinal tract
    4. The archicerebellum and fastigial nucleus of the cerebellum via the juxtarestiform body (medial part of the inferior cerebellar peduncle)
    5. Brainstem reticular formation
    6. Hair cells of the macula and crista of the membranous labyrinth via the vestibular nerve as the vestibular efferent fibers
D. ALL CONTRALATERAL PROJECTIONS from the vestibular nuclei to the MLF decussate at the level of the vestibular nuclei
E. THE MLF IS COMPOSED of many ascending and descending fibers from the vestibular nuclei and is important because it helps to integrate the vestibular system with movements of the eyes, neck, and head

II. **Eye movements and the MLF:** functional interaction of the vestibular receptors resulting in eye movements is mediated via ascending fibers that project from the vestibular nuclei to the MLF of both sides and ascend in the brainstem
A. THE SUPERIOR VESTIBULAR NUCLEUS PROJECTS ascending uncrossed fibers; the inferior and medial nuclei project both crossed and uncrossed fibers; and the lateral nucleus projects largely crossed fibers via the MLF
    1. The pathway is called the vestibulomesencephalic pathway and connects with the nuclei of cranial nerves III, IV, and VI to form the basis of many coordinated (conjugate) eye movements, synchronizing movements of both eyes to the side, upward, and downward (i.e., stimulation of the nerve innervating the crista ampullaris of the lateral canal on one side results in lateral conjugate deviation of the eyes to the contralateral side as a result of contractions of the lateral rectus m. of the rt. eye and the medial rectum m. of the lt. eye)
B. VESTIBULAR PROJECTIONS to the nuclei of the extraocular muscles are very precise
    1. The most potent vestibular projections are primarily to horizontal eye movements (lateral and medial recti mm.). Effects of vestibular influences on elevation and depression of the eyes are minimal

PRIMARY VESTIBULAR FIBERS AND NUCLEI

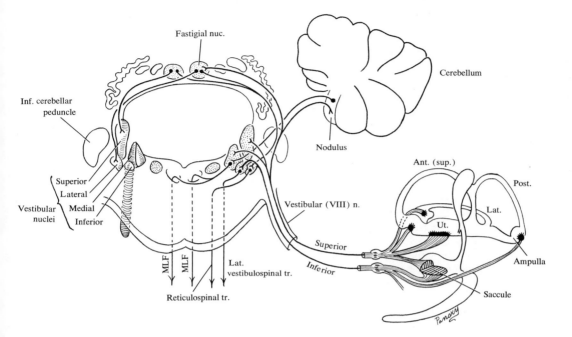

VESTIBULAR CONNECTIONS TO EYE MOVEMENTS

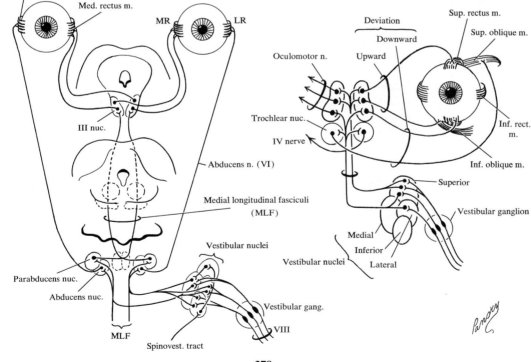

172. VESTIBULAR SYSTEM: EYE, HEAD, AND BODY MOVEMENTS

I. Eye movements and the MLF (cont.)

A. THE INTERSTITIAL NUCLEUS OF CAJAL AND THE NUCLEUS OF DARKSCHEWITSCH are midbrain nuclei in the ventrolateral periaqueductal gray at the lower level of the IIIrd nucleus
   1. Fibers from the interstitial nucleus descend (some cross in the posterior commissure) bilaterally in the MLF and terminate in the IIIrd, IVth, and medial vestibular nuclei of the brainstem and in laminae VII and VIII of the cord
      a. Lesions of these nuclei result in impaired vertical eye movement
   2. The role of the nucleus of Darkschewitsch and its fibers is unknown. It does not project through the MLF

B. NYSTAGMUS: as the head and body pivot and circle, the eyes attempt to fix on an object in space (slow component); as the head and body continue to circle, the eyes snap quickly in the direction in which the head is circling (fast or quick component or compensating component)
   1. Nystagmus is named by the direction of the fast component
   2. Following a sudden stop after spinning, the direction of nystagmus is reversed (postrotatory nystagmus) because the semicircular canals stop when the head stops, but the endolymph, with its momentum, continues to move and thus pushes the cupula and hairs of the hair cells in the opposite direction from that of the spin
   3. Postrotating nystagmus is associated with a fast component and sensation of vertigo or dizziness, which are in the opposite direction, and a slow component, tendency to fall, pastpointing, and endolymph movement, all in the same direction
   4. Pastpointing is the phenomenon of missing an object when reaching to touch it
   5. The plane of the head in the spin determines the nystagmus: horizontal, oblique, vertical, rotatory, etc.

II. The vestibular system and head and body movements

A. THE VESTIBULAR NUCLEI project crossed and uncrossed fibers (largely from the medial and inferior vestibular nuclei) that descend in the MLF as the vestibulospinal pathway which terminates in cervical segments of the spinal cord and conveys influences of inhibition to extensor tonus

B. THE LATERAL VESTIBULAR NUCLEUS projects uncrossed fibers to all levels of the cord via the lateral vestibulospinal tract which is somatotopically organized; the rostroventral part goes to cervical spinal levels; the dorsocaudal part of the nucleus projects to lumbosacral spinal levels
   1. The archicerebellum has some influence via the lateral vestibular nucleus on the lateral vestibulospinal tract and to lower motor neurons in the cord. The lateral vestibular nucleus can be thought of as a displaced cerebellar nucleus since it is the only nucleus receiving direct input from the Purkinje cells of the cerebellar cortex

C. THE SOMATIC VESTIBULOSPINAL PATHWAYS play a role in muscular activities of the body and extremities associated with postural movements and balance
   1. The vestibulospinal pathways exert facilitatory influences on extensor tonus via the alpha and gamma motor neurons of the spinal nerves
   2. In addition, spinovestibular fibers project ascending influences to the vestibular nuclei from the cord

Area 19   Area 8

Interstitial nucleus of Cajal

MLF MLF

Oculomotor (III) nucleus

Oculomotor (III) nerve Superior oblique muscle

Superior rectus muscle

III
IV

Trochlear (IV) nucleus

Trochlear (IV) nerve

Lateral rectus muscle

Inferior rectus muscle

Medial longitudinal fasciculus (MLF)

Inferior oblique muscle

Vestibular nuclei { Superior
Medial
Lateral
Inferior

Abducens nucleus

Abducens (VI) nerve

EYE MOVEMENT CONTROL

VI

Pansky

L SLOW PHASE ◄ **NYSTAGMUS** ► R FAST PHASE

Lat. rectus m. Med. rectus m. Med. rectus m. Lat. rectus m.

**VESTIBULAR
NUCLEI**

Nuc. III

Dextrorotary cells

MLF

Levorotary
cells

Paramedian pontine reticular formation

LEFT
Horizontal canal

Nuc. VI Nuc. VI

RIGHT
Horizontal canal

Pansky

-381-

173. VESTIBULAR SYSTEM: PATHWAYS AND SPECIAL FEATURES

I. Vestibular connection with cerebellum and reticular formation

A. PRIMARY FIBERS of the vestibular nerve and secondary fibers from the inferior and medial vestibular nuclei pass through the juxtarestiform body and course as mossy fibers to the ipsilateral flocculonodular lobe and uvula (the flocculonodular lobe and adjacent cortex are often called the vestibulocerebellum)
   1. In addition, fibers from these vestibular nuclei project to the fastigial nuclei
B. THE MACULAE OF THE UTRICLE AND SACCULE are the major sources of vestibular input to the cerebellum
C. THE PURKINJE CELLS OF THE VESTIBULOCEREBELLUM AND VERMIS relay inhibitory influence to the fastigial nuclei and lateral vestibular nuclei
   1. The fastigial nucleus projects fibers conveying excitatory influences to part of each vestibular nucleus and to the reticular formation (lateral reticular nucleus and nucleus reticularis pontis caudalis)
   2. The vestibular nuclei probably receive most of their input from the cerebellum
D. THE VESTIBULOCEREBELLUM exerts its influence on the coordination of the axial muscles of the neck and vertebral column in balance and posture through
   1. The lateral vestibular nucleus and lateral vestibulospinal tract
   2. The medial vestibular nucleus and the medial vestibulospinal tract
   3. The caudal pontine reticular nucleus and the pontine reticulospinal tract
   4. Fibers from the vestibular pathway to the pontine and medullary reticular formation are integrated in the reticular system and are involved in producing nausea, vomiting, and sweating after vestibular stimulation

II. Vestibular efferent pathway to the membranous labyrinth

A. EFFERENT FIBERS from the brainstem pass through the vestibular nuclei and end in the vestibular receptors of the membranous labyrinth (the nuclei of origin and the role of these fibers are controversial). 1. They probably exert inhibitory influences and ameliorate the effects of motion sickness and its aftersensation nystagmus

III. Functional features of the vestibular system: the labyrinth plays a significant role in maintenance of head and neck position and body posture

A. VERTIGO: the dizziness or sensation associated with lack of equilibrium involves the subjective sense of rotation, either of the surroundings or of the subject
   1. It is a cardinal sign of labyrinthine and vestibular dysfunction, but may occur in normal subjects with excessive stimulation of the semicircular canals
   2. In vestibular disease, nystagmus occurs and is a consequence of influences conveyed from the vestibular system via the MLF to the extraocular muscles
B. DIZZINESS, HEADACHE, NAUSEA, AND VOMITING are symptoms of motion sickness and are primarily due to the excessive stimulation of the utricle and saccule and involve the vestibulocerebellar projections
   1. Drugs like Dramamine raise the threshold of vestibular stimulation and help prevent motion sickness*
C. MÉNIÈRE'S DISEASE is a disorder probably resulting from abnormal circulation of the vestibular endolymph. Its symptoms are tinnitus (ringing in the ear), attacks of vertigo, pallor, vomiting, nausea, and increased respiration. The cristae ampullaris are normal
D. LOSS OF BOTH LABYRINTHS is not followed by vertigo or nystagmus. Normal locomotion under these circumstances requires visual cues
E. MAN IS NOT DEPENDENT ON HIS VESTIBULAR SYSTEM as on his visual and general proprioceptive systems. Injury to the labyrinths, vestibular nuclei, or vestibular pathways may lead to nystagmus, tendency to fall to one side, pastpointing, and some erect posture problems, but these attenuate in time as other proprioceptive cues are used

*Because the labyrinth is not functional during the 1st year of life, infants do not get motion sickness.

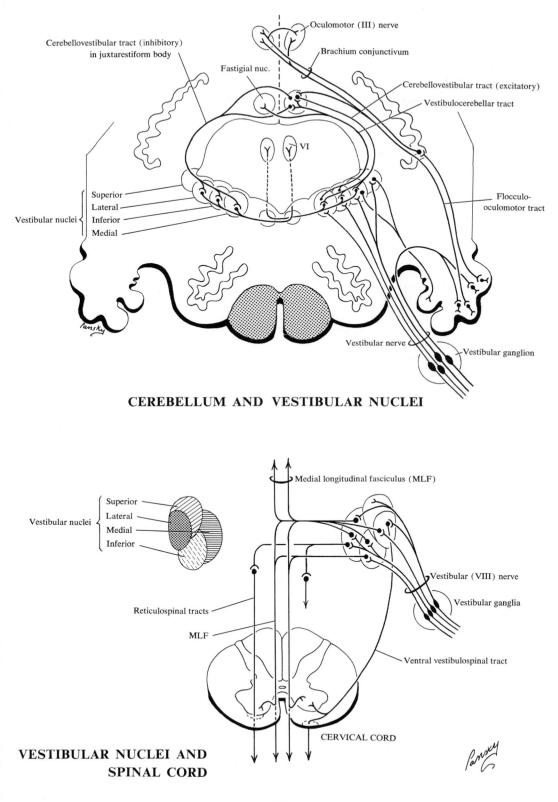

CEREBELLUM AND VESTIBULAR NUCLEI

Oculomotor (III) nerve

Cerebellovestibular tract (inhibitory) in juxtarestiform body

Brachium conjunctivum

Fastigial nuc.

Cerebellovestibular tract (excitatory)

Vestibulocerebellar tract

VI

Flocculo-oculomotor tract

Superior
Lateral
Inferior
Medial

Vestibular nuclei

Vestibular nerve

Vestibular ganglion

Medial longitudinal fasciculus (MLF)

Superior
Lateral
Medial
Inferior

Vestibular nuclei

Vestibular (VIII) nerve

Vestibular ganglia

Reticulospinal tracts

MLF

Ventral vestibulospinal tract

CERVICAL CORD

VESTIBULAR NUCLEI AND SPINAL CORD

– 383 –

174. ANATOMY OF THE EYE

I. **Anatomy of the eye:** the eye is an organ capable of transducing light to visual impulses and consists of three layers of tissue

A. THE OUTER FIBROUS TUNIC is composed of the sclera and cornea

   1. The sclera or white of the eye: tough, fibrous protective structure affording attachment for the eye muscles that rotate the eyeball
   2. The cornea: avascular, anterior, transparent portion which serves as a fixed lens that initially bends the light rays to a near focus on the retina

B. THE MIDDLE VASCULAR TUNIC OR UVEAL TRACT (UVEA) is composed of

   1. The choroid: highly vascularized, richly black-pigmented posterior layer which absorbs stray light rays and is coexistent with the photoreceptive retina which it supplies with nutriment and oxygen
   2. The ciliary body contains the ciliary muscles for fine focusing of the lens, and the iris, a thin, muscular diaphragm (circulating and radiating fibers) that regulates the opening (pupil) which controls light entering the eye
   3. The lens and supporting zonular fibers (zonules): biconvex, elliptical in shape

C. THE INNER TUNIC OR RETINA: a 10-layered structure containing the photosensitive or visual receptors, the rods and cones, and neurons leading to the optic nerve

   1. Is an extension of the forebrain and divided into a photoreceptive layer (neuro-retina) and pigment epithelium to the ora serrata, an unpigmented layer (ciliary retina) over the ciliary body, and a pigmented epithelium (iridic retina) over the posterior iris
   2. Its posterior portion is modified into 3 concentric regions: the central 5–6 mm of retina, the macula lutea (3 mm) with a pit in its center, the fovea centralis (0.4 mm), and the blind spot or optic disk just medial to the posterior pole (site of emergence of the optic nerve)
   3. The functional retina terminates anteriorly at the ora serrata

D. ADDITIONAL ESSENTIAL STRUCTURES

   1. The eye is protected by *the eyelids* and lubricated by the lacrimal apparatus
   2. The aqueous humor (anterior) chamber between the cornea and the lens is filled with an alkaline, watery solution secreted by the ciliary processes
      a. Glaucoma: leading cause of blindness in U.S. develops commonly when the aqueous humor fails to drain properly; pressure builds up in the eyeball and can, if untreated, lead to optic nerve damage
   3. The large vitreous humor (posterior) chamber: behind the lens, a viscous and gelatinous-filled chamber
   4. Iridocorneal filtration angle: lateral circular border of the anterior chamber
   5. Canal of Schlemm (aqueous vein) drains anterior chamber into venous system

II. **Visual sensation:** the receptors of the eye are sensitive to only that small part of the spectrum of electromagnetic radiation called light

A. ELECTROMAGNETIC RADIATION has both wavelength and particle-like properties

   1. The radiant energy is propagated as small discrete "packets" called photons
   2. Radiant energy, however, is described in terms of wavelengths and frequencies
      a. Wavelength is the distance between 2 successive wave peaks; varies from fractions of a millimeter (bottom of spectrum) to several miles (top of spectrum)
      b. Frequency is the number of wave peaks (cycles) that pass a point in a given time and is expressed in cycles per second or herz (Hz)
         i. Photons of visible light oscillate from 4×10^{14} to 7×10^{14} times per second
      c. Light of different wavelengths is associated with different color sensations: a wavelength of 540 nm looks green; one of 565 nm looks red
      d. Wavelength times frequency equals velocity: velocity of light in free space is a fundamental constant of nature, thus as wavelength decreases, frequency increases

EYE STRUCTURE

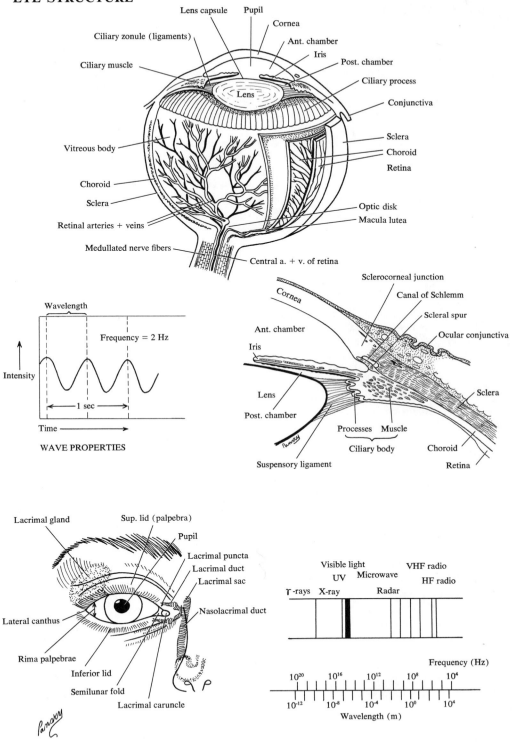

Lens capsule

Pupil

Cornea

Ant. chamber

Ciliary zonule (ligaments)

Iris

Post. chamber

Ciliary muscle

Ciliary process

Conjunctiva

Lens

Sclera

Vitreous body

Choroid

Retina

Choroid

Sclera

Optic disk

Retinal arteries + veins

Macula lutea

Medullated nerve fibers

Central a. + v. of retina

WAVE PROPERTIES

Wavelength

Frequency = 2 Hz

Intensity

1 sec

Time

Sclerocorneal junction

Canal of Schlemm

Cornea

Scleral spur

Ant. chamber

Ocular conjunctiva

Iris

Lens

Sclera

Post. chamber

Processes   Muscle

Choroid

Ciliary body

Retina

Suspensory ligament

Lacrimal gland

Sup. lid (palpebra)

Pupil

Lacrimal puncta

Lacrimal duct

Lacrimal sac

Lateral canthus

Nasolacrimal duct

Rima palpebrae

Inferior lid

Semilunar fold

Lacrimal caruncle

Visible light

UV   Microwave

VHF radio

HF radio

γ-rays   X-ray

Radar

Frequency (Hz)

10^{20} 10^{16} 10^{12} 10^{8} 10^{4}

10^{-12} 10^{-8} 10^{-4} 10^{0} 10^{4}

Wavelength (m)

175. VISUAL OPTICS

I. **The optics of vision:** the image of the object being viewed must be focused on the retina of the eye, a thin layer of neural tissue lining the back of the eyeball, where light-sensitive receptor cells of the eye are found. When a light wave hits a boundary between 2 substances, like the cornea of the eye, the rays are bent and travel in a new direction. The degree of bending depends on the frequency of the light and the angle at which it enters the 2nd medium

  A. THE LENS AND CORNEA are the optical systems that focus the image of the object on the retina. The shape of both, as well as the length of the eyeball, determines the point where the light rays converge on the retina

    1. The cornea plays a greater role than the lens in focusing, because light rays are bent more in passing from air into it than pass into and out of the lens

      a. The corneal surface is curved so that light rays from a single point source hit it at different angles and are bent different amounts. However, all are directed to a point after leaving the lens

      b. When the object has more than one dimension, the image to the retina is upside down to the original light source, and is also reversed left to right

    2. The lens makes all adjustments for distance by changing its shape (accommodation), although the cornea does most of the focusing on the retina

      a. The shape of the lens is controlled by a muscle (ciliary muscle) that flattens the lens when distant objects are to be focused, and allows it to become more spherical to provide for more bending of light for near objects

      b. Cells are added to the outer surface of the lens throughout life. Cells at the center are the oldest and farthest from nutrient fluids that bathe the lens exterior. They age and die first, making the lens more stiff and less able to accommodate

      c. Decrease in refractive power of lens (reduced elasticity with age) until it is almost nonaccommodating is called presbyopia (corrected with convex lenses)

      d. Cells of the lens can become opaque and impair detailed vision, forming a cataract (defective lens can be removed and vision restored by compensating lenses)

  B. VISUAL DEFECTS occur when the lens or cornea does not have a smooth spherical surface, resulting in astigmatism (improper eyeball, corneal, or lens irregularities can be compensated for by corrective lenses)

  C. THE EYEBALL, when normal, focuses sharp images on the retina (emmetropic)

    1. Defects in vision occur if the eyeball is too long in relation to lens size, thus, images of far objects are focused in front of the retina resulting in nearsightedness or myopia (cannot see distant objects clearly)

    2. If the eyeball is too short for the lens, distant objects focus well, but near objects are focused behind the retina resulting in farsightedness or hypermetropia (hyperopia)

  D. THE IRIS controls the amount of light entering the eye by means of its ringlike sphincter and dilator pupillae muscles. In addition, it contains connective tissue and pigmented cells which give it its color

    1. The pupil (hole in the center of the iris) allows light to enter the eyeball

    2. The iris sphincter muscle reflexly contracts in bright light, decreasing the pupil diameter, not only reducing the amount of light entering the eye, but also directing it to the central and most optically accurate part of the lens

      a. The sphincter pupillae muscle conversely relaxes in dim light and, with help of dilator pupillae muscle, allows for maximum sensitivity and opening of the pupil

II. **Glaucoma** is a dangerous disorder of the aqueous system of the anterior chamber of eye

  A. FLUID may be secreted excessively by ciliary body (open-angle glaucoma)

  B. DILATED IRIS may fold into area of trabeculae, blocking drainage into canal of Schlemm (closed-angle glaucoma)

  C. PRESSURE IN EYE becomes excessive, altering circulation in retina, and results in permanent blindness, unless treated

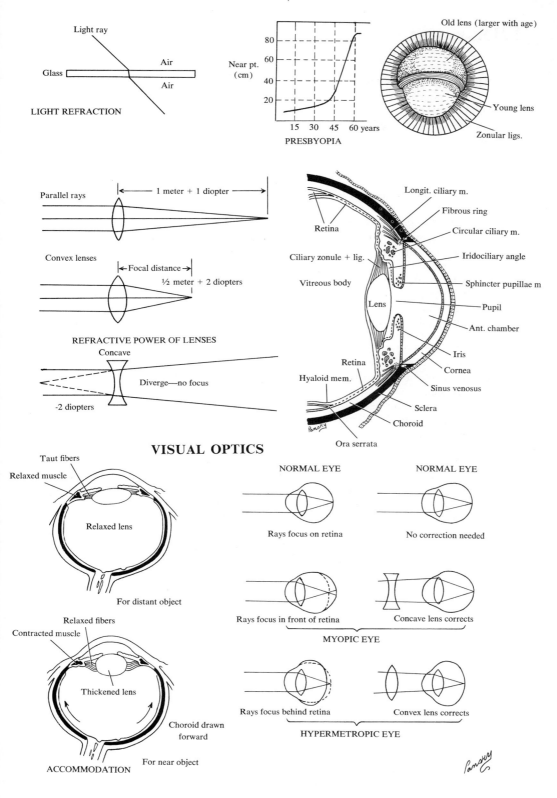

Light ray

Glass

Air

Air

LIGHT REFRACTION

PRESBYOPIA

Near pt. (cm)

15 30 45 60 years

Old lens (larger with age)

Young lens

Zonular ligs.

Parallel rays

1 meter + 1 diopter

Convex lenses

Focal distance

½ meter + 2 diopters

REFRACTIVE POWER OF LENSES

Concave

Diverge—no focus

-2 diopters

Longit. ciliary m.

Fibrous ring

Circular ciliary m.

Retina

Ciliary zonule + lig.

Iridociliary angle

Vitreous body

Sphincter pupillae m

Lens

Pupil

Ant. chamber

Retina

Iris

Hyaloid mem.

Cornea

Sinus venosus

Sclera

Choroid

Ora serrata

VISUAL OPTICS

Taut fibers

Relaxed muscle

Relaxed lens

For distant object

Relaxed fibers

Contracted muscle

Thickened lens

Choroid drawn forward

ACCOMMODATION

For near object

NORMAL EYE

Rays focus on retina

NORMAL EYE

No correction needed

Rays focus in front of retina

Concave lens corrects

MYOPIC EYE

Rays focus behind retina

Convex lens corrects

HYPERMETROPIC EYE

-387-

176. RETINA

I. The retina contains the receptor cells, called either rods or cones because of their microscopic appearance. Only the rods and cones respond directly to light; all other components of the pathway are influenced only by synaptic input to them. Each retina contains about 120 million rods and 7.0 million cones, but only about 1.0 million optic nerve fibers leading from it

A. BOTH CELL TYPES contain light-sensitive molecules called photopigments whose major function is to absorb light

    1. Light energy causes the photopigments to change their molecular configuration which then alters the properties of the receptor-cell membranes in which they are located

        a. Unlike other receptor cells, in response to stimulation, the membrane decreases its permeability to sodium ions, which hyperpolarizes the receptor-cell membrane. This seemingly releases other neurons in the visual pathway from inhibition

B. FOUR KINDS OF PHOTOPIGMENTS

    1. Rhodopsin is very sensitive to low levels of illumination and is found in the rods, in man, with a maximum scotopic sensitivity of 500 nm

    2. Erythrolabe, chlorolabe, and cyanolabe are found in the cones and are sensitive to light wavelengths of the 3 primary colors: red, green, and blue, respectively

        a. Colorblindness: genetic, sex-linked trait in which cones that respond to red or green light are missing and the person is blind to these colors

C. COMPOSITION OF PHOTOPIGMENTS: all four are made up of a protein (opsin) bound to a chromatophore molecule (retinal or retinaldehyde or retinene or vitamin A aldehyde)

    1. All four photopigments possess the 11-*cis* retinaldehyde as the chromophore but are united to four different opsins, which confers the specific light sensitivities on the photopigment, i.e., whether it responds to all light or selectively to red, blue, or green

    2. Light acts on the chromatophore and it splits away from the opsin, changing its molecular configuration and generating a nerve impulse. Retinal eventually breaks down into vitamin A. Most of the latter is transformed back to retinal, which combines, in the dark, with opsin to regenerate rhodopsin

        a. After the breakdown of the photopigment in the presence of light, the chromatophore molecule is rearranged and rejoined to opsin to restore the photopigment

        b. Thus, the only action of light in vision is to change the chromatophore; everything else in the sequence leading to vision (chemical, physiologic, or psychologic) is a "dark" consequence of the light reaction

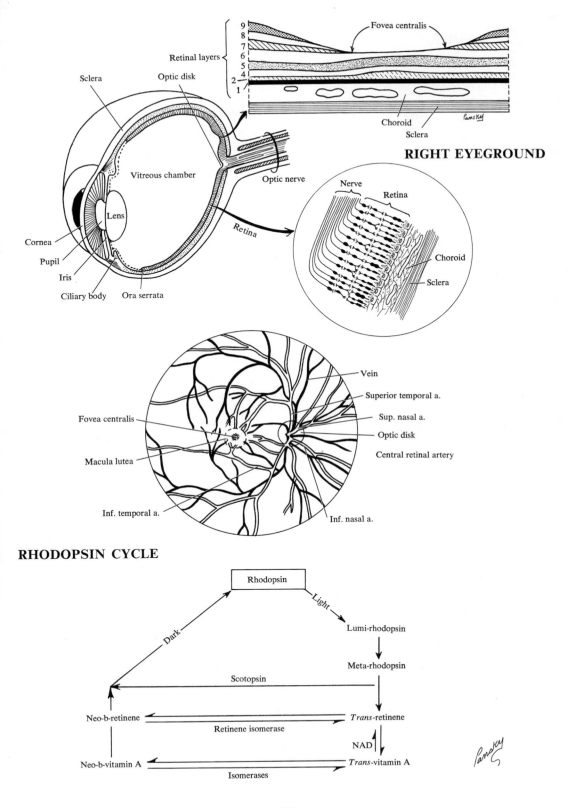

Fovea centralis

Retinal layers
9
8
7
6
5
4
2
1

Sclera

Optic disk

Choroid

Sclera

RIGHT EYEGROUND

Vitreous chamber

Optic nerve

Cornea

Pupil

Iris

Ciliary body

Ora serrata

Lens

Retina

Nerve

Retina

Choroid

Sclera

Vein

Superior temporal a.

Sup. nasal a.

Optic disk

Central retinal artery

Fovea centralis

Macula lutea

Inf. temporal a.

Inf. nasal a.

RHODOPSIN CYCLE

Rhodopsin

Light

Dark

Lumi-rhodopsin

Meta-rhodopsin

Scotopsin

Neo-b-retinene

Trans-retinene

Retinene isomerase

NAD

Neo-b-vitamin A

Trans-vitamin A

Isomerases

177. RODS AND CONES

I. **The rod receptor cell** contains rhodopsin, a pigment called visual purple, which is very sensitive and can detect small amounts of light (scotopic vision). The rods act as photo-receptors during conditions of poor lighting and for night vision
   A. THE ROD RESPONSES do not indicate color, show only shades of gray, do not indicate brightness, and their acuity is very poor
   B. THE RODS ARE MOST NUMEROUS in the peripheral retina, nearest the lens, and are absent from the retinal center (the fovea)
      1. There is greater sensitivity of the peripheral retina to weak light where as many as 600 rods can converge on the same optic nerve fiber with marked summation of the peripheral ganglion cell

II. **The cones** are of 3 types, each with one of the 3 photopigments for color vision. They operate only at high illumination levels (photopic vision), are the photoreceptors for day vision, and have a very high visual acuity
   A. THE CONES are concentrated in the center of the retina (the fovea) which we use for finely detailed vision
      1. Near the fovea, fewer and fewer rods and cones converge on each fiber and both become more slender which progressively increases the acuity of vision near the center of the retina

III. **Synapses:** each receptor cell in the retina (rod or cone) synapses on a 2nd neuron, a bipolar cell, which in turn synapses on a ganglion cell. The axons of the latter form the optic nerve, which passes out of the eye directly to the brain
   A. CONE RECEPTOR CELLS have relatively direct lines to the brain because each bipolar cell receives synaptic input from few cones, and each ganglion cell receives synaptic input from few bipolar cells. In the fovea, where only cones exist, there is little convergence; instead, the cones are represented by about equal numbers of bipolar and ganglion cells. The number of optic nerve fibers leading from the fovea is almost equal to the number of cones there
      1. The above results in a lack of convergence which provides for precise information about the retinal areas stimulated, but little chance for summation of subthreshold events to fire the ganglion cell
   B. MANY ROD CELLS, conversely, converge on bipolar and ganglion cells so that spatial and temporal summation are very good although acuity is poor. Thus, a low intensity of light stimulus that would cause only a subthreshold response in a cone ganglion can cause an action potential in a rod ganglion cell, explaining why objects in a darkened room are indistinct and appear only in shades of gray
   C. SENSITIVITY OF THE EYE improves after being in the dark for some time, as a result of dark adaptation
      1. The theory states that the excitability of the rod vision pathways depends on the number of intact rhodopsin molecules in the rods. In bright light, so many rod rhodopsin molecules break down that they are ineffective and vision is due to cone activation. If one moves to a darkened room from a lighted one, there are relatively few intact rhodopsin molecules initially available, but as the rhodopsin gradually regenerates in the dark, visual sensitivity improves

Sclera

Choroid

1. Pigment epithelium

2. Bacillary layer (rod + cone outer segment)
   - Cone cell
   - Rod cell
   - 3. Outer limiting membrane

4. Outer nuclear layer

5. Outer plexiform layer — Horizontal cell

6. Inner nuclear layer
   - Bipolar cell
   - Amacrine cell

7. Inner plexiform layer

8. Ganglion cell layer — Ganglion cell

9. Nerve fiber layer
   - Optic n. fibers
   - 10. Inner limiting membrane
   - 11

DIAGRAM OF RETINA

Light

Rod

Cone

Outer segment

Outer segment

Connecting structure

Cilia

Ellipsoid zone

Ellipsoid zone with mitochondria

Myoid zone

Nucleus

Nucleus

Müller's cell

Ribbon filament   Spherules   Pedicle

Visibility coefficient

Scotopic (rods)

Photopic (cones)

1.0
0.8
0.6
0.4
0.2

400   500   600   700

Wavelength mμ

ABSORPTION CURVES

Sensitivity %

100
80
60
40
20

Blue   Green   Red

3 types of cones

400   450   500   550   600   650   700

Wavelength mμ

ULTRASTRUCTURE OF A ROD AND CONE

178. VISUAL (RETINOGENICULOSTRIATE) PATHWAYS— PART I

I. **General summary of visual pathways:** impulses from the retinae pass backward through the optic nerves. At the optic chiasma, all fibers from the opposite nasal halves of the retinae cross and join fibers from the temporal retinae of the same side to form the optic tracts. The fibers of each tract synapse in the lateral geniculate body, and from here, the geniculocalcarine fibers pass through the optic radiation or geniculocalcarine tract to the optic or visual cortex in the calcarine area of the occipital lobe. In addition, visual fibers pass to the lower areas of the brain, into the hypothalamus, superior colliculi, and pretectal nuclei

  A. THE BOUNDED SPACE OF THE ENVIRONMENT SEEN BY ONE EYE (monocular vision) while that eye is fixed on a stationary or fixed point is called the field of vision or monocular field of that eye. Normally we see with both eyes (binocular vision) and have a binocular field of vision formed by the almost complete overlapping of the monocular fields of both eyes
    1. The temporal hemiretina of one eye sees the same half of the visual field as the nasal hemiretina of the other eye
    2. When the 2 hemiretinas do not see precisely the same part of the visual fields, we have diplopia or double vision

II. **Stimulation of rods and cones with light** results in a photochemical decomposition and acts on the membrane of the receptor to cause a receptor potential that lasts as long as the light continues. It is a hyperpolarization signal rather than a depolarization signal

  A. RECEPTOR POTENTIALS are transmitted unchanged through the bodies of the rods and cones since neither generate action potentials. The receptor potentials themselves, however, acting at the synaptic bodies, induce signals in the successive neurons by electronic current flow across the synaptic gaps

III. **Stimulation of the bipolar, amacrine, and horizontal cells:** these are all cells seen in the inner nuclear layer of the retina

  A. THE SYNAPTIC BODIES OF THE RODS AND CONES make intimate contact with the dendrites of both bipolar and horizontal cells and the hyperpolarization is transferred directly to both with the excitation signal summating in both. (*Note:* the action potentials are not transmitted)

  B. THE BIPOLAR CELLS are the main transmitting link for the visual signal from the rods and cones to the ganglion cells and they are entirely excitatory

  C. THE HORIZONTAL CELLS spread horizontally and transmit signals laterally as far as several hundred microns. They transmit signals mainly to bipolar cells in the areas lateral to the excited rods and cones. They are inhibitory, enhance contrasts in the visual scene, and are important in helping differentiate colors

  D. THE AMACRINE CELLS inhibit the ganglion cells, but their response is transient rather than steady

Binocular field

Monocular field

Monocular field

Macula lutea

VISUAL FIELDS

Nasal retina

Temporal (lateral) retina

Optic nerve

Temporal (lateral)

Nasal

Left visual field

a b
1 2
3 4
c d

a b
1 2
3 4
c d

Rt. visual field

Optic tract

Med. genic. body

Left retina

d c
3 4
1 2
b a

d c
3 4
1 2
b a

Rt. retina

Optic n.

Optic chiasma

Optic tract

Pul.

P

G

M

Mesen.

Optic chiasma

Lateral geniculate body

Sup. colliculus

Lat. vent.

CENTRAL CONNECTIONS OF OPTIC NERVE FIBERS

Pulvinar

Pin.

Geniculocalcarine (optic) radiations

Area 17—cacarine cortex (occipital lobe)

Temporal (lateral) fiber

Nasal fiber

Macular fiber

Optic n.

Optic tract

Inf. accessory optic tract

Hypothalamus

1
2
3
4
5
6

Lat. geniculate body

1, 2, 4 receive fibers from nasal ½ of opposite retina

3, 5, 6 ipsilateral temporal fibers

Pretectum

Sup. colliculus

III (Edinger-Westphal) nuc.

Tectum

To pupilloconstrictor mm.

Pupillodilator mm.

MLF (to III, IV + VI cranial nuc.)

Lat. gray matter (T1–T3) → Sup. cerv. gang.

Tectospinal tr. (to neck mm.)

179. VISUAL (RETINOGENICULOSTRIATE) PATHWAYS— PART II

I. **Stimulation of the ganglion cells:** bipolar, horizontal, and amacrine cells all stimulate the ganglion cells

A. THE GANGLION CELLS TRANSMIT THEIR SIGNALS via the optic nerve fibers in the form of action potentials, and they transmit continuous nerve impulses (5 per second), even when not stimulated. Thus, the visual signal is superimposed on this basic level of stimulation and can be either excitatory or inhibitory

B. THE GANGLION CELLS TRANSMIT luminosity signals (intensity), contrasts in the visual scene, changes in intensity, and color signals

1. Thus color analysis begins in the retina and is not entirely a brain function

C. FOUR TYPES OF GANGLION CELLS are present according to the manner in which their discharge is initiated: classified as belonging to an on-system (discharge when light is on), off-system (discharge at end of light stimulation), on-off system (discharge at the beginning and end of light), and a steady-background system (discharge spontaneously with no light needed*

II. **Function of the lateral geniculate bodies:** each of which is composed of 6 nuclear layers

A. LAYERS 1, 4, AND 6 receive signals from the nasal retina of the opposite eye; layers 2, 3, and 5 receive signals from the temporal part of the ipsilateral retina

B. ALL LAYERS OF THE LATERAL GENICULATE BODY relay visual information to the visual cortex through the geniculocalcarine tract (optic radiations)

C. THE PAIRING OF LAYERS FROM THE 2 EYES plays a major role in the fusion of vision because corresponding retinal fields in the 2 eyes connect with respective neurons that are approximately superimposed over each other in the successive layers

D. THE SIGNALS RECORDED IN THE RELAY NEURONS OF THE GENICULATE BODY are similar to those recorded in the ganglion cells of the retina; a few transmit luminosity but the majority transmit signals of contrasting borders in the visual image, and many are responsive to movement of objects across the visual scene

1. The signals in the geniculate body differ from the retina in that a greater number of complex interactions are found as a result of convergence of excitatory and inhibitory signals from 2 or more ganglion cells on the relay neurons of the geniculate body

III. **Function of the primary visual striate (area 17) cortex:** the ability to detect spatial organization of the visual scene, namely, forms of objects, brightness, shading, etc. Is dependent on the primary visual cortex (mainly in the calcarine fissure) located bilaterally on the medial aspect of each occipital cortex

A. SPECIFIC POINTS OF THE RETINA connect with specific points of the visual cortex; i.e., the right halves of the 2 respective retinae connect with the right visual cortex and the left with the left

1. The macula is represented at the occipital pole, and the peripheral regions of the retina are represented in concentric circles farther and farther forward from the occipital pole

2. The upper part of the retina is seen superiorly in the cortex, the lower part inferiorly

B. COLOR ANALYSIS: there are specific cells in the primary cortex that are stimulated by color intensity or by contrasts of the opponent colors. These effects are identical to those seen in the lateral geniculate body except that the proportion of cells excited by opponent color contrasts is less here

1. Since neuronal excitation by color contrasts is a means of deciphering color, it appears that the cortex is concerned with an even higher order of detection of color than simply deciphering color itself, a process mainly completed by the time the signals pass through the geniculate bodies

---

*The on-sets, off-sets, and on-off sets can also be identified in the lateral geniculate body and the primary visual cortex of the optic pathways.

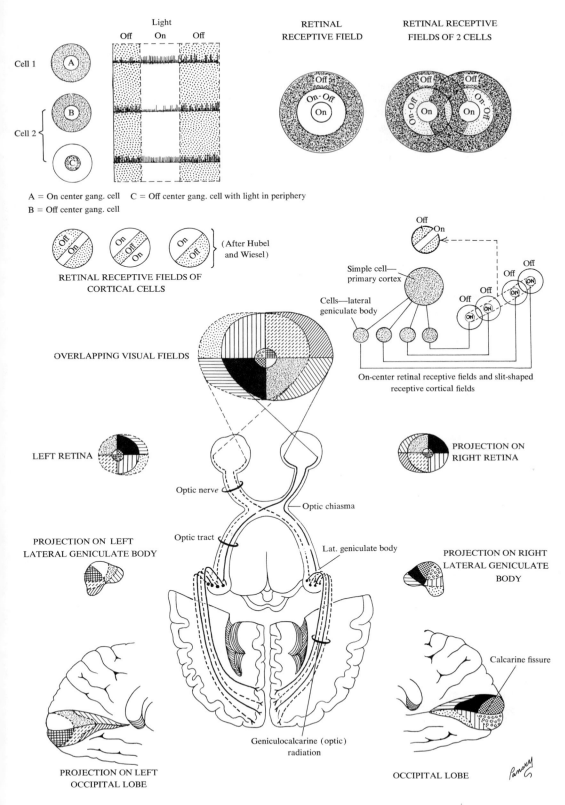

Light

Off On Off

Cell 1

A

Cell 2 { B

C

A = On center gang. cell   C = Off center gang. cell with light in periphery
B = Off center gang. cell

RETINAL
RECEPTIVE FIELD

Off
On-Off
On

RETINAL RECEPTIVE
FIELDS OF 2 CELLS

Off   Off
On-Off   On-Off
On   On

Off
On
On
Off
On
Off

(After Hubel
and Wiesel)

RETINAL RECEPTIVE FIELDS OF
CORTICAL CELLS

Off
On

Simple cell—
primary cortex

Off
Off
Off
Off
ON
ON
ON
ON

Cells—lateral
geniculate body

OVERLAPPING VISUAL FIELDS

On-center retinal receptive fields and slit-shaped
receptive cortical fields

LEFT RETINA

PROJECTION ON
RIGHT RETINA

Optic nerve

Optic chiasma

Optic tract

Lat. geniculate body

PROJECTION ON LEFT
LATERAL GENICULATE BODY

PROJECTION ON RIGHT
LATERAL GENICULATE
BODY

Calcarine fissure

Geniculocalcarine (optic)
radiation

OCCIPITAL LOBE

PROJECTION ON LEFT
OCCIPITAL LOBE

–395–

180. OTHER VISUAL PATHWAYS

I. Signals from the primary cortex project laterally in the occipital cortex into visual association areas (areas 18, 19) which are loci for additional processing of visual information

A. HERE THE NEURONS RESPOND TO MORE COMPLEX PATTERNS such as geometrical design, curving borders, angles, etc.

B. DESTRUCTIVE LESIONS OF THE VISUAL ASSOCIATION AREAS result in difficulty with certain types of visual perception and learning such as dyslexia or word-blindness (difficulty in understanding the meaning of words one sees)

II. The tectal system or secondary visual system is difficult to define and comprises a sequence of retina to superior colliculus, to pulvinar of the thalamus, to the inferotemporal cortex

III. Other pathways from the retina include those involved with various reflexes and the pineal gland

A. SUPERIOR COLLICULUS AND PRETECTUM are important nuclear structures intercalated in the pupillary reflex, accommodation, and conjugate eye movements

B. THE ACCESSORY OPTIC SYSTEM involves several small nuclei in the midbrain tegmentum and has a significant role in the function of the pineal gland

IV. The superior colliculus serves the reflex control and regulation of many movements of the eyes and head and certain aspects of vision. It has a significant function in attention and perception associated with vision. It is more than a reflex center but is an integrative center subserving visual perception

V. Accommodation is the process by which the refractory power of the lens is changed and is essential to a high degree of visual acuity

A. THIS REFLEX IS MEDIATED BY THE CEREBRAL CORTEX

  1. One can voluntarily focus on a nearby or faraway object

  2. It is regulated by a negative feedback mechanism that automatically adjusts the focal power of the lens

  3. The pathway includes the optic pathways from the eye to the cerebral cortex; in turn, a projection back from cortical area 19 (and from frontal eye field area 8) via the optic radiation through the brachium of the superior colliculus to terminate in the nucleus of the superior colliculus (interneurons interconnect these bilateral nuclei)

    a. Through a chain of several interneurons, the collicular nuclei are connected to the preganglionic neurons in the accessory motor nucleus of cranial nerve III (Edinger-Westphal). The latter project via the IIIrd nerve and synapse in the ciliary ganglion with postganglionic fibers innervating the ciliary muscles in the ciliary body. Thus, pupillary constriction and convergence of the eye accompany accommodation to near vision

VI. Depth perception: the most precise and accurate means for localizing objects in space utilizes the images seen simultaneously by both eyes, called stereopsis or solid vision

A. BASED ON THE FACT that the visual cortex on one side receives input from the contralateral nasal hemiretina and the ipsilateral temporal hemiretina, and the presence of an interhemispheric link through the corpus callosum

B. THE SOLE BASIS FOR STEREOPSIS is the horizontal disparity between the 2 hemiretinal images; the horizontal disparities between the 2 hemiretinas viewing the same visual field obtain the cues of depth and, in turn, the processing in the neurons of the visual pathways lead to depth perception

ACCOMMODATION REFLEX

181. VISUAL REFLEXES

I. Pupillary constriction (direct light) and consensual reflexes
A. THE PUPILS CONTRACT when the eyes are exposed to bright light and reduce the intensity of light reaching the retina. This involuntary light reflex is in response to direct light stimulation and its pathway is
1. Retinal ganglion cell axons via the optic nerve, optic chiasma (some fibers decussate), optic tract, and brachium of superior colliculus (bypassing the lateral geniculate body) to terminate in the pretectum. Interneurons project to the contralateral pretectum via the posterior commissure. Interneurons interact with other interneurons which, in turn, synapse with preganglionic neurons of the oculomotor nerve (arising in the Edinger-Westphal nucleus). These preganglionic neurons project via the IIIrd nerve and synapse in the ciliary ganglion from which postganglionic neurons innervate the constrictor muscle of the iris
2. When only one eye is exposed to bright light, the pupils of both eyes constrict (consensual reflex). This is a result of crossing of some fibers in the optic chiasma and interconnections across the midline of the bilateral pretectum

II. Pupillary dilatation
A. THE PUPILS DILATE when the eyes are exposed to dim light. Optic pathways act through the reticular formation and descending sympathetic pathways (in the dorsolateral tegmentum of the brainstem and in the anterior half of the cervical spinal cord) to stimulate preganglionic neurons of the sympathetic intermediolateral cell column at C8 and T1 levels. The preganglionic fibers ascend in the sympathetic chain, synapse in the superior cervical ganglion, and postganglionic neurons follow adjacent blood vessels to innervate the dilator muscle of the iris. Interruption of these fibers may result in Horner's syndrome

III. Corneal reflex: touching the cornea causes the eyelids to close reflexly. The afferent fibers are in the ophthalmic nerve (V), and the efferent fibers of the reflex arc are in the facial nerve (VII) to the orbicularis oculi muscle

IV. Argyll-Robertson pupil may occur in syphilis of the CNS. In this syndrome, the pupil is small in dim light and does not constrict farther when the eye is exposed to bright light. This same pupil will respond by constricting farther during accommodation-convergence reaction
A. THE USUAL STATED CAUSE for this dissociation of pupillary reflex is a lesion in the pretectum

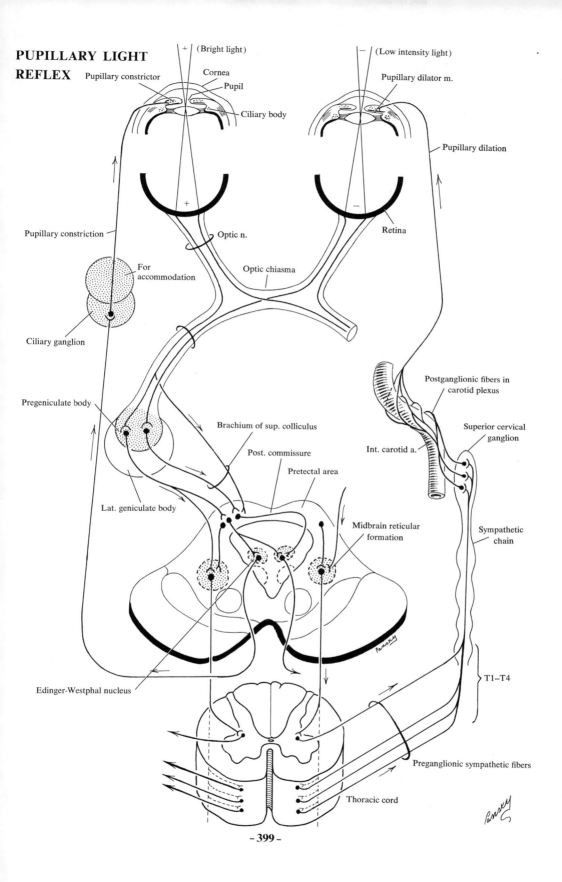

PUPILLARY LIGHT REFLEX

+ (Bright light)

− (Low intensity light)

Pupillary constrictor

Cornea

Pupil

Ciliary body

Pupillary dilator m.

Pupillary dilation

Pupillary constriction

Optic n.

Retina

For accommodation

Optic chiasma

Ciliary ganglion

Pregeniculate body

Postganglionic fibers in carotid plexus

Brachium of sup. colliculus

Superior cervical ganglion

Post. commissure

Int. carotid a.

Pretectal area

Lat. geniculate body

Midbrain reticular formation

Sympathetic chain

Edinger-Westphal nucleus

T1–T4

Preganglionic sympathetic fibers

Thoracic cord

182. EYE MOVEMENTS AND THEIR CONTROL

I. **Binocular vision:** the same points in the field of vision common to the two eyes must be focused on corresponding loci in the two retinas. The simultaneous movement of both eyes in the same direction is called conjugate movement (convergence is a disconjugate movement, when the eyes look at a close-up vision together)

A. THE OCULOMOTOR SYSTEM regulates these movements and comprises several central pathways and the lower motor neurons of cranial nerves III, IV, and VI, innervating the extraocular muscles

1. Each set of muscles (medial and lateral recti, superior and inferior recti, and the superior and inferior oblique) is reciprocally innervated so that the contraction of one muscle of each pair is synergistically synchronized with the relaxation of the other muscle, in order to direct the gaze to any position

2. The oculomotor system comprises four subdivisions
   a. The saccadic system (search for visual targets): the movement is a small rapid jerk called a saccade. Saccades search the visual field, move the visual image over the receptors and prevent adaptation. Saccades are the fastest movement in the body. They occur during sleep with the eyes closed. If stopped, all color detail fades in seconds.
   b. The smooth pursuit system (tracking of visual objects) are smooth movements causing the eye to follow objects in the visual fields. They require continual feedback of visual information about the moving object
   c. The vestibular system (compensation for head movements): if a stationary visual object is focused on the fovea and the head is moved, the eyes must be moved an equal distance for the object to remain focused on the fovea (i.e., if the head moves down, the eyes must move up)
   d. The vergence system (convergence): used to track a visual object in depth through the visual field by turning the eyes inward as the object gets closer and outward as it moves away

3. The four movements appear to be controlled by separate neurologic systems, yet they cooperate in all movements
   a. The control system depends on information from the retina
   b. When the eyes and head are fixed, the image of a moving object moves across the receptors and gives rise to retinal signals
   c. When the eyes follow a moving object, however, the image remains stationary on the retina and the retina cannot signal movement, but we see movement because the rotation of the eye in the head can apparently give rise to perception of movement, even estimate velocity
   d. In movement, one must decide what is moving and what is stationary with regards to a reference frame. If the information about movement is by vision alone (and not due to body movement), we generally assume that the larger objects in the visual field are stationary

B. VOLITIONAL MOVEMENTS OF THE EYE: the frontal eye field in area 8 is the cortical center which influences voluntary eye movements mediated through cranial nerves III, IV, and VI and is called the voluntary eye field

C. AUTONOMIC EYE MOVEMENTS AND EYE FIXATION: our eyes can scan a line or follow a moving object without any volitional effort

1. The neuronal center and pathways involved with locking the eyes on an object once it has been located comprise the involuntary fixation mechanism and include reflex pathways from retina to primary visual cortex to association visual cortex and then via corticotectal fibers to the superior colliculus, from which connections are made with cranial nerve nuclei of III, IV, and VI. Area 19 may play a role in this mechanism

Sup. rectus m. (SR)

Levat. palp. sup. m.

Sup. oblique (SO) m.

Med. rectus m. (MR)

Lateral rectus m. (LR)

Inf. oblique m. (IO)

Inf. rectus m. (IR)

SR   SO   SO   SR

LR   MR   LR

IR

RIGHT EYE   IO   IO   LEFT EYE

IO

Inward rotation

Elevation

SR

SO

Down and out

L.
(Lateral) abduction ←————
R.

M.
————→ Adduction (medial)
R.

Up and out

IO

Depression

IR

Inward rotation

**EYEBALL AND MUSCLE
MOVEMENTS**

Medial rectus m.

Lateral rectus m. ←

IIIrd nerve

Cortex

III

Oculomotor (III) nucleus

IV

Abducens (VI) nerve

Contralateral pontile center of conjugate lateral
ocular deviation (parabducens nucleus)

Medial longitudinal fasciculus

VIth nucleus (abducens)

Ipsilateral parabducens nucleus

Ipsilateral VI nucleus

Pansky

CONJUGATE LATERAL DEVIATION OF EYES

183. LESIONS OF THE OPTIC PATHWAYS
AND RELATED PHENOMENA

I. **Impairment of a small area of the retina** results in scotoma (blind spot) in the field viewed by the eye

II. **Complete interruption of the optic nerve** results in permanent blindness; in one eye it is called monocular blindness
  A. THE BLIND EYE can still accommodate and show consensual light reflex because the normal eye activates the intact reflex arcs through the pretectum and superior colliculus and the optic nerve to the blind eye

III. **A fixed dilated pupil** may be a sign of pressure on the oculomotor nerve

IV. **A midline lesion of the optic chiasma** may interrupt the decussating retinofugal fibers resulting in a loss of reception in the nasal hemiretina of each eye, accompanied by loss of vision in the temporal half of the visual field of each eye resulting in bitemporal hemianopia (hemianopsia)
  A. SUCH A DEFECT IS HETERONYMOUS: portions of both retinas that view different areas of the fields of vision are involved
  B. INTERRUPTION OF THE NONDECUSSATING FIBERS on both sides of the optic chiasma produces a binasal hemianopia: loss of reception in the temporal hemiretina of each eye accompanied by loss of vision in the nasal half of the visual field of each eye

V. **Complete interruption of the optic tract or lateral geniculate body or the entire primary visual cortex on one side (e.g., right)** results in a left contralateral homonymous hemianopia or blindness in the field of vision on the side opposite the lesion
  A. HOMONYMOUS refers to corresponding regions of both retinas, hence to a single visual field

VI. **Partial lesions of the visual pathways** produce partial defects in the visual fields
  A. A LOWER QUADRANTIC HOMONYMOUS ANOPSIA results from a contralateral lesion in the upper (superior) half of the optic radiation or the entire primary visual cortex (area 17) above the calcarine fissure
    1. This field defect occurs because the pathways from the upper temporal quadrant of the ipsilateral retina and upper nasal quadrant of the contralateral retina are interrupted
  B. A LESION OF THE LOOP OF MEYER will produce a contralateral upper quadrantic anopsia because pathways from the lower temporal quadrants of the ipsilateral retina and the lower nasal quadrant of the contralateral retina are interrupted

VII. **Lesions of a portion of the primary visual cortex** may result in a homonymous field defect in which the field defect of one eye is exactly superimposed on that of the other eye; the loss is said to be congruous
  A. A LESION IN THE OPTIC TRACT OR RADIATION may result in a field defect in which the fields from each eye are not exactly superimposed; the loss is said to be incongruous

VIII. **A blow or injury that damages the choroid layer of the eye** may result in sympathetic ophthalmia; not only may sight be impaired in the injured eye but the other eye also gradually loses its sight. The cause is that the body responds to certain substances released from the injured choroid layer resulting in an immunologic reaction that destroys both eyes

IX. **Absence of pigment:** true albinos shun light and have difficulty seeing because pigment is absent in the iris, choroid, and retinal pigment layers. Visual acuity is rarely better than 20/100 to 20/200

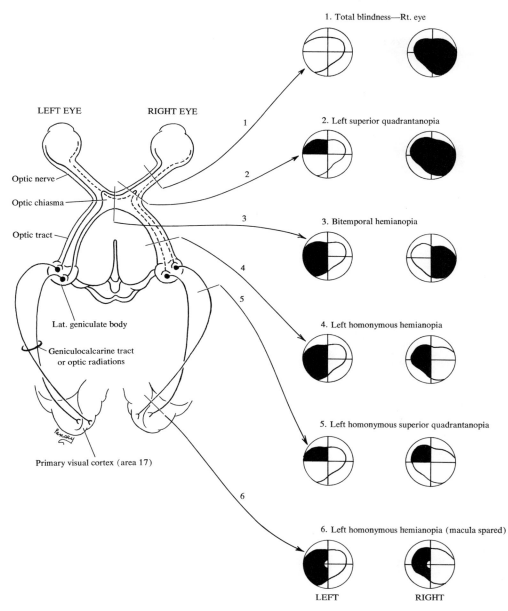

VISUAL FIELDS

1. Total blindness—Rt. eye

LEFT EYE   RIGHT EYE

2. Left superior quadrantanopia

Optic nerve

Optic chiasma

3. Bitemporal hemianopia

Optic tract

4. Left homonymous hemianopia

Lat. geniculate body

Geniculocalcarine tract
or optic radiations

5. Left homonymous superior quadrantanopia

Primary visual cortex (area 17)

6. Left homonymous hemianopia (macula spared)

LEFT   RIGHT

**DEFECTS IN VISUAL FIELDS
PRODUCED BY LESIONS IN
REGIONS OF THE VISUAL PATHWAYS**

PART III

Neurology and Neuropathology

Unit Thirteen

INTRACRANIAL VASCULAR ACCIDENTS AND TRAUMA

184. COMPUTERIZED TOMOGRAPHY (CT) OF THE BRAIN

I. Introduction

The development of the CT scanner has had a great impact on the fields of neuroanatomy, neuroradiology, neurosurgery, neuropathology, and neurology. At the same time, computerized cerebral scanning has clearly demonstrated and emphasized the need to understand the anatomic minutiae of the brain and its coverings (usually taught in neuroanatomy) in order to interpret the details and relationships now possible with the CT scanner. In addition to revealing the extent and exact location of intracranial space-occupying lesions, cerebral hemorrhage, cerebral edema, infarcts, malformations, and demyelinating diseases, it can serve also as an invaluable tool for studying normal neuroanatomy. The student of neuroanatomy is able to obtain a better three-dimensional image when coronal and horizontal CT scans of the same brain are compared

II. Some characteristics and major advantages of the CT scanner
A. NONINVASIVE CHARACTER makes it practically free of morbidity
B. REPEATED SCANS without exposure to x-radiation are possible
C. SAFER THAN ANGIOGRAPHY AND PNEUMOENCEPHALOGRAPHY
D. PAINLESS, COMFORTABLE, AND RELATIVELY FAST PROCEDURE
E. AMBULATORY UTILIZATION makes it more practical

III. Basis of computer tomography
A. CT UTILIZES DIFFERENTIAL ABSORPTION of head and neck (or body) obtained by axial circumferential measurements which are mathematically arranged by a computer into a matrix (grid) of picture elements on an oscilloscope
B. EVEN SOFT TISSUES have slightly different absorption coefficients, and by taking a number of measurements of the amount of x-ray absorption and systematically altering the angle of the passage of the x-ray beam through the head, the absorption coefficient can be calculated
C. THE IMAGE that is formed can be manipulated by using the computer's data bank to suit the particular organ or pathology in the area of interest which includes magnification, density, and area of measurements
D. HIGHER DENSITIES appear light (bone), while lower densities are darker
E. CERTAIN TISSUES (abscess margins, tumors, edematous fluids, etc.) and intravascular blood can show increased CT number density when intravenous contrast materials (diatrizoate, iothalamate ions, etc.) are utilized
F. THE DEGREE OF INCREASE IN BLOOD CT DENSITY is related to the dose and rate of intravenous infusion (e.g., fast high dose causes more intravascular CT density enhancement)

IV. Procedure used for neuroradiologic application
A. X-RAY BEAM that is finely focused is rotated around the head numerous times
B. AMOUNT OF X-RADIATION EMITTED is known, and that passing through the skull is determined by means of scintillation crystals
C. SECTIONS (slices) of the head are scanned in this fashion
D. THE BRAIN CAN BE PROGRESSIVELY SCANNED, in this manner, from its base to its vertex when the positions of the slices are varied
E. DATA FROM THE SCANS are processed by the computer which in turn displays the absorption coefficients of the tissues in each section on a cathode ray tube. The intensity of the beam is proportional to the absorption coefficient
F. SOME OF THE STRUCTURES (both normal and pathologic) that can be recognized are shown on the CT scans that are included herein

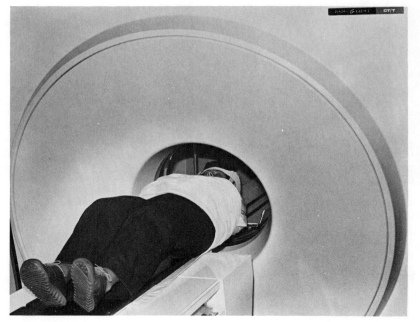

Patient in position in CT scanner

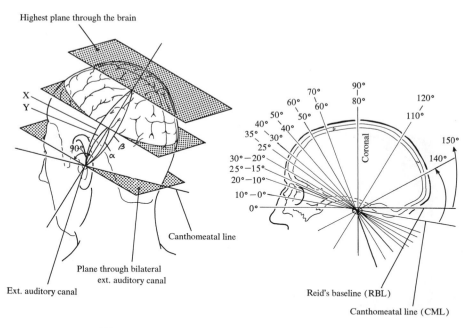

Highest plane through the brain

X
Y

90°
β
α

Ext. auditory canal

Plane through bilateral
ext. auditory canal

Canthomeatal line

90°
80°
70°
60°
60°
50°
50°
40°
40°
35°
30°
25°
30° — 20°
25° — 15°
20° — 10°
10° — 0°
0°

Coronal

120°
110°
150°
140°

Reid's baseline (RBL)

Canthomeatal line (CML)

CT SCAN

Horizontal CT scan of normal adult brain showing lateral ventricles (LV), falces (F), and sulci (arrows). Radiopaque medium used for contrast.

Horizontal CT scan of normal 7-yr-old brain. Note difference between white and gray matter.

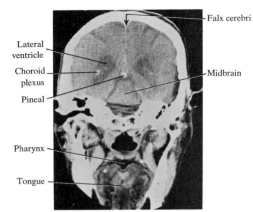

Falx cerebri

Lateral ventricle

Choroid plexus

Pineal

Midbrain

Pharynx

Tongue

Coronal CT scan of normal adult brain.

Lateral ventricle

3rd ventricle

Coronal CT scan of normal adult brain.

Rectus muscles

Eye

Optic nerve

Temporal lobe

Cerebellum

Horizontal CT scan of normal brain at midorbit level.

Optic chiasm

Middle cerebral artery

Cerebral peduncle

Horizontal CT scan of normal brain at level of optic chiasm.

FIGURE 13. **Normal CT scans.**

–410–

A

Normal ventricles in seven-yr-old.
Compare with fig. *E* below.

B

Normal ventricles in seven-yr-old appear slitlike
(arrows).

C

Brain atrophy (mild) in senile patient. Lateral
fissures are widened (arrows) and ventricles appear
enlarged.

D

Adult hydrocephalus.

Dilated anterior horn

Shunt
tube

Dilated
post. horn

E

Infant hydrocephalus with shunt tube as seen in
CT scan.

F

Multiple cerebellar metastases.

FIGURE 14. **CT scans.** *A* and *B*, **normal;** *C*, **brain atrophy;** *D* and *E*, **hydrocephalus;** *F*,
cerebellar metastases.

Three aneurysms (subtraction film).

CT scan of same patient showing the three aneurysms:
1. Lt. middle cerebral artery.
2. Ant. communicating art.
3. Rt. middle cerebral art.

Arteriovenous (AV) malformation (angiogram) feeding vessel indicated by arrow.

CT scan showing AV malformation.

FIGURE 15. **CT scans.** Aneurysms and arteriovenous malformations.

Excessive fluid around falx

Brain atrophy (mild) in 88-yr-old patient. Note widened sulci.

Bilateral frontal lobe contusion. Note fractures (1) and hematoma (2) with spectrum density.

Epidural hematoma (arrow). Note anterior and medial displacement of rt. choroid plexus.

Hematoma   Edema

Thalamic glioma

Infarct

Chronic subdural hematoma (subacute), thalamic glioma, edema, and displacement of midline structures.

Multiple cerebral metastases.

Multiple cerebral metastases (arrows) with surrounding edema.

FIGURE 16. **CT scans.** Atrophy, contusion, hematomas, and cerebral metastases.

185. INTRACRANIAL VASCULAR ACCIDENTS: GENERAL INFORMATION

I. Introduction: "stroke," the best-known disorder of the nervous system, is not primarily a neurologic disease, but one in which nervous tissue is damaged as a result of catastrophe affecting blood vessels carrying oxygen. Under normal body temperature conditions, interruption of oxygen supply for more than 3–4 minutes results in widespread damage and death to nerve cells. With hypothermia or reduced body temperatures, O_2 lack can be withstood for considerably longer

A. STROKE OR CEREBROVASCULAR ACCIDENT refers to the blood supply to the brain being seriously disturbed spontaneously and not due to surgery or trauma. Though some vascular accidents inside the skull do not damage brain tissue, we tend to group all strokes together as "intracranial vascular accidents"

   1. In general, thrombosis, leading to gradual occlusion, will permit the adaptation of an anastomotic substitution network of supply. By contrast, emboli produce massive and sudden occlusion after which reirrigation is inadequate and the resulting infarct is usually extensive

II. Characteristics of a cerebrovascular accident: the onset is sudden; the paralysis is maximum at onset; and with survival, the disability tends to improve

III. Etiologic factors

A. ATHEROSCLEROSIS is the chief factor in production of cerebral infarction

B. CARDIAC EMBOLI are frequent causes of arterial occlusion

C. ARTERITIS is a rare cause of cerebral infarction

D. TRAUMA: injuries to the neck or in the mouth may occasionally give rise to internal carotid occlusion

E. VASCULAR MALFORMATIONS, e.g., arterial aneurysms

IV. Types of strokes (clinically)

A. CEREBRAL THROMBOSIS: a clot or thrombosis forms at a diseased point in a cerebral artery

   1. The disease in the artery is an atheroma, although inflammation of the wall may be responsible. An embolism, in most cases, is of cardiac origin

   2. The clot forms, blocks the vessel and cuts off the flow of blood (and O_2) to a particular area of the brain, and produces a zone of damage, an infarct

     a. The clinical result is that the part of the body normally controlled by this zone loses its function, e.g., becomes paralyzed

     b. Not all the cells in the zone are permanently destroyed, and after a period of time some cells begin to function again and other blood vessels open up to take over the job of the blocked artery

     c. The paralysis then begins to recover and may do so partly or completely according to the degree of the original damage. (*Note:* nerve cells never regrow, but others may take over their function)

   3. Patients tend to be over 50 years old (since atheroma is a major cause)

   4. There is a slight tendency for occlusion to occur during periods of relative immobility, e.g., during sleep or bedrest

   5. The onset is sudden or very rapid, with little headache, and there may not be a loss of consciousness

   6. The form of the stroke varies with the vessel affected and certain clinical syndromes have been recognized (see next page)

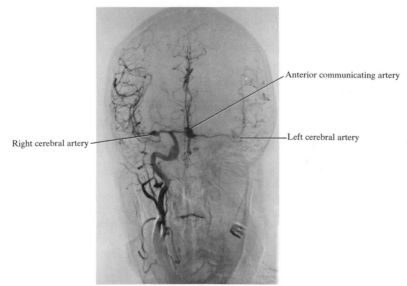

Anterior communicating artery

Right cerebral artery

Left cerebral artery

Angiogram showing three aneurysms.

Ruptured berry aneurysm (arrow) resulting in massive bleeding.

186. INTRACRANIAL VASCULAR ACCIDENTS: CEREBRAL THROMBOSIS

I. Cerebral thrombosis

A. CLINICAL SYNDROMES

1. Middle cerebral artery thrombosis: less common than once thought and less than internal carotid thrombosis
   a. Paralysis of the opposite side of the body, mainly affecting face and arm
   b. If the patient is right-handed and the left artery is affected, there is also a loss of speech (aphasia)
   c. Limbs are usually flaccid at first, but as the "shock" passes off, they become spastic or recover completely

2. Anterior cerebral artery thrombosis: relatively uncommon
   a. The opposite leg is principally affected

3. Posterior cerebral artery thrombosis is seldom an isolated event and usually occurs as a result of anterograde extension of thrombosis of the basilar artery
   a. Sudden loss of opposite half of visual fields: homonymous hemianopia

4. Basilar artery thrombosis
   a. The brainstem is disturbed: coma and death very often follow
   b. Gross disturbances of eye movements, paralysis of swallowing, weakness of all 4 limbs, and often bilateral blindness

5. Internal carotid artery thrombosis: symptoms similar to those of middle cerebral artery involvement, and the two may be confused
   a. Is suspected when there are repeated strokes that may affect the same side, with paralysis lasting only a few hours or days
   b. The episodes are due to a thrombus formed in the vessels near its origin in the neck and from here, small emboli are carried up and temporarily block the branches of the middle cerebral artery and then pass on
   c. The carotid need not be completely blocked, but may be markedly narrowed or stenotic as a result of an atheroma patch and overlying thrombus

6. Vertebral artery thrombosis
   a. One artery may block without any symptoms (blood going up the other side and thus it may be clinically and pathologically silent)
   b. May have attacks of vertigo, faintness, and vague sensory disturbances throughout the body (often when neck is turned or bent)
   c. Symptoms arise from inadequate supply to the brainstem since these two vessels join to form the basilar artery
   d. One syndrome involving this artery is related to thrombosis of the posterior inferior cerebellar artery
      i. Attack of vertigo is followed by a loss of sensation on one side of the face and opposite side of the rest of the body due to arrangement of the sensory fibers at this part of the brainstem. Facial pain, which is very troublesome, may also follow

7. Subclavian artery thrombosis may result in ischemic lesions of the vertebrobasilar artery territory after diversion of the arterial flow (also called subclavian steal syndrome)

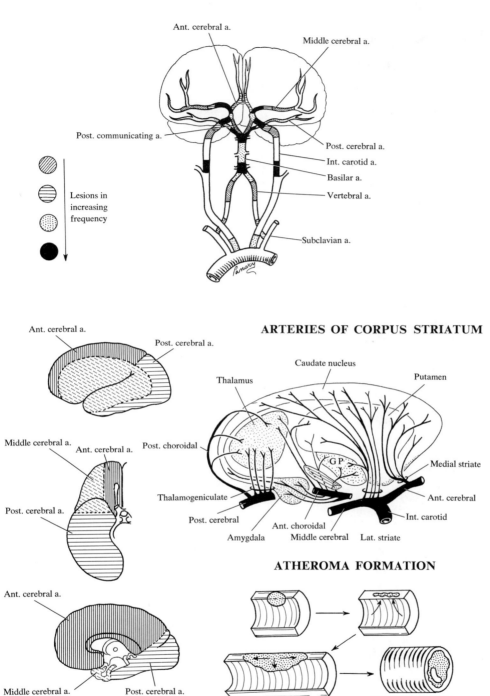

Ant. cerebral a.

Middle cerebral a.

Post. communicating a.

Post. cerebral a.

Int. carotid a.

Basilar a.

Vertebral a.

Subclavian a.

Lesions in
increasing
frequency

ARTERIES OF CORPUS STRIATUM

Ant. cerebral a.

Post. cerebral a.

Caudate nucleus

Putamen

Thalamus

Post. choroidal

Middle cerebral a.

Ant. cerebral a.

G P

Medial striate

Post. cerebral a.

Thalamogeniculate

Ant. cerebral

Int. carotid

Post. cerebral

Ant. choroidal

Amygdala

Middle cerebral

Lat. striate

Ant. cerebral a.

ATHEROMA FORMATION

Middle cerebral a.

Post. cerebral a.

Thrombotic occlusion

187. INTRACRANIAL VASCULAR ACCIDENTS: CEREBRAL EMBOLI, ARTERIAL INSUFFICIENCY, AND VENOUS THROMBOSIS

I. Cerebral embolism

A. THE CLOT is carried to the cerebral vessel, which may itself be healthy but is too narrow to allow the clot to pass from some other vessel

B. THE SYMPTOMS AND SIGNS are similar to those of thrombosis except that the onset is usually quite abrupt

C. THE EMBOLUS must come from somewhere, and this may be the heart (cardiac embolism) whose valves have been diseased by rheumatic fever or whose walls have been damaged by a coronary thrombosis

    1. Emboli may also arise from patches of thrombus formed in the large vessels in the neck and chest (carotids, vertebrals, brachiocephalics, and even the aorta). The latter is apparently even more common than primary intracranial thrombosis of such vessels as the middle cerebral artery

II. Cerebral arterial insufficiency

A. SYMPTOMS AND SIGNS of a cerebral thrombosis may develop after a short period of time (only minutes) without any actual thrombosis having occurred—transient ischemic attacks, which may occur as a result of

    1. Minute emboli may break off from a diseased patch in a carotid or vertebral artery, temporarily block a small cerebral or retinal artery, fragment, and pass on into the veins; whereupon, flow is restored and symptoms recover

    2. In others, there may be a marked narrowing of a vessel by arteriosclerosis or suffering a sudden loss of blood due to an ulcer which results in a fall of blood pressure which is then inadequate to force blood through the narrowed vessel

        a. The patient develops signs of insufficient blood in the area supplied by the vessel. When the blood pressure is restored, the symptoms recover

        b. If one large vessel is blocked, the other may work for both sides

        c. Merely turning the head sharply and kinking the vessel may cause temporary symptoms

        d. If any one of the above is of long duration, infarction may result without an actual thrombosis, and this is not an uncommon clinical happening

III. Venous thrombosis: thrombi can form in the veins as well as in arteries

A. VENOUS THROMBOSIS is rare but may occur during severe illness, malnutrition, dehydration, or after major surgery or confinement to bedrest

B. PATCH AREAS OF THE CEREBRAL CORTEX are involved, since the veins lie on the surface of the brain

C. IRRITATED CELLS result in recurrent fits usually followed by weakness of the involved parts of the body

    1. If the thrombus spreads, the fits will also spread to affect more and more of one side of the body and may even spread to the other side as well

Hemorrhagic infarct in midbrain.

Cerebral infarct following arterial insufficiency.

188. INTRACRANIAL VASCULAR ACCIDENTS: CEREBRAL HEMORRHAGE

I. **Intracranial hemorrhage:** a blood vessel ruptures allowing blood to rush through cerebral tissue, destroying it, or forming a collection in the brain (a hematoma) or permitting it to escape into the subarachnoid space and into the CSF (a subarachnoid hemorrhage). Consists of 2 types: cerebral and subarachnoid hemorrhage

A. CEREBRAL HEMORRHAGE: a term used when a cerebral vessel ruptures, resulting in blood escaping through brain substance and leading to an immediate destruction of much nervous tissue

   1. Hemorrhage usually results from a weakness in a vessel wall (i.e., due to a patch of atheroma, an abnormally high pressure inside the vessel as a result of high blood pressure, or both)
   2. Tends to occur in hypertensive patients over the age of 55 and often during periods of activity
   3. Abrupt onset, a bursting headache, vomiting, and rapid loss of consciousness
   4. Breathing is stertorous, there is profound paralysis of one side of the body, and the paralyzed cheek flaps in and out with each breath
   5. Many die rapidly; others remain unconscious for days and then die; still others recover consciousness and may use their limbs, but are left with serious disability and speech defects
   6. Prognosis is much worse than for a thrombosis or embolism

B. DIFFERENTIATION BETWEEN THROMBOSIS AND HEMORRHAGE: arteriography and other techniques have shown that it is difficult to be certain whether there is a massive cerebral thrombosis (may mimic hemorrhage) or a small hemorrhage which may, in every respect, resemble a cerebral thrombosis

   1. Neurologists are reluctant to prescribe anticoagulants without some certainty of what has happened and thus patients must be carefully investigated

C. TREATMENT OF INTRACRANIAL VASCULAR ACCIDENTS is far from satisfactory and consists of management of the patient following the event and prevention of further similar episodes

   1. The stroke damages the brain in seconds and nothing can alter its course at that moment, but one treats the resultant coma and paralysis

D. PREVENTION OF FURTHER ACCIDENTS

   1. Little can be done to stop the thrombosis of small brain vessels
      a. Anticoagulants are dangerous because by reducing the clotting tendency of the blood one may convert a small infarct into a massive hemorrhage
   2. If the main thrombus lies in a large vessel in the neck, from which small emboli are being thrown off into the intracranial vessels, prolonged treatment by anticoagulants is often helpful
      a. It must be known that this is the case since nothing could be more dangerous if it turned out to have been a small hemorrhage
      b. When symptoms suggest a carotid artery lesion (i.e., transient ischemic episodes, recurrent small strokes), arteriography is used if the patient isn't too old and can tolerate the procedure
      c. If the carotid is stenosed but not fully blocked, surgery may help to restore full blood flow
   3. In young people, emboli often come from a diseased heart, although this is uncommon. In this case, the heart condition (i.e., atrial fibrillation, bacterial endocarditis) must be treated as well as possible, but anticoagulants are also indicated
   4. In cerebral hemorrhage due to hypertension, one must lower the blood pressure. Without a definite hematoma present, it is of little value to operate on a hypertensive cerebral hemorrhage

A series of sections through the brain showing how extensive intracranial hemorrhage may become.

Cerebral hemorrhage.

189. INTRACRANIAL VASCULAR ACCIDENTS: SUBARACHNOID HEMORRHAGE

I. **Subarachnoid hemorrhage** is a type of hemorrhage where bleeding occurs into the CSF between the arachnoid and pia. It is most often due to a weakness in a blood vessel wall at the base of the brain (in the circle of Willis) or a rupture of abnormally developed vessels lying anywhere. Two major causes are

A. CEREBRAL ANEURYSMS: a weakness in the wall of a pulsating cerebral artery at a point where it branches, resulting in a bulge or aneurysm. Called a "berry" aneurysm because of its appearance

   1. A weak-walled sac which ruptures sooner or later. The weakness is present at birth and is never due to syphilis

   2. In older patients, may develop due to degenerative changes weakening the vessel wall

   3. Referred to by name of the vessel from which they originate

   4. Patient may be young with a normal blood pressure and no other vascular disease

   5. Rapidly develops intense headaches that spread to the back of the head and down the neck and the neck becomes stiff

   6. Usually no paralysis, but if force of rupture damages the brain itself, results may resemble an ordinary cerebral hemorrhage

   7. Repeated vomiting and loss of consciousness [or a dislike for light (photophobia) and a restlessness and irritability develop]

   8. With complications, nearby vessels go into spasm resulting in cerebral infarction and hemiplegia a few days after the rupture

   9. Without complications, the irritant blood is slowly absorbed from the CSF, the headaches improve in a few days, the neck stiffness decreases, and the patient may recover without any residual trace of disability

   10. Hemorrhages do tend to recur, particularly in the 1st months, even the 1st fortnight, and, unless treated, further rupture may prove to be fatal

   11. Aneurysms may press on nearby cranial nerves prior to rupture (3rd nerve commonly) resulting in symptoms, i.e., ptosis or paralysis of eye movements

B. CEREBRAL ANGIOMA (vascular anomalies or arteriovenous aneurysms) consist of a mass of abnormal arteries, capillaries, and veins which have very fragile walls and, if lying inside the head, are liable to rupture and produce subarachnoid hemorrhage

   1. Are usually found on or near the cerebral cortex and act as an irritant and thus produce fits (not seen with "berry" aneurysms)

   2. Blood rushes through these widened vessels, and one may hear a pulsating noise in the head with a stethoscope—an intracranial bruit—diagnostic of an angioma in the adult

C. TREATMENT OF SUBARACHNOID HEMORRHAGE

   1. Patient should lie in a position of comfort until headache subsides

   2. If diagnosis is suspected, a lumbar puncture will confirm blood in the CSF

   3. Avoid any unnecessary movement for the first two days after hemorrhage

   4. Analgesics given as needed (vomiting reduced with appropriate drugs)

   5. Strict bedrest for 4–5 weeks with further convalescence for 4–5 weeks. No special precautions after this stage reduce the risk of further hemorrhage, but one avoids excessive straining or repeated head injury

   6. Surgical treatment depends on surgeon and the patient's condition

      a. Prognosis has been improved by advances in neurosurgery

      b. Use angiography to demonstrate aneurysm or angioma (both size and site)

      c. Clips or ties may be used or even splints in order to remove the weak points from the circulation

      d. If surgery is successful, it relieves the patient from fear of further hemorrhage

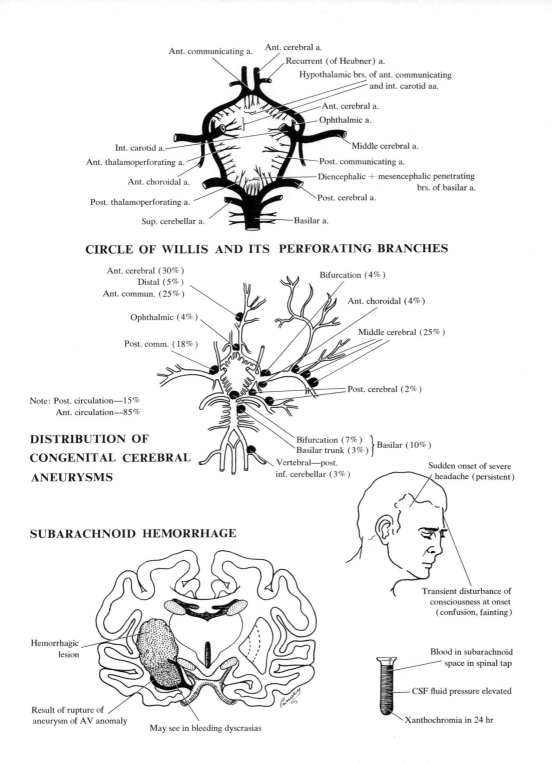

CIRCLE OF WILLIS AND ITS PERFORATING BRANCHES

Ant. communicating a.
Ant. cerebral a.
Recurrent (of Heubner) a.
Hypothalamic brs. of ant. communicating and int. carotid aa.
Ant. cerebral a.
Ophthalmic a.
Middle cerebral a.
Int. carotid a.
Ant. thalamoperforating a.
Ant. choroidal a.
Post. communicating a.
Diencephalic + mesencephalic penetrating brs. of basilar a.
Post. thalamoperforating a.
Post. cerebral a.
Sup. cerebellar a.
Basilar a.

DISTRIBUTION OF CONGENITAL CEREBRAL ANEURYSMS

Ant. cerebral (30%)
Distal (5%)
Ant. commun. (25%)
Bifurcation (4%)
Ant. choroidal (4%)
Ophthalmic (4%)
Middle cerebral (25%)
Post. comm. (18%)
Post. cerebral (2%)
Note: Post. circulation—15%
Ant. circulation—85%
Bifurcation (7%)
Basilar trunk (3%) } Basilar (10%)
Vertebral—post. inf. cerebellar (3%)

SUBARACHNOID HEMORRHAGE

Sudden onset of severe headache (persistent)

Transient disturbance of consciousness at onset (confusion, fainting)

Hemorrhagic lesion

Result of rupture of aneurysm of AV anomaly

May see in bleeding dyscrasias

Blood in subarachnoid space in spinal tap
CSF fluid pressure elevated
Xanthochromia in 24 hr

190. Cerebral Arteriosclerosis and Other Vessel Diseases

I. **Cerebral arteriosclerosis:** the cerebral arteries narrow with age, reducing the flow of blood through the brain

A. AREAS DEPRIVED OF OXYGEN lead to a slow death of nervous tissue with resultant shrinkage or atrophy of the brain

1. Early symptom is gradual loss of memory which at first is mainly for recent events with memory for early years remaining good
2. If the process is severe, the patient also becomes unsteady, tends to take short, shuffling steps, and have attacks of giddiness and sudden falls
3. Epileptic attacks may occur in the early morning hours
4. Patients complain of continuous headaches, become irritable, and may even complain of being persecuted or neglected and are hard to live with
5. May progress to complete loss of mental and physical faculties leading to arteriosclerotic dementia
6. There is no treatment to prevent or reverse the inevitable deterioration

II. **Oral contraceptives:** cerebral thrombosis is normally rare in young women, but recent high incidence has been related to women on "the pill"

A. THE CHANCE OF A THROMBOSIS is 3 to 6 times as great if one of the oral contraceptives contains high doses of estrogen

1. Low-dose estrogen pills are less likely to cause trouble, though not totally free from risk

III. **Other diseases of the blood vessels**

A. POLYARTERITIS NODOSA is a rare inflammatory or possibly allergic disease of the arteries, affecting all vessels including those of the brain, spinal cord, and peripheral nerves

1. Repeated vascular catastrophes occur in different parts of the nervous system
2. Small vessels supplying different peripheral nerves become blocked, and the nerves become paralyzed
3. One of the so-called collagen diseases (includes disseminated lupus erythematosus)
4. Blood sedimentation rate is high, blood proteins are abnormal, and diagnosis is confirmed by muscle biopsy

B. TEMPORAL ARTERITIS is an inflammation of the walls of the arteries of the scalp

1. Limited to elderly people
2. Vessels outside of the skull become so inflamed and so intensely painful and tender that the patient cannot bear to touch the head, wear a hat, or even rest head on a pillow
3. Condition subsides eventually, but unfortunately the ophthalmic artery is often affected resulting in blindness on both sides
4. Treatment: improvement is seen with the use of ACTH and cortisone, which are also used in polyarteritis.

C. SPINAL VASCULAR DISEASE: spinal intramedullary infarcts are rarer than central infarcts

1. Thrombosis or rupture of the spinal arteries is rare
2. If the above do occur, they are usually a complication of another disease
   a. Pressure on the arteries as a result of prolapsed disk or spinal tumor producing thrombosis; or anomalies of the vessels producing hemorrhage
3. A diagnosis of spontaneous spinal thrombosis is almost invariably incorrect

Cerebral arteriosclerosis.

Cerebral arteriosclerosis.

191. TRAUMA: HEAD INJURIES

I. **Introduction:** the skull and vertebral column offer such good protection for the CNS that, although the nervous tissue is delicate, a severe injury is needed for either the brain or spinal cord to be damaged. However, bone strength, thickness, and resilience vary in different people and are influenced by age and disease

II. **Head injuries:** in an open head injury, the brain or meninges are exposed; in a closed head injury, though the skull may be fractured, the coverings of the brain remain intact

A. FRACTURES OF THE SKULL may occur either in the vault or across the base. The fracture itself is of little importance. The danger lies in the damage to the underlying structures such as the meninges, blood vessels, and the brain. Thus, the position of the fracture is important. Examples
   1. A frontal fracture may pass through the frontal sinuses and breach the dura, creating a pathway for CSF to escape down the nose (CSF rhinorrhea). It also allows air or infection to enter skull with the danger of recurrent attacks of meningitis
   2. Base of the skull fracture may injure the petrous bone which contains the inner ear, allowing both blood and CSF to escape through the ears (CSF otorrhea). The inner ear is also subject to infection

B. TYPES OF FRACTURES
   1. Simple fissure fractures: cracks with no breaks in the skin
   2. Compound fractures: when the scalp is breached
   3. Comminuted fractures: when the bone is broken into several pieces
   4. Depressed fractures are most troublesome, and a piece of bone is driven inward

C. EXTRADURAL HEMATOMA: bleeding between the bone and dura which usually collects and enlarges fairly rapidly
   1. Injury to the frontoparietal region may rupture the middle meningeal artery
   2. History of injury
      a. Patient has head injury and is knocked unconscious or recovers consciousness rapidly. In 1 or 2 hours, there is an increased drowsiness, paralysis develops down one side, and there is dilatation of one pupil—all due to pressure of the expanding hematoma on the brain
   3. Treatment requires an emergency operation. If left alone, patient will die

D. SUBDURAL HEMATOMA: injury of the fine arteries and veins between arachnoid and dura
   1. Collects slowly so that signs and symptoms are like extradural bleeding, only they may be delayed for days or even weeks and come on more slowly
   2. Great variation from ·hour to hour in the degree of drowsiness. As clot irritates cortex, the patient may have fits and complain of increasing headache
   3. In the young, the skull is usually fractured; in the elderly, milder trauma may be the cause of this injury due to the presence of more brittle vessels
   4. Treatment: surgical evacuation of the clot

E. TRAUMATIC SUBARACHNOID BLEEDING: tearing of the vessels between the arachnoid and the pia resulting in bleeding into the subarachnoid space
   1. This is not uncommon in head injuries
   2. After the patient regains consciousness, he has severe headaches for many days with marked stiffness of the neck which gradually improves
   3. There is no further loss of consciousness and no hemiplegia
   4. The patient normally recovers spontaneously

Traumatic injury with subarachnoid space bleeding.

192. TRAUMA: CONCUSSION, CONTUSION, LACERATION, AND CONSCIOUSNESS

I. **Concussion, contusion, and laceration:** the obvious danger from a head injury is damage to the brain, e.g., an object piercing the skull to form a penetrating injury or a depressed fracture. Most head injuries are not penetrating wounds, but the whole brain can be severely "shaken up" without the skull actually being fractured

A. CONCUSSION: sudden acceleration of the skull by a blow or a sudden deceleration as a result of a fall on the head, resulting in a temporary loss of function of the brain cells and temporary loss of consciousness
   1. Need not result in permanent damage to any part of the brain
   2. Patient is stunned or dazed or actually unconscious for some minutes or even hours, and may be confused for several hours after injury, but usually recovers with no abnormal signs

B. CONTUSION: the injury is more severe and the brain tissue is bruised and many cells may die
   1. Depending on the area damaged, the signs will vary from mild transient weakness of a limb to prolonged unconsciousness and even paralysis

C. LACERATION: brain tissue is actually torn, as in the case of a compound depressed fracture
   1. There is usually bleeding into and around the tear
   2. Results in more severe prolonged unconsciousness and paralysis coming on at once and leaving permanent scarring and some degree of permanent disability

II. **Consciousness and the brainstem:** in addition to cranial nerves and fibers in the brainstem, there is a network of cells and fibers called the reticular formation. Impulses from it spread to the cerebral cortex and keep the brain in a state of activity and in a state of wakefulness. Damage to the RF cuts these impulses and results in a loss of consciousness

A. MANY CONDITIONS, SUCH AS HEMORRHAGE AND TUMORS, can cause this damage, as can head injuries
   1. Contusion of the brainstem causes the profound loss of consciousness often seen in cases where there is no sign of severe cerebral damage and patients may lie in a coma for weeks or months or even years
   2. In lesser degree, consciousness will return after a few days or weeks and recovery may be complete

B. FRACTURES IN THE REGION OF THE BRAINSTEM are more likely to be associated with injury to it because the brainstem is so closely applied to the base of the skull

III. **Levels of consciousness:** in head injuries, steadily increasing lowering of consciousness is a serious sign, and failure to regain consciousness that was lost immediately at the time of an accident need not have the same importance, for though it means cerebral concussion or contusion, it does not necessarily mean increasing bleeding inside the skull. Each of the following states or levels of consciousness merges into the other when unconsciousness is deepening or lightening

A. DROWSINESS resembles normal sleep, and the patient can easily be awakened and cooperates fully

B. STUPOR: state of unconsciousness in which stimulation produces only restlessness, noisiness, and irritability but no true cooperation

C. COMA: deepest degree when no response occurs even to the most painful stimulus

Sections of brain showing extent of hemorrhage in cerebrum and brainstem after traumatic injury.
Note petechial hemorrhages (arrows).

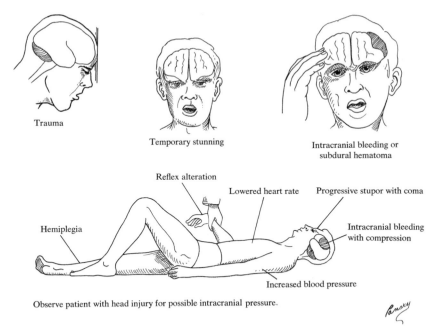

Trauma

Temporary stunning

Intracranial bleeding or
subdural hematoma

Reflex alteration

Lowered heart rate

Progressive stupor with coma

Hemiplegia

Intracranial bleeding
with compression

Increased blood pressure

Observe patient with head injury for possible intracranial pressure.

193. TRAUMA: POSTTRAUMATIC EVENTS

I. **Tentorial herniation and the dilated pupil:** the tentorium separates the cerebral hemispheres from much of the brainstem and cerebellum with the upper brainstem passing through the tentorial hiatus or notch
A. PRESSURE ABOVE THE TENTORIUM as a result of "something" expanding inside the brain or pressing on it from the outside forces the nearest part of the cerebral hemispheres, namely, the uncus of the temporal lobe, through the hiatus since it represents the only "escape" valve
   1. Herniation through the hiatus alongside the brainstem squeezes it and damages its blood supply resulting in hemorrhage or softening in the midbrain and pons which affects the reticular formation and causes rapid deepening coma
   2. Simultaneously, the IIIrd cranial nerve, which passes near the opening, is displaced and stretched by the tentorial herniation. Its autonomic fibers are damaged, and the pupil dilates first on the side of the herniation and then on the other side as well
B. INCREASING COMA AND DILATING PUPILS are also seen with late stages of cerebral tumors and abscesses as well as with extradural and subdural hematomas
C. SURGICAL TREATMENT is essential before both pupils dilate and cease to react to light so the conscious level and pupil size must be watched in all injuries

II. **The aftereffects of head injury:** most patients recover from a head injury without serious ill effects. A few severe injuries damage the brainstem so that consciousness never returns. Others recover consciousness, but damage to the brain is so great that they remain mentally abnormal and permanently unable to look after themselves. The latter are in the minority, but a larger number suffer from symptoms which, although not incapacitating, may profoundly change their way of life
A. POSTTRAUMATIC EPILEPSY: laceration of the brain may result in a scar which later acts as an irritating focus capable of producing epileptic fits
   1. Convulsions may occur shortly post-head injury (in 24 hours), especially in children, as a result of direct brain contusion or hemorrhage around it
   2. Some patients, on first getting out of bed after a head injury, have a tendency to lose consciousness easily. Convalescence clears up these simple "faints"
   3. More severe head injuries tend to develop true epileptic convulsions which usually occur in the first 2 years afer injury and need the same treatment as ordinary epilepsy
   4. Epilepsy is most common after a penetrating wound through the skull, dura, and brain (50% of cases) and is rare following an injury which has produced only a short loss of consciousness and no fracture or intracranial bleeding
B. POSTTRAUMATIC VERTIGO: a feeling as if the body or outside world were spinning around. Often accompanied by vomiting and unsteadiness, but no loss of consciousness
   1. Some degree of vertigo on change of position is common after head injury and may last for a few days or weeks only or persist in diminishing degree for many months, but finally clears up completely
C. POSTCONCUSSIONAL SYNDROME (now called accident neurosis) rarely follows severe injury and is common after minor injury
   1. Patient may complain of constant headaches, impairment of memory, increased irritability, intolerance to noise, inability to concentrate on reading or television, insomnia, and inability to work
   2. It is not due to brain damage and usually clears up when compensation problems are settled. Its exact nature is a matter of controversy

Section taken just above the floor of frontal (anterior) cranial fossa.

Higher section showing diminishing hematoma.

Progressively higher sections showing the deformed ventricular system with the anterior horns of the lateral ventricles pushed straight backward.

BILATERAL INTRACRANIAL FRONTAL LOBE HEMATOMA SHOWN WITH COMPUTERIZED TOMOGRAPHY

194. TRAUMA: SPINAL AND PERIPHERAL INJURY

I. Spinal trauma: fracture of the bones around the spinal cord is not sufficient for the cord to be damaged, but they must be displaced as well

A. THE CORD may be crushed, overstretched, or even completely divided

    1. The result is that below the level of the lesion, all neurons carrying motor impulses are cut off from the muscles they innervate, and all neurons carrying sensation from parts below the lesion are unable to transmit to the brain

    2. A severe cord lesion will cause complete paralysis and complete loss of sensation below a clear-cut lesion

        a. A neck injury produces paralysis of the arms, trunk, and legs

        b. A thoracic injury paralyzes the lower part of the trunk and legs

        c. Injury to the lumbar spine may probably miss the spinal cord, which ends at just below L1, but will involve the cauda equina and produce paralysis of and loss of sensation in the legs and the urinary bladder

    3. Immediately after spinal injury at any level, there is usually a flaccid paralysis of the limbs and complete retention of urine as a result of spinal shock which usually passes off within 3 weeks

        a. Flaccidity then changes to spasticity if the injury is to the cord itself, as a result of damage to the upper motor neurons

        b. If the lesion is in the cauda equina, the legs remain flaccid because in this case, it is the lower motor neurons that are destroyed

        c. In any case, there is marked loss of skin sensation, the upper level corresponding to the level of the damage

    4. In partial lesions of the cord, the paralysis and sensory loss show a peculiar distribution called the "Brown-Sequard syndrome" (described under clinical signs of damage to the nervous system in another unit)

II. Peripheral nerve injuries: as most peripheral nerves carry both motor and sensory fibers, injury results in weakness of the muscles they supply and loss of all forms of sensation over the particular area of the skin. Vulnerable nerves are

A. ULNAR NERVE INJURY: the nerve lies very near the surface at the elbow and is easily damaged, resulting in a wasting of the hand muscles and loss of sensation over the little finger

B. MEDIAN NERVE INJURY is usually involved as it passes through a narrow tunnel at the wrist, resulting in wasting of the thumb muscles and loss of sensation over the thumb and first finger. This has been called carpal tunnel syndrome

C. SCIATIC NERVE may be injured in the buttock by badly placed injections which results in paralysis of the foot and loss of sensation over the front and back of the lower leg and foot

D. LATERAL POPLITEAL NERVE may be injured just below the knees by the patient's wearing high boots or by skin traction as a result of disk lesions or by allowing the patient to lie on the nerve

    1. Injury results in paralysis of dorsiflexion of the foot and loss of sensation between the big toe and 2nd toe

E. OTHER PERIPHERAL NERVE INJURIES are discussed in detail elsewhere

III. Summary: in all traumatic lesions, more nervous tissue is "knocked out" of action immediately after the injury than is actually destroyed, and there is a remarkable degree of recovery possible. It is not unusual for many weeks, months, or even years to pass before one can even venture to confidently state that no further recovery is possible. However, by the end of the 2nd year, the maximum degree of improvement has usually been reached

GUIDE TO CORD LEVEL DAMAGE

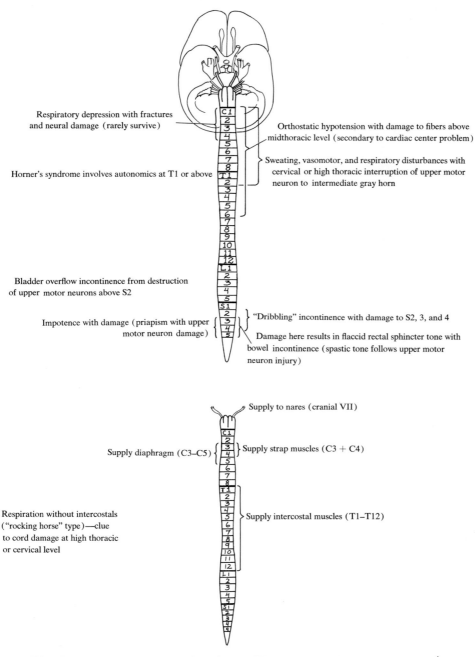

Respiratory depression with fractures and neural damage (rarely survive)

Orthostatic hypotension with damage to fibers above midthoracic level (secondary to cardiac center problem)

Sweating, vasomotor, and respiratory disturbances with cervical or high thoracic interruption of upper motor neuron to intermediate gray horn

Horner's syndrome involves autonomics at T1 or above

Bladder overflow incontinence from destruction of upper motor neurons above S2

Impotence with damage (priapism with upper motor neuron damage)

"Dribbling" incontinence with damage to S2, 3, and 4

Damage here results in flaccid rectal sphincter tone with bowel incontinence (spastic tone follows upper motor neuron injury)

Supply to nares (cranial VII)

Supply diaphragm (C3–C5)

Supply strap muscles (C3 + C4)

Respiration without intercostals ("rocking horse" type)—clue to cord damage at high thoracic or cervical level

Supply intercostal muscles (T1–T12)

RESPIRATION AS CLUE TO CORD LEVEL DAMAGE

CONGENITAL MALFORMATIONS
AND PERINATAL PATHOLOGY

195. PATHOLOGY OF SPINAL CORD AND BRAIN

I. **Introduction:** the cause of abnormal development is not always clear. If the developing fetus is damaged, there is usually some abnormality at birth, but this abnormality depends on the site of the damage and the stage of development at which it takes place. Malformations of the nervous system, like the rest of the body, may be due to 3 groups of factors

A. EXOGENOUS: such as infections, especially rubella and toxoplasmosis; radiation (x-rays, atomic exposure); and chemical agents (especially various drugs)

B. GENETIC AND CHROMOSOMAL FACTORS: in particular, trisomy 21 or mongolism (Down's syndrome) and trisomy 13–15

C. AN INTERACTION OF THE ABOVE TWO: the relatively late completion of brain development in fetal life accounts for the protracted effects of some of these teratogenic factors

II. **Disturbances as a result of defective closure of the neural groove (dysraphic state):** these constitute the most frequent malformations of the nervous system. If in the course of neural tube formation there is defective closure of the neural groove, various nervous system malformations will appear, depending on the localization and extent of the closure defect. These malformations to some extent will be accompanied by developmental defects affecting the subjacent tissue planes (meninges, posterior arch of the vertebrae, dermis) whose development is normally subject to inductive influences from the neural tube

A. SPINAL CORD: malformations are most often localized to the lumbar and lumbosacral regions and are listed under the generic term of spina bifida (during development of the spine, the neural arch which is formed by the two laminae and the spinous processes, fail to join in the center). (See also developmental anatomy unit)

1. Spina bifida aperta (myeloaraphia) is a major form of dysraphia
   a. The open neural groove is immediately exposed to the surface, with its borders in continuity with the adjacent epiblast
   b. Because of the nonclosure of the neural groove, no leptomeninges are formed, the vertebral arch does not close, and the related skin does not develop

2. Myelomeningocele is a lesser form of dysraphia
   a. The neural tube has undergone closure, while the posterior arch of the vertebrae has not, with the result that the meninges protrude directly under the skin
   b. There is also faulty development of the dermis
   c. Various nervous tissue elements, i.e., spinal cord parenchyma and nerve roots, are present in the sac thus formed, which is filled with CSF
   d. Fairly often, such myelomeningoceles are associated with hydrocephalus due to malformation involving the medulla and the cerebellar tonsils which impedes the passage of CSF at the level of the 4th ventricle resulting in the Arnold-Chiari deformity
      i. With compression of the upper cervical cord one sees ataxia and weakness of the arms and legs
      ii. Surgical decompression may relieve a progressive disability

Presence of cysts in the brain cortex known as porencephaly.

Arnold-Chiari syndrome (cerebellum and medulla oblongata protruded down into spinal cord).

196. PATHOLOGY OF SPINAL CORD AND BRAIN

II. Disturbances as a result of defective closure of the neural groove (dysraphic state):

A. SPINAL CORD (cont.):

   3. Meningocele is essentially similar to myelomeningocele but the sac does not contain any nervous tissue

     a. Can occur in any part of the spine but is most common at or just below the occipital region and in the lumbar region

     b. Treatment: temptation to excise the sacs and close the skin over the defect is great and may be done to avoid rupture and meningeal infection, but hydrocephalus develops sometimes after this procedure

       i. If the myelocele is excised, there may be permanent paralysis of the legs, but this is often accepted and the child is trained at a paraplegic center from early age

       ii. Some claim good results simply from closing the defect and correcting the deformities of the limbs

       iii. The problem is in deciding whether to allow children with some of these defects to deteriorate rapidly or whether to undertake a procedure which allows the infant to survive but leaves him mentally defective or paralyzed

   4. Congenital dermal sinus: the meninges are linked to the epidermal tissues by a narrow aperture across the incompletely closed posterior vertebral arch

   5. Spina bifida occulta is a highly frequent malformation

     a. It does not give rise to any clinical manifestations and is usually discovered incidentally on radiologic examination of the spine (occulta meaning hidden)

     b. Consists essentially in a closure defect of the posterior arch of the vertebrae, sometimes accompanied by discrete abnormalities in the overlying skin, i.e., tufts of hair

     c. It affects one or several spines

     d. Occasionally there is a failure of development of the nervous tissue underneath with the result that there are deformities of the feet (pes cavus) or difficulties in the control of urination

     e. Some patients with the above condition have a nevus on the overlying skin and even tufts of hair growing on it. There may be a tract leading from this into the theca and ending in a lipoma or dermoid. The latter may compress the cord or cauda equina and act like a spinal tumor or allow infection to enter and recurrent meningitis to occur. Tracts can be surgically removed

B. BRAIN (also see developmental anatomy unit)

   1. Anencephaly is due to failure on the part of the neural tube to close at the level of the encephalon

     a. There is a lack of development of the meninges, of the cranial vault, and of the skin

     b. The open nervous system is exposed to the air

     c. This malformation is fatal within a few hours after birth

   2. Encephalocele is a condition where part of the brain protrudes under the skin through a bony defect in the vault of the skull

   3. Meningocele consists of a meningeal sac filled with CSF and protruding under the skin through a bony defect in the vault of the skull

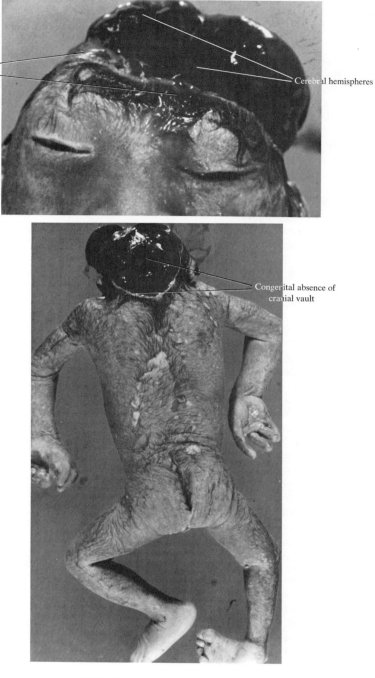

Scalp margin

Cerebral hemispheres

Anterior view
of head

Congenital absence of
cranial vault

Posterior view of
body and head

ANENCEPHALY

197. AGENESES, DYSGENESES, AND CORTICAL ANOMALIES

I. Ageneses and dysgeneses

A. MICROENCEPHALY describes various destructive processes which occur before birth and result in a decrease in volume of the brain associated with other anomalies
   1. True microencephaly (small brain) is characterized by a decrease in size of the cranial perimeter and is used only when referring to a brain of small size (less than 900 grams) in the adult, with no degenerative lesions

B. MEGALENCEPHALY may be seen in normal subjects as often as in mentally deficient subjects
   1. The increase in volume of a brain weighing more than 1800 grams may be diffuse or localized
   2. The brain sometimes may be structurally normal, but most often there is either a diffuse glial proliferation or a variable pathologic process

C. CYCLOCEPHALY: rare malformation affecting the tissues derived from the prosencephalon
   1. The telencephalon is not divided, and there is a single central ventricle which communicates with the third ventricle
   2. There is a single median eye (cyclopia), absence of the pituitary, and aplasia of the nasal structures
   3. Due to an induction defect of the prochordal plate as a result of a chromosomal abnormality

D. PORENCEPHALY consists of cerebral cavitation due to localized agenesis of the cortical mantle, leading to the formation of a cavity or a lateral slit through which the lateral ventricle communicates with the convexity

E. HYDRANENCEPHALY is generally the result of destructive processes
   1. It is similar to porencephaly and presents as a large intracerebral cavitation which may at first mimic ventricular dilatation but, in contrast to hydrocephalus, the cavity is not lined by ependyma but by a glial border, and the thin external wall is formed by leptomeninges and by a thin layer of superficial cortex
   2. Generally involves the anterior part of the cerebral hemispheres

F. AGENESIS AND COMMISSURAL MALFORMATIONS include septal formations (septal agenesis and cysts) and agenesis of the corpus callosum

G. ARHINENCEPHALY may involve only the olfactory bulbs and is associated with other malformations, especially of the cerebellum
   1. Is typical of trisomy 13–15
   2. Indicates a severe disturbance in development of the fore-end of the telencephalon

H. CEREBRAL AND CEREBELLAR AGENESES are rare

II. Cortical anomalies are related to a disturbance in the maturation of the germinative layer around the 2nd month of fetal life

A. AGYRIA (lissencephaly) is the normal state up to 7 months of fetal life but remains with the absence of fissures and convolutions. May be total or involve only a part of the hemisphere

B. PACHYGYRIA: the convolutions are abnormally wide and thick, and the cortex has a four-layer type of lamination

C. HETEROTOPIAS: seen as isolated phenomena in the shape of masses of gray matter in the centrum ovale near the caudate nucleus, in the cerebellar white matter, in the brainstem, and in the inferior olives. Associated with other malformations

D. MICROGYRIA AND POLYMICROGYRIA are characterized by the grouping of small convolutions that are reduced in size, are irregular, malformed, and with a verrucous appearance. Show large numbers of sulci, tend to involve cerebral and cerebellar cortex, and are frequently bilateral and symmetrical

MICROCEPHALUS

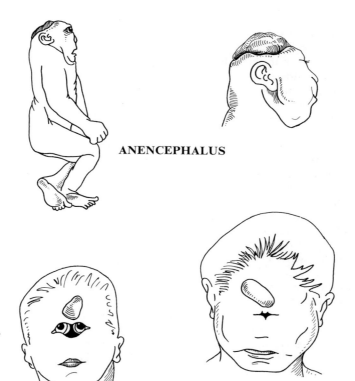

ANENCEPHALUS

Fused globe
Double retina, lens,
iris, and pupil

CYCLOCEPHALUS AT TERM

Pansky

198. ARACHNOID CYSTS AND HYDROCEPHALUS

I. **Arachnoid cyst** refers to the formation of intracranial extraparenchymatous cysts which are lined internally by arachnoid tissue and contain clear fluid. They are most frequently seen in children and are sometimes associated with underlying cerebral abnormalities

II. **Hydrocephalus,** strictly speaking, is an increase in cerebral mass due to the excessive amounts of CSF, but in practice, it corresponds to a variable degree of dilatation of the ventricular system. It is often spoken of as "water on the brain" since it is due to the presence of too much CSF at too high a pressure resulting, in children, in the enlargement of the head since the bones are free to separate at the sutures. In the adult, this cannot happen since the sutures are fixed, thus we would have raised intracranial pressure instead.

A. TYPES: in either, the ventricles eventually enlarge
1. Obstructive noncommunicating: if the CSF, after being formed in the lateral ventricles cannot pass through the ventricular system due to an obstruction in the 3rd ventricle, aqueduct of Sylvius, or 4th ventricle, the lateral ventricles enlarge and "blow" up the brain like a balloon, stretching the cortex. The pressure is highest in the ventricles
2. Communicating: if CSF gets out of the ventricular system, but is blocked in the subarachnoid space and cannot circulate to reach the surface of the brain where it is absorbed, the pressure builds up outside the brain
   a. Is the result of adhesions and thickening in the subarachnoid spaces due to past inflammation, such as infantile meningitis
   b. May also be a result of hypersecretion of CSF from a choroid plexus papilloma
B. HYDROCEPHALUS may be present at birth, resulting in mechanical difficulties, or it may develop later
C. APPEARANCE: the head is large, the fontanelles remain wide open, the sutures can be felt separated, the forehead is bossed, and the eyes may be pushed downward
D. MENTAL RETARDATION is frequent, but high intelligence is sometimes possible
E. THE CONDITION MAY BE ARRESTED if a balance develops between the formation of CSF and the defect in its absorption, and the pressure does not go on increasing in which case, normal intelligence is possible. Thus one attempts to treat the condition before irreparable damage is done by stretching and thinning of the brain
F. TREATMENT
1. The site of the block of CSF must first be discovered (use dyes, air, or opaque fluid)
2. Treatment consists of bypassing the block and directing the fluid from the lateral ventricle out of the skull, under the skin, to the pleural or peritoneal cavities or into the spinal CSF pathways (*note:* this tube may become blocked or too short as the child grows). One may even make an opening in the walls of the 3rd ventricle to allow the fluid to drain directly into the subarachnoid space (these holes frequently close by themselves)
3. A tube may be led from the lateral ventricles, through a burr-hole under the scalp, and into the cisterna magna (Torkildsen's procedure)
4. Use of the Spitz-Holter valve: a tube is led from the ventricle out through the skull and under the skin to empty the CSF into the jugular vein in the neck. The tube has a valve with a small pumping device to allow for intermittent flow from the ventricles into the bloodstream (infection is not uncommon)
5. Note that the above treatments are of value in hydrocephalus that is increasing. An arrested condition should be left alone since these children may develop normally

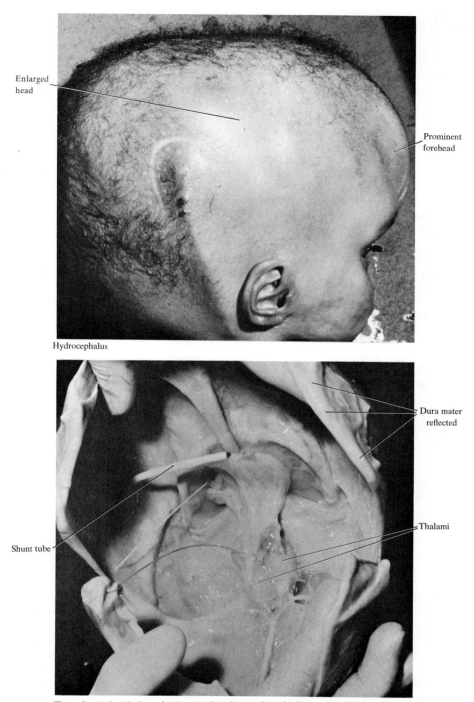

Enlarged
head

Prominent
forehead

Hydrocephalus

Dura mater
reflected

Thalami

Shunt tube

Floor of cranial vault above showing complete absence of cerebral hemispheres (hydranencephaly).

199. SYRINGOMYELIA

I. Syringomyelia is the result of a cavitation of the spinal cord (syrinx), which extends over several metameric segments, most often involving the cervicothoracic segments. If it forms in the lower brainstem it is called syringobulbia

A. DESCRIPTION
1. The spinal cord is widened and swollen at its level, in the living, but at autopsy, the cord appears to be flattened
2. The syringomyelic cavity, most often single, can be multiple, occupies the center of the cord, and is situated behind the ependyma, involving in the midline, the crossing fibers of the ascending pain and temperature fiber tracts
3. It may involve the gray matter and fuse laterally with the entering sensory nerve roots or affect to a variable extent the anterior horns, resulting in a lower motor neuron paralysis with amyotrophy
4. It may extend transversely into the white matter of the lateral and posterior columns, as well as caudally, but seldom involves the lumbar segments
5. It is limited by highly fibrillary glial tissue which often includes blood vessels with hyalinized walls

B. THEORIES OF DEVELOPMENT
1. The dysraphic theory states that there is a closure defect of the neural tube at the level of the posterior raphe
2. The hydrodynamic theory states it is due to a disturbance in the outflow of CSF from the 4th ventricle as a result of failure of part of the foramina of Luschka and Magendie to open, resulting in hydrocephalus accompanied by herniation of the medulla and of the cerebellar tonsils (Chiari malformation). This is later followed, as a result of increased CSF pressure, by dilatation of the ependymal canal (hydromyelia). A syringomyelic cavity then might develop as a result of a rupture in the wall of the dilated ependymal canal

C. SYMPTOMS
1. The patient is usually between 20 and 45 years of age
2. Patient first notices either that his hands are becoming thinner or that he is burning or cutting his fingers without noticing the heat or pain
3. The hands and arms become wasted, cold, and blue
4. Wasting and sensory loss may extend to the rest of the body, including tongue
5. The legs may become stiff as the pyramidal tracts are later involved

D. SIGNS
1. Wasting is usually present in the hand muscles, the forearms, and shoulder girdles, and the hands show scars from old burns and cuts
2. There is a loss of sensation to pain and temperature over the arms, shoulders, upper chest, and often the occiput, but above and below these, the sensations are normal
3. Despite sensory loss, light touch can still be felt in the affected areas, resulting in what is called dissociated sensory loss
4. Reflexes are absent in the arms, but if the cavity enlarges and the pyramidal tract is involved, the legs will show increased tone, increased reflexes, and extensor plantars

E. TREATMENT
1. No drug therapy is available
2. After careful myelography, surgical decompression of the foramen magnum and upper cervical region, with or without incision of a large syrinx, may halt what is an inevitable progressing disease

SYRINGOMYELIA

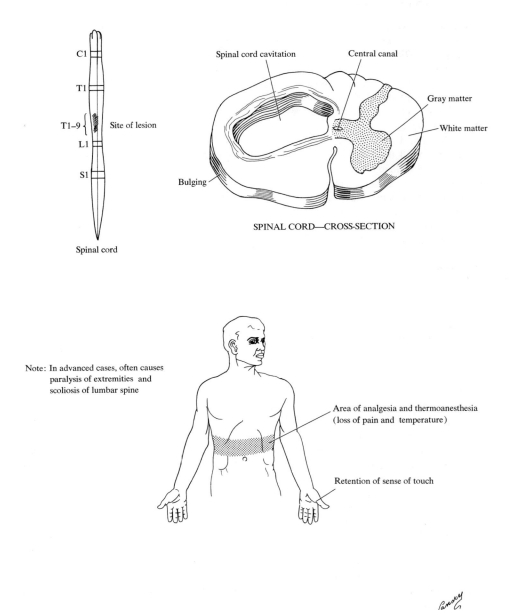

C1

T1

T1–9 { Site of lesion

L1

S1

Spinal cord

Spinal cord cavitation   Central canal

Gray matter

White matter

Bulging

SPINAL CORD—CROSS-SECTION

Note: In advanced cases, often causes
paralysis of extremities and
scoliosis of lumbar spine

Area of analgesia and thermoanesthesia
(loss of pain and temperature)

Retention of sense of touch

200. CRANIOSTENOSIS, BLASTOMATOUS DYSPLASIAS, AND PHAKOMATOSES

I. **Craniostenosis:** the skull sutures fuse prematurely, the fontanelles close very early, and the skull becomes a rigid box too soon, since the brain is still growing rapidly
   A. DESCRIPTION
      1. The brain becomes compressed, pressure rises and is transmitted to the orbits, resulting in the eyes being pushed forward, even to a gross degree
      2. The skull develops a pointed shape (turricephaly or oxycephaly), and the fused sutures can be felt as hard ridges
   B. SYMPTOMS
      1. The child may be mentally deficient and epileptic
      2. The child may have other abnormalities (congenital) such as webbed fingers and bronchiectasis
   C. TREATMENT
      1. Attempts have been made to split the sutures or break up the skull bones (morcellation) to allow the brain to grow—not too successful
      2. Surgeons have tried placing tantalum foil between the sutures to prevent fusion, but success is limited and variable

II. **Blastomatous dysplasias and phakomatoses:** in the context of malformations and neuroectodermal dysplastic processes, there are a group of diseases that have in common an association of distinctive malformations of the neuraxis with small tumors (phakomas or lentil-like neoplasia) which involve the skin, the nervous system, or the eyes
   A. VON RECKLINGHAUSEN'S NEUROFIBROMATOSIS is a common familial disorder showing
      1. Cutaneous lesions: café-au-lait spots and pigmentary nevi, pendulated tumors, and nodular subcutaneous neurofibromas, and so-called "royal tumors" which may become very extensive
      2. Neural tumors most frequently involve nerve roots and may present as either neurofibromas or as schwannomas (or neurilemomas)
         a. They may involve the cranial nerves, particularly the VIIIth, giving rise to bilateral acoustic schwannomas
         b. Multiple meningiomas and a form of meningiomatosis are also characteristic
      3. Nervous system malformations consist of heterotopias, cortical dysplasias, stenosis of the aqueduct of Sylvius, or syringomyelia
         a. In association with bony malformations of the spine, they indicate the very dysplastic nature of this process
   B. TUBEROUS SCLEROSIS (BOURNEVILLE'S DISEASE) is less common and shows an association of the following
      1. Skin lesions, including adenoma sebaceum, involving the skin of the nasolabial fold, angiofibromas, and cutaneous and periungual fibromas
      2. Cortical cerebral malformations: circumscribed nodules with voluminous cells, cerebral tumors of variable size, and visceral tumors (skeletal and cardiac muscle, endocrine, and pulmonary)
   C. VON HIPPEL-LINDAU'S DISEASE (retinocerebellar angiomatosis) is familial and characterized by the presence of multiple hemangioblastomas, localized most often in the retina and cerebellum
   D. ENCEPHALOTRIGEMINAL ANGIOMATOSIS OR STURGE-WEBER'S DISEASE: a flat extensive unilateral cutaneous angioma of the face and a leptomeningeal encephalic angiomatosis, most often in the parieto-occipital region

SCAPHOCEPHALY $\left\{\begin{array}{l}\text{craniosynostosis}\\\text{(craniostenosis)}\end{array}\right.$

Premature closure of sagittal suture

Note: wedge-shaped skull

Premature closure of coronal suture

Oxycephaly or acrocephaly
"Tower-like skull"

Note: skull asymmetry

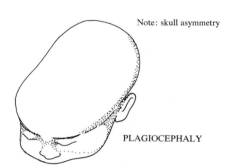

PLAGIOCEPHALY

Premature closure of coronal and lambdoid sutures

NEUROFIBROMATOSIS

Fibrotic tumors of various sizes: von Recklinghausen's disease

201. PERINATAL AND POSTNATAL PATHOLOGY

I. Perinatal pathology

A. INTRODUCTION: at birth and during the 1st months of life, various etiologic factors may result in definite cerebral lesions and account for various neurologic syndromes. The most common are
   1. Infantile cerebral hemiplegia
   2. Infantile cerebral diplegia or Little's disease
   3. Some forms of congenital epilepsy

B. THE LESIONS ARE TYPICAL for the conditions, but their supervention on the immature brain gives rise to a highly distinct pathologic picture. These lesions are obviously more frequent and more severe in premature infants. One separates the cerebral damage related to the birth process itself (neonatal pathology) from those that are secondary to disturbances in the 1st months of life (postnatal), even though the sequelae are often identical and the lesions of the same type
   1. Neonatal pathology: the pathogenic mechanism consists of
      a. Traumatic mechanism: in which obstetric injury results in cerebral lesions that are direct or indirect, the effects being usually hemorrhagic as a consequence of blood vessel rupture (often venous)
      b. Circulatory mechanism: either venous and related to increase of intracerebral venous pressure at the time of birth, or arterial and secondary to a fall of blood flow resulting from circulatory arrest
      c. Anoxic and asphyxial mechanisms
   2. The lesions may be acute, leading to death, or cicatrical, resulting in a picture of secondary encephalopathy of neonatal origin
      a. Acute lesions: essentially hemorrhagic, such as subdural hematomas, subarachnoid hemorrhage, and intracerebral hemorrhage
      b. Secondary lesions: responsible for the group of infantile encephalopathies, the sequel of hemorrhage or necrotic lesions
         i. Extensive lobar or hemispheric atrophies suggest vascular occlusion
         ii. Cortical changes: in which ulegyria is the most common appearance, consisting of cortical atrophy involving deeper parts of the cortex of the cerebral gyrus with relative sparing of its convexity. Also involves the white matter and basal ganglia

II. Postnatal pathology: the cerebral lesions may be secondary to numerous etiologic and pathologic mechanisms

A. EARLY VASCULAR ENCEPHALOPATHIES: some are related to venous lesions, such as thrombosis of the vein of Galen or of the superior longitudinal sinus. Are often secondary to early meningeal infections
   1. Others are related to arterial occlusions whose emboli or thrombic origin is hard to prove. May produce a picture of lobar or hemispheric atrophy

B. POSTEPILEPTIC ENCEPHALOPATHY is characterized by gliosis with atrophy of Ammon's horn, secondary episodes of anoxia due to repeated epileptic fits, or status epilepticus
   1. Believed by some as being of circulatory origin, due to ischemia of the territory of the posterior cerebral artery at the time of birth and as being the cause of the epileptic fits

C. NUCLEAR JAUNDICE: result of hemolytic neonatal anemia due to Rh incompatibility
   1. The lesions are characterized by their topographic distribution, involving the globus pallidus, the subthalamic nuclei, and to a lesser extent, Ammon's horns, the dentate nuclei, and the inferior olives
   2. Their yellow appearance is due to bilirubin

Spontaneous depression

Depressed parietal fracture in newborn

Passage through pelvis

Parietal cephalohematoma

Parietal subperiosteal hemorrhage

Periosteum

Bilateral subdural hematoma (seen in 1st yr of life)

Uncommon in older children

HEAD INJURY SEQUELAE

Scalp: Contusion, hematoma; open wound

Skull: Linear fracture, diastatic fracture, other simple fracture, CSF otorrhea
Depressed fracture, growing fracture, compound fracture, pneumocephalus, fistula

Meninges: Subarachnoid hemorrhage
Epidural hematoma, acute and chronic subdural hematoma, subdural hygroma

Cranial nn.: Paralysis I–XII

Brain: Concussion, edema, contusion, laceration, hematoma, penetrating wound

Unit Fifteen

DEMYELINATING DISEASES

202. DEMYELINATING DISEASES (OF WHITE MATTER) AND MULTIPLE SCLEROSIS

I. **Introduction:** myelin forms the sheaths that cover the majority of the nerve fibers, whether they lie in the brain and cord or run in the cranial and spinal nerves. When myelinated fibers are collected together, they appear white and form the white matter of the brain and cord. Any damage to nervous tissue can cause the myelin sheaths to degenerate. Nevertheless, there are certain diseases in which this occurs spontaneously without a known cause. Loss of myelin is called demyelination. Without its myelin, the nerve fiber does not conduct impulses properly, and there is a loss of function in that part being supplied. Demyelination is one feature of massive white matter lesions such as cerebral infarcts, hemorrhages, and cerebral tumors or it may be secondary to degeneration.

If only the myelin sheath is damaged (often the axis cylinder is spared), it can re-form so that function can be restored to the nerve fiber. If the nerve fiber is also destroyed, recovery is less likely. Nerve fibers can grow again, as is often the case in peripheral nerve injuries, but often scar tissue forms and prevents proper growth and recovery. Scar tissue formation in the CNS is formed by the glial cells and is called gliosis. In a healing process, the scar tissue gradually shrinks, but here, too, it may involve nearby healthy tissue resulting in a slowly progressive disability.

Diseases of the white matter result from pathologic processes dissimilar to each other. Some by virtue of their genetic character belong to the leukodystrophies and show metabolic changes in the myelin sheaths while others, like multiple sclerosis, have inflammatory features

II. **Classification of demyelinating disease:** 2 types are distinguished in etiologic and histopathologic views
A. ETIOLOGIC
   1. Multiple sclerosis and Schilder's disease are the result of myeloclastic processes. The etiology is unknown, but an immunopathologic process is suspected
   2. The leukodystrophies are familial and are enzymatic disorders of genetic origin (see Genetic and Familial Diseases)
B. HISTOPATHOLOGIC: are distinguished according to whether or not they show inflammatory cellular lesions or not (these may, however, be seen only in the early stages of the disease and disappear in the course of the process)

III. **Disseminated or multiple sclerosis:** the most common form of demyelinating disease
A. SCLEROSIS means "hardening." At autopsy, hard areas of gliosis are found scattered or disseminated throughout the CNS
   1. The cause of the disease remains unknown (probably immunologic or viral)
   2. Familial incidence of the disease is not uncommon
   3. Starts in young adults, occasionally in adolescence or young children, infrequent in middle life, and rarely begins over the age of 50
   4. Women are slightly more often affected than men; country dwellers more than town dwellers
   5. It is a disease of temperate climates (exceedingly rare in the tropics)
   6. Patches of demyelination, called plaques, develop in some part of the cord, cerebellum, brainstem, cerebral hemispheres, or optic pathways
      a. The patches flare up, take weeks or months to settle down, and after an interval of months or years, flare up again in another site. After several attacks, the healing process of gliosis upsets all attempts at remyelination and permanent sclerosis and function loss result
   7. The peripheral nerves are not affected
   8. Inflammatory cellular lesions are seen in recent lesions but are absent in old ones
   9. Coexistence of lesions of different ages are seen at various levels of the neuraxis

Sections through the brain showing patches of demyelination.

An enlargement of Fig. *B* in the top photograph to show demyelination throughout white matter of the cerebrum.

MULTIPLE SCLEROSIS

203. MULTIPLE SCLEROSIS: SYMPTOMS, SIGNS, AND TREATMENT

I. Disseminated or multiple sclerosis

A. SYMPTOMS: the course of the disease is characterized by a series of relapses and remissions

1. The symptoms of each relapse depend on the site where the plaques of demyelination take place; e.g., blurring or complete loss of vision in one eye; paralysis of eye movement and double vision; numbness, weakness, and clumsiness of one or both arms or legs; slurring of speech; unsteadiness of the hands or legs; and urgency and increased frequency

2. The symptom may occur by itself or together with any or all of the others

3. The symptoms may last several weeks or even months (never several hours) and then clear up, and the patient may be for months or years in remission. There is then another episode or relapse which may involve the same or entirely different parts of the nervous system

4. Recovery during a remission may not be so complete and, following the next relapse, the patient may make only partial recovery

5. The progressive disability picture is not necessarily always true
   a. The disease may consist of mild relapses and long remissions throughout life with very little permanent effects
   b. Others have a series of relapses with some residual disability and further deterioration stops
   c. There are, of course, cases in which relapses worsen the condition until the patient becomes bedridden and incontinent in early middle life; and others are so severely affected from the very beginning that they progress very rapidly and become grossly disabled in a year. But even the latter may remain in this state for many years
   d. It is not in itself a killing disease, and the prognosis in many cases is much better than is usually believed. Thus, it has been called "a benign relapsing variety" of disease in many cases

B. SIGNS: many more abnormal signs are often found on examination than the symptoms indicate

1. The optic disks may be abnormally pale, even without a history of visual disturbance
2. One or the other of the eye muscles may be weak, causing diplopia
3. The eyes often show rhythmical oscillation seen in cerebellar and brainstem disease, called nystagmus
4. The speech is slurred (dysarthria)
5. The limbs show a mixture of cerebellar signs, producing ataxia and upper motor neuron signs, such as weakness, spasticity, exaggerated reflexes, and extensor plantars
6. Euphoria exists where the patient appears cheerful and unconcerned about his apparent gross disability

C. TREATMENT

1. No curative treatment has yet been found
2. Since relapses tend to occur spontaneously, it is difficult to know if any drug has been helpful
3. In the acute stages of relapse, a course of ACTH injections will more often than not produce a rapid and marked improvement
   a. Long-term use of ACTH or cortisone, however, does not prevent relapses

Gross section through brain showing patches and large plaques of demyelination throughout the white matter.

Histologic section through cerebral white matter showing extensive demyelination throughout entire area.

204. SCHILDER'S DISEASE AND RETROBULBAR (OPTIC) NEURITIS

I. Schilder's disease, from a histologic point of view, consists of lesions whose structure is like multiple sclerosis

A. DISTINCTION

1. By its more frequent occurrence in children and its continuous evolution
2. Characteristic distribution of lesions
   a. Large areas of bilateral and symmetrical demyelination in the cerebral hemispheres involving the corpus callosum—"butterfly appearance"
   b. Predominantly in the parieto-occipital regions, sparing the gray matter, also involves the optic pathways
   c. In some cases, endocrine disturbances are seen, especially adrenal lesions with cutaneous pigmentation
3. Results in acute onset of fits, rapidly developing blindness, and often spastic paralysis
4. The entire picture is of a progressively slow total disability
5. Several varieties are seen with slight clinical differences
6. ACTH given in the acute stages may help prevent gross disability, but many cases have had too extensive a lesion for useful improvement

II. Retrobulbar neuritis or optic neuritis: a condition usually due to a patch of demyelination developing in one optic nerve and most commonly an early stage of disseminated sclerosis. Later, the other nerve may be affected, occasionally both at the same time

A. SYMPTOMS

1. Pain develops in one eye, and it hurts to move it or even press on it
2. Vision very quickly becomes blurred, particularly its central part, and this may progress to a total loss of vision. One may recover after a few days or within weeks or months and vision may return to normal
3. In some cases, a little area of permanently impaired vision remains at the center of the field which may interfere greatly with reading, writing, etc.

B. SIGNS

1. If the patch develops in the nerve just behind the globe (retrobulbar), mild swelling of the optic disk can be seen—papillitis
2. After the acute stage, some nerve fibers atrophy and the disk looks pale or even dead white (optic atrophy)
3. If the patch develops farther back along the nerve, there may be no swelling, but the atrophy can still develop later
4. By charting the visual fields, the blind area can be shown to be in the center and is referred to as central scotoma

C. TREATMENT

1. ACTH and cortisone often hasten the improvement and are worth trying
2. The sooner the acute stage settles down, the less chance there is of permanent visual damage

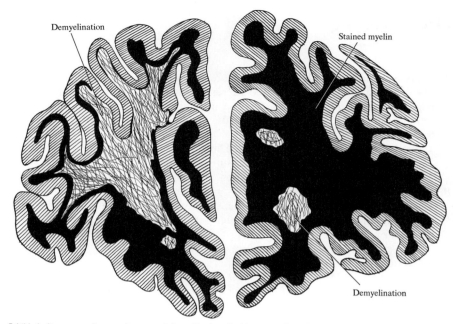

Demyelination

Stained myelin

Demyelination

Schilder's disease revealing massive areas of demyelination of white matter throughout the cerebrum.

Myelinated fibers

Multiple sclerosis: note corpora amylacea (arrows).

Demyelinated fibers

205. MYELITIS

I. Transverse myelitis: a plaque of myelination may extend across the spinal cord during one of the relapses of disseminated sclerosis, producing a transverse cord lesion. This may also occur without any evidence of disseminated sclerosis. The lesion is usually found in the thoracic region and its cause is unknown
A. SYMPTOMS
1. Preliminary period of pain around the lower thoracic or abdominal regions which has been mistaken for pleurisy, cholecystitis, or appendicitis
2. Rapid weakness and loss of sensation in the legs, with urine retention
B. SIGNS
1. Initially, particularly in children, there may be a flaccid paralysis of the legs which soon becomes spastic with exaggerated reflexes and extensor plantars
2. All forms of sensation are lost below a clear-cut level on the trunk
3. Bed sores and urinary infection are likely to occur
C. TREATMENT
1. ACTH may help in the acute stages
2. Treatment is usually careful nursing of a paraplegic patient
3. Prognosis varies with different patients from recovery to complete paralysis

II. Bilateral retrobulbar neuritis and transverse myelitis: seen in a condition also called neuromyelitis optica (Devic's disease) which is a rare condition
A. SIGNS AND SYMPTOMS
1. Patients develop retrobulbar neuritis in both eyes (not necessarily together) followed soon by a transverse myelitis
2. This is probably a variety of disseminated sclerosis
3. If recovery occurs from the acute stage, there is not the same tendency for relapse to occur

III. Cerebrospinal fluid in disseminative disease
A. THE CSF may be normal in the chronic stages of these diseases
B. A RISE IN LYMPHOCYTE COUNT (usually less than 100 per cu ml) and a slight rise in protein is often seen in the acute stages of the disease
C. THE LANGE CURVE may show a paretic pattern, but the WR is always negative
D. IN DISSEMINATED SCLEROSIS, even when the cells and protein are normal, the gamma globulin portion may rise to over a quarter of the total protein

IV. Influence of pregnancy on demyelinative disease
A. IN DISSEMINATED SCLEROSIS, there is often a relapse shortly after the end of pregnancy
B. THERE IS NO EVIDENCE that pregnancy really makes things worse in the long run
C. IT HAS NOT AS YET BEEN DEFINED whether or not oral contraceptives have any influence on the course of disseminated sclerosis

TRANSVERSE MYELITIS

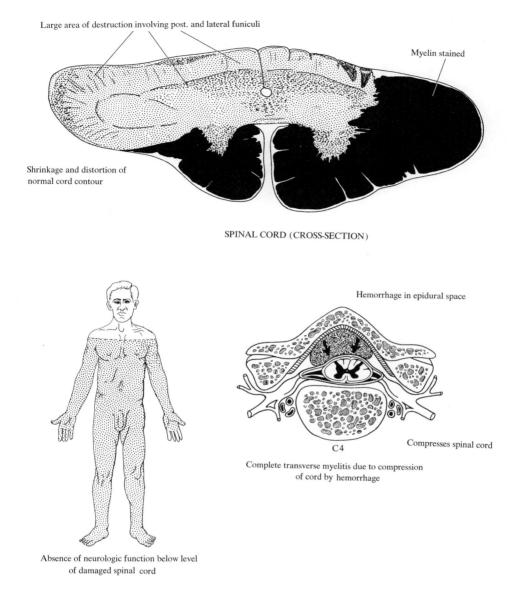

Large area of destruction involving post. and lateral funiculi

Myelin stained

Shrinkage and distortion of
normal cord contour

SPINAL CORD (CROSS-SECTION)

Hemorrhage in epidural space

Compresses spinal cord

C4

Complete transverse myelitis due to compression
of cord by hemorrhage

Absence of neurologic function below level
of damaged spinal cord

HEREDITARY AND
FAMILIAL DISEASES

206. THE HEREDOFAMILIAL ATAXIAS

I. **Introduction:** a hereditary disease is passed on from generation to generation. Familial disease affects several members of the same family. The term "heredofamilial" is common; they are diseases of the nervous system which can be traced in several relatives, though the relationship is sometimes a distant one. Most neurologists use the term to refer to the group of uncommon diseases that are degenerative in type and tend to produce abnormalities either of coordination or of the peripheral nerves and thus are classified under the heading of spinal medullary degenerative diseases

II. **Heredofamilial ataxias**
  A. GENERAL FEATURES
    1. Group of diseases in which family members develop a progressive ataxia
    2. In any family, the symptoms and signs are much the same, but in different families they vary as to age of onset and the part of the system affected
    3. On the basis of the above, they are given names, usually after the physician who described them, e.g., Marie's ataxia, Charcot-Marie-Tooth disease, etc.
    4. Differences are not of practical importance; certain features are common to all
  B. SYMPTOMS
    1. Increasing clumsiness and unsteadiness of the arms and legs
    2. There may be defective vision and difficulty in speech
    3. No headache, vomiting, vertigo, or diplopia, as in posterior fossa tumors
    4. Condition progresses steadily and not by remissions and relapses (as seen in disseminated sclerosis)
  C. SIGNS
    1. Ataxia is the major sign, as a result of cerebellar disease
    2. Slurring of speech, nystagmus, and unsteadiness of arms, legs, and gait are often seen
    3. Tendon reflexes are usually absent, but as the pyramidal tracts are affected, in some cases, there may be extensor plantar responses
    4. Pes cavus is common and often found in relatives who are not otherwise affected
    5. Optic atrophy is often seen, but patients rarely complain of visual loss
  D. SPECIAL VARIETIES
    1. Hereditary degenerative spinocerebellar disease (Friedreich's ataxia)
      a. The most frequent and best characterized of these diseases
      b. Involves the spinocerebellar tracts, Clarke's column, and posterior columns. Neuronal cell loss from some cranial nerve nuclei (XII, XI, VIII) is frequent and in some cases optic atrophy has been seen
      c. Develops in adolescence
      d. Shows a marked loss of position and vibratory sense, with a postural ataxia due to posterior column degeneration
      e. Weak back muscles and severe scoliosis are common
    2. Olivopontocerebellar degeneration develops in middle life
      a. Dysarthria is a major symptom, as is cerebellar ataxia, and there are spasticity and exaggeration of reflexes
      b. Resembles disseminated sclerosis, but it advances without remissions, is familial, and dysarthria is outstanding
    3. Hereditary spasmotic (spastic) paraplegia (Strumpell-Lorraine disease) is rare
      a. Lesion consists of bilateral atrophy of the pyramidal tracts throughout the length of the cord associated with involvement of the posterior columns and spinocerebellar tracts but is distinct because there are usually no cerebellar signs and no true ataxia, though reflexes are exaggerated

A

Loss of myelin in dorsal funiculus (*DF*), spinocerebellar (*SC*), lateral corticospinal (*LCS*), and anterior corticospinal (*ACS*) tracts.

B

C

Olivopontocerebellar degeneration as evidenced by the absence of a Purkinje cell layer in the cerebellar cortex. Compare with normal cerebellar cortex shown in *C*.

Normal cerebellum showing molecular layer, Purkinje layer (arrows) and granular layer.

207. HEREDITARY NEUROPATHIES

I. **The hereditary neuropathies:** rare diseases due to peripheral nerve degeneration in mainly motor nerves in some cases and in sensory nerves in other cases

A. RADICULOSPINAL DEGENERATIVE DISORDERS
   1. Peroneal muscular atrophy (Charcot-Marie-Tooth disease)
      a. Involves both lower motor and peripheral sensory neurons
      b. This is a neuropathy, not purely a degeneration of muscle
      c. It affects either sex and develops in adolescence or early adult life
      d. Symptoms: walking becomes increasingly difficult because the legs have to be lifted high to avoid catching the toes; wasting is first seen in the legs and later in the hands and forearms
      e. Signs
         i. There is a high stepping gait due to bilateral footdrop, each leg being slapped down in turn. There is also pes cavus
         ii. Hands and feet are cold and blue; little glove-and-stocking sensory loss
         iii. All tendon reflexes are absent, and vibration sense is impaired
         iv. Wasting is the most characteristic feature because, though very marked, it stops suddenly half-way up the thighs and forearms—shoulder girdles, upper arms, hip girdles, and upper thighs are normal—giving the extremities the appearance of "inverted bottles"
         v. Condition progresses slowly without incapacity until late in life
   2. Hypertrophic polyneuritis has all the features of a polyneuritis but progresses slowly over many years. May resemble (1) but there is great thickening of the peripheral nerves which can be felt at the elbows, below knees, and top of feet
   3. Hereditary sensory neuropathy (of Denny-Brown)
      a. Associated with picture of acropathia ulceromutilans, is familial, and involves peripheral sensory neurons. Lesions implicate the spinal ganglia, dorsal nerve roots, and posterior columns
      b. Sensation is severely impaired so that painless sores develop on fingers and toes which may need amputation or may even drop off by themselves
   4. Spinal hereditary amyotrophy (Werdnig-Hoffmann disease)
      a. Lesions involve anterior horns of cord and nuclei of certain cranial nerves (VII, XII). Ventral roots are atrophied and demyelinated but posterior columns are not involved
      b. Appears in infancy and one sees a progressive spinal muscle atrophy
      c. May be fatal because of involvement of respiratory muscles. Death usually occurs by the 5th year of life
   5. Chronic anterior poliomyelitis produces a degenerative lesion which evolves slowly in the adult, and changes are caused by massive neuronal cell loss and degeneration of ventral nerve roots. May be hard to tell from amyotrophic lateral sclerosis

B. AMYOTROPHIC LATERAL SCLEROSIS (CHARCOT'S DISEASE) produces degenerative lesions of the pyramidal tracts associated with lesions of the lower motor neurons, atrophy of anterior horn cells, and nerves XII, XI, and X
   1. Is a chronic, progressive disease of unknown cause in which there is degeneration of both upper and lower motor neurons
   2. More common in males (4:1) and appears in 4th and 5th decades
   3. There are weakness, wasting, and fine tremors in the thenar, hypothenar, and inter-ossei muscles of the hands; then arms and shoulders are involved. Spastic paraplegia, increased deep reflexes, and (+) Babinski are seen. Death is due to bulbar paralysis within 3 years of onset

Transverse section through spinal cord showing degeneration of lateral corticospinal tracts on both sides and the anterior corticospinal tract on one side.

Anterior gray horn from amyotrophic lateral sclerosis patient showing loss of neurons.

Anterior horn cells from normal anterior gray horn.

AMYOTROPHIC LATERAL SCLEROSIS

208. LEUKODYSTROPHIES

I. **General comments:** it is not uncommon to find examples of more than one of these diseases in the same family and, in fact, there may even be two types in the patient. The diseases in some way may all be linked together. It is very common, on examining relatives, to find unmistakable signs in someone who is entirely symptom-free (called "formes frustes" of the disease). All the diseases progress slowly and no drug therapy is available. Physiotherapy aimed at educating coordination of movements can be helpful but, sooner or later, the disability becomes severe, even though comparatively rarely are the patients bedridden

II. **The leukodystrophies** are diseases affecting children, show a lack of inflammatory cellular lesions and a frequency of genetic factors (familial). Most of them are, apparently, largely the result of a disturbance of myelin synthesis and are often classified under the dysmyelinating diseases

A. DEGENERATIVE CHANGES take place, particularly in the white matter of the brain, with frequent involvement of the cerebellum

B. THERE ARE FREQUENT IRREGULAR CHARACTER OF DEMYELINATION and sparing of U fibers

C. THE PRESENCE OF MACROPHAGIC REACTION whose distribution is diffuse and not perivascular

D. DISTINCT TYPE OF MYELIN CATABOLISM with formation of nonsudanophilic and metachromatic prelipids

E. MOST OF THESE DISEASES are the result of a known disturbance of myelin metabolism (see also Metachromatic Leukodystrophy and Krabbe's Disease), and others are of uncertain origin. Thus, the term "dysmyelinating disease" has been used

F. THE LEUKODYSTROPHIES appear early in childhood and may cause successive children in the family to become increasingly backward, with a loss of intellect, have ataxia, and develop fits and weakness of the limbs which progress to complete paralysis

G. METACHROMATIC LEUKODYSTROPHY is the best known of the group
   1. Can be diagnosed in life by finding substances called sulphatides in the urine or in biopsies of the kidneys, brain, or peripheral nerves
   2. There is no known method of interfering with the disease process
   3. Others in this group are less common, all are rare, and many are not diagnosed until after death

H. SUDANOPHILIC LEUKODYSTROPHIES encompass a number of disorders grouped together because of the absence of inflammatory features and the sudanophilic character of the lipid found in the white matter

I. DIFFUSE OR SIMPLE TYPE ORTHOCHROMATIC LEUKODYSTROPHY: comparable to Schilder's disease. No inflammatory cellular infiltrates but frequent axonal lesions

J. TIGROID FORMS (E.G., PELIZAEUS-MERZBACHER'S DISEASE): seen very early in life and affects only male children. Has a prolonged clinical course

K. MASSIVE CONGENITAL FORMS: virtually total absence of central myelin with normal cranial and spinal nerves

L. SPONGY DEGENERATION OF NEURAXIS (CARAVAN'S DISEASE): spongy aspect of subcortical layers of cerebral and cerebellar white matter

Coronal section through brain presenting with enlarged ventricles and sulci, narrow gyri, and corpus callosum and grayish-white matter.

Some loss of myelin

Lipoprotein residue

Complete loss of myelin

Section through grayish-white matter shown in gross brain.

LEUKODYSTROPHY

NEUROLOGIC DISTURBANCES
WITH SKELETAL ABNORMALITIES

209. INTERVERTEBRAL DISK DISEASE

I. **Introduction:** since the skull and spine closely encase the brain and spinal cord, many peripheral nerves run very close to bones, joints, and ligaments. Thus, skeletal abnormalities may involve nerve tissue. Certain commonly recurring conditions are seen in both neurologic and orthopedic departments

II. **Intervertebral disk disease:** the disks are "cushions" of cartilage separating the vertebral bodies and are held in place by a fibrous ring, the anulus fibrosus, which joins the bodies of the vertebrae together

A. ACUTE DISK PROLAPSE: if the anulus ruptures, the disk will bulge through the gap (prolapse)
   1. In the cervical region, will compress the cord causing an acute quadriplegia
      a. This is uncommon and the protrusion more likely will be to the side and compress a nerve root causing severe pain down the arm
      b. Requires urgent surgery
   2. Thoracic disk prolapse: since there is little room here, it usually compresses the cord and results in paraplegia
      a. Requires urgent surgery
   3. Lumbar disk prolapse is very common
      a. It misses the lower end of the cord usually and presses on the lumbar and sacral nerves resulting in sciatica
         i. Results in pain in the back, passing down the back of the thigh and either over the front of the lower leg toward the big toe (L4, L5 lesions) or down the calf to the little toe (L5, S1 lesions)
         ii. Muscle spasms make the back rigid, scoliosis develops, and the straight-leg-raising test (Lasègue's sign) is limited to the affected side
      b. 95% of lumbar disk lesions settle down after complete bedrest for about 4 to 5 weeks. With severe symptoms, about 5% require removal
      c. Massive prolapse can compress the entire cauda equina resulting in flaccid paralysis with urinary retention and requires urgent surgery

B. CERVICAL SPONDYLOSIS
   1. X-rays of the necks of middle-aged and older patients have shown degenerative changes in the disks with overgrowth of bone (osteophytes) forming hard bars
   2. The bars compress nerve roots, thus these spondylitic changes are responsible for most cases of "fibrositis," "rheumatism," or "neuritis" in the arms and shoulders
      a. The bars may press on the blood vessels supplying the anterior horn cells and pyramidal tracts in the cord which results in a wasting and numbness in the arms with progressive spastic paraparesis
   3. Treatment
      a. Immobilization of the neck in a well-fitting collar for many months
      b. Neck manipulation is highly dangerous
      c. Few are suitable for surgery
      d. Immobilization can be made permanent with a bone graft

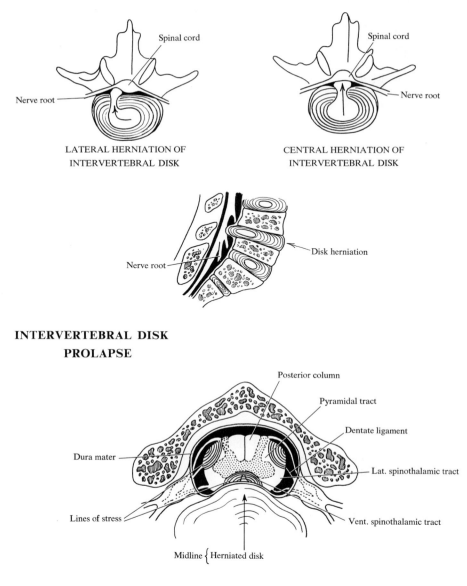

LATERAL HERNIATION OF
INTERVERTEBRAL DISK

Spinal cord

Nerve root

CENTRAL HERNIATION OF
INTERVERTEBRAL DISK

Spinal cord

Nerve root

Nerve root

Disk herniation

INTERVERTEBRAL DISK
PROLAPSE

Posterior column

Pyramidal tract

Dentate ligament

Dura mater

Lat. spinothalamic tract

Lines of stress

Vent. spinothalamic tract

Midline { Herniated disk

Note: Strain greatest ventral (not clinically tested); secondary
stress on pyramidal tracts—hands usually spared since
leg area is most lateral

210. CARPAL TUNNEL SYNDROME AND PRESSURE NEUROPATHIES

I. The carpal tunnel syndrome

A. AS THE MEDIAN NERVE enters the wrist to supply the muscles of the thumb and give sensation to the thumb and first two fingers, it passes through the carpal tunnel where soft tissue and fat may become swollen and thickened and compress the nerve
1. Carpal tunnel compression is common in middle-aged women resulting in severe pain, burning, and tingling of the hand, especially at night
2. Muscles may waste after a long time

B. Treatment
1. Splinting the wrist during the night offers some relief
2. Injections of ACTH or local infiltration of hydrocortisone may help
3. To cure, the tunnel must be opened surgically and the pressure relieved

II. The thoracic outlet syndrome: as the nerves pass through the root of the neck to reach the arms, they are compressed by tight bands of muscle in the region. These patients, predominantly women, after carrying heavy weights or after an increase in body weight with a drooping of the shoulders, develop tingling in the entire arm

A. TREATMENT: physiotherapy to correct posture and division of the muscle bands

III. Cervical ribs are congenital anomalies consisting of small ribs attached to the 7th cervical vertebrae and, as a result of their position, they raise and stretch the nerves that run over the 1st rib

A. THE CONDITION results in wasting of the small muscles of the hand and loss of sensations over the little and ring fingers and part of the forearm

B. TREATMENT: if symptoms are severe, the small ribs can be removed surgically

IV. Pressure neuropathies: certain nerves, especially the ulnar at the elbow and the popliteal outside the knee, lie superficial and can be compressed by resting on them for periods of time (result of bedrest, application of a poorly padded cast, or even use of adhesives for skin traction)

A. THE HAND CAN WASTE AND SENSATION IS LOST over the ring and little fingers, if the ulnar nerve is involved

B. IN THE FOOT, if the popliteal nerve is involved, it can lead to footdrop

C. TREATMENT: adequate protection for the nerves involved

V. Bony abnormalities at the base of the skull

A. THE SKULL BASE AND POSTERIOR FOSSA may be abnormally flat (platybasia)
1. Associated with basilar invagination or basilar impression, resulting in the first cervical vertebrae being almost inside the posterior fossa so that parts of the cerebellum (cerebellar tonsils) are squeezed down through the foramen magnum
2. These patients usually have short necks and low hairlines and become increasingly ataxic

B. THE BONY ABNORMALITY may be congenital or as a result of bone softening due to Paget's disease

C. BASILAR IMPRESSIONS are also associated with Arnold-Chiari deformity and syringomyelia

Flex. pollicis long.

Radial a.

Flexor tendons (deep + superficial)

MEDIAN NERVE

Artery
Nerve } Ulnar

Muscular branch (recurrent branch of median n.)

Flexor retinaculum

CARPAL TUNNEL

Superficial arterial arch

Palmar digital branches of median n.

Palmar digital branches

Pronator teres m.

Carpal tunnel

Palmar cutaneous

Motor br. to thenar mm.

Transverse intermetacarpal ligs.

Cutaneous innervation to hand

Cutaneous interdigital

Median nerve supply to back of fingers (stippled)

CARPAL TUNNEL SYNDROME
Pain + paresthesias in hand + wrist very often most troublesome in hours of sleep

SITES OF MEDIAN NERVE COMPRESSION

Unit Eighteen

DEGENERATIVE DISEASES

211. DEGENERATIVE CORTICAL DISEASE

I. **Introduction:** when the cells and fibers cease to function, shrink, and gradually disappear for no known reason, the process is vaguely called degeneration. The degenerative diseases of the CNS form a heterogeneous group of disorders in which changes are mainly neuronal and occur sporadically, independent of any inflammatory, toxic, or metabolic factor that we know of as yet. These changes can occur as a result of ischemia from the narrowing of blood vessels in arteriosclerosis, but often all other systems are normal; yet, certain portions of the brain or cord die. Genetic factors determine a large number of these conditions which frequently have a familial character, but the pathogenic mechanisms are unknown. The degenerative process may be diffuse or it may select particular parts of the nervous system, e.g., cerebral cortex (giving rise to demential disorders), basal ganglia (causing disturbances in which the symptomatology is extrapyramidal), or a number of spinocerebellar systems (seen in the group of hereditary degenerative diseases). Often the symptoms and signs, when collected together, form a syndrome that is recognizable, e.g., parkinsonism motor-neuron disease or the cerebral atrophies. The diseases occur generally late in adult life and often in the presenile age group.

II. **Degenerative cortical disease:** cortical degenerative processes are the most frequent neuropathologic causes of organic dementia, although other disorders can give these symptoms. Senile dementia and Alzheimer's disease account for about 80% of cases in this group of diseases

A. CEREBRAL (CORTICAL) ATROPHY: atrophy or shrinkage of brain tissue may be the result of injury, inflammation, or interference with blood supply, but the tissue cells and fibers can atrophy for reasons yet unknown; and with a loss of cerebral tissue, the gyri thin out, the sulci widen, and the ventricles enlarge

  1. Senile dementia causes only discrete cerebral atrophy with symmetric thinning of the convolutions, predominantly in the frontal lobes and a moderate degree of ventricular dilatation

     a. The lesions include

        i. Senile plaques are occasionally seen in physiologic senescence but are abundant in senile dementia. Are seen mainly in the superficial cortical layers

        ii. Neuronal cell changes: neurofibrillary degeneration and tangles, granulovacuolar degeneration, excess of lipopigments, and disappearance of Nissl bodies

        iii. Vascular changes: thickening of the capillaries and terminal arteriolar vessel walls, deep arteriolar hyalinosis, and atherosclerosis of arteries

  2. Alzheimer's disease causes a striking cerebral atrophy and diffusely affects all the cerebral convolutions. May sometimes show a predilection for the parieto-occipital regions where gross atrophy is very striking, accounting for early symptoms of aphasia, apraxia, and agnosia with dementia. May also give rise to circumscribed lobar atrophies and clinically resemble Pick's disease

     a. Microscopic lesions are those of senile dementia, including senile plaques and neurofibrillary and granulovacuolar degeneration, but no vascular changes are noted. The changes are, however, more severe than is seen in senile dementia

     b. Occurs earlier than senile dementia, even seen in young adults and often on a familial basis. Neuronal cell loss is considerable

ALZHEIMER'S DISEASE

Coronal section showing narrowing of gyri and widening of sulci (arrows).
Note also the enlarged lateral ventricles (LV).

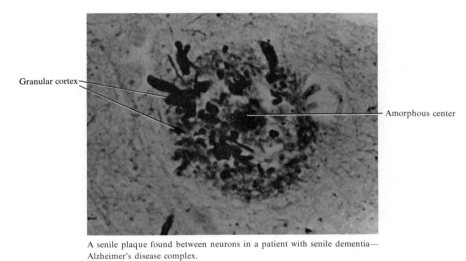

A senile plaque found between neurons in a patient with senile dementia—
Alzheimer's disease complex.

212. CEREBRAL ATROPHY

II. Degenerative cortical disease (cont.)

A. CEREBRAL ATROPHY (cont.)

3. Pick's disease is frequently hereditary
   a. Presents a picture of circumscribed frontotemporal cerebral atrophy which spares the posterior $\frac{1}{3}$ of the superior temporal gyrus, thus accounting for the rarity of aphasic phenomena. The parietal cortex is seldom involved and the occipital never
   b. There is an extreme degree of neuronal loss associated with dense astrocytic gliosis and sometimes a spongy state
   c. There are white matter lesions and basal ganglia changes involving the head of the caudate nucleus
4. Alzheimer's and Pick's diseases, together, have been called the presenile dementias
5. Symptoms of the cortical diseases
   a. Memory is most noticeably disturbed: at first for recent events only (childhood being remembered clearly), but gradually all memory is lost, even of words just recently spoken
   b. Patients become unable to look after normal affairs of life, and later help is needed even to dress and find one's way about the house
   c. The physical state may remain good throughout, and the patient appears normal on a casual visit
   d. Some patients are subject to recurrent fits
6. Signs of the cortical diseases
   a. Patients are confused and disoriented, not knowing their own names, age, or those of their nearest relatives; do not know where they are, nor can they identify those around them
   b. The physical examination is usually within normal limits
7. Treatment of cortical diseases
   a. Fits are controlled by anticonvulsants and investigations are necessary to be certain that the dementia is not due to an operable tumor, vitamin B_{12} deficiency, folic acid deficiency, or neurosyphilis
   b. Generally, these cases deteriorate and eventually need institutional care
8. Low pressure hydrocephalus is a form of cerebral atrophy in which the ventricles are very large, but the cortex shows little atrophic change
   a. Some feel that if the CSF absorption is improved, the patient's mental state may improve, and the patient is treated as a hydrocephalic child with a Spitz-Holter valve. The results of these treatments are still controversial

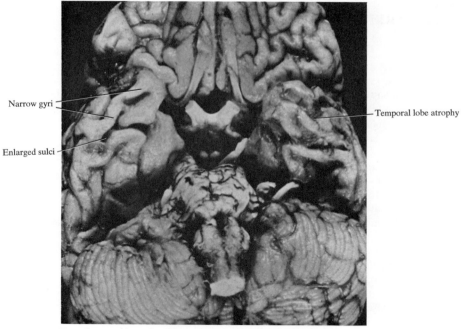

Narrow gyri

Enlarged sulci

Temporal lobe atrophy

Ventral aspect of brain showing atrophy associated with Pick's disease.

Histologic section showing neurons with cytoplasmic Pick inclusions (arrows).

PICK'S DISEASE

213. PARKINSONISM (DISEASE AND SYNDROMES)

I. **Subcortical degenerative diseases** involve the basal ganglia to a variable degree and account for many of the extrapyramidal syndromes

A. PARKINSONISM (DISEASE AND SYNDROMES): in addition to Parkinson's disease proper, which is of degenerative origin, a number of processes of different etiology involve identical anatomic structures, resulting in a series of parkinsonian syndromes that are clinically related

1. Parkinson's disease, once called paralysis agitans or shaking palsy, is the most frequent of the subcortical degenerative disorders
   a. Lesions involve pigmented nuclei of the brainstem, especially the substantia nigra (origin of the dopaminergic pathway) and nuclei in the upper brainstem. Sympathetic ganglia neurons are often involved

2. Nondegenerative parkinsonian syndrome: to a variable extent shows lesions in the same neural structures
   a. Postencephalitic parkinsonism syndrome develops years after an encephalitis, particularly encephalitis lethargica
      i. Involves the substantia nigra chiefly and is diffuse
      ii. Lesions of the pontine and midbrain reticular formation are frequent
      iii. Affects younger people and is likely to produce mental changes
      iv. Often mainly unilateral
      v. May have attacks called oculogyric crises: eyes (sometimes the head) appear to be forced to turn upward for minutes at a time without loss of consciousness
      vi. History of encephalitis or a serious illness is prevalent
   b. Arteriosclerotic parkinsonism: due to defective blood supply to basal nuclei
      i. Onset is often abrupt and affects the later age groups
      ii. There is less rigidity than in above groups
   c. Syphilitic parkinsonism is a rare complication of syphilitic infection and has seldom been systematically investigated from a pathologic viewpoint
      i. Diagnosed by a clear history of syphilis or a positive WR
   d. Posttraumatic parkinsonism
   e. Vascular origin of parkinsonism is only exceptionally pure
   f. Parkinsonism following large doses of certain tranquilizer drugs

3. Symptoms: three major forms are tremor, muscle rigidity, and akinesia (inability to initiate movements or perform them quickly enough). Any one of these may predominate
   a. Paralysis agitans begins in middle life with a slowing up of all movements and aching in the shoulders and arms. As rigidity and akinesis increase, it becomes difficult to raise arms, get out of a chair, or even turn suddenly
   b. Writing becomes progressively small
   c. Speech becomes monotonous and indistinct
   d. The first symptom may be a tremor, starting in a thumb and hand on one side, gradually involving the wrist, other arm, and then the legs. The tremor is worse when the patient tries to remain still, but gradually increases until it becomes impossible to control or conceal
   e. In most cases, tremor, rigidity, and akinesia develop together
   f. The autonomic system is also disturbed as seen by excessive formation of saliva, sweating, and greasiness of skin
   g. There is intractable constipation
   h. In late stages, the memory deteriorates and obsessional trends develop
   i. Patients become irritable, querulous, and never seem satisfied

PARKINSON'S DISEASE

Section through midbrain showing decreased neuromelanin pigment in the substantia nigra.

Substantia nigra from Parkinson's disease patient showing reduction in number of pigmented neurons.

Normal substantia nigra to show neuromelanin pigment in neuron.

214. PARKINSONISM (SIGNS AND TREATMENT) AND HUNTINGTON'S CHOREA

I. Subcortical degenerative diseases (cont.)

A. PARKINSONISM (cont.)

4. Signs: a fully developed case is unmistakable
   a. The face shows no change of expression and blinking is frequent—masklike appearance
   b. Due to muscular rigidity, the patient walks with small shuffling steps, stooped forward, arms and legs bent, and the arms held across the body instead of swinging in the normal manner
   c. The rapid movements needed to correct a sudden change of position are impossible, so that, if given a slight push, the patient will appear to trot forward more and more quickly. This is called propulsion
   d. The rigidity can be felt by moving a joint at which time it gives a jerky sensation, like a cogwheel
   e. Tremor is most obvious in fingers, hands, and wrist, but may affect any body part, including the head, chin, and tongue. Tremor is rhythmical and regular but stops as limb is put into motion—begins again at rest
   f. Speaking results in a rapid, monotonous run of words, often indistinguishable
   g. The mind remains clear

5. Treatment
   a. Use of antispasmodics, e.g., Artane, lysivane, Kemadrin, Disipal, etc.
   b. The drugs appear to improve rigidity but have little effect on the tremor, and it is the tremor that worries the patient the most
   c. L-Dopa has been used and is of value in the akinetic form of the disease having much less effect on the tremor, but it has troublesome side-effects and the patients must be watched carefully during the early stages of its use
   d. There have been some surgical operations aimed at relieving both tremor and rigidity, including the cutting out of areas of cortex, cutting pyramidal tracts, tying off blood vessels to the basal nuclei and a small lesion in the basal nuclei (usually the thalamus) using stereotaxis. The results have varied, but the natural course of the disease does not appear to be altered, and further trouble usually develops in time

B. HUNTINGTON'S CHOREA OR CHRONIC CHOREA is a familial disease of genetic origin chiefly involving the corpus striatum and to some extent the cerebral cortex (frontal and parietal regions show neuronal cell loss). Atrophy of the putamen and caudate nucleus preferentially selects the head of the caudate and is accompanied by ventricular dilatation of the frontal horns

1. Involuntary movements
   a. Athetosis consists of slow writhing movements of the limbs, face, and tongue, seen most frequently after brain damage at birth, or after rhesus incompatibility
   b. In chorea the movements are very rapid, never repeated, and accompanied by flapping of the tip of the tongue and grimacing
   c. When both a and b are present, we speak of choreoathetosis
   d. Chorea may follow rheumatic fever and is called Sydenham's chorea
      i. It may occur in pregnancy
      ii. Seen as a rare complication of the use of oral contraceptives
      iii. May occur as a familial disturbance in middle life
      iv. Usually accompanied by mental deterioration (called Huntington's chorea)
   e. Hemiballismus is a wild violent flinging movement of a limb, usually as a result of a vascular accident in the subthalamic nucleus

Atrophy of
caudate nucleus

HUNTINGTON'S CHOREA. There is a diffuse atrophy, but caudate nuclei are most affected

ATHETOTIC HAND MOVEMENTS

ATHETOTIC HAND MOVEMENTS

Congenital choreoathetosis
due to degeneration involving
striatum

215. RARE SUBCORTICAL, MOTOR NEURON, AND CEREBELLAR DEGENERATIVE DISEASES

I. Rarer forms of subcortical degeneration
   A. STRIATONIGRAL DEGENERATION: rare, with clinical features of extrapyramidal rigidity involving the substantia nigra and atrophy of corpus striatum
   B. PURE PALLIDAL ATROPHY: often familial, rare, and involves subthalamic nuclei
   C. ISOLATED ATROPHY OF SUBTHALAMIC NUCLEI: often associated with other pallidonigral degeneration
   D. HALLERVORDEN-SPATZ DISEASE: neuronal cell loss with deposition of iron pigment
   E. SHY-DRAGER DISEASE: extrapyramidal syndrome associated with orthostatic hypotension
   F. PROGRESSIVE SUPRANUCLEAR PALSY (STEELE-RICHARDSON-OLSZEWSKI DISEASE): clinical picture of hypertonia, supranuclear ocular palsies, and disturbance of wakefulness

II. Motor neuron disease: degenerative changes here affect the motor system only, namely, the anterior horn cells, the nuclei of some cranial nerves, and the fibers of the pyramidal tracts. Three syndromes develop
   A. PROGRESSIVE MUSCULAR ATROPHY
   B. PROGRESSIVE BULBAR PALSY
   C. AMYOTROPHIC LATERAL SCLEROSIS
   D. ALL OF THE ABOVE ARE PART OF THE SAME DISEASE, the difference in their signs depends on the part of the motor system affected, thus, they are grouped together
      1. The cause is unknown
   E. SYMPTOMS
      1. The disease most frequently develops between the ages of 40 to 60
      2. Weakness may first be noticed in the hands, shoulders, or neck muscles
      3. Occasionally the legs are first affected, usually as a painless footdrop
      4. The hand and shoulder girdle muscles begin to waste, and gradually the weakness and wasting spread throughout the body until the trunk muscles no longer give support. Weakness of the tongue and throat muscles makes speech difficult and swallowing impossible. When the latter symptoms begin, we speak of the disease being progressive
      5. In amyotrophic lateral sclerosis, pyramidal tract degeneration causes a spasticity that may predominate over wasting and weakness at first, but not for long
   F. SIGNS
      1. Muscle wasting is usually first seen in the shoulder girdles and hands, where the weakness appears greater than the wasting suggests. Later, the wasting becomes widespread and the tongue becomes small, wrinkled, and paralyzed
      2. Bundles of muscle show a constant jumping and flickering (spoken of as fasciculation) which are clearly seen in the tongue
      3. The tendon reflexes are greatly exaggerated, despite muscle wasting, and are combined with fasciculation and normal sensation
   G. Treatment: there is no effective treatment, and the disease progresses inevitably to death, usually from respiratory infection in from 6 months to 2 years

III. Cerebellar degenerative diseases: the complex group of cerebellar atrophies comprises all the degenerative disorders that involve cerebellar cortex and its efferent and afferent pathways. They may be focal or diffuse
   A. FOCAL CEREBELLAR ATROPHY
   B. DIFFUSE CEREBELLAR ATROPHY
   C. OLIVOPONTOCEREBELLAR ATROPHY

Caudate nucleus

Globus pallidus

Putamen

DISTRIBUTION OF LESIONS

HUNTINGTON'S DISEASE

Histologic preparation showing loss of neurons and increased glial cells in basal ganglia.

Unit Nineteen

EPILEPSY

216. EPILEPSY—PART I

I. **Introduction:** nerve cells function by discharging small electrical waves. Discharges arising from the brain can be measured by the electroencephalograph (EEG). If the brain cells suddenly produce a burst of larger electrical waves than normal, an epileptic fit may result. Patients whose brains have a tendency to do this repeatedly are said to suffer from epilepsy. This tendency is sometimes weakly hereditary or may develop from an area of damaged tissue. These two types are called idiopathic and symptomatic epilepsy, respectively

A. IDIOPATHIC EPILEPSY
   1. The exact cause for the electrical abnormality in this group is unknown
   2. Microscopic examination of the brain at postmortem shows no abnormality, but there may be some biochemical abnormality in the cells
   3. The tendency seems to be passed on through certain families for many generations. Some members develop fits; most have none
   4. Fits begin in childhood, adolescence (usually around puberty), or early adult life and take the form of generalized convulsions or petit mal attacks
   5. The patients are usually normal between attacks

B. SYMPTOMATIC EPILEPSY: the fits are one of the symptoms of brain damage or disease
   1. The suspicion that epileptic attacks are symptomatic arises when the fits begin later in life (over age 30), the attacks are focal, and there are abnormal signs in between attacks or shortly after an attack
   2. There is no family history and patients do not have true petit mal attacks
   3. When other symptoms of brain disease are also present, one is alerted to the possibility of a tumor, because removal may cure the patient

II. **Types of epileptic fits:** in either of the two major types of epilepsy, several types of fits may take place

A. GRAND MAL ATTACKS (major fits)
   1. May start abruptly with loss of consciousness, or may be preceded by a short warning or aura. The type of aura is important since it shows where in the brain the abnormal discharge begins
   2. In a generalized fit, there is often a vague indescribable sensation, followed by a loss of consciousness and rigid contraction of all muscles
      a. The arms are flexed, legs are extended, head thrown back or turned to one side, teeth are clenched tightly (tongue or cheek may be bitten), and respiratory movements are checked so the patient becomes cyanosed—this is the tonic phase of the fit
      b. After a few moments, the clonic phase begins—the muscles of the face, limbs, and trunk develop a coarse jerking, which becomes violent. There is often incontinence of urine and occasionally of feces
      c. Breathing becomes stertorous, the face flushes, muscles relax, but the patient remains unconscious, unarousable, with absent corneal and pupillary reflexes and extensor plantars—a postepileptic coma lasting 10 minutes to several hours
      d. This is followed by a period of confusion, the patient being restless and may wander about doing things automatically—postepileptic automatism—where the patient knows nothing of it and may endanger himself or others
   3. All grand mal attacks do not show all the features described. There may merely be a sudden loss of consciousness, a loss of consciousness with rigidity, but no clonic movements, or any combination of symptoms
      a. Symptoms may be so mild as to consist only of incontinence of urine at night or so severe that the patient suffocates and dies
   4. Generalized attacks may be idiopathic or symptomatic

At rest (not asleep)

Alpha waves (9–11 cycles/sec)

Normal waking adult

Beta waves (25–35 cycles/sec)

Normal

1 sec

GRAND MAL

Aura may precede ictus

Stare

1 sec

Clonic eye movements

1 sec

Sudden loss of consciousness

General tonic attack

1 sec

Convulsion
generalized with tonic (or clonic) spasms of
extremities

Regaining consciousness,
patient is confused and drowsy
and may lapse into a long sleep

217. EPILEPSY—PART II

II. Types of epileptic fits (cont.)

B. JACKSONIAN ATTACKS
1. Begin with a twitching of the mouth, thumb, or big toe and spread to other areas of the same side
2. They then pass over to the other side and end in a loss of consciousness
3. The attacks are symptomatic of irritation of the cerebral cortex by a tumor, scar tissue, or some form of inflammation

C. FOCAL EPILEPTIC ATTACKS affect just one part of the body (jacksonian attacks are a form of focal epilepsy, but many focal attacks show no spread, never become generalized, or cause loss of consciousness)
1. They are always symptomatic of a localized cerebral lesion, and the form that the attack takes depends on where the lesion lies, thus an accurate account of the fits is essential
   a. A lesion in the motor cortex will cause convulsive twitching of part of the opposite side of the body. This is often followed by a temporary paralysis of the parts involved (Todd's paralysis)
   b. Causes tingling and numbness of arm, leg, or face if in the sensory cortex
   c. In the occipital cortex, flashes of light in the opposite visual field result
   d. In the temporal lobe, vivid scenes, curious sensations of smell or of taste, and often a feeling that this is all very familiar—the so-called déjà-vu phenomenon—may occur
   e. The above may be the only symptoms of the fit or they may be followed by unconsciousness and a generalized fit

D. PSYCHOMOTOR EPILEPSY: abnormal discharge results in a sudden disturbance of behavior
1. The patient may start running around in a circle, or fighting, or just rearranging objects with the behavior stopping as suddenly as it began and with knowledge of what has taken place
2. This type of attack originates in the temporal lobe and may be preceded by a hallucination of taste, smell, or vision
3. This attack may be dangerous, but rarely is

E. PETIT MAL ATTACKS have often been called "minor epilepsy," but these attacks are not minor grand mal attacks, are quite different, and are always idiopathic
1. Begin in childhood and adolescence; only occasionally persist into adult life
2. Many attacks occur each day, sometimes even hundreds, consisting of
   a. A momentary loss of touch with one's surroundings ("absence") so that the patient "pauses" in conversation, e.g.
   b. The eyelids may flicker, the head may nod, and something held may be dropped
   c. Words or figures may be repeated
   d. Then the patient picks up the conversation and only a close observer may be aware of anything being wrong
   e. These attacks show a characteristic EEG abnormality (see later), and their treatment differs from that of grand mal epilepsy (see later)

F. MYOCLONIC ATTACKS
1. Myoclonus is a shocklike jerking of the muscles
   a. It may happen in the arms in many epileptics, especially first thing in the morning
   b. It sometimes occurs without other signs of epilepsy
2. Similar jerking sometimes occurs in normal people just after they drop off to sleep
3. Myoclonus of a repeating kind is seen in several diffuse diseases of the brain including subacute encephalitis

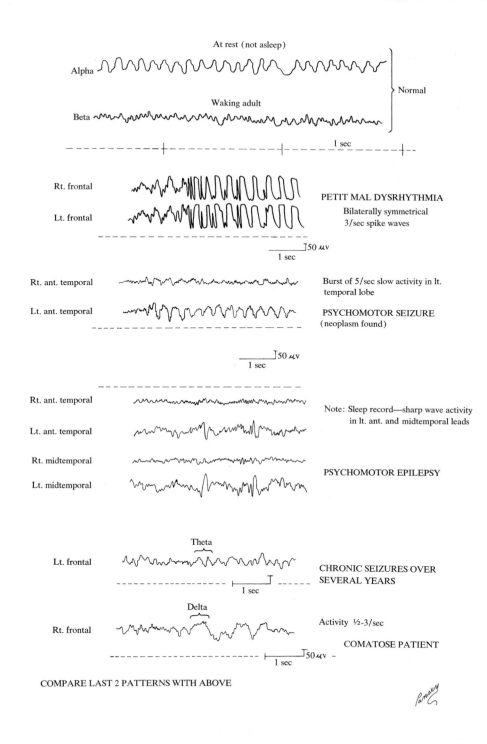

At rest (not asleep)

Alpha

Waking adult

Beta

Normal

1 sec

Rt. frontal

Lt. frontal

PETIT MAL DYSRHYTHMIA

Bilaterally symmetrical
3/sec spike waves

50 μv

1 sec

Rt. ant. temporal

Lt. ant. temporal

Burst of 5/sec slow activity in lt.
temporal lobe

PSYCHOMOTOR SEIZURE

(neoplasm found)

50 μv

1 sec

Rt. ant. temporal

Lt. ant. temporal

Rt. midtemporal

Lt. midtemporal

Note: Sleep record—sharp wave activity
in lt. ant. and midtemporal leads

PSYCHOMOTOR EPILEPSY

Theta

Lt. frontal

CHRONIC SEIZURES OVER
SEVERAL YEARS

1 sec

Delta

Rt. frontal

Activity ½-3/sec

COMATOSE PATIENT

50 μv

1 sec

COMPARE LAST 2 PATTERNS WITH ABOVE

218. EPILEPSY—PART III

II. Types of epileptic fits (cont.)

G. HYPSARRHYTHMIA is a name given to a gross EEG abnormality seen in infants subject to fits, especially the so-called "salaam" seizures in which the child throws his arms back and repeatedly bends his head and body forward

H. POSTTRAUMATIC EPILEPSY: in epilepsy that follows a serious injury to the brain, all types of attack may be seen (that is, all but petit mal)

1. Fits just after the injury, due to cerebral bruising, are relatively unimportant, but when they develop later, usually within the first 2 years, they are probably due to cerebral scarring and may recur for years
2. They are most common if the injury has penetrated the dura and brain, but are rare if there is no loss of consciousness and no fracture

I. STATUS EPILEPTICUS

1. Epileptic attacks usually occur at well-spaced intervals, but at times, one fit may pass into the next so quickly that consciousness is not regained and the attacks may last for many hours
2. The temperature rises to hyperpyrexial levels and, unless vigorously treated, the patient may become exhausted and die
3. It is more common in idiopathic epilepsy, but may occur as a result of structural disease
4. An epileptic may develop this condition during some intercurrent illness, but it is a particular danger if treatment is suddenly stopped

III. Treatment of a fit

A. ONCE A FIT BEGINS, nothing will work quickly enough to stop it, so that immediate treatment is used to prevent the patient from injuring himself and to ensure that there is no breathing obstruction

1. A firm object is placed between the teeth as soon as the fit begins
2. The patient should be removed from water, fire, traffic, or any other hazard where he might injure himself during the convulsive period
3. Clothing should be loosened and collars undone
4. The tongue must not be allowed to fall back into the throat, and the head should not be cushioned or pillowed to prevent suffocation

B. ON RECOVERY, the patient must not be left alone in case of postepileptic automatism

C. PREVENTION OF FURTHER FITS: symptomatic epilepsy has a demonstrable cause, and attempts should be made to eliminate the cause, e.g., removal of a benign tumor. If it is impossible to correct the underlying condition, e.g., in cerebral atrophy, then the treatment is the same in the symptomatic and idiopathic types

1. For grand mal: barbiturates form the basis for treatment, and the effect is greatly increased if hydantoins are given at the same time
   a. Primidone is another effective drug; there are many more, most of which are variations of the above. None has any specific advantage
2. The above drugs do not influence petit mal
   a. The "-diones" or suximides are indicated, e.g., ethosuximide
   b. Tridione and Paradione may be beneficial, but they may have toxic side effects
3. Absolute regularity and adequate dosage are essential
4. No reduction of treatment should be made until there has been complete freedom from fits for an absolute minimum of 2 years, however long this may take. Sudden stoppage of treatment can precipitate status epilepticus, which can be fatal

RECURRENT CONVULSIONS SEEN IN DIFFERENT AGE GROUPS

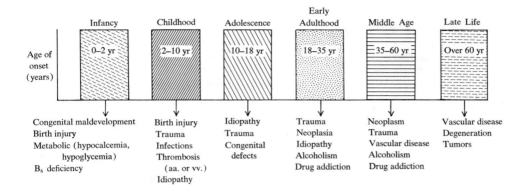

	Infancy	Childhood	Adolescence	Early Adulthood	Middle Age	Late Life
Age of onset (years)	0–2 yr	2–10 yr	10–18 yr	18–35 yr	35–60 yr	Over 60 yr

Congenital maldevelopment	Birth injury	Idiopathy	Trauma	Neoplasm	Vascular disease
Birth injury	Trauma	Trauma	Neoplasia	Trauma	Degeneration
Metabolic (hypocalcemia, hypoglycemia)	Infections	Congenital defects	Idiopathy	Vascular disease	Tumors
B_6 deficiency	Thrombosis (aa. or vv.)		Alcoholism	Alcoholism	
	Idiopathy		Drug addiction	Drug addiction	

SPREAD OF EPILEPTIC DISCHARGES

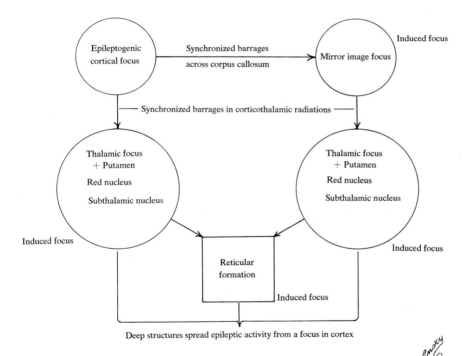

Deep structures spread epileptic activity from a focus in cortex

219. EPILEPSY—PART IV

III. Treatment of a fit (cont.)

D. TREATMENT OF STATUS EPILEPTICUS

1. Instructions regarding injury and suffocation are even more important and the patient should not be left alone for a moment
2. Paraldehyde remains the most useful tool, 10 ml given intramuscularly, which can be repeated in half an hour if the fits continue
   a. Additional doses of 5 ml can be given hourly and at more widely spaced intervals until oral dosage of anticonvulsants has once again been established efficiently
   b. If the above fails, paraldehyde can be given in a saline drip in 10% solution
   c. An injection of 180 mg epanutin given at the same time builds up the blood level of anticonvulsants more slowly
3. Diazepam (Valium) given IV may stop fits quickly but only transiently, and it is necessary to watch for a dangerous drop in blood pressure
4. Occasionally it is necessary to give an IV anesthetic such as Pentothal
5. Rarely, curare is used to paralyze the patient completely until the epileptic storm is over. Artificial ventilation and the constant presence of expert staff are needed for this procedure
6. The temperature must be watched and hyperpyrexia treated at once by tepid sponging, spraying, or nursing under a moistened sheet on which a fan is blowing a constant current of air
7. All the care of the unconscious patient applies and continues even after the fits have stopped
8. The danger of status epilepticus comes predominantly from failure to maintain an adequate ventilation and resultant damage to brain cells from lack of oxygen

E. TOXIC EFFECTS OF ANTIEPILEPTIC DRUGS

1. Barbiturates and Mysoline cause drowsiness, which is overcome by persisting with treatment (Mysoline sometimes causes abdominal pain, vomiting, and dilated pupils)
2. Large doses of panurin cause hypertrophy of the gums
3. Heavy combinations of drugs may cause ataxia and dysarthria
4. Methoin and the -diones occasionally cause disturbances of blood cell formation, sometimes agranulocytosis; barbiturates and hydantoins disturb folic acid metabolism and may cause macrocytic anemia
5. The above listed effects should never prevent adequate dosage (the dosage needed to effectively prevent the fits), providing it is not producing toxic effects.

IV. Surgery in epilepsy has a place apart from the removal of a tumor

A. TO EXCISE RAGGED SCAR TISSUE in posttraumatic epilepsy
B. IN TEMPORAL LOBE EPILEPSY, when removing part of the temporal lobe may stop the fits completely
C. BEFORE SURGERY, it must be certain that the fits arise from a single focus, that they are so frequent that they disrupt the patient's life, and that other adequate treatment has completely failed

V. Prognosis

A. EPILEPTIC FITS usually diminish with age and often stop altogether in middle life
B. A FEW PATIENTS, unfortunately, are completely uncontrollable, deteriorate steadily, and may require institutional care
C. PETIT MAL rarely troubles a patient in adult life

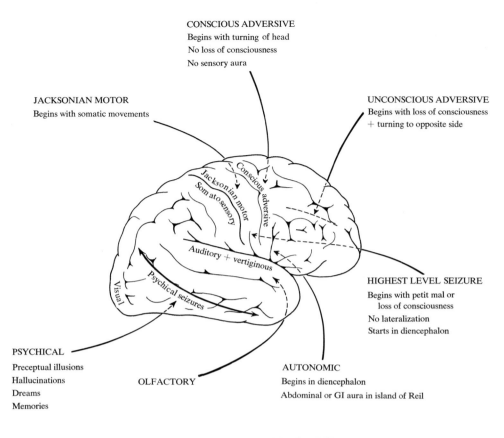

CONSCIOUS ADVERSIVE
Begins with turning of head
No loss of consciousness
No sensory aura

JACKSONIAN MOTOR
Begins with somatic movements

UNCONSCIOUS ADVERSIVE
Begins with loss of consciousness
+ turning to opposite side

HIGHEST LEVEL SEIZURE
Begins with petit mal or
    loss of consciousness
No lateralization
Starts in diencephalon

PSYCHICAL
Preceptual illusions
Hallucinations
Dreams
Memories

OLFACTORY

AUTONOMIC
Begins in diencephalon
Abdominal or GI aura in island of Reil

SEIZURE PATTERNS AND
THEIR LOCATIONS

Unit Twenty

DEFICIENCY, METABOLIC, AND TOXIC DISEASES

220. POLYNEURITIS AND POLYNEUROPATHY

I. **Introduction:** the diseases discussed here are caused by a poison or toxin acting on the nervous system, a shortage of some substance essential for nervous tissue to function properly, or both. The poison may be produced by abnormalities inside the body (e.g., uremia) or it may be taken in from the outside (e.g., drug addiction). A vital substance may become deficient as a result of not enough being taken into the body (starvation), or because it is not absorbed properly (as in pernicious anemia) or because the body fails to use it properly once it is absorbed. The peripheral nerves and the posterior columns of the cord are particularly vulnerable to chronic toxic and deficiency states, thus, the initial discussion deals with polyneuritis and polyneuropathy before the more common diseases are discussed

II. **Polyneuritis and polyneuropathy**
A. DEFINITIONS: neuritis means inflammation of a nerve; polyneuritis refers to inflammation of many nerves. However, nerves do not often become inflamed in the true sense of the word, and the term "polyneuropathy" is preferred by many. "-Opathy" means something wrong but not necessarily an inflammation, and exactly what is wrong may not be certain
   1. In polyneuritis or polyneuropathy, the symptoms usually begin in the fingers and hands, toes, and feet on both sides (thus, the alternative terminology of peripheral neuritis)
B. SYMPTOMS consist of numbness, pins and needles, or burning; or they can be motor such as weakness and wasting (discussed later)
   1. Symptoms spread upward to involve the calves, thigh and hip muscles, and the arms and shoulder girdles. In severe cases, even the trunk, facial, and respiratory muscles are involved
   2. Reflexes are characteristically absent, and there is a loss of sensation in the hands and forearms as well as feet and lower legs (glove-and-stocking anesthesia)
   3. The condition may range from a mild tingling in all 4 limbs to total paralysis of all muscles, requiring a tracheotomy and artificial respirator
C. THE ABSENCE OF MENINGEAL IRRITATION SIGNS, the slow onset, the symmetry of the paralysis, and the presence of sensory loss distinguish polyneuritis from poliomyelitis
D. THE EXACT CAUSE of many cases has never been determined, but fortunately recovery is quite common
E. ACUTE "INFECTIVE" POLYNEURITIS is a well-known syndrome but may not be of toxic or metabolic origin
   1. Develops rapidly, often after a few weeks of vague pyrexial illness and sometimes as a complication of known infection
   2. Symptoms are of polyneuritis, but the motor paralysis may be severe and bilateral facial paralysis is very characteristic
   3. Diagnosis is confirmed by a very high protein in the CSF (over 250 mg%) but no cell rise
   4. After 2 or 3 weeks of great disability, the patient slowly recovers and usually makes a complete recovery
   5. It may represent one of the autoimmune diseases where the patient's own body defenses react to damage his own tissue
   6. Early use of ACTH may help in the acute but not in the chronic stage
   7. Condition has been called the Guillain-Barré syndrome (after 2 French neurologists who described it during World War I)

Guillain-Barré syndrome: polyneuritis.

Guillain-Barré syndrome: demyelinating at node of Ranvier.

221. DEFICIENCY DISEASES—PART I

I. Deficiency diseases

A. SUBACUTE COMBINED DEGENERATION is usually associated with pernicious anemia, the underlying cause of which is atrophy of the mucous membrane of the stomach, resulting in the lack of absorption of factors for the development of the RBC and for the nutrition of nervous tissue

1. The vital substance of deficiency is vitamin B_{12} (isolated from liver)
   a. In its absence, the number of RBCs that develop is small, yet they are larger than normal and carry more hemoglobin than usual, resulting in macrocytic anemia
2. The deficiency also results in a gradual degeneration of the posterior columns and pyramidal tracts in the spinal cord, together with a polyneuropathy, thus accounting for its name of "subacute combined" degeneration
3. Symptoms: patient is usually middle-aged, white-haired, and often a woman who complains of numbness and tingling in the fingers and toes, soreness of the calves (due to peripheral neuritis), and unsteadiness in walking, particularly in the dark (result of posterior column degeneration). They also admit to soreness of the tongue
4. Signs: with no treatment, there are a mild pallor, slightly yellowish tinge to the tongue, and the tongue is shiny and sore (superficial glossitis)
   a. The limbs are weak and reflexes may be absent if the polyneuritis predominates
   b. Reflexes may be exaggerated if the pyramidal tract degeneration predominates
   c. Plantar reflexes are usually extensor; position sense and vibration sense are impaired in the legs and there is postural ataxia
   d. The calves are tender and there may be a slight impairment of sensation of the glove-and-stocking distribution
   e. The blood count and sternal marrow show characteristics of pernicious anemia
   f. As a result of stomach atrophy, hydrochloric acid is not secreted so that a fractional test meal shows no free acid even after the injection of histamine (a histamine-fast achlorhydria). Serum vitamin B_{12} is very low, but the CSF is normal
5. Treatment
   a. Vitamin B_{12} is given by injection. A small dose corrects the vitamin deficiency, but to treat the neurologic lesion, much higher doses are needed (250 micrograms daily, at first, then twice weekly for many months and less frequently on a permanent basis)
   b. The polyneuritis should clear up, but the symptoms of spinal cord disease, if present for a long time, may only partially recover
   c. Relapse may occur with inadequate treatment

B. OTHER VITAMIN B_{12}–DEFICIENCY SYNDROMES: absorption of B_{12} may be defective in a variety of intestinal diseases

1. Steatorrhea: where the absorption of fat is deficient
2. Surgical removal of the stomach
3. Crohn's disease (regional enteritis)
4. Chronic diarrhea
5. In the above conditions there are deficiencies of other vitamins as well, such as vitamin B_1, pyridoxine, nicotinic acid, and folic acid
6. The symptoms and signs may be suggestive of polyneuritis or of subacute combined degeneration, but they are not associated with a histamine-fast achlorhydria or the typical marrow of pernicious anemia

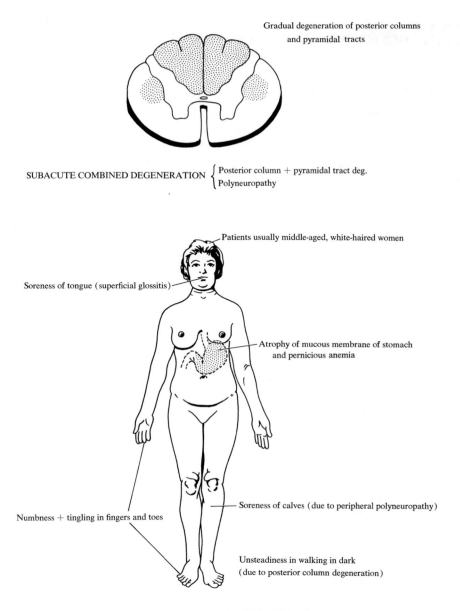

Gradual degeneration of posterior columns
and pyramidal tracts

SUBACUTE COMBINED DEGENERATION {
Posterior column + pyramidal tract deg.
Polyneuropathy

Patients usually middle-aged, white-haired women

Soreness of tongue (superficial glossitis)

Atrophy of mucous membrane of stomach
and pernicious anemia

Soreness of calves (due to peripheral polyneuropathy)

Numbness + tingling in fingers and toes

Unsteadiness in walking in dark
(due to posterior column degeneration)

VITAL DEFICIENCY OF VITAMIN B$_{12}$

222. DEFICIENCY DISEASES—PART II

I. Deficiency diseases (cont.)
C. VITAMIN B$_1$ (THIAMINE) DEFICIENCY leads to beriberi
   1. In conditions of starvation, inadequate diets or severe gastric disturbances, not enough vitamin B$_1$ may be absorbed, and a polyneuritis may develop
   2. In severe cases, the heart enlarges and fails, and edema of the limbs develops
   3. If changes take place in the upper brain stem resulting in paralysis of eye movement, ataxia, confusion, and coma, we call it Wernicke's encephalopathy
   4. Treatment: large doses of vitamin B$_1$ by injection (100 mg daily)
D. NIACIN (NICOTINIC ACID) DEFICIENCY leads to pellagra (B-group deficiency)
   1. Symptoms may be those of polyneuropathy, or a syndrome similar to subacute combined degeneration occurs. However, in addition, there is a scaly brown rash on those parts of the skin exposed to light
   2. Consists of isolated neuronal changes characterized by cellular chromatolysis without associated glial or vascular changes
E. FOLIC ACID DEFICIENCY: since folic acid is concerned with blood formation, deficiencies cause anemias and polyneuropathies. Severe mental changes also occur
   Treatment: restoration of blood folic acid levels

II. Neurologic complications of certain metabolic diseases: used clinically to describe a number of cerebral disturbances secondary to general metabolic disorders. Lesions are nonspecific and related to visceral lesions or anoxic disturbances
A. DIABETES MELLITUS
   1. Polyneuritis: a common complication of poorly controlled diabetes
      a. Symptoms: particularly sensory with severe burning sensation in the hands and feet, numbness of extremities, and increasing difficulty in walking
      b. Exact cause is unknown and response to treatment not very good
      c. Complete control of the diabetes is essential to prevent worsening
   2. Diabetic coma: a danger present in poorly controlled patients
      a. General malaise for some hours or days, increased thirst, followed by vomiting. There are dehydration, drowsiness, then passes into a coma if not treated
   3. An additional infection (e.g., a carbuncle) may lead to coma even in a well-controlled patient since control breaks down, urine is loaded with sugar and acetone, and the blood sugar is high
      a. Treatment: insulin, probably IV, balanced with IV glucose together with saline to correct the dehydration
B. HYPOGLYCEMIC COMA is the result of a too low blood sugar and may lead to coma
   1. Patient begins to sweat profusely, becomes ataxic, confused, and may have convulsions and lose consciousness
   2. Prolonged hypoglycemia may cause permanent brain damage
   3. Treatment: sugar by mouth or IV or give an injection of adrenalin, which causes sugar to pour into the blood from the liver
C. SPONTANEOUS HYPOGLYCEMIA: very low blood sugar in patients not on insulin
   1. Symptoms vary from transient disturbances of behavior, episodes of ataxia and drowsiness, to fits and loss of consciousness with profuse sweating
   2. Most common cause is a tumor of cells in pancreas which secretes insulin
      a. Can be diagnosed by starving patient 24 hr or more. Blood sugar will fall below 40 mg%. Some of the symptoms described will develop, and if an EEG is taken at the same time, it will be grossly abnormal
      b. Diagnosis is confirmed by giving glucose orally or by injection (blood sugar and EEG should return to normal in 15 minutes)
      c. Treatment: explore pancreas and remove islet-cell tumor

HYPOGLYCEMIC STATE—MECHANISMS AND SYMPTOMS

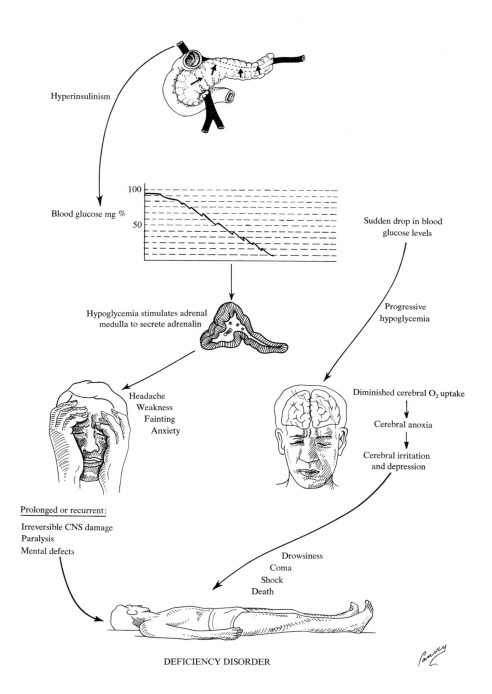

Hyperinsulinism

Blood glucose mg %

100

50

Sudden drop in blood
glucose levels

Hypoglycemia stimulates adrenal
medulla to secrete adrenalin

Progressive
hypoglycemia

Headache
Weakness
Fainting
Anxiety

Diminished cerebral O$_2$ uptake

Cerebral anoxia

Cerebral irritation
and depression

Prolonged or recurrent:

Irreversible CNS damage
Paralysis
Mental defects

Drowsiness
Coma
Shock
Death

DEFICIENCY DISORDER

223. DEFICIENCY DISEASES—PART III

II. Neurologic complications of certain metabolic diseases (cont.)

D. POTASSIUM DEFICIENCY
1. If the blood potassium level falls very low, a weakness of the limbs can develop leading to total flaccid paralysis
2. Can occur in patients receiving cortisone and in some forms of nephritis when the potassium is excreted in excess
3. A similar paralysis may occur spontaneously, however, in cases of familial periodic paralysis
   a. Occurs repeatedly while the patient is asleep, especially after heavy meals
   b. Several members of the family are affected
   c. Severity of attacks can be reduced by taking potassium regularly by mouth
4. Potassium should also be monitored very carefully during renal dialysis
5. Treatment: give potassium chloride or citrate

E. CARCINOMATOUS NEUROPATHY: cause is unknown
1. A slow degenerative process of the sensory fibers in the peripheral nerves and the posterior columns which can occur in patients who have carcinoma of the bronchus (occasionally with other carcinomas)
2. Patients develop progressive numbness, weakness, and wasting of the legs with increasing ataxia due to loss of position sense. Symptoms may be present long before the carcinoma is discovered
3. Treatment: to no avail, although occasionally, successful removal of the primary tumor may lead to recovery
4. A similar mysterious degeneration of the cerebellum produces cerebellar ataxia and occurs in patients with unsuspected carcinoma. Again, the bronchus is the most common primary site

F. RESPIRATORY ENCEPHALOPATHIES: secondary to chronic bronchopulmonary disease and attributable to hypoxia and hypercapnia

G. HEPATIC INSUFFICIENCY: severe hepatic insufficiency as a result of terminal coma in hepatic cirrhosis and severe hepatitis, portacaval encephalopathy, and Wilson's disease (hepatolenticular degeneration)
1. Lesions predominate in the globus pallidus but may be seen in the dentate nucleus and cerebral cortex

H. DISORDERS OF IRON METABOLISM: seen in primary or secondary hemochromatosis. The blood-brain barrier provides an efficient protection against the diffusion of iron pigment into the CNS and consequently cerebral lesions are very constricted, thus hemosiderin is seen in the choroid plexuses, area postrema of the medulla, the pineal gland, and other vestigial remnants such as the paraphysis and subfornical organ

I. DISORDERS OF CALCIUM METABOLISM
1. Hypoparathyroidism may lead to Fahr's disease with a massive perivascular deposit of pseudocalcium in the basal ganglia, especially the caudate nucleus
2. When the serum calcium levels approach the critical level, increased neuromuscular excitability develops, leading to muscle twitching, and occasionally to a convulsive disorder resembling epilepsy, although there is usually no loss of consciousness. This excitable state, known as tetany, can be relieved by the IV administration of calcium

Spider angioma

Confused state

Jaundice and multiple spider angiomas

Asterixis + palmar erythema

HYPOCALCEMIA

Wide range of mental aberrations
Seizures
Coma occasionally develops
Tetany (cardinal finding)
Increased deep tendon reflexes
ECG changes

HEPATIC FAILURE

Hepatic encephalopathy secondary to severely
impaired liver function or portacaval
shunting of blood into peripheral
circulation—may progress to coma

Encephalopathy evolves over days or weeks—
varies with pathogenesis

Mental confusion with increased or decreased
psychomotor activity prior to progressive
drowsiness, stupor, and coma

Neurologic changes: Asterixis
Increased muscle tone
Hyperreflexia with clonus
Ataxia
Babinski signs
Focal or generalized
seizures

Chvostek's sign elicited

Numbness + tingling,
then stiffness

Trousseau's sign (latent tetany)

Pansky

224. INBORN ERRORS OF METABOLISM

I. **Inborn errors of metabolism** (genetic metabolic diseases due to an enzyme defect): some people are born with a tendency for one of the vital processes of the body to go wrong. As a result, some substance, normally broken down and excreted, builds up in the body to toxic levels. There are many such diseases. Fortunately all are rare and few affect the nervous system

A. DISORDERS OF METAL METABOLISM
   1. Wilson's disease (hepatolenticular degeneration): due to a deficiency of a substance called ceruloplasmin, which normally carries copper in the blood
      a. Copper cannot be maintained at its normal blood level, excessive amounts are excreted in the urine and, more important, are deposited in various organs, including the liver and basal ganglia (lenticular nucleus)
      b. Copper is also deposited in the cornea of the eye, producing a diagnostic golden brown ring, the Kayser-Fleischer ring
      c. Several members of a family are affected
      d. Symptoms
         i. Severe involuntary movements develop in youth with progressive disabling and mental deterioration
         ii. Liver failure later develops
      e. Treatment: Penicillamine is used which extracts copper from the CNS and increases kidney secretion of copper. Striking improvement is seen in the clinical condition, which appears to be maintained for many years in cases that are treated early
   2. Kinky hair disease (Menkes' disease): defect of intestinal copper absorption. Abnormalities in hair and neuropsychiatric manifestations

B. DISORDER OF PIGMENT METABOLISM: porphyria (hematoporphyrinuria)
   1. Porphyrins are substances concerned in the formation of blood pigments. If metabolism is disturbed, large amounts are excreted and detected in the urine. In some cases, the urine may turn to a port wine color after exposure to light and in other cases special tests are needed
   2. Symptoms
      a. The skin of the patient blisters easily on exposure to sunlight
      b. Some patients have abdominal pain attacks and mental disturbances lasting several weeks, during which time they develop a polyneuritis. The attacks often vanish spontaneously and may follow administration of barbiturates
   3. Treatment: none is known

C. DISORDERS OF AMINO ACID METABOLISM: cause of many syndromes of mental retardation in childhood which may be associated with various neurologic manifestations
   1. Phenylketonuria: due to the absence of phenylalanine hydroxylase
      a. The neurologic findings are variable and poorly defined
      b. Can be detected by adding a few drops of ferric chloride to the urine
      c. Treatment: special diet free from phenylalanine has been used with good results. This condition is the only type of mental deficiency known that is preventable
   2. Hartnup's disease: due to a disorder of tryptophan absorption (resembles pellagra)
   3. Homocystinuria: due to a deficit of cystathione synthetase
      a. Disease causes alterations in blood vessel walls, with fibrosis of the intima and degeneration of elastic fibers. Also see foci of cerebral necrosis
   4. Leucinosis or maple sugar urine disease: identified by presence of alphahydroxybutyric acid in urine with characteristic odor. Due to decrease in decarboxylase activity
   5. Tyranosis: related to deficiency in oxidation of parahydroxyphenylpyruvic acid

Kayser-Fleischer ring (pathognomonic)

Discoloration of
lunnulae due to
copper accumulation

WILSON'S DISEASE (HEPATOLENTICULAR DEGENERATION)

Golden or green-brown Kayser-Fleischer rings at periphery of cornea

Serum ceruloplasm decreased; serum + urinary copper increases

Choreoathetosis and staggering possible

Blue-brown discoloration of lunulae

Tremors, incoordination, speech difficulty, and deteriorating handwriting

Personality changes—episodes of acute schizophrenia

Dysphagia, rigidity, transient hemiparesis

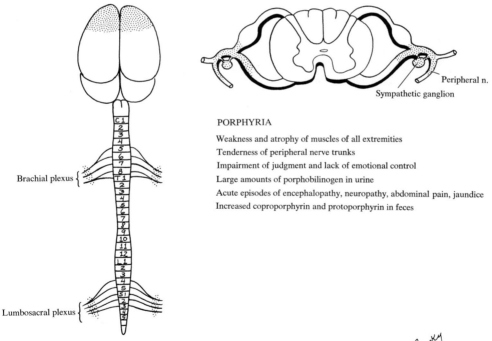

Peripheral n.

Sympathetic ganglion

Brachial plexus {

Lumbosacral plexus {

PORPHYRIA

Weakness and atrophy of muscles of all extremities

Tenderness of peripheral nerve trunks

Impairment of judgment and lack of emotional control

Large amounts of porphobilinogen in urine

Acute episodes of encephalopathy, neuropathy, abdominal pain, jaundice

Increased coproporphyrin and protoporphyrin in feces

METABOLIC BRAIN DISEASES

225. DISORDERS OF CARBOHYDRATE, MUCOPOLYSACCHARIDE, AND LIPID

I. Disorders of carbohydrate metabolism

A. GLYCOGENOSES: some forms of glycogen storage disease (MacArdle's and Forbes' diseases) involve the skeletal musculature with an exceptional CNS involvement. Pompe's disease, caused by acid maltase deficiency, does implicate the CNS

B. GALACTOSEMIA: deficit of galactose-1-phosphate-N-uridyltransferase (CNS lesions)

C. LAFORA'S DISEASE: exact metabolic disturbance uncertain. May be due to excessive storage of amylopectin. Picture of progressive familial myoclonic epilepsy

D. SUBACUTE NECROTIZING ENCEPHALOPATHY (LEIGH-FEIGIN DISEASE) involves basal ganglia, tegmentum of brainstem, corpora quadrigemina and inferior olives. Neurons are spared and there is gliosis. Metabolic disorder poorly understood

II. Disorders of mucopolysaccharide metabolism: a systemic disturbance of acid mucopolysaccharides or glycosaminoglycans (excreted in urine) is accompanied by a lipid neuronal storage disorder that essentially implicates the gangliosides

A. HURLER'S DISEASE (TYPE I MUCOPOLYSACCHARIDOSIS): with excessive chondroitin sulfate B and heparin sulfate

B. HUNTER'S DISEASE (TYPE II MUCOPOLYSACCHARIDOSIS) differs from Hurler's by its sex-linked recessive character and absence of corneal opacities

C. SANFILIPPO'S DISEASE (TYPE III MUCOPOLYSACCHARIDOSIS): only heparin sulfate accumulates in excessive amounts

III. Disorders of lipid metabolism: an abnormal accumulation, mostly in neurons of the CNS, of a specific lipid metabolite. May or may not be associated with PNS lesions and visceral lesions. Classifications refer to the lipids involved and enzyme(s) defect(s) rather than the clinical features. The chief lesions are of intense neuronal swelling with distention of the cell bodies. Cerebellar lesions are common

A. THE SPHINGOLIPIDOSES: are the most important group among the neurolipidoses and are characterized by an excess of sphingolipids

B. THE GANGLIOSIDOSES (AMAUROTIC IDIOCIES) are characterized by accumulation of gangliosides
  1. GM2 gangliosidoses: classified according to their specific enzyme deficiency
     a. Tay-Sachs disease: due to accumulation of GM2 gangliosides and ceramide trihexoside as a result of deficiency of hexosamidase A
     b. Sandhoff disease: due to deficiency of both hexosamidases A and B
  2. GD3 gangliodisosis: congenital amaurotic idiocy
  3. GM1 gangliosidosis: late amaurotic infantile idiocy

C. THE CEREBROSIDOSES
  1. Gaucher's disease: deficiency of glucocerebroside-B-glucosidase. In acute stage one sees macrophages (Gaucher cells) in the CNS
  2. Krabbe's disease: deficiency seems to be a galactoside-B-galactosidase. Characterized by conspicuous involvement of myelin of the white matter

D. THE SULFATIDOSES: characterized by a deficit of arylsulfatase
  1. Metachromatic leukodystrophy: deficit of arylsulfatase A
  2. Austin's disease: deficit of arylsulfatases A, B, and C

E. NIEMANN-PICK'S DISEASE displays several forms both clinically and genetically
  1. All show an accumulation of sphingomyelin with a sphingomyelinase deficit
  2. CNS involvement is accompanied by an infiltration of foamy cells and changes in the endothelial and perithelial elements of blood vessel walls

F. REFSUM'S DISEASE: excessive storage of phytanic acid due to a deficit of phytanic oxidase. The PNS is affected by a demyelinating neuropathy

G. THE LIPOPROTEINOSES: acanthocytosis with a deficit of β-lipoprotein and Tangier's disease with a deficit of α-lipoprotein

Gross section of the brain showing widened sulci, narrowed gyri, a thinned corpus callosum, and enlarged ventricles typical of brain atrophy observed in Tay-Sachs disease and other ganglioside storage diseases.

Histologic section showing swollen neuron and increased number of glial cells (arrows) characteristic of Tay-Sachs disease.

TAY-SACHS DISEASE

226. ALCOHOLISM, BARBITURATES, AND ANTIEPILEPTIC DRUGS

I. Toxic diseases

A. ALCOHOLISM

1. When a person is drunk, he develops a disturbance of coordination, resulting in double vision, slurring of speech (dysarthria), unsteadiness of the arms and legs, and ataxia of gait
   a. These develop during a bout of drinking, but usually clear up some hours afterward, usually with a "hangover"
2. Excessive alcohol intake over a long period of time, however, has other side effects
   a. The toxin acts not only directly on the brain and liver (cirrhosis), but also causes sufficient damage to the stomach to prevent absorption of vitamins, especially vitamin B_1
   b. The direct effects are gross tremulousness of the entire body, followed a few days later by wild confusion and agitation with terrifying visual and auditory hallucinations, and absolute wakefulness—delirium tremens (DTs). Seen in chronic alcoholics who have had a heavy drinking bout and then stopped
   c. Treatment: sedation and no alcohol
3. After years of heavy drinking (chronic alcoholism), the deficiency of vitamin B_1 causes severe damage to the brainstem leading to mental confusion and paralysis of eye movements, double vision, ataxia of the arms and legs, and in addition a polyneuropathy—Wernicke's encephalopathy
   a. Korsakoff's syndrome: in alcoholics is regarded as a late chronic stage of Wernicke's encephalopathy. Predominant involvement of the mamillary bodies accounts for memory disturbances with fixation amnesia, confabulation, and temporospatial disorientation with polyneuritis
   b. Treatment: massive doses of vitamin B_1
4. Alcoholic polyneuritis is not uncommon and responds to vitamin therapy, provided alcohol is avoided completely

B. BARBITURATES AND ANTIEPILEPTIC DRUGS

1. If a large dose of barbiturates is suddenly ingested (e.g., in an attempted suicide), coma develops rapidly with complete flaccidity of the limbs, absent reflexes, and often collapse of the circulation. The patient can be saved if the stomach is pumped out at once, and intravenous cerebral stimulants (e.g., megimide) are given until there are signs of arousal
2. Excessive amounts of some drugs taken over a long period of time produce slurring of speech, ataxia of the limbs, and recurrent drowsiness
3. Many poorly controlled epileptics develop toxic effects from a combination of too much phenobarbitone, epanutin, and primidone. They also develop drowsiness, gross ataxia, nystagmus, and dysarthria, but often with sufficient headache and vomiting to arouse suspicions of a cerebellar tumor. Folic acid deficiency also occurs
   a. The problem of treatment is that if anticonvulsants are stopped suddenly, status epilepticus can occur
   b. The reduction in medication is essential but it must be gradual and take place where treatment is readily available if fits should occur

Wernicke's encephalopathy—thiamine deficiency.

Alcoholic neuropathy resulting from nutritional deficiency. Note scattered areas of demyelination along nerve fibers (arrows).

227. OTHER TOXIC DISEASES

I. Toxic diseases (cont.)

C. OTHER TOXIC CONDITIONS AND DISEASES
  1. Tranquilizers
     a. May cause slowing reactions, even in moderate doses; defective judgment and slight unsteadiness; and even these signs may be more marked if the drug is mixed with alcohol
     b. Some cause troublesome polyneuropathy
     c. Large doses over a long period of time produce a syndrome almost indistinguishable from parkinsonism
  2. Amphetamines (and to a lesser extent belladonna)
     a. Cause excitement, tremor, rapid pulse, collapse, and delirium
     b. Belladonna or atropine also causes dilatation of the pupils
  3. Lead
     a. Produces a polyneuropathy, especially wristdrop
     b. The type of lead used in paints and gasolines can cause a cerebral disturbance called lead encephalopathy with headaches, mental changes, fits, and even coma
  4. Mercury
     a. This is rarely a problem these days, but when it happens it causes severe tremors of the arms and legs called hatter's shakes
     b. Certain plant sprays contain a form of mercury which causes degeneration of the posterior columns with severe postural ataxia
  5. Arsenic
     a. Taken by mouth over a long period of time causes polyneuropathy
     b. The form given by injection for the treatment of syphilis can cause an encephalopathy similar to that of lead poisoning
  6. Oral contraceptives
     a. May cause cerebral thrombosis, papilledema, and chorea. Migraines may also increase

D. GENERAL COMMENTS
  1. Most of the toxic effects of any substance can be cured providing the substance is withdrawn completely
  2. If intoxication has been present for a long time, permanent damage may have occurred in the nervous system and cure cannot be complete
     a. This applies to drugs as well as to alcoholic addiction
     b. One cannot be certain that the patient is still not secretly obtaining supplies

Astrocytosis: a histologic preparation of cerebral cortex showing an increase in astrocyte population due to lead poisoning.

Normal astrocyte population in normal cerebral cortex.

Unit Twenty-One

DISEASES OF MUSCLE

228. THE MYOPATHIES OR PROGRESSIVE MUSCULAR DYSTROPHIES

I. **Introduction:** diseases of the muscles have always come under the province of the neurologists since muscles are a vital portion of the motor system, even though the nervous system is normal. It is not always easy to tell whether or not muscle weakness and wasting are due to a lesion of the motor nerves or to disease limited to the muscle itself. Muscle diseases may be degenerative and are called myopathies or progressive muscular dystrophies, or they may be inflammatory and the term "myositis" or "polymyositis" applies. In addition, there is a condition called myasthenia gravis in which impulses cannot "get across" from the nerve into the muscle to be supplied

II. **The myopathies or progressive muscular dystrophies** are heredofamilial diseases in which the muscle fibers slowly degenerate and waste, eventually producing severe disability. Since the muscles are often replaced by fat, the patient superficially may not appear "wasted." Several types exist, varying in the age groups and particular muscles affected as well as the gravity of prognosis

A. PSEUDOHYPERTROPHIC MUSCULAR DYSTROPHY (DUCHENNE TYPE) is the most severe and serious type, so named because the muscles are unusually large or hypertrophied, though very weak
1. Almost, though not completely, confined to boys
2. Progression is such that very few reach adult life, and its rapid course is usually fatal by the age of 20
3. Mode of inheritance: though mainly affecting boys, the disease is passed on by apparently unaffected female carriers (recessive, sex-linked). It can be detected by a combination of chemical, electrical, and biopsy examination of muscles. There is a 50% chance of any son of a carrier to be affected, and a daughter may become a carrier
4. Symptoms: child walks badly and has difficulty in picking himself up after falling and in climbing stairs. Early onset in muscles of pelvic girdle
5. Signs
   a. Child walks with a waddle and has marked lordosis
   b. Calves, buttocks, and shoulder girdle muscles look much larger than expected for the age, but have a tough, rubbery feel and may be weak
   c. Back muscles are weak
   d. Uses "climbing up the legs" technique to get up if lying on the floor: rolls onto face, then half kneels, places hands first on ankles and then on shins, his knees, his thighs, and gradually pushes himself up
   e. Facial muscles are rarely affected; tendon reflexes are usually absent
   f. Pseudohypertrophy gives way to wasting, and patients die of chest infection
   g. Diagnosis: blood level of muscle enzyme creatine phosphokinase is elevated in the disease. This is also true in apparently unaffected boys who later develop the disease and may also be the case in female carriers

B. FACIOSCAPULOHUMERAL DYSTROPHY (OF LANDOUZY-DÉJERINE)
1. Wasting and weakness affect the face, shoulder girdles, and upper arms (often one side more than the other)
2. Of slow evolution, compatible with a relatively long life. Onset in middle childhood or sometimes in the 2nd decade of life and affects both sexes
3. The least progressive of the dystrophies
4. The muscles do not hypertrophy
5. Inherited from an affected individual, not via a carrier, and is transmitted as an autosomal dominant trait

Duchenne-type muscular dystrophy showing great variation in size of muscle fibers.

Progressive muscular dystrophy showing atrophy and wasting of muscle fibers.

229. MYOPATHIES AND MYOTONIA

II. The myopathies or progressive muscular dystrophies (cont.)

C. LIMB GIRDLE MYOPATHIES OR DYSTROPHIES involve muscles of the hip and shoulder girdles or both and affect older patients
1. Weakness and wasting progress slowly and disability is great in 10 years
2. Usually no history in preceding generations and transmission is recessive but not sex-linked

D. OCULAR MYOPATHIES
1. Degenerative changes develop in the muscles of the lids and eyes in late middle life (occasionally earlier)
2. Progressive ptosis and difficulty in moving eyes; finally fixed in midline
3. The lesion is now known to lie in the muscles
4. There is no treatment, but plastic surgery may be needed to raise eyelids

E. TREATMENT OF MYOPATHIES: no curative treatment is known
1. Physiotherapy is of little use and may even overtire weakened muscles
2. Patient is prone to chest infections which may prove to be fatal

III. Myotonia is a name given to a condition in which muscles continue to contract for some time after the need has passed. Seen in two major diseases: dystrophia myotonia and myotonia congenita

A. DYSTROPHIA MYOTONIA (MYOTONIA ATROPHICA): a heredofamilial disorder which affects men more than women and appears to begin in young adult life
1. Symptoms
   a. Patients complain of increasing weakness of the hands and arms, and occasionally of muscular wasting
   b. Occasionally have difficulty in "letting go" after picking up an object or when shaking hands and may realize that their vision is failing
2. Signs
   a. Appearance is very characteristic and all patients look alike
      i. Have long, thin faces, with drooping eyelids and corners of mouth
      ii. Go bald early and remaining hair is unkempt
      iii. Sternomastoids are wasting, giving the neck a scraggy appearance, and patients raise their heads from a pillow with great difficulty
   b. Most muscles show some wasting, but especially the hand and thigh muscles
   c. The myotonia is seen when the patient smiles, because the smile remains after the face has relaxed
   d. After gripping something, unable to release grip and hand must be dragged away
   e. Tapping the tongue or other muscles causes a dent that remains for long time
   f. Other abnormalities may include cataracts, testicular atrophy, endocrine disturbances, and even cardiac abnormalities
3. Treatment: nothing stops the steady progression of weakness and wasting, but the course varies greatly so that if the myotonia can be relieved, it is very helpful
   a. Injection of long-acting preparations of ACTH will reduce the myotonia in 4 hours, although the effects last only 24 hours to 3 days

B. MYOTONIA CONGENITA is a hereditary condition present from infancy
1. Dystrophic features (muscle wasting, baldness, cataracts, testicular atrophy, or ovarian atrophy) are usually absent
2. The patient complains of muscle stiffness rather than muscle weakness

Myotonic dystrophy: low-power cross-section of muscle fibers showing their variation in size.

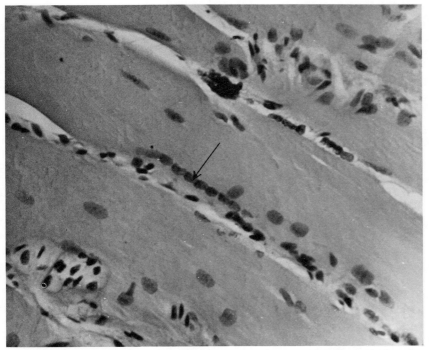

Myotonic dystrophy: high-power long. section of muscle fibers. Note characteristically chained nuclei along cell membrane.

230. MYOSITIS AND MYASTHENIA GRAVIS

IV. Other muscle disorders

A. MYOSITIS: acute myositis (viral or infectious) is rare; parasitic myositis is less exceptional and is common in the tropics and subtropics
1. Muscles can be infected from dirty wounds or careless injections and they become swollen and tender as inflammatory cells enter from the bloodstream
2. Whether allergic or a toxic state it can occur with inflammatory lesions of the skin (dermatomyositis) or of the blood vessels (polyarteritis nodosa)
3. When many muscles are involved, we speak of polymyositis
4. Symptoms: the patient (of any age) feels generally unwell and complains of increasing weakness of the limbs with swelling and tenderness of the muscles
5. Signs
   a. Slight fever and a raised pulse rate
   b. Affected muscles are weak and tender and overlying skin may be slightly swollen, and there are absent tendon reflexes
   c. High blood sedimentation rate with elevation of serum aldolase and creatine kinase
   d. Muscles may waste if the disease is present for a long time
   e. Diagnosis is confirmed by biopsy of affected muscle
6. Treatment: ACTH and cortisone may improve situation and must be taken until all signs settle down

B. MYASTHENIA GRAVIS: myasthenia means weakness of muscles. The term usually is used for a weakness that worsens with use and rapidly recovers when muscle is at rest
1. Mainly affects young women but may occur in either sex at any age
2. Caused by an abnormality preventing impulses from branches of motor nerves from passing across nerve endings in the muscle fibers at the neuromuscular junctions
3. Symptoms
   a. The eyes are often affected: eyelids droop, eyes squint, and there is double vision (symptoms vary from hour to hour)
   b. Patients may be normal on awakening, but get worse as day wears on: voice gets weak and hoarse near the end of each sentence; chewing and swallowing become more difficult as meal progresses; arm and back muscles become weak after use (varies greatly); and recovery is complete on resting
      i. Months may pass with no symptoms at all
4. Signs: any muscle may be affected, so signs will vary greatly
   a. Most commonly seen are ptosis, squinting, hoarseness, and a nasal voice, all varying from time to time during the examination
   b. One major sign is the ease with which the muscles become fatigued
   c. Diagnosis can be confirmed by giving subcutaneous injection of neostigmine (Prostigmine), 2.5 mg with $\frac{1}{100}$ gr atropine. The weak muscles become nearly normal in about 45 minutes and this lasts for about 4–5 hours
   d. A quick test is using edrophonium chloride (Tensilon), 1 ml (10 mg) given IV. Restores power to normal in 1 minute, but effect usually lasts for less than 5 minutes or just slightly longer
5. Treatment: prostigmine, or longer acting pyridostigmine, given regularly by mouth. The dose may be large and taken frequently to keep patient well
   a. Most patients respond well except for some eye muscle problems
   b. In some cases, nothing seems to stop the disease, paralysis of throat and respiratory muscles develops, and drug injections every 1–2 hours are needed, or even a tracheotomy or respirator becomes necessary
   c. Thymus removal helps some (young women) if medical treatment fails, but results are unpredictable. ACTH may also help in some cases

Inflammation (myositis) of a voluntary muscle. Note cellular infiltrate in center of field (arrow).

Multiple myositis or polymyositis showing cellular infiltration.

Unit Twenty-Two

TUMORS

231. TUMORS OF THE CENTRAL NERVOUS SYSTEM: INTRODUCTION AND CLASSIFICATION

I. Introduction and classification: tumors or neoplasms that affect the nervous system are said to be primary if they grow from cells of the nervous system itself; or secondary (metastases) if they are carried to the nervous system by the blood from a primary tumor elsewhere (e.g., the bronchus, breast, etc.). Nervous system tumors are classified into 2 groups: those of the central nervous system and those of the peripheral nervous system. Tumors of the PNS are situated outside the cranial and spinal cavities and include tumors of the peripheral nerve trunks, sympathetic ganglia, adrenal medulla, and chemoreceptors (these lie generally outside the context of traditional neuropathology). Tumors of nerve cells and fibers are also very rare. Neoplasms grow, instead, from the supporting tissue: gliomas from the glia, meningiomas from the meninges, ependymomas from the lining of the ventricles, neurofibromas from the sheaths of nerves, and pituitary tumors

CNS tumors can further be divided into two groups: intracranial or intrinsic (erroneously called cerebral tumors) and extrinsic or tumors causing spinal root and cord compression. Most extrinsic tumors are benign, slow growing, and compress the nervous tissue while remaining separate from it. They do not spread elsewhere. Most intrinsic tumors are malignant, grow more rapidly, and destroy nervous tissue by spreading through it to a point where one cannot tell where tumor ends and normal tissue begins. Metastases to other parts of the body from even malignant cerebral tumors are rare. Some intrinsic tumors might be called "relatively benign" since they grow slowly and do not produce much destruction for years. Some extrinsic tumors, on the other hand, may become malignant. All metastases (extrinsic or intrinsic) are malignant. Finally, malignant tumors of nearby structures may also invade the skull and spinal cord. In addition, tumors are classified topographically and histologically

II. Cerebral tumors

A. GENERAL FEATURES: intracranial tumors produce symptoms in three ways
1. Space-occupying lesion: since the skull is a rigid box almost filled with brain, any new mass fills up the remaining space and then causes pressure inside the skull to rise
2. Tumors may compress the ventricles or the narrow parts of the CSF pathways and dam up the flow. Since CSF is still produced, pressure is built up
3. The tumor presses on, or destroys, some part of the brain itself directly, producing symptoms and signs that gradually become worse and more extensive as the tumor grows (in contrast to a sudden onset of a vascular accident)

B. SYMPTOMS AND SIGNS OF INCREASED INTRACRANIAL PRESSURE
1. Headaches: especially after lying flat. Are increased by coughing and stooping
2. Vomiting: usually at the height of the headache
3. Diplopia
4. Blurring of vision: on movement of the head
5. Drowsiness: increasing and progressive and leading to coma
6. Slowing of the pulse
7. Dilatation of the pupils with failure to react to light
8. Papilledema: swelling of the optic disk as seen through an ophthalmoscope (an important vital sign). The disk is where the optic nerve fibers leave the eyeball. Elevated pressure inside the skull is transmitted along the nerve as far as this point and elevates the disk

Intracranial tumors: a glioblastoma shown in gross brain.

Histologic preparation of a glioblastoma.

232. EFFECTS OF TUMORS ON THE BRAIN

II. Cerebral tumors (cont.)
C. DIRECT EFFECTS OF A TUMOR ON THE BRAIN
1. The two most common effects of local damage to the brain are weakness of the opposite side and fits
   a. The fits may take the form of convulsions affecting all parts of the body, or they may be focal fits affecting, e.g., one forearm alone
   b. The nearer the surface the tumor lies, the more likely it is to produce focal fits, but as it grows, the amount of the limb affected becomes greater and it is common to find that later, after the twitching has stopped, part of the body remains paralyzed for some time. This is referred to as Todd's paralysis
   c. Fits need not consist of convulsions
      i. Temporal lobe tumors cause sudden hallucinations of taste, smell, or vision. The patient may see a clear picture, smell some vague aroma before losing consciousness, or often have a strong feeling of déjà-vu sensation (this has all happened before sensation)
      ii. Occipital lobe fits may be also seen as blobs and flashes of light
2. In addition, special features develop according to the lobe affected
   a. If in the frontal lobe: there are also loss of memory and personality changes
   b. If in the temporal lobe or occipital lobe: can damage the optic radiations so that vision is lost in the half of each visual field opposite the lesion (homonymous hemianopia)
   c. Tumors growing in the dominant hemisphere (left, in most) cause progressively increasing speech disturbances (dysphasia)
   d. Tumors arising in the brainstem are more common in children than in adults
      i. They cause increasing paralysis of eye movements and of other cranial nerves, with a loss of sensation of the face, and later difficulty in swallowing
      ii. Intrinsic brainstem tumors are invariably fatal (the stem compressed by benign tumors growing nearby can, however, be helped by surgery)
   e. Tumors in the cerebellum cause unsteadiness (ataxia) of the limbs on the same side as the tumor and, as a result of IVth nerve involvement, produce very early signs of elevated intracranial pressure
      i. Tumors in the vermis of the cerebellum can cause severe ataxia when the patient walks, but this disappears when he is lying down so it appears that the patient is fabricating (lying)

Gross section through cerebrum showing meningioma. Note compression rather than invasion of cerebral tissue.

Histology of a meningioma: note whorl at arrow.

233. PRIMARY NEOPLASMS: CEREBRAL TUMORS—PART I

I. Primary neoplasms
A. CEREBRAL TUMORS
1. Meningiomas: benign extrinsic tumors seen between 20–60 years with a maximal incidence at 45 years, accounting for 13–18% of primary intracranial tumors and about 25% of intraspinal tumors
   a. Predominantly seen in the female
   b. Grow from arachnoid cells to irritate and compress the brain surface
   c. Patients have long history of headaches, focal fits, and focal paralysis. In time they produce all general signs of a space-occupying lesion
   d. Can be completely cured by surgical removal
2. Gliomas: most cerebral tumors, unfortunately, fall into this group
   a. Develop from glial cells (supporting tissue) such as astrocytes, oligodendrocytes, and microglia
   b. Are always intrinsic: produce signs by working through nerve cells and fibers and compressing surrounding tissue they have not destroyed
   c. They produce all and any of the general features of a cerebral tumor but do so more rapidly than do the meningiomas
   d. Astrocytomas: histologically benign, but a fair number of cases, and after a variable interval are susceptible to malignant changes
      i. Account for 20–30% of tumors in the glioma group
      ii. Are of varying malignancy, some being relatively benign, and grow slowly over many years. Others grow rapidly, taking only a few weeks or months. The latter are called glioblastomas, which are malignant, rapidly fatal astrocytic neoplasms, accounting for about 50% of tumors in the glioma group
         (a) Occur at any age, maximally between 45–55 years and in any region of the CNS but have a predilection for the frontal or temporal lobes, the basal ganglia, and commissural pathways
         (b) Multiple localizations are seen in 5% of cases
      iii. Cysts form in some astrocytomas and drainage is helpful
      iv. Types: seen in cerebral hemispheres, floor of 3rd ventricle, chiasma and optic nerves, brainstem, cerebellum, and spinal cord
   e. Oligodendrogliomas or oligodendrocytomas account for 5% of the gliomas
      i. Are seen between the ages of 30–50 years and involve the cortex and white matter of the cerebral hemispheres
      ii. A very slow-growing tumor which often has calcium in it (seen on x-ray)
   f. Microglias are very rare
   g. Ependymoma: like a medulloblastoma but is usually seen in childhood and adolescence, and results of excision often better (6% of the gliomas)
   h. Choroid plexus papilloma: 0.5% of intracranial tumors and 2% of glioma group
      i. Seen in first decade of life in the ventricular system where the choroid plexus is found
      ii. These tumors are histologically benign
   i. Colloid cysts of 3rd ventricle account for 2% of intracranial gliomas and are seen in young adults, rarely in children
      i. Always seen at anterior end of 3rd ventricle near foramen of Monro
      ii. Cyst wall has a fibrous capsule attached to the stroma of the choroid plexus and is lined by cuboidal or columnar epithelium

Cerebral tumor: meningioma.

Cerebral tumor: glioblastoma.

234. PRIMARY NEOPLASMS: CEREBRAL TUMORS— PART II

I. Primary neoplasms (cont.)

A. CEREBRAL TUMORS (cont.)

3. Hemangioblastomas: a tumor of immature blood vessels, invariably in the cerebellum
   a. Produces signs of raised intracranial pressure and marked ataxia
   b. Tends to form a cyst with tumor itself remaining as a nodule in cyst wall
   c. Removal of cyst and nodule results in complete cure

4. Medulloblastoma is a malignant neoplasm accounting for $\frac{1}{3}$ of the posterior fossa tumors and is almost always confined to children under 12 years of age
   a. Develops from primitive cells in the roof of the 4th ventricle and spreads into the cerebellum (vermis) and brainstem
   b. Quickly blocks the CSF pathways and causes all the signs of high intracranial pressure. The children become increasingly ataxic
   c. Surgical removal followed by deep x-ray therapy (tumor is very radiosensitive) can give years of survival, but most tumors eventually recur

5. Ganglioneuromas and gangliogliomas are rare; small neoplasms seen in children and young adults
   a. Site is the floor of the 3rd ventricle or the temporal lobe
   b. Show two types of cells: neuronal and astrocytic
   c. Prognosis after surgical removal is favorable

6. Pineal tumors are a variety of tumors arising in the pineal region

7. Schwannomas (neurilemomas or neurinomas) are benign tumors arising from Schwann cells and are found on cranial nerves, spinal nerve roots, peripheral nerve trunks, and at nerve endings
   a. The most frequent kind is the acoustic neuroma which grows from the sheath of the 8th cranial nerve and lies in the angle between the cerebellum and pons (cerebellopontine angle)
      i. Because of their position, they compress the 8th nerve and produce deafness on that side; 7th nerve involvement produces weakness of facial muscles; and 5th nerve involvement produces loss of sensation over the forehead and eye
      ii. They compress the cerebellum and produce ataxia
      iii. They soon compress the brainstem, causing high intracranial pressure and later bulbar paralysis and death
      iv. If diagnosed early, they can be completely removed, although the operation leaves a facial paralysis and deafness on that side
      v. Other tumors may occur in this site as well, like neurofibromas of the 5th cranial nerve or a meningioma

8. Neurofibromas form multiple tumors of nerve roots and peripheral nerves as part of the picture of von Recklinghausen's neurofibromatosis
   a. Are a kind of hyperplasia of the Schwannian and fibroblastic supporting elements of the nerve

9. Melanomas are a wide variety of disorders ranging from a simple increase in normal leptomeningeal pigmentation to highly malignant melanomas
   a. Whether benign or malignant, primary melanomas of the CNS are rare
   b. Since meningiomas sometimes contain melanin pigment, diagnosis may be rather difficult

Schwannoma.

Neurofibroma.

235. PRIMARY NEOPLASMS: CEREBRAL TUMORS—
PART III

I. Primary neoplasms (cont.)

A. CEREBRAL TUMORS (cont.)

10. Sarcomas: intracranial sarcomas are very rare and account for 1–3% of all intracranial tumors. Two types are recognized
    a. Fibrosarcoma, derived from fibroblasts, are well circumscribed but nonencapsulated and are firm tumors seen in the dura, leptomeninges, perivascular spaces, tela choroidea, or stroma of the choroid plexuses
    b. Reticulum-cell sarcomas are of controversial origin
        i. Are not rare and are seen between the ages of 50–70 years
        ii. More invasive and less circumscribed than fibrosarcomas and most often involve the cerebral hemispheres with a predilection for the temporobasal regions and midline structures of the brainstem

11. Pituitary adenomas form a special group. The gland lies in the sella turcica and is attached to the brain by the pituitary stalk. The gland contains 3 major cell types: basophil (stains blue), eosinophil (stains red), and the chromophobe (does not stain). Each cell type can produce a tumor
    a. Basophil tumor: small tumors that are probably the result of a disturbance of the close relationship that exists between the pituitary and adrenal glands— found in association with tumors or hypertrophy of the adrenals
        i. Results in Cushing's syndrome: obesity, hairiness of face, stretching of skin to produce striae, amenorrhea, hypertension, and glycosuria
        ii. Are not space-occupying tumors
        iii. Treatment is directed at the adrenal glands
    b. Eosinophil adenomas cause an increased output of growth hormone
        i. They develop before adult life and produce a very feeble giant
        ii. If bone growth has stopped, they cause acromegaly: face, nose, and jaws enlarge; skin coarsens and folds; hands and feet become progressively larger; and wide spaces develop between the teeth
        iii. These tumors can increase in size until they burst out of the sella and press on the optic chiasma causing bitemporal hemianopia (loss of the outer halves of both visual fields)
        iv. Tumors can be removed, treated by implantation of radioactive material or with deep x-rays
    c. Chromophobe adenomas are quite common and grow to considerable size
        i. Their cells have no special function, but because of their size, they compress the rest of the pituitary gland causing loss of hormones and a condition of hypopituitarism: patients become feeble, with pale, round, yellowish, plump faces; there is body hair loss; patients feel excessively cold; there is a loss of sexual capacity and desire; and women develop amenorrhea
        ii. The tumors expand upward and compress the optic chiasma, resulting in bitemporal hemianopia and, if not treated, blindness follows
        iii. Tumors can be removed surgically, treated with deep x-rays, or implantation of radioactive substance (with some danger)

Ventral aspect of brain with pituitary adenoma.

Histologic section of pituitary adenoma (eosinophilic).

I. Primary neoplasms (cont.)

A. CEREBRAL TUMORS (cont.)

12. Craniopharyngiomas account for 3% of all intracranial tumors
    a. Are encapsulated, solid, and/or cystic and are related to the pituitary gland and stalk and often arise from Rathke's pouch
    b. They are seen in children and adolescents and are the most common of the supratentorial tumors seen in childhood
    c. They are histologically benign but, due to extensions, are difficult to remove surgically

13. Cholesteatomas ("pearly tumors") are cystic tumors resulting from the inclusion of epiblastic elements in areas from which they are normally absent (take place in fetal life or as a result of trauma)
    a. Are very rare tumors seen in the posterior fossa, intra- and suprasellar regions, and the lumbosacral spinal region
    b. Histologically two groups are distinguished: epidermoid and dermoid cysts

14. Lipomas are rare, benign growths that favor the corpus callosum, suprasellar and pineal regions. In the spinal cord, they are intradural and extramedullary and at the thoracic level of the cord. In $\frac{1}{3}$ of cases, they are associated with other congenital anomalies

15. Teratomas are very rare within the CNS and account for only 0.1% of primary intracranial growths
    a. Found mostly in the 1st decade of life and tend to favor the midline
    b. Are composed of various derivatives of the 3 primitive germ cell layers: i.e., epidermal, dermal, glandular, muscular, vascular, cartilaginous elements

16. Hemangioblastomas account for 1 to 2.5% of all intracranial tumors and are seen at any age, but most frequently in young and middle-aged people
    a. Most often seen in the cerebellum (represent about 7% of the primary tumors of the posterior fossa)
    b. Are most often solitary but fairly frequently multiple, and in the latter case, fall into the definition of Hippel-Lindau's disease (familial; retinal and cerebellar hemangioblastoma with visceral lesions, particularly renal and/or pancreas tumors or cysts)

17. Chordomas originate from interosseous notochordal remnants. In the cranial cavity they are seen at the level of the sella turcica and clivus; in the spine at the sacrococcygeal region

B. DIAGNOSIS OF CEREBRAL TUMORS

1. The CSF is usually elevated (though this can be shown by lumbar puncture, it is highly dangerous if a tumor is present and should rarely be performed in this case)
2. Arteriography is the important investigation, for the pattern of blood vessels on x-ray will show the position and often the type of tumor
3. If the tumor is thought to be in the brainstem, cerebellum, or cerebellopontine angle (posterior fossa tumor), ventriculography is the procedure most often used
4. Brain scanning is an important diagnostic tool
5. After locating the exact position, operative exploration is needed in most cases to make certain of the type of tumor and see whether removal is at all possible

Craniopharyngioma (low power).

Craniopharyngioma (high power).

237. SECONDARY AND SPINAL TUMORS

I. Secondary tumors: metastatic neoplasms from primary visceral cancer are some of the most frequent histologic types of intracranial and intraspinal tumors. They are found in any region of the cranial cavity or spinal canal, and may involve the cerebral hemispheres, the cerebellum, the brainstem, or less often, the spinal cord, spinal or cranial nerve roots, and meningeal coverings

A. METASTASES may be solitary, but most often are multiple, and their size varies from a very small size to that of a large egg. They are generally well-circumscribed, either firm or soft, and may show necrosis or cystic degeneration

B. THE TWO CHIEF PRIMARY SITES are bronchopulmonary carcinoma in the male and mammary carcinoma in the female
  1. Other sites may include malignant melanoma, renal carcinoma, and GI carcinoma

C. HISTOLOGICALLY they resemble their primary sources, but atypical features are not rare

II. Spinal tumors: most are benign, whereas most brain tumors are malignant

A. FIVE MAJOR TYPES are seen
  1. Neurofibromas grow on the nerve roots
  2. Meningiomas grow from the arachnoid
  3. Hemangioblastomas grow from immature blood vessels
  (*Note:* types 1–3 are all benign, extrinsic, and compress the cord. They can be removed with complete cure)
  4. Astrocytomas*
  5. Ependymomas*

B. ALL SPINAL TUMORS produce similar symptoms which depend on the level of the spine at which they develop
  1. Spinal tumors are suspect with slowly progressive paralysis of the lower body and a clear-cut level on the skin, below which sensation is lost
     a. Neurofibromas may cause pain in the nerve root distribution for months before it is large enough to compress the cord. It then causes a Brown-Sequard syndrome and later, as it grows larger, a total transverse lesion
     b. Meningiomas and hemangioblastomas behave like neurofibromas but with more cord signs and less primary root pain

C. THE INTRINSIC TUMORS OF THE CORD are common in the cervical region
  1. They first damage those sensory fibers that are crossing each other in the center of the cord and thus produce a large patch of loss of sensation to pain and temperature over the arm and upper part of body (like syringomyelia)
  2. They later spread to involve the motor fibers in the pyramidal tracts and cause paralysis of the lower limbs

D. TUMORS OFTEN DEVELOP BELOW THE CORD itself among the fibers of the cauda equina. Due to their position, they cause a progressive flaccid paralysis of the legs with loss of sensation over the buttocks and the back of the legs, and paralysis of the bladder and rectum (if the tumor is benign and not left too long before removal, all these signs are capable of recovery)

E. THE DIAGNOSIS OF SPINAL CORD COMPRESSION is confirmed by finding a block on the Queckenstedt test on lumbar puncture and usually a high CSF protein level. Surgical exploration is necessary to determine the exact position of the tumor following a myelogram—necessary because an examination alone finds that a curable extrinsic tumor may masquerade as an inoperable intrinsic one

---

*Types 4 and 5 grow in the cord substance, are intrinsic, and cannot be removed, but their growth rate is slow. Secondary deposits grow in spaces about the cord, develop rapidly, and removal is unsuccessful.

Gross section through brain showing metastatic tumors.

Histologic section of a metastatic carcinoma.

238. OTHER SPINAL TUMORS

I. **Peripheral nerve tumors** rarely cause much trouble and consist entirely of neurofibromas
   A. THESE TUMORS MAY CAUSE PAIN AND SENSORY LOSS in the skin area that the nerve is supplying, but they are usually symptomless
   B. THESE TUMORS CAN OFTEN BE REMOVED WITHOUT NERVE SACRIFICE, but it is rarely necessary

II. **von Recklinghausen's disease**
   A. THE OCCURRENCE OF LARGE NUMBERS OF NEUROFIBROMAS throughout the body, involving cranial, spinal, and peripheral nerves
   B. THE SKIN IS DARK IN COLOR, there are brown so-called "café-au-lait" patches and blue nevi, together with numerous little wartlike tags called mollusca fibrosa
   C. THE SKIN FEATURES may be present by themselves, and the disease is often familial

III. **Congenital tumors** may occur anywhere in the nervous system (also see Primary tumors) and consist of 3 major types
   A. EPIDERMOIDS contain cheesy keratin
   B. DERMOIDS contain hair or teeth
   C. TERATOMAS may have any type of human tissue
   D. ALL THESE TUMORS GROW SLOWLY and, if conveniently situated, may be completely removed
   E. OTHER SPACE-OCCUPYING LESIONS due to aberrations in development include
      1. Chordomas grow at the base of skull, compressing the brainstem and cranial nerves
      2. Colloid cysts, usually of the 3rd ventricle, cause blockage of CSF flow due to its position, and symptoms of high pressure buildup develop, at first intermittently on change of position and then constantly
         a. There are often mental changes and curious episodes in which the legs suddenly give way (drop attacks)
      3. These tumors are very rare

IV. **Bone tumors involve the skull and/or the spine**
   A. BENIGN: osteomas, chondromas, aneurysmal bone cysts, cholesteatomas, etc.
   B. MALIGNANT
      1. Primary: osteosarcoma, myeloma, etc.
      2. Secondary: metastases from mammary gland, prostate, kidney, lungs, or thyroid gland

V. **Structural changes from expanding intracranial space-occupying lesions**
   A. FOCAL CHANGES
      1. Changes in the tumor mass with an increase in volume with local extension, hemorrhage, or cyst formation
      2. Changes involving the adjacent neural parenchyma resulting in compression and displacement of the parenchyma and/or destruction and invasion of the neural parenchyma
   B. REGIONAL CHANGES
      1. Cerebral edema as a result of circulatory disturbances
      2. Disturbances of CSF circulation
      3. Cerebral herniations

Chordoma: a malignant tumor arising from the embryonic remains of the notochord (low power).

Chordoma: high power of an area of tumor showing physaliphorous or bubblelike appearance of cytoplasm.

Unit Twenty-Three

INFECTION AND INFLAMMATION

239. BACTERIAL INFECTIONS

I. Introduction: the same variety of pathogenic organisms (bacterial, fungal, parasitic, rickettsial or viral) that affect the organs of the body can be involved in the nervous system and its coverings. Some are selective, but not all. Illnesses resulting may be acute (poliomyelitis), subacute (some forms of encephalitis), or drawn out and chronic (neurosyphilis). Infection first leads to an inflammation of the tissues. Inflammation of the meninges (most commonly pia and arachnoid) is called meningitis; of the brain, encephalitis; and of the spinal cord, myelitis. The terms "meningism" or "meningismus" are often used to indicate signs of irritation of the meninges without necessarily meaning they are directly infected (merely refers to signs suggestive of meningitis, although the infection may be elsewhere in the body)

Inflammatory lesions consist of either disseminated or localized perivascular cuffings and inflammatory infiltrates of variable cell types depending on the etiology. Their presence does not necessarily indicate an infectious etiology, since an inflammatory infiltrate of reactive, or symptomatic, character is well known to occur in other pathologic processes (polymorphonuclear leukocytes in early stages of cerebral infarction or lymphocytic infiltrations are seen in some demyelinating diseases)

II. Bacterial infections

A. PYOGENIC INFECTIONS, caused by pus-forming bacteria, the commonest being meningococci, pneumococci, and the influenza bacillus. Less often the agent is staphylococcus, *Listeria monocytogenes,* enterococcus, streptococcus, or others. Some pyogenic infections may be localized in the epidural space (epidural abscess) or to the subdural space (subdural abscess or empyema), but these are exceptional. Most often they involve the leptomeninges (purulent meningitis) or the cerebral parenchyma (cerebral abscesses).

1. The organism reaches the meninges directly after open skull fractures or after head injury with fracture of base of skull, spreads from an adjacent focus of suppuration (otitis, sinusitis, mastoiditis), or as a result of blood-borne dissemination from an infective focus (lung, GI tract, etc.)

2. Meningitis

    a. Meningococcal meningitis is spread by nasopharyngeal infection and occurs in epidemics

    b. Pneumococcal meningitis often follows pneumonia, or when a fracture of the skull involves the paranasal sinuses

    c. Influenzal meningitis especially appears in infants

    d. Symptoms of meningeal infection: certain features are seen in all types

     i. The illness starts eith a few days of pyrexia (fever), vague influenzalike state, a cold, or sore throat

     ii. Followed by a rapidly developing headache, which spreads backward to give pain in the neck and back, accompanied by vomiting, photophobia (abnormal eye sensitivity to light), drowsiness with restlessness and irritability and signs of meningeal irritation

     iii. Regardless of cause, neck stiffness or neck rigidity is seen: inability to bend the head forward to bring the chin onto the chest

     iv. Kernig's sign is positive (hip is first flexed and an attempt is made to straighten the knee. This is restricted and painful in a positive test)

     v. In infants, the neck spasms may be so great that the neck is bent backward (neck retraction) or the entire body may be arched (opisthotonos)

Hemophilus meningitis.

Purulent meningitis.

240. PYOGENIC INFECTIONS

I. Pyogenic infections (cont.)
A. PYOGENIC INFECTIONS (cont.)
  1. Meningitis (continued)
    e. Diagnosis: confirmed by doing a lumbar puncture: one sees an elevated CSF pressure, cloudy or milky fluid containing thousands of polymorphonuclear leukocytes (i.e., pus cells), and also a much reduced sugar
      i. CSF is placed on slide and stained for the responsible organism
      ii. Other specimens are placed on culture plates for sensitivity drug testing
    f. Treatment: tests take time and treatment must begin immediately. Patients are put on penicillin (assuming no sensitivity) and sulfonamides. In suspected pneumococcal infection, diluted penicillin may also be placed directly into the CSF. Treatment may be altered with identification of the organism
  2. Pyogenic encephalitis, as with meningitis, is caused by pyogenic bacteria or viruses (see later). If the meninges are involved, it is called meningoencephalitis
    a. General features of encephalitis: involvement of neuronal cell bodies, resulting in their destruction and being engulfed by macrophages; predominantly perivascular infiltrates; microglial proliferation with nodules
      i. Symptoms consist of moderate headaches, vomiting, confusion, delirium, increasing drowsiness (which passes into coma), and epileptic fits
      ii. Unless the meninges are involved, there is little neck stiffness, Kernig's sign is negative, and the CSF may be under normal pressure and show a moderate rise in lymphocytes only
      iii. The seriousness of the disease depends on the organism involved: if due to pyogenic bacteria, it will kill the patient unless treated; if viral, the patient may recover without serious side effects. The preceding are, of course, uncertain. In addition, some patients show personality changes with tendency toward outbursts of temper, fits, and difficulty in learning which may alter their entire lives
    b. Specific features
      i. Bacterial infection can spread from the nasal sinuses (or infected ears) or may be carried by the blood from infected lungs (i.e., bronchiectasis)
      ii. Part of the brain first becomes edematous (swollen), then the inflamed area liquefies and white blood cells pour in until pure pus is formed, resulting in a cerebral abscess. The latter are most common in the frontal and temporal lobes
        (a) Initially (presupperative cephalitis or cerebritis), the lesion is ill-defined and characterized by early necrosis of cerebral tissue, vascular congestion, petechial hemorrhage, microthromboses, perivascular exudates, and cellular infiltration with polys. Edema and cerebral herniation may also be seen
        (b) Gradually, in a few weeks, the abscess becomes circumscribed and a shell is progressively organized around the central, purulent, necrotic focus. In later stages, the abscess is truly encapsulated
        (c) The 2 major complications of abscesses are raised intracranial pressure with risk of cerebral herniation and rupture into ventricles
      iii. Symptoms, at first, are the same as for encephalitis, but are more severe and acute, and paralysis of the limbs is common
      iv. Treatment is surgical. A needle is passed into center of abscess, pus is withdrawn, slides and cultures made, and antibiotic along with a radiopaque fluid is placed into abscess. Aspiration and replacement by antibiotic are done frequently at first and abscess size followed by x-ray. When small enough, it may be excised fully.

Leptomeningitis or meningitis proper (low power).

Inflammation of meninges and subarachnoid space (between membranes).

MENINGITIS

241. TUBERCULOSIS AND SYPHILIS

II. Bacterial infections (cont.):

B. TUBERCULOSIS is caused by the tubercle bacillus, is difficult to diagnose, consists of tumorlike masses of cells on the center of which are the bacilli. May behave like a tumor or may rupture into the meninges to cause a meningitis. The masses are called granulomas

  1. Epidural tuberculous abscess: usually a complication of tuberculosis of the spine (Pott's disease) involving either the vertebral bodies or the disks
  2. Subdural tuberculous abscess: less frequent and usually discovered incidentally after surgery or at necropsy
  3. Tuberculous meningitis differs from purulent meningitis in that the meninges over the skull base are predominantly involved; the exudate is essentially made up of lymphocytes, mononuclear cells, and tubercles with areas of caseous necrosis; arterial lesions are constant and responsible for foci of ischemic necrosis
     a. Onset is much less acute, and patients may have been vaguely unwell for weeks with gradually increasing headaches and listlessness
     b. Eventually the same symptoms and signs as in other cases of meningitis develop, but the CSF shows fewer cells; polys and lymphocytes are seen in equal numbers; and both sugar and chlorides are very low
     c. Once a fatal disease, but now treated by prolonged course of streptomycin injections, together with isonicotinic acid hydrazide (INAH) given orally. Majority respond to streptomycin given IM; others may need injection directly into CSF. Treatment takes about 3 months
  4. Cerebral tuberculoma; a spherical or multiloculated lesion which may be single or multiple with sites of predilection being the cerebellum, pontine tegmentum, and paracentral lobule. May spontaneously become cystic, fibrous, and calcified, but their danger lies in spilling into the meninges

C. SYPHILIS: CNS involvement is a sequel of primary luetic disease that has escaped notice or been inadequately treated and appears years after the original infection. The organism is a spirochete, *Treponema pallidum,* and causes a chronic inflammation of the nervous tissue and its coverings

  1. Subacute secondary syphilitic meningitis: leptomeningeal invasion resulting in meningitis with lymphocytes and plasma cells and perivascular infiltrates
  2. Tertiary syphilis
     a. Chronic meningitis leads to fibrous organization and ultimate occlusion of the CSF pathways of resorption with concomitant infiltration of arterial walls (syphilitic arteritis)
        i. Episodes of thrombosis occur resulting in sudden paralysis of one of the cranial nerves or of a limb, and the clinical picture is varied since so many parts of the CNS can be involved
     b. General paresis (general paralysis of insane): inflammatory meningovascular lesions associated with parenchymatous changes of encephalitic type
        i. Characterized by much cortical atrophy, neuronal depopulation, and proliferation of rod-shaped microglia are seen as scattered foci of different ages
        ii. Symptoms are of slow mental deterioration accompanied by grandiose delusion
        iii. There are often fits and signs of upper motor neuron disease such as spasticity, exaggerated reflexes, and extensor plantars
        iv. Argyll Robertson pupils are common
        v. The WR is positive in 100% of untreated cases (in CSF) and the Lange curve is of the paretic type.

BACTERIAL CAUSES OF FEVER IN CHILDREN

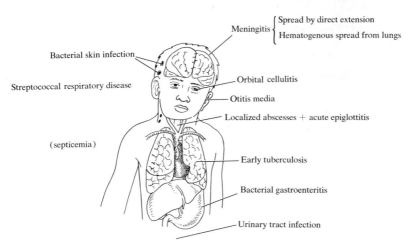

Meningitis ⎰ Spread by direct extension
⎱ Hematogenous spread from lungs

Bacterial skin infection

Streptococcal respiratory disease

Orbital cellulitis

Otitis media

Localized abscesses + acute epiglottitis

(septicemia)

Early tuberculosis

Bacterial gastroenteritis

Urinary tract infection

Convulsions not uncommon and justify lumbar puncture in child with febrile illness

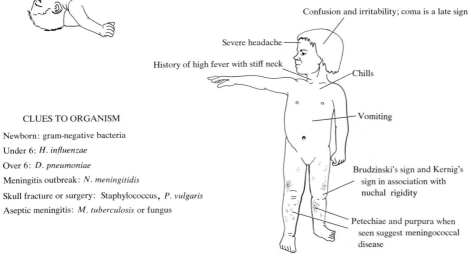

Confusion and irritability; coma is a late sign

Severe headache

History of high fever with stiff neck

Chills

Vomiting

Brudzinski's sign and Kernig's sign in association with nuchal rigidity

Petechiae and purpura when seen suggest meningococcal disease

CLUES TO ORGANISM

Newborn: gram-negative bacteria

Under 6: *H. influenzae*

Over 6: *D. pneumoniae*

Meningitis outbreak: *N. meningitidis*

Skull fracture or surgery: Staphylococcus, *P. vulgaris*

Aseptic meningitis: *M. tuberculosis* or fungus

CLINICAL PICTURE OF
BACTERIAL MENINGITIS
OVER AGE 2

242. TERTIARY SYPHILIS AND OTHER BACTERIAL INFECTIONS

II. Bacterial infections (cont.)

C. SYPHILIS (cont.)

2. Tertiary syphilis (cont.)

c. Tabes dorsalis consists of degeneration of the posterior columns and spinal roots with involvement of dorsal root ganglia (tabes = wasting)

i. The result of meningovascular lesions localized to the subarachnoid portion of the dorsal nerve roots

ii. Spinal cord involvement presents as demyelination and gliosis of the posterior columns secondary to radiculoganglionic lesions

iii. There is no inflammatory process and the treponema are absent

iv. Root irritation causes "lightning" pains like a sudden outburst of needlelike pins in the shins, calves, and arms as well as "girdle pains" which resemble a constricting band around the waist

v. Lesions result in a loss of position sense so patient becomes ataxic and walks with legs wide apart and a stamping gait—helpless in the dark

vi. Sensory fiber degeneration results in an upset reflex arc so that tendon reflexes are absent and bladder sensations are disturbed leaving no desire to pass urine resulting in overfilling or overflow incontinence

vii. Patchy loss of skin sensation may result in ulcers on feet

viii. Gross destructive changes occur in knees and other joints but due to sensory loss, are painless—the disorganized joint of Charcot

ix. Most characteristic sign of neurosyphilis is in the eyes—the so-called Argyll Robertson pupil (small irregular pupil which does not react to light but does constrict on looking at near objects)

x. In active disease, the CSF shows a raised lymphocyte count and a tabetic (luetic) Lange curve

xi. Other tests: treponema immobilization or fluorescent antibody tests

3. Treatment of neurosyphilis

a. Penicillin is treatment of choice: 20,000,000 units given in divided doses over about 10–12 days

i. Usually cures meningovascular syphilis and improves GPI but is disappointing in tabes dorsalis

ii. Courses have to be repeated at 3–6 month intervals until all signs of activity have cleared up

iii. Wide use of penicillin has made diagnosis of neurosyphilis more difficult and classical pictures are much less common

D. SARCOIDOSIS: origin is uncertain

1. Granulomas develop that are similar to tuberculomas but do not contain bacilli
   a. Form in brain, meninges, around peripheral nerves, and in glands of body

2. Diagnosis is by biopsy of the above glands

3. Treatment: use of cortisone

E. LEPROSY: leprosy bacillus causes peripheral nerve thickening (seen and felt)

1. Sensation is lost in parts supplied by the nerves and trophic changes eventually take place: areas of skin and bone slough off resulting in mutilation of hands, feet, and face

2. It is only slightly infectious and modern treatment with sulfones offers a considerable hope of cure

F. BRUCELLOSIS (MALTA FEVER): leptomeningeal involvement common in the septicemic phase of this disease. Different forms of neurobrucellosis correspond to a variety of clinicopathologic pictures which frequently involve the 8th (acoustic) nerve.

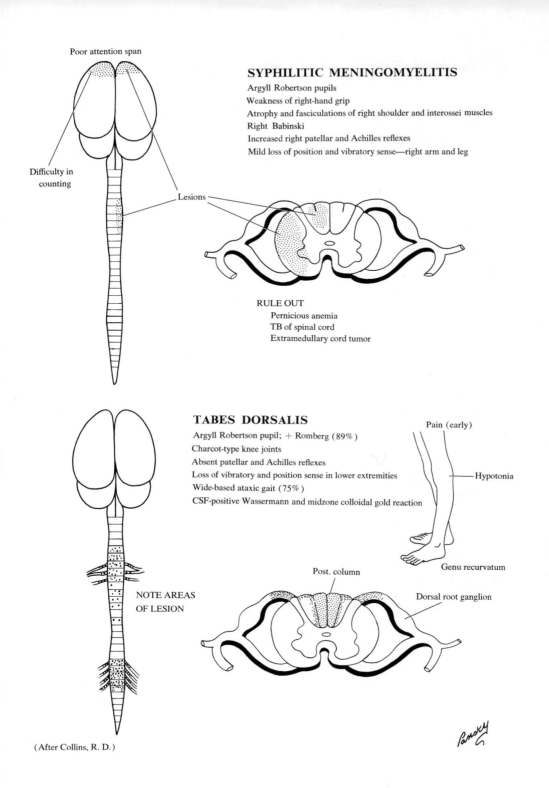

Poor attention span

SYPHILITIC MENINGOMYELITIS

Argyll Robertson pupils
Weakness of right-hand grip
Atrophy and fasciculations of right shoulder and interossei muscles
Right Babinski
Increased right patellar and Achilles reflexes
Mild loss of position and vibratory sense—right arm and leg

Difficulty in
counting

Lesions

RULE OUT
  Pernicious anemia
  TB of spinal cord
  Extramedullary cord tumor

TABES DORSALIS

Argyll Robertson pupil; + Romberg (89%)
Charcot-type knee joints
Absent patellar and Achilles reflexes
Loss of vibratory and position sense in lower extremities
Wide-based ataxic gait (75%)
CSF-positive Wassermann and midzone colloidal gold reaction

Pain (early)

Hypotonia

Genu recurvatum

Post. column

Dorsal root ganglion

NOTE AREAS
OF LESION

(After Collins, R. D.)

243. OTHER INFECTIONS AND INFESTATIONS

III. Mycoses are fungal infections of the CNS that are seen mainly in special circumstances, such as following prolonged antibiotic treatment, in the course of hematologic or lymphoreticular disease, or as a complication of immunosuppressive therapy
A. THE MOST FREQUENT FORMS are candidiasis and torulosis (or cryptococcosis)
B. SOME HAVE CHARACTERISTIC APPEARANCE; others are difficult to identify
C. RARE FORMS are aspergillosis, mucormycosis (or phycomycosis), and coccidioidomycosis

IV. Rickettsial infections: some forms of such infections (murine or endemic typhus, exanthematic or epidemic typhus, Rocky Mountain spotted fever) may cause nervous system lesions involving the gray matter of the cerebral hemispheres and brainstem

V. Parasitic infections: various parasites may affect the nervous system but vary greatly in different geographic areas. The chief causative parasites are protozoa and metazoa. A few examples are listed below
A. TAPEWORM has an intermediate or larval form, which usually develops in other animals. The human tapeworm, *Taenia solium,* can at times develop its larval stage not, as normally, in the pig, but in human muscle and brain
   1. These larvae are called cysticerci and the condition called cysticercosis
   2. The muscle cysts are harmless, but they do calcify and are seen on x-rays of the thighs and pelvis
   3. The cysts of the brain cause epilepsy and sometimes mental changes
B. *Trichinella spiralis* is another parasite that affects man if eaten in uncooked sausage meat
   1. Usually develops in the muscles and the eyelids, but it can produce a serious encephalitis
   2. Condition is called trichiniasis and is common in America among Negro children
C. HYDATID DISEASE
   1. Cysts formed by the parasite *Echinococcus*
   2. Very rare and may develop in the brain, skull, or meninges and behaves like a brain tumor
   3. Common in Australia and Middle East. Also seen in South Wales and some areas of England
D. LEPTOSPIROSIS
   1. Caused by an organism carried by rats and dogs
   2. The dog variety, *Leptospira canicola,* causes high fever, meningitis, and inflammation of the eyes, which is known as canicola fever

A form of mycotoxicosis of the leptomeninges caused by a member of the *Aspergillus* genus.

Mycotoxicosis of the subarachnoid space accompanied by secondary bacterial infection.

244. VIRAL DISEASES

I. Viral diseases: lesions show variable forms

A. SOME LESIONS are nonspecific, related to immunoallergic reactions, and are secondary to viral infection, involving the meninges and white matter; whereas, other lesions are directly related to penetration of the CNS by the virus and involve mainly the gray matter (polioencephalitis)

1. Some of the above forms have an acute course (in viral encephalidites); but others show an immunologic phenomenon and development of latent infections (herpes zoster)

B. ONE GROUP shows animal transmissibility and resembles some comparative neuropathology suggesting a viral mechanism related to the "slow-virus diseases"

C. VIRAL INFECTIONS with nonspecific nervous system manifestations

1. Acute viral lymphocytic meningitis: response common to viral infections of a number of different viruses

2. Postinfectious perivenous encephalitis may complicate a variety of viral diseases, particularly, exanthemas such as measles, chickenpox, rubella, smallpox, or even follow smallpox vaccination

   a. Produces a clinical picture of an acute disseminated encephalomyelitis that may show up during the convalescence from the viral infection

   b. The basic mechanism appears to be immunoallergic, but this is questionable since the finding of viral antigen in glial cells of the lesion

D. VIRAL ENCEPHALITIS is directly related to viral penetration of the CNS as a result of direct spread along the peripheral nerves (particularly the olfactory), or more commonly in the course of a viremia. The inflammatory processes often involve the meninges (meningoencephalitis), the cord (encephalomyelitis), or both (meningoencephalomyelitis) as well as the nerve roots. Diagnosis and treatment of these conditions are often very difficult because a virus is rarely isolated and the diagnosis is usually presumptive. Much more insight is obtained from showing changes in the antibodies in the blood than from attempting to find the organism. The symptoms (described under pyogenic encephalitis) apply to all forms of encephalitis, although the pattern of development varies

1. Encephalitides of RNA viruses: due to enteroviruses (Coxsackie group and echoviruses) present as a lymphocytic meningitis. The CNS is, however, more often involved by a poliovirus

   a. Poliomyelitis: most frequent form of the disease; does not correspond to a typical encephalitis, but rather to an acute anterior poliomyelitis

      i. The lesions selectively involve the motor neurons of the anterior horns

      ii. Also involve other regions: frontal gyri, hypothalamus, reticular formation, and posterior horns (polio encephalitis)

      iii. The inflammatory infiltrates, edema, vasodilatation, microglial and macrophage proliferation can be severe

      iv. Following resolution, the residual lesions consist of atrophy of the anterior horns with neuronal cell loss and gliosis

      v. About 50% of the patients are adults

      vi. A few cases appear each year, but epidemics do develop from time to time

      vii. Onset is acute with pyrexia, headache, vomiting, neck stiffness, and pain in the back and limbs—may last 2–4 days and may subside completely and is then called nonparalytic poliomyelitis

      viii. However, the above are the major sources of spread of infection and, except during an epidemic, they are difficult to distinguish from any other viral meningitis

Superior long. fissure

Encephalitis: note necrosis in gross brain (arrows).

Encephalitis: histologic preparation showing reactive astrocytes.

245. ENCEPHALITIDES OF RNA VIRUSES—PART I

I. Viral diseases (cont.)

D. VIRAL ENCEPHALITIS (cont.)

  1. Encephalitides of RNA viruses (cont.)

    a. Poliovirus (cont.)

      ix. In a few cases, after the nonparalytic period, there is a 48-hour interval and then the paralytic stage begins with pain and extreme tenderness of various muscles. The muscles rapidly become weak or totally paralyzed, and the danger lies with involvement of respiratory (diaphragm and intercostal) or throat muscles preventing swallowing, coughing, or proper breathing and may require the use of an artificial "iron" lung

      x. When the acute stage settles down, many muscles may recover completely, but others may waste, at times almost completely. In the latter case, some function may be restored by physiotherapy and rehabilitation

      xi. The polio virus is detected in the feces of patients with the nonparalytic type and in early stages of the paralytic variety. By EM, viral particles may be seen in the cytoplasm of the nerve cells

      xii. Antibody tests can also be used for blood studies

      xiii. Treatment

        (a) The pre- or nonparalytic stage requires no treatment other than rest and analgesics for headache. Rest is essential since physical exercise during this stage increases the chances of paralysis developing

        (b) The limbs are supported in positions to avoid overstretching of muscles. Joints also must be protected

        (c) Hot packs may ease muscle spasms, but movement should be avoided for several days

        (d) Respiration and pulse rate must be watched carefully since changes can take place rapidly

        (e) When pain subsides, each joint must be moved through its full range of motion, twice daily, to avoid fixed joints, and care must be taken with pressure points, urine, and bowels

        (f) Treatment of paralyzed patients is described in other texts

        (g) Modern immunization has greatly changed the world picture of this disease entity. Paralysis is less common in immunized people. However, the incidence of nonparalytic polio has not apparently changed

    b. An identical or fairly similar picture has been seen with other enteroviruses, in particular with Coxsackie virus, possibly as the result of polio immunization

  2. Encephalitides due to arboviruses (arthropod-borne) have a distinct geographic distribution, as often indicated by their names. The best known of these are

    a. Mosquito-borne encephalitides: St. Louis encephalitis, eastern and western equine encephalitis, Japanese B encephalitis seen in the Far East, and Murray Valley encephalitis seen in Australia

    b. Tickborne encephalitis: Russian spring-summer encephalitis and Central European encephalitis

VIRAL SYNDROMES IN A FEBRILE CHILD

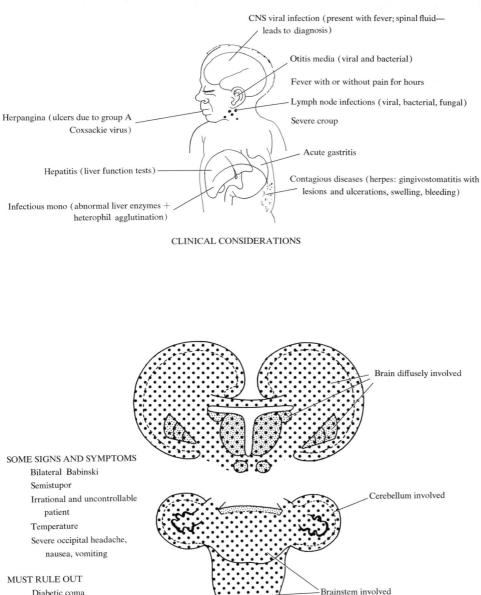

CNS viral infection (present with fever; spinal fluid—leads to diagnosis)

Otitis media (viral and bacterial)

Fever with or without pain for hours

Lymph node infections (viral, bacterial, fungal)

Severe croup

Herpangina (ulcers due to group A Coxsackie virus)

Acute gastritis

Hepatitis (liver function tests)

Contagious diseases (herpes: gingivostomatitis with lesions and ulcerations, swelling, bleeding)

Infectious mono (abnormal liver enzymes + heterophil agglutination)

CLINICAL CONSIDERATIONS

Brain diffusely involved

Cerebellum involved

Brainstem involved

SOME SIGNS AND SYMPTOMS
Bilateral Babinski
Semistupor
Irrational and uncontrollable
    patient
Temperature
Severe occipital headache,
    nausea, vomiting

MUST RULE OUT
Diabetic coma
Lead encephalopathy
Meningitis
Barbiturate intoxication
Lesion
Ruptured aneurysm

ST. LOUIS ENCEPHALITIS

246. ENCEPHALITIDES OF RNA VIRUSES—PART II

I. Viral diseases (cont.)

D. VIRAL ENCEPHALITIDES (cont.)

    1. Encephalitides of RNA viruses (continued)

        c. Human rabies is due to a rhabdovirus and is transmitted to man by the bite of an infected animal (cat, dog, bat, wild animal, etc.) and is always fatal once it has declared itself

            i. Identified by the presence of characteristic cytoplasmic inclusions in the neurons, e.g., Negri bodies, accompanied by discrete inflammatory cellular infiltrates

            ii. Polyneuritis and encephalomyelitis have been reported after rabies vaccination with either the attenuated or the inactivated virus and are probably related to inadequate inactivation of the virus or to an immunopathologic process

        d. Encephalitides due to paramyxoviruses: in addition to influenza and mumps, which may be associated with different forms of encephalitis, measles infection can cause 2 types of encephalitis: acute postinfectious encephalitis and subacute sclerosing panencephalitis

            i. Mumps encephalitis tends to occur during the worst stage of the acute illness and is often accompanied by signs of meningitis as well

                (a) It lasts only about 2–3 days and is not usually followed by serious side effects

            ii. Measles encephalitis, in general, usually occurs as the child is getting over the worst and is beginning to recover, when signs of cerebral disturbances develop and the child becomes very ill again. There is a danger of serious sequelae

            iii. Subacute sclerosing panencephalitis is seen in children, several years after a known episode of measles

                (a) The lesions involve the gray matter (cortex and basal ganglia) where inclusion bodies are seen in neuronal and glial nuclei as well as in the cytoplasm of neurons (inclusion-body encephalitis)

                (b) Lesions may involve the white matter where one sees a diffuse demyelination with marked astrocytic proliferation

                (c) One sees perivascular inflammatory infiltrates in both white and gray matter

                (d) Although the measles virus has been implicated, the precise mechanism of this prolonged viral infection is not understood

                (e) Is a rare condition and usually begins as a slowly progressive mental deterioration followed by fits, tremulousness, ataxia, and finally by extreme spasticity and rigidity, accompanied by attacks of drooping of the head and limbs followed by tremor

                (f) EEG gives an almost diagnostic abnormality

                (g) No treatment exists to check its fatality and long illness

Note hypertrophied astrocytes (arrows) and perivascular cuffing in white matter.

High power of one perivascular cuff and reactive hypertrophied astrocytes (arrows).

SUBACUTE SCLEROSING PANENCEPHALITIS

247. ENCEPHALITIDES OF DNA VIRUSES

I. Viral diseases (cont.)

D. VIRAL ENCEPHALITIDES (cont.)

2. Encephalitides of DNA viruses

a. Encephalitides due to herpes virus group

i. Herpes simplex encephalitis is due to a subgroup A virus and topographically involves predominantly the temporal lobes and limbic areas

(a) Severe inflammatory infiltrates which may result in hemorrhagic necrosis with marked cerebral edema

(b) Intranuclear inclusions may be seen in the neurons and in some glial cells

(c) Virus is identified by EM and by immunofluorescence and is becoming more and more recognized

(d) Clinical picture develops rapidly over 3–4 days and may be accompanied by focal fits and paralysis of one part of the body. Speech disturbances and visual field loss also may be seen

(e) Patients are gravely ill with high fever and many cells in the CSF and signs show resemblances to a cerebral abscess in many ways

(f) Can be diagnosed by increasing levels of herpes antibody in the blood but more directly and quickly by taking a biopsy of the affected brain and examining this with special fluorescent stains for antibodies and for virus particles in the electron microscope

(g) Since the brain disturbances may be an allergic response to the virus, some cases are helped with ACTH treatment

ii. Cytomegalovirus encephalitis is caused by a herpes virus of subgroup B and may be seen as part of a multisystem viral infection or as an isolated CNS infection associated with chorioretinitis

(a) Usually caused by early fetal infection and may lead to microcephaly and hydrocephalus with intracranial calcifications correlated with subependymal and subcortical lesions

iii. Herpes zoster varicella: the virus of shingles, which is the same one that causes varicella, selectively involves the neurons of one or more spinal root ganglia and the posterior horn of the corresponding metameric segment. Trigeminal nerve ganglia are also involved

(a) The inflammatory cellular infiltrates also tend to involve the adjacent leptomeninges and spinal medullary levels

(b) Illness begins with severe pain in the distribution of the nerve root, often around the trunk, or above the eyes in the case of ophthalmic herpes, and red patches appear on the skin. The latter change to vesicles which run together, crust over and heal, leaving pale depressed scars like those seen in chickenpox

(c) In the elderly, the pain may persist in the affected areas for many years and become a major obsession in their lives

(d) No treatment thus far has been found to be consistently helpful

(e) In geniculate herpes, the vesicles develop in the ears, and often facial paralysis and deafness result

3. Encephalitides due to papoviruses: progressive multifocal leukoencephalopathy is due to a papovirus, as are complications of some diseases associated with immunodeficiency, e.g., chronic lymphatic leukemia, Hodgkin's disease, etc.

a. The white matter, particularly of the occipital lobe, is the site of limited foci of demyelination

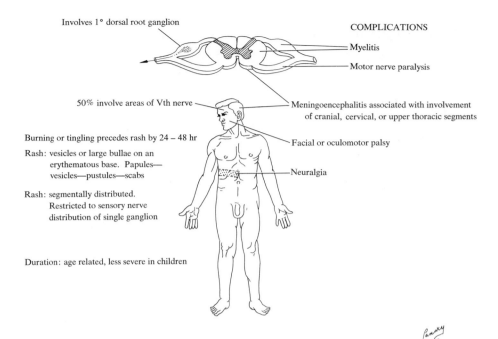

Involves 1° dorsal root ganglion

COMPLICATIONS

Myelitis

Motor nerve paralysis

50% involve areas of Vth nerve

Meningoencephalitis associated with involvement of cranial, cervical, or upper thoracic segments

Burning or tingling precedes rash by 24 – 48 hr

Facial or oculomotor palsy

Rash: vesicles or large bullae on an erythematous base. Papules—vesicles—pustules—scabs

Neuralgia

Rash: segmentally distributed. Restricted to sensory nerve distribution of single ganglion

Duration: age related, less severe in children

HERPES ZOSTER: AN ACUTE INFECTION BY A FILTERABLE VIRUS

Histologic preparation showing cellular infiltration.

248. ENCEPHALITIDES OF NONIDENTIFIED VIRUSES AND SLOW VIRUSES

I. Viral diseases (cont.)

D. VIRAL ENCEPHALITIDES (cont.)

   3. Due to nonidentified viruses

      a. Encephalitis lethargica: known as epidemic encephalitis of von Economo or "sleeping sickness"

        i. Was rampant in the European epidemic between 1916 and 1930, but the viral agent was never demonstrated

        ii. Characterized by preferential involvement of the midbrain and basal ganglia

        iii. Is rare, but isolated cases are seen

          (a) Brainstem is affected so that, along with general symptoms, there can be paralysis of eye movements, and the patients are drowsy by day and wakeful by night

          (b) Face becomes rigid and there is considerable personality disturbance

        iv. Recovery usually occurs but with an important aftereffect, namely, a type of parkinsonism called postencephalitic parkinsonian syndrome which does not necessarily appear for many years

      b. Uveomeningoencephalitides: seen as an inflammatory encephalitic, meningitic and uveal (choroid, ciliary body, and iris) lesions of unknown etiology (seen in Behçet's disease)

E. CEREBRAL INFECTIONS DUE TO TRANSMISSIBLE "SLOW VIRUS" are characterized by animal transmissibility after the injection of cerebral biopsy fragments and include 2 known major disorders

   1. Creutzfeldt-Jakob disease: clinically characterized by dementia associated with various neurologic manifestations and fatal in a few months

      a. See neuronal cell depopulation with spongiosis and dense astrocytic gliosis of the cerebral cortex, especially the occipital lobes

      b. Resembles some animal virus diseases (scrapie in sheep and mink encephalopathy)

   2. Kuru: a human disease seen in New Guinea

      a. A progressive, fatal encephalopathy

   3. The pathogenic agent for both of the above has not as yet been isolated, and the usual histopathologic picture of encephalitis is entirely lacking

ENCEPHALITIS LETHARGICA

Scattered lesions

Caudate nuc.

T T

Basal ganglia lesions

Lenticular nuc.

Cerebellum

Brainstem lesions

Dilated areflexic pupils

Paralysis of convergence

Masklike facies

Generalized rigidity

Tremor: rhythmic of eyes,
tongue, hands

Patient treated for parkinsonism for years
(postencephalic parkinsonian syndrome)

LESIONS

Cortex

Basal ganglia

Thalamus

Cerebellum

CREUTZFELDT-JAKOB DISEASE

Dementia associated with various neurologic
manifestations

Fatal in a few months

Neuronal cell depopulation + spongiosis
and astrocytic gliosis

Slow virus infection

Anterior horn

APPENDIXES

APPENDIX I. GLOSSARY

Abdominal reflexes. Abdominal muscle contractions on stroking the skin

Adiadochokinesia. Inability to perform rapidly alternating movements. Also called dysdiado-chokinesia

Agnosia. Lack of ability to recognize significance of sensory stimuli (auditory, visual, tactile, etc.)

Agraphia. Inability to express thoughts in writing, due to a central lesion

Air encephalography (pneumoencephalography). X-raying the ventricles and subarachnoid spaces after introduction of air by lumbar or cisternal puncture

Akinesia. Inability to start a movement; slowness in movement

Ala cinerea. Vagal triangle in floor of fourth ventricle

Alexia. Loss of power to grasp the meaning of written or printed words and sentences

Allocortex. Phylogenetic older cerebral cortex, consisting of three layers. Includes paleocortex and archicortex

Alveus. Thin layer of white matter covering ventricular surface of hippocampus

Amacrine. A long fiber (amacrine nerve cells of retina)

Ammon's horn. The hippocampus has the configuration of a ram's horn (also called cornu ammonis)

Amygdala. Like an almond. Amygdaloid nucleus or amygdala in temporal lobe of cerebral hemisphere

Analgesia. Inability to feel pain

Anarthria. Inability to pronounce words

Anesthesia. Inability to feel touch

Aneurysm. A weak bulge in an arterial wall

Angioma. A collection of abnormal arteries, capillaries, and veins

Anopsia. A defect of vision

Anosmia. Loss of sense of smell

Anterior horns. Part of the spinal gray matter containing motor nerve cells

Anticonvulsants. Drugs used to control epileptic fits

Aphasia. A defect of the power of expression by speech or of comprehending spoken or written language

Aphonia. Inability to make sounds

Apoplexy, cerebral. Old term for massive cerebral hemorrhage

Apraxia. Inability to carry out purposeful movements in the absence of paralysis

Arachnoid mater. The weblike middle layer of the meninges forming outer boundary of subarachnoid space

Archicerebellum. Phylogenetically old part of cerebellum

Archicortex. Three-layered cortex included in limbic system; located mostly in hippocampus and dentate gyrus of temporal lobe

Argyll Robertson pupil. An irregular pupil reacting to convergence but not to light

Arteriography (-gram). X-raying blood vessels after injecting an artery with opaque dye

Association fibers. Nerve fibers connecting different areas of the cerebral cortex in the same hemisphere.

Association areas. Areas whose neurons integrate sensory and other information. In parietal lobe, they integrate sensory data from somesthetic, auditory, visual, and taste areas. In temporal lobe, post. part, they integrate visual and auditory data; in anterior part, concerned with past experiences. In frontal lobe, they deal with judgment, reasoning, abstraction, planning, behavior patterns, etc.

Astereognosis. Inability to recognize objects by shape and size by touching or feeling them

Astrocyte. A neuroglial cell of ectodermal origin

Astrocytoma. A tumor arising from astrocytes

Asynergy. Disturbance of proper association in contraction of muscles that assumes the different components of an act follow in correct sequence, at the right moment, and correct degree for accurate execution of act

Ataxia. Unsteadiness. A loss of power of muscle coordination with irregularity of muscle action

Athetosis. Slow writhing involuntary bizarre movements, especially of fingers and toes

Auditory area. Located in superior temporal gyrus of temporal lobe

Aura. Warning symptoms (of an attack of epilepsy or migraine)

Autonomic. Autonomic system. Efferent or motor innervation of smooth muscle, glands, and heart muscle

Axolemma. The plasma membrane of an axon

Axon. The longest process of a nerve cell. Efferent process conducting impulses to other neurons or muscle fibers and gland cells.

Axon hillock. Region of nerve cell body from which axon arises. Contains no Nissl material

Axon reaction. Changes in cell body of a neuron after damage to its axon

Axoplasm. Cytoplasm of the axon

Babinski reflex (sign). Extension of the great toe on scratching the sole

Baroreceptor. A sensory nerve terminal that is stimulated by changes in pressure

Basal ganglia. Islands of gray matter in each cerebral hemisphere. They are associated with the postural aspect of voluntary activity and habitual activity.

Betz cells. Large pyramidal cells of the precentral gyrus associated with motor activity

Bitemporal hemianopia. Loss of the outer halves of both visual fields

Brachium. Denotes a large bundle of fibers connecting one part with another (in CNS)

Bradykinesia. Abnormal slowness of movements

Brainstem. Portion of brain comprising midbrain, pons, and medulla. Contains vital centers, ascending and descending tracts, nuclei of cranial nerves, and reticular formation.

Broca's area. Area for speech in frontal lobe

Brown-Sequard syndrome. Signs produced by damage to one half of the cord

Bruit (intracranial). A sound in time with the heart beat

Bulbar. Concerning the medulla

Burr-hole. A hole drilled in the skull

Café-au-lait patches. Brown skin marks in neurofibromatosis

Calamus scriptorius. An area in the caudal part of the floor of the fourth ventricle shaped like a pen point.

Calcar. Any spur-shaped structure. Calcar avis, an elevation on medial aspect of lateral ventricles, at junction of posterior and inferior horns. Calcarine sulcus is responsible for calcar avis

Capsule. Fibrous barrier around a tumor or abscess

Carpal tunnel. Channel in the wrist through which the median nerve passes

Cauda equina. Lumbar and sacral spinal nerve roots in lower part of spinal canal below spinal cord

Caudate nucleus. Part of corpus striatum named because of its long extension or tail

Central nervous system (CNS). Part of nervous system consisting of the spinal cord and brain (cerebrum, cerebellum, midbrain, pons, and medulla)

Cephalgia. Headache

Cerebellum. Second largest part of brain, located in postcranial fossa, below tentorium

Cerebral. Pertaining to the brain

Cerebral aqueduct (aqueduct of Sylvius). Connection between 3rd and 4th ventricles

Cerebral hemorrhage, massive. Massive bleeding into an area of brain substance with resultant loss of function of that part of the brain

Cerebrovascular accident (CVA). Same as massive cerebral hemorrhage

Charcot's joints. Painless disorganized joints in tabes

Chiasma (-al). The X-like crossing of the two optic nerves

Chordotomy. Division (cutting) of lateral spinothalamic tract for intractable pain

Chorea. Very rapid irregular spasmotic involuntary movements of limbs or facial muscles attributed to degenerative changes in neostriatum

Choroid. Forming a delicate membrane. Choroid or vascular eye coat; choroid plexuses in brain ventricles

Chromatolysis. Dissolution. Dispersal of Nissl material of neurons after axon section or in viral infections of nervous system

Cinerea. Gray matter. Tuber cinereum in basal portion of hypothalamus; tuberculum cinereum in elevation on medulla due to spinal V tract and nucleus; and ala cinerea in vagal triangle in floor of fourth ventricle

Cingulum. Bundle of association fibers in white matter of cingulate gyrus on medial surface of cerebral hemisphere

Circle of Willis. The arteries at the base of the brain

Cisterns. Collection of CSF (e.g., cisterna magna)

Claustrum. Thin sheet of gray matter (unknown function) between lenticular nucleus and insula

Clava. Slight swelling on dorsal surface of medulla due to gracile nucleus

Clonus (-ic). Rhythmical contractions of a spastic muscle on stretching

Coma. Deep loss of consciousness

Commissural fibers. Nerve fibers traveling from one side of the brain to the other

Concussion (of brain). Temporary impairment of brain function, due to trauma; often characterized by loss of memory of the traumatic incident, unconsciousness, temporary visual disturbances, and loss of equilibrium

Coordination. Smooth, efficient movement

Corneal reflex. Normal blinking on touching of the cornea

Corona. A crown. Corona radiata are fibers radiating from internal capsule to cerebral cortex

Corpus callosum. Large mass of commissural fibers located at the bottom of the longitudinal fissure and connecting one hemisphere to the other

Corpus striatum. Furrowed or striped. A mass of gray matter, or motor function, at base of cerebral hemisphere

Cortex. The surface layer (gray matter) of the cerebral and cerebellar hemispheres

Cortical areas. Areas of the cerebral cortex that have specific functions

Cranial nerves. Twelve (12) pairs arising from or running to the brain and brainstem

CSF. The cerebrospinal fluid, bathing the brain and spinal cord

Cuneus. A wedge-shaped gyrus on medial surface of cerebral hemisphere

Deficiency disease. One due to lack of a normal food substance

Degeneration. Death of tissue, often of unknown origin or cause

Déjà-vu. A sense of familiarity

Dementia. Deterioration of intellect

Demyelinative. Due to loss of the myelin sheath

Dendrite. A nerve cell process on which axons of other neurons terminate. Also peripheral process of a primary sensory neuron

Dentate. A nucleus of the cerebellum or a gyrus in the temporal lobe

Diabetes insipidus. Passage of large amounts of sugar-free urine

Diencephalon. Part of cerebrum consisting of epithalamus, thalamus, subthalamus, and hypothalamus. Formed from embryonic prosencephalon

Diplopia. Seeing double

Discriminitive sense. Ability to detect difference in size, shape, texture, and number of stimuli

Disk (optic). The optic nerve leaving the eye, seen through an ophthalmoscope

Disorientation. Confusion as to time, place, and person

Dissociated sensory loss. Loss of pain sense with light touch preserved

Dominant hemisphere. The cerebral hemisphere controlling speech

Dorsiflexion. Bending backward (usually wrist or ankle)

Drop attacks. Sudden falls without loss of consciousness

Dura mater. The outer, toughest sheet of the meninges

Dural sinuses. Spaces in layers of the dura mater that contain venous blood

Dys-. Difficulty in (e.g., dysphasia, dysarthria (difficulty in expressing thoughts), dysmetria: disturbance of power to control range of movement in muscle action

Dystrophy. Degeneration with loss of function

Electrocorticography. Recording electrical activity direct from the cortex

Electroencephalography (EEG). Recording the electrical activity of the brain

Electromyography (EMG). Recording the electrical activity of muscle

Emboliform. In form of a plug. Nucleus of cerebellum

Encephalitis. Inflammation of the brain

Encephalopathy. A disorder of brain function

Endoneurium. Delicate connective tissue sheath around an individual nerve fiber of a peripheral nerve

Engram. A latent memory picture (psychology)

Entorhinal. This area is the anterior part of parahippocampal gyrus of temporal lobe. Functions as olfactory association cortex

Ependyma. Lining epithelium of ventricles of brain and central canal of spinal cord

Euphoria. A casual cheerfulness

Epidural (extradural) hematoma. A hematoma is a somewhat localized mass of blood that has escaped from a blood vessel; the blood is usually partly clotted and discolored. An epidural hemorrhage becomes an epidural hematoma

Epilepsy. Chronic disorder of brain function, often associated with seizures and loss of consciousness

Epineurium. Connective tissue sheath around a peripheral nerve

Epithalamus. Region of diencephalon above thalamus (includes pineal gland)

Extensor plantar response. The big toe goes up when the sole is scratched

Exteroceptor. Sensory receptor serving to acquaint one with environment

Extracranial. Outside the skull

Extradural. Between the dura and the skull

Extrapyramidal. Motor fibers arising from cells other than the pyramidal cells (often basal ganglia)

Extrinsic. Outside and separate from nervous tissue

Falx cerebri. Sheet of dura mater separating the 2 cerebral hemispheres

Fasciculation. Flickering contractions of muscle fibers

Fasciculus. A bundle of nerve fibers

Fastigial. Top of a gabled roof. Nucleus of cerebellum

Fimbria. Band of nerve fibers along medial edge of hippocampus, continuing as fornix. A fringe

Flaccid. Limp, floppy, loss of tone

Flexor plantar response. The normal downward movement of the big toe when the sole is scratched

Flexor spasm. Painful contraction of muscles in spastic limbs

Focal. Arising from, or limited to, one part

Focal epilepsy. A fit affecting one part of the body, arising from one point in the brain (an attack which begins with specific muscles and then spreads to adjacent ones)

Foramen. An opening

Foramen magnum. Opening in base of skull (occipital bone) where brain becomes the spinal cord

Foramina of Luschka. There are 2: they connect the 4th ventricle and the subarachnoid space

Foramen of Magendie. One of the 3 connections between the 4th ventricle and the subarachnoid space

Forceps. U-shaped configuration of fibers constituting anterior and posterior portions of corpus callosum (minor and major)

Fornix. Efferent tract of hippocampus "arching" over thalamus and ending in mamillary body

Fortification spectra. Zigzag patterns seen in migraine

Fossa. A compartment of the skull holding a part of the brain

Fovea. Pit or depression. Fovea centralis is depression in center of macula lutea of retina

Functional. Due to a disturbance in the working, not the structure, of some part

Funiculus. Area of white matter of cord containing several functionally different fasciculi

Ganglia. Collections of nerve cells outside the central nervous system

Generalized fits. Convulsions affecting all parts of the body

Genu: Knee. Anterior end of corpus callosum: genu of facial nerve

Girdle pains. Constricting pains around the trunk

Glia. The supporting cells and fibers (neuroglia) of the nervous system

Glioblastoma. The most malignant glioma

Glioma. A tumor arising from one of the neuroglial cells, usually from astrocytes. Tumors growing from the supporting cells

Gliosis. Overgrowth of glial fibers

Globus pallidus. A pale ball. Medial part of lenticular nucleus of corpus striatum

Glove-and-stocking anesthesia. Impaired sensation over the periphery of all four limbs.

Gracilis. Slender. Fasciculus gracilis of cord and medulla; nucleus of medulla

Grand-mal. Epilepsy characterized by major fits (an epileptic attack characterized by loss of consciousness and generalized seizures)

Granule. Grain. Small neurons of cerebellar cortex and stellate cells of cerebral cortex

Gray matter. Nervous tissue containing nerve cells (neuron cell bodies, fibers, and neuroglial cells)

Gyri. The folds of the cerebral cortex (elevated areas of cerebral cortex between the sulci)

Habenula. Small swelling in epithalamus near the posterior end of the roof of the third ventricle

Hallucinations. Sensory impressions of something that is not there

Hemianopia. Loss of half of the visual field

Hemiballismus. Half-jumping. Violent form of motor restlessness

Hemiplegia (-paresis). Paralysis (weakness) of one half of the body

Heredofamilial. Passed both from generation to generation and to several members of one family

Herpes simplex. The virus of "cold sores"

Herpes zoster. Shingles

Hiatus. A large gap

Hippocampus. "Seahorse." A gyrus that is an important part of limbic system. Produces elevation on floor of inferior horn of lateral ventricle

Homeostasis. Tendency toward stability in internal environment of organism

Homonymous. The same on both sides

Hydrocephalus. "Water head." Excessive accumulation of cerebrospinal fluid in ventricles

Hyper-. Increased

Hypertrophy. Enlargement

Hyperventilation. Excessively deep and fast breathing

Hypo-. Decreased

Hypoglycemia. An abnormally low blood sugar

Hypopituitarism. Loss of pituitary function

Hypothalamus. A region of diencephalon below thalamus serving as main controlling center of autonomic nervous system

Idiopathic. Of unknown cause

Impulses. Electrical waves traveling along nerves

Inborn. Part of an individual's "makeup"

Indusium. To put on. Indusium griseum is a thin layer of gray matter on upper surface of corpus callosum (gray tunic)

Infarct. A zone of tissue deprived of blood supply

Infundibulum. Funnel-shaped stalk connecting hypothalamus with neural lobe of pituitary gland

Insula. An island. Cerebral cortex concealed from surface and lying at bottom of lateral fissure. Island of Reil

Interceptor. One of sensory end organs in viscera

Intracranial. Inside the skull

Intracranial hypertension. High pressure inside the skull (not high blood pressure)

Intrinsic. Inside the substance of the nervous system

Involuntary movement. Muscular activity not under the patient's control

Ischemia. Shortage of blood supply

Isocortex. Cerebral cortex having six layers (neocortex)

Jacksonian fits. Convulsions starting at one point and spreading to involve wider areas

Kernig's sign. Inability to straighten the knee with the hip flexed, in meningeal irritation

Kinesthesia. Sense of perception of movement

Konicortex. Areas of cerebral cortex containing large numbers of small neurons (typical of sensory areas)

Labyrinth. The semicircular canals of the inner ear

Lange curve. A test carried out by adding CSF to colloidal gold solution

Lasègue's sign. Limitation of straight leg raising

Lemniscus. A ribbon or band. A bundle of nerve fibers in CNS (medial and lateral lemniscus)

Lenticular. Shape of lentil or lens. A nucleus of the corpus striatum

Leptomeninges. Arachnoid and pia mater

Leukodystrophy. Degeneration of white matter

Lightning pains. Needlelike pain in the limbs in tabes

Limbus. C-shaped configuration of cortex on medial surface of cerebral hemisphere consisting of cingulate and parahippocampal gyri. Limbic system: limbic lobe, hippocampal formation, and parts of diencephalon

Limen. Threshold. Limen insulae is basal part of insula

Lobe. A major division of the cerebral hemispheres or cerebellum based on surface marking

Local, localize. One point affected; to determine the exact point affected

Locus ceruleus. Small dark spots on either side of floor of fourth ventricle marking position of nerve cells that have melanin pigment

Longitudinal fissure. Cleft that separates the cerebrum into right and left cerebral hemispheres

Lower motor neuron. Cells and fibers of motor cranial and spinal nerves

Macroglia. Larger types of neuroglial cells (astrocytes, oligodendrocytes, and ependymal cells) of ectodermal origin

Macrosmatic. Sense of smell strongly or acutely developed

Macula. A spot. Macula lutea is a spot of posterior pole of eye having yellow color when seen with red-free light. Maculae sacculi and utriculi are sensory areas in vestibular part of membranous labyrinth

Mamillary. Breast-shaped like nipple. Mamillary bodies are small swellings on basal surface of hypothalamus

Massa intermedia. Bridge of gray matter connecting thalami of two sides across third ventricle (in 70% of brains)

Medulla. The lowest portion of the brainstem

Meninges. The three membranes clothing the brain and cord and lining the skull and vertebral canal

Meningioma. A benign tumor growing from the arachnoid

Meningism. Signs of irritation of the meninges, not due to infection

Meningitis. Inflammation of the meninges (usually bacterial in origin)

Meningocele. A bulge of the meninges through a breach in the bony coverings

Meningoencephalitis. Inflammation of both the brain and meninges

Meningomyelocele. A meningocele containing spinal cord tissue

Mesencephalon. The midbrain. Second of three primary brain vesicles

Mesocortex. Cerebral cortex intermediate between three-layered and six-layered cortex (in cingulate gyrus)

Mesoglia. Microglial cells of mesodermal origin

Metathalamus. Medial and lateral geniculate bodies (nuclei)

Metencephalon. Pons and cerebellum. Anterior division of rhombencephalon or posterior brain vesicle

Microglia. Neurological cells of mesodermal origin belonging to reticuloendothelial system of cells of body

Microsmatic. A sense of smell of relatively poor development

Mitral. Bishop's miter. Mitral cells of olfactory bulb

Mixed nerve. One containing both motor and sensory fibers

Molecular. Denotes tissue containing large numbers of fine nerve fibers. Has a punctate appearance in silver-stained sections. Layers of cerebral and cerebellar cortices

Monoplegic (-paresis). Paralysis (weakness) of one limb

Motor. Concerned with movement

Motor areas. Located in precentral gyrus of frontal lobe; control specific skeletal muscles of opposite side of body

Multiple sclerosis. Disseminated sclerosis

Myasthenia. Weakness of muscle

Myelencephalon. Medulla oblongata. Posterior of two divisions of rhombencephalon or posterior primary brain vesicle

Myelin. The white sheath of nerve fibers

Myelitis. Inflammation of the spinal cord

Myelography (myelogram). X-raying the vertebral canal by introducing opaque fluid into the CSF

Myoclonus. Shocklike muscle contractions

Myopathy. Degenerative disease of muscle (muscular dystrophy)

Myositis. Inflammation of muscle

Myotonia. Contraction of muscle persisting after the need for it is passed

Neck stiffness, rigidity, retraction. Signs of irritation or infection of the meninges

Neocerebellum. Newest part of cerebellum seen in mammals (especially well developed in man). Ensures smooth muscle action in finer voluntary movements

Neocortex. Six-layered cortex, characteristic of mammals. Most of cerebral cortex in man

Neostriatum. Striped or grooved. Newer part of corpus striatum consisting of caudate nucleus and putamen; the striatum

Nervous system. All nerve tissue of the body and associated supporting tissue. Divided into a central nervous system (CNS) and a peripheral nervous system (PNS)

Neuralgia. Pain in the distribution of a nerve

Neuraxes. A short form referring to brain and spinal cord. May refer to an axon

Neurofibril. Small nerve. Delicate filaments in cytoplasm of neurons

Neuroglia. Accessory or interstitial cells of CNS include astrocytes, oligodendrocytes, microglial cells, and ependymal cells

Neurolemma. Delicate sheath around a peripheral nerve fiber consisting of a series of neurolemma cells or Schwann cells

Neuromuscular junction. A point where a nerve fiber ends in a muscle

Neuron. The nerve cell, its fibers, and all its branches

Neuropil. Complex net of nerve cell processes occupying the intervals between cell bodies in gray matter

Neurosyphilis. Involvement of the nervous system in the later stages of syphilis

Nociceptive. Responsive to injurious stimuli

Nucleus (-ei). A collection(s) of nerve cell bodies

Nystagmus. Rhythmic involuntary oscillation of the eyes

Obex. Barrier. Small transverse fold overhanging the opening of the fourth ventricle into central canal of closed position of medulla

Oculomotor. Concerned with eye movement

Olfactory. Concerned with the sense of smell

Oligodendrocyte. A neuroglial cell of ectodermal origin. Forms myelin sheath in CNS

Oligophrenia. Mental retardation

Operculum. Cover or lid. Frontal, parietal, and temporal opercula bound the lateral fissure of cerebral hemisphere. Conceal insula

Ophthalmoplegia (-ic). Paralysis of eye movement

Opisthotonos. Backward arching of the whole body

Optic. Concerned with the eyes or visual pathways

Organic. Due to structural disease

Otorrhea. Running from the ear

Overflow incontinence. Constant dribbling due to overdistention of an insensitive bladder

Pachymeninx. Thick membrane. The dura mater

Paleocerebellum. Phylogenetically old part of cerebellum functioning in postural changes and locomotion

Paleocortex. Ancient, three-layered olfactory cortex

Paleostriatum. Ancient, striped, or grooved phylogenetically older and efferent part of corpus striatum. Globus pallidus or pallidum

Pallidum. Pale. Globus pallidus of corpus striatum. Medial part of lenticular nucleus.

Pallium. Cloak. Cerebral cortex with subjacent white matter. Sometimes used synonymously with cortex

Panencephalitis. Inflammation of the whole brain

Papilledema. Swelling of the optic nerve seen with an ophthalmoscope

Para-. Alongside (e.g., paraventricular)

Paraplegia (paraparesis). Paralysis (weakness) of both legs and lower part of trunk

Parkinsonism. The tremor and rigidity of Parkinson's disease

Perikaryon. Cytoplasm around the nucleus. Sometimes refers to cell body of a neuron

Perineurium. Connective tissue sheath surrounding a bundle of nerve fibers in a peripheral nerve

Peripheral nervous system (PNS). Nerves, ganglia, and special receptors

Pes cavus. Very high arches of the feet

Petit mal. Minor epilepsy; frequent episodes of detachment from one's surroundings (an epileptic attack characterized by brief loss of consciousness and sometimes mild facial twitchings)

Photophobia. A dislike for light

Pia mater. Thin innermost layer of meninges, attached to surface of brain and spinal cord. Forms inner boundary of subarachnoid space

Pineal gland. A structure lying in the center of the skull, frequently calcified and visible on x-ray

Plantar reflex. Toe movement when the sole is scratched

Plaques. Patches; usually applied to areas of demyelination in disseminated sclerosis

Pneumoencephalography. *See* Air encephalography

Polioencephalitis. Brainstem infection of poliomyelitis virus

Poly-. Many (e.g., polyneuritis, inflammation of many nerves)

Pons. Bridge. Part of brainstem lying between medulla and midbrain. A bridge between right and left halves of cerebellum

Position (postural) sense. Knowledge of where each part of the body is without looking at it

Posterior columns. Spinal cord tracts carrying position, vibration, and discriminative sensations

Posterior horns. Part of the spinal gray matter receiving sensory roots

Premotor areas. Areas in frontal lobe just anterior to motor areas; control general muscular activity of opposite side of body

Presenile dementia. Cerebral atrophy in middle age

Pressure cone. The forcing of brain tissue through a foramen, as a result of high pressure above, e.g., tentorial or cerebellar pressure cones

Projection fibers. Nerve fibers traveling to or from lower parts of the brain and spinal cord

Proprioceptor. One of sensory endings in muscles, tendons, and joints. Gives information about movement and position of body parts

Prosencephalon. Forebrain consisting of telencephalon (cerebral hemispheres) and diencephalon. Anterior primary brain vesicle

Pseudohypertrophy. Apparent, but not true, enlargement

Psychomotor epilepsy. Disturbance of behavior due to epileptic discharges

Psychoneurotic. Of psychologic, not disease, origin

Ptosis. Drooping of the eyelid

Pulvinar. Posterior projection of thalamus beneath which the medial and lateral geniculate bodies are found

Putamen. Shell. The larger and lateral part of lenticular nucleus of corpus striatum

Quadriplegia. Paralysis of all four limbs

Queckenstedt test. Rise in lumbar CSF pressure when the jugular vein is compressed

Reflex. An automatic response to a stimulus

Relapse. A further attack of a disease

Remission. A period of recovery from a disease

Restiform. Rope-form. Restiform body: main mass of inferior cerebellar peduncle containing afferent fibers to cerebellum

Retrobulbar. Behind the eye

Rhinal. Related to nose

Rhinencephalon. In man, refers to components of olfactory system

Rhinorrhea. Running from the nose

Rhombencephalon. The pons and cerebellum (metencephalon) and medulla (myelencephalon). Posterior primary brain vesicle

Rigidity. Stiffness due to equal resistance in all muscles

Roots. Nerve fibers as they leave or enter the stem or cord

Rostrum. Recurved part of corpus callosum passing backward from genu to lamina terminalis

Satellite. Satellite cells: flattened cells of ectodermal origin, forming a capsule for nerve cell bodies in dorsal root ganglia and sympathetic ganglia. Satellite oligodendrocytes adjacent to nerve cell bodies in CNS

Scanning (brain-). Detecting the distribution of a radioactive substance in the brain

Sclerosis. Hardening

Scotoma. A patch of blindness

Sella turcica. A saddle-shaped cavity at the base of the skull containing the pituitary gland

Sensory. Concerned with feeling

Sensory level. The point where sensation changes from abnormal to normal

Septal area. Cortex and underlying septal nuclei, the latter extending into septum pellucidum beneath genu and rostrum of corpus callosum on medial side of frontal lobe. Also called medial olfactory area

Septum pellucidum. Triangular double membrane separating anterior horns of lateral ventricles. Fills interval between corpus callosum and forms in the median plane

Small areas. Located in gyrus on medial surface of temporal lobe

Somatic. Denotes the body, exclusive of viscera

Somesthetic. The consciousness of having a body. Somesthetic senses are general senses of pain, temperature, touch, pressure, position and movement, and vibration

Space-occupying lesion. A tumor or other growing lesion

Spastic ataxia. A combination of spasticity and unsteadiness

Spinal cord. Part of the CNS in the spinal canal. *Structure:* gray matter on inside, white matter on outside

Spinal cord segments. There are 8 cervical, 12 thoracic, 5 lumbar, 5 sacral, and 1 coccygeal (31 in all). Each segment gives rise to a pair of spinal nerves

Splenium. Thickened posterior extremity of corpus callosum

Spondylosis. Degenerative changes in bones and the disk of the spine

Status epilepticus. Fits following each other in rapid succession

Stenosis. A narrowing

Stereotaxis. Accurately placing small lesions in the depths of the brain

Strabismus. A squint

Stretch reflex. Contraction of muscle following its sudden stretching

Stria terminalis. Furrow limit. A slender strand of fibers running along medial side of tail of caudate nucleus originating in amygdaloid nucleus and mostly ending in septal area and hypothalamus

Striatum. Furrowed. More recent part of corpus striatum (neostriatum) consisting of caudate nucleus and putamen

Stroke, cerebral. Same as massive cerebral hemorrhage

Stupor. Unconsciousness, but rousable

Subarachnoid space. Between arachnoid and pia mater containing CSF

Subdural. Between dura and arachnoid

Subdural hemorrhage. Bleeding, usually due to trauma (blow to the head), in which veins entering into sinuses are broken, e.g., cerebral veins entering the sagittal sinus. Subdural hemorrhage is usually venous in origin

Subiculum. A layer. Transitional cortex between parahippocampal gyrus and hippocampus

Substantia gelatinosa. Column of small neurons of apex of dorsal gray horn throughout spinal cord

Substantia nigra. A large nucleus, with motor functions, in midbrain. Many cells contain melanin

Subthalamus. Region of diencephalon beneath thalamus, containing fiber tracts and sub-thalamic nucleus

Sulci. Furrows on the surface of the brain

Superficial reflex. Various muscle contractions following stimuli to the skin surface

Symptomatic. Representing a disease process

Synapse. Site of contact between neurons, at which site one neuron is excited or inhibited by another neuron

Syndrome. A collection of signs and symptoms recurring frequently enough to be recognizable

Syringomyelia. A condition characterized by central curvature of spinal cord and gliosis around the cavity

Syrinx. A cavity in the brainstem or spinal cord

Tapetum. A carpet of fibers of the corpus callosum sweeping over the lateral ventricle and forming lateral wall of its posterior and inferior horns

Taste area. Located on ventral aspect of postcentral gyrus of parietal lobe

Tectum. Roof of midbrain consisting of paired superior and inferior colliculi

Tegmentum. To cover. The dorsal portion of pons. Also the major part of cerebral peduncle of midbrain between substantia nigra and tectum of midbrain

Teichopsia. Flashes of light in migraine

Tela choroidea. Like a membrane. The vascular connective tissue, continuous with that of pia mater, which continues into core of choroid plexuses

Telencephalon. Cerebral hemispheres. Anterior of the two divisions of prosencephalon or anterior primary brain vesicle

Telodendrea. Terminal branches of axons

Tentorial herniation. Forcing part of the brain through the tentorial hiatus

Tentorium (cerebelli). Tentlike sheet of dura separating the cerebellum from the cerebral hemispheres (occipital lobes)

Thalamus. Large mass of gray matter of diencephalon in each cerebral hemisphere, bordering on 3rd ventricle with sensory and integrative functions

Theca. The space formed by the meninges containing CSF

Tic. A recurrent spasm (e.g., tic douloureux)

Tinnitus. Ringing in the ears

Todd's paralysis. Temporary paralysis of a part following a fit

Tone. The tension present in a muscle at rest

Torcular. To twist. The confluence of dural venous sinuses at internal occipital protuberance

Tracts. Collection of nerve fibers in the CNS having similar functions

Trapezoid body. Transverse fibers of auditory pathway found at junction of dorsal and basal portions of pons

Trophic. Changes occurring in tissues that have lost their nerve supply

Uncinate. Hook-shaped. Uncinate fasciculus: association fibers connecting cortex of basal surface of frontal lobe with temporal pole; a bundle of fastigiobulbar fibers (of Russell) that curves over superior cerebellar peduncle on way to inferior cerebellar peduncle

Uncus. A hook. Portion of rostral end of parahippocampal gyrus of temporal lobe. A landmark for lateral olfactory area

Upper motor neuron. The motor cells and fibers running from the cerebral cortex to cranial nerve nuclei and anterior horn cells

Uvula. Portion of inferior vermis of cerebellum

Vallecula. Midline depression on inferior aspect of cerebellum

Vagus. Tenth (X) cranial nerve. Once called the pneumogastric nerve.

Ventricles. Cavities in the brain containing CSF [fluid-filled (CSF) interconnected spaces in the brain, 4th, 3rd, and 2 lateral ventricles]

Ventriculography. X-raying the ventricles by injecting air directly into them

Vermis, worm. Midline potion of cerebellum

Vertigo. A sense of rotation

Vestibular. Concerned with the inner ear labyrinth and its connections

Vibration sense. Ability to detect tuning fork vibration

White matter. The part of the brain and spinal cord that has a light appearance and contains myelinated fibers and neuroglial cells

Xanthochromic. Yellow colored

Zona incerta. Gray matter in subthalamus, representing rostral extension of reticular formation of brainstem

APPENDIX II. DISORDERS OF FUNCTION: MIGRAINE

I. **Introduction:** the majority of people who have symptoms related to the nervous system do not have anything wrong with its structure but with its function. Symptoms as headache and giddiness fall into this group. Unfortunately, the term "functional" is often used to suggest an imaginary disturbance since psychological upsets disturb function rather than structure. However, many disorders of the normal functioning of the brain, its vessels, and its coverings produce symptoms for which an explanation is hard to come by, or which may have a psychological basis, but nonetheless, there is no doubt they are real. The skull and brain are incapable of feeling pain. It is from the scalp, its muscles and blood vessels, and from the dura mater and its venous sinuses as well as blood vessels at the base of the brain that pain can arise. Headaches most commonly originate in the muscles, blood vessels, and dura mater and are produced by tension in, or stretching of, these structures

II. **Migraine** is due to a disturbance in the behavior of the cranial blood vessels. What causes an attack is not known and there may be many causes. It seems relatively certain that chemical changes in and around the blood vessel walls play an important role in its development
   A. Symptoms are produced by spasms of vessels inside the skull and dilation of vessels outside of the skull (theory): the spasm produces the warning of the attack aura, and the dilation produces the headache
   B. Migraine attacks sometimes bear an obvious relationship to anxiety, menstruation, and excitement, but may come "out of the blue" and often on weekends
      1. Some foods are liable to produce attacks, particularly chocolate
   C. Attacks usually begin in youth, even in childhood, affect women more than men, and are common in "perfectionists" living under much pressure
   D. Migraine is strongly hereditary, and allergic disorders (asthma, eczema) are frequently found in the same families
   E. THE AURA
      1. The 1st symptoms depend on the blood vessel that goes into spasm
      2. Not all attacks have a clear aura. Many begin with flashing, scintillating lights zig-zagging across the visual fields, sometimes forming colored patterns with a dark center and jagged edges, called teichopsia or fortification spectra
      3. Other visual auras include loss of half of the visual field (homonymous hemianopia), loss of the outer fields of each eye (bitemporal hemianopia), or complete loss of both eyes or small patches of loss in both eyes (scotoma)
         a. The above symptoms are due to spasm affecting the vessels supplying the optic nerves, the chiasma, or the occipital visual cortex
      4. If the middle cerebral artery is involved, the patient develops numbness of one hand and arm, one half of the face and tongue, and even dysphagia
      5. If the basilar artery is involved, there may be bilateral blindness, vertigo, tinnitus, tingling in both hands and legs, and dysarthria
      6. The symptoms last from a few minutes to as long as an hour, die away, and are replaced by the headache
   F. THE HEADACHE: migraine headache is said to be one-sided, but often affects the entire head or just the occipital region
      1. Because of dilatation of the superficial arteries it throbs violently and the superficial temporal vessels may even be seen to stand out
      2. It increases to a maximum when vomiting occurs and then there is improvement, and the patient goes to sleep to probably wake up free from pain
      3. Occasionally lasts 24–48 hours; vomiting is sometimes prolonged and constant
      4. Occasionally, just at the stage when aura changes to headache, the patient may even lose consciousness for a few moments

G. TREATMENT
  1. The aspirin-containing analgesics are the only drugs that will influence a migraine once the headache is developed
  2. With a long aura, ergotamine preparations taken early may prevent the headache from developing
      a. They are not very effective by mouth, though cafergot in tablets, allowed to dissolve under the tongue, or used as suppositories, may help
      b. Ergotamine tartrate by injection is the most satisfactory, but may make some patients sick and should be preceded by Largactil or Dramamine
  3. The aura, however, may be short or absent and then long-term prevention is needed
      a. Regular sedation with phenobarbitone or prochlorperazine (Stemetil) or methysergide, used for months or years, often reduces the frequency and severity of the attacks, but there is no wonder drug available
  4. Attacks tend to subside with age and many women improve with menopause. In some cases, development of hypertension with age worsens and prolongs attacks

III. **Specific types of migraine**
  A. OPHTHALMOPLEGIC MIGRAINE: due in most cases, not to migraine, but to pressure on the 3rd cranial nerve by an aneurysm on the circle of Willis
      1. Severe pain over one eye, sometimes followed by double vision, ptosis, and paralysis of the 3rd nerve which lasts for a week or more
  B. FAMILIAL HEMIPLEGIC MIGRAINE: mechanism is obscure and can be diagnosed only if other members of the family have identical attacks
      1. Numbness of $\frac{1}{2}$ of the body may occur in ordinary migraine, but here a frank paralysis of one arm and one leg may persist for hours or days
  C. COUGH HEADACHES: due to vascular distention during coughing
      1. Very severe pain develops over one eye or in the head generally, lasting some moments after the cough has stopped
      2. It is not serious and the attacks usually stop after several months
  D. PERIODIC MIGRANOUS NEURALGIA (histamine cephalgia or cluster headaches)
      1. Migraine rarely disturbs patient's sleep; this condition always does
      2. The attacks occur every night at about 2–3 A.M., for weeks on end, followed by months or years without any symptoms
      3. The pain is behind the same eye, accompanied by reddening and running of the eye, blocking of the nose, and flushing of side of face
      4. Lasts $\frac{1}{2}$ to 2 hours and is very severe
      5. Attacks may occur in the daytime, often at the same time each day
      6. Injection of ergotamine tartrate, just before sleep, given by the patient to himself (like insulin), will stop most attacks
  E. MUSCLE TENSION HEADACHES: commonest cause of all chronic headaches
      1. Many people get very tense at one time or another. In some, this tension affects the muscles of the scalp, resulting in headaches that are much more prolonged than migraines, lasting days or weeks
      2. Described as "pressure" or a "tight band" round the head and down the neck
      3. There is no vomiting, aura, or sleep disturbance
      4. People helped by an explanation of the cause and the use of regular doses of prochlorperazine (Stemetil) or diazepam (Valium)

I. **Neuralgias,** strictly speaking, are pains in the distribution of a nerve, but the term is often vaguely used with some recurring severe pains which do not fall into a definite pattern that can be recognized and separated from other types of pain. Some definitive types are

A. TRIGEMINAL NEURALGIA OR TIC DOULOUREUX is an intense spasmotic pain affecting the areas supplied by the branches of the 5th cranial nerve
  1. May affect all ages, but is less common below the age of 50
  2. Symptoms
      a. Intense stabs of pain, even lasting only a few seconds, but often frequently repeated, shoot upward from the lower lip, or upward from the eyebrows, or downward into the point of the jaw and side of the tongue
      b. Pain may also start from the same trigger point and can be brought on by touching this point, brushing the face, shaving, exposure to cold wind, eating, drinking, or even talking
      c. The face is contorted during attacks, but is otherwise free of pain
  3. Treatment
      a. The trigeminal nerve can be injected with alcohol, or the nerve root divided by intracranial surgery. Both leave that side of the face numb, and there may be some troublesome tinglings after injection
      b. Carbamazepine (Tegretol) has been found to suppress the pain, if taken regularly, in many patients. In others, the pain breaks through the medication and surgery is necessary
  4. Spontaneous remissions are common
B. ATYPICAL FACIAL NEURALGIA
  1. Many patients, particularly women, complain of a constant burning pain that lasts for months or years, affecting one part of the face, but often spreading somewhat to the other side
  2. Its onset is often dated to some dental treatment
  3. The patients are depressed and there is usually some deficiency in their lives, e.g., a loss of a husband or a failure to have children
  4. Treatment is difficult and unsatisfactory and drug addiction is a common danger
C. POSTHERPETIC NEURALGIA
  1. After an attack of herpes zoster, neuralgic pain may persist for months or years
  2. This condition is rare in the young, but common in the aged, especially affecting those who have had ophthalmic herpes (herpes of the 1st division of the trigeminal nerve)
  3. Despite their complaints, the patients do not seem to be in pain
  4. The pain seems to become part of their personalities and begins to control their lives
  5. Even root section does not successfully cure the condition
  6. All forms of treatment appear to fail
  7. The greatest danger for the patient is drug addiction

I. Vertigo: a feeling that one's surroundings are spinning round, either horizontally or vertically. "Giddiness" may or may not mean vertigo since the word is often used incorrectly for a sense of instability or attacks of faintness. True vertigo is due to a disturbance of the vestibular system, which consists of the semicircular canals of the inner ear, vestibular fibers of the 8th cranial nerve and its nuclei in the brainstem, and fibers that pass from these to the cerebellum, temporal lobes, and spinal cord. Lesions producing vertigo may occur at many sites, the most common being the semicircular canals

A. ACUTE LABYRINTHITIS is a commonly used term, but it is doubtful that the condition really exists, i.e., as a true inflammation of the semicircular canals
  1. The symptoms begin suddenly with severe vertigo, vomiting, and unsteadiness, but no deafness. It lasts a few days and clears up spontaneously
  2. The patient is kept at strict rest, given chlorpromazine (Largactil) or dimenhydrinate group of drugs (i.e., Dramamine), and then gradually mobilized
  3. Syndrome seen in epidemics suggesting a relation to infection

B. MÉNIÈRE'S DISEASE: the labyrinth, for some reason, sometimes becomes swollen and produces a characteristic sequence of events
  1. Symptoms
    a. Patient is prone to sudden, often severe, attacks of vertigo
    b. During the attack, which may last several hours, the patient vomits repeatedly and is too unsteady to stand
    c. Between attacks, tinnitus develops in one ear with gradual increasing deafness
    d. The other ear finally may also be affected
    e. Eventually the vertigo ceases, and the patient is left severely deaf
  2. Signs: nystagmus and unsteadiness are present during an attack and the patient invariably lies flat and is afraid to move
    a. Between attacks, increasing nerve deafness (one or both ears) is only sign
  3. Treatment
    a. Dramamine or Largactil should be given by injection in an acute attack
    b. Above medication should also be taken regularly between attacks
    c. In severe cases, a decompression or destruction of the labyrinth may be done, or the 8th nerve may be divided. In these cases, the patient should be grossly disabled by the attacks before surgery is attempted because a sense of instability may be a troublesome result of surgery

C. ARTERIOSCLEROTIC VERTIGO
  1. Sudden attacks of vertigo on movement are common in elderly people with atheroma of the vertebral and basilar arteries
  2. If small vessels supplying the vestibular system become blocked, severe vertigo may persist for hours or even days
  3. One may also see attacks after spine manipulations (osteopathic or otherwise)

D. OTHER CAUSES OF VERTIGO
  1. Tumors of the brainstem and 4th ventricle, especially in childhood, may disturb the vestibular nuclei and cause vertigo
  2. An acute relapse of disseminated sclerosis affecting the brainstem may cause vertigo and vomiting lasting for days
  3. May occur as part of an attack of temporal lobe epilepsy
  4. Vertigo can be a symptom of a psychoneurosis

APPENDIX II (cont.). DISORDERS OF FUNCTION:
DISTURBANCES OF SLEEP

I. Disturbances of sleep; precisely how normal sleep and waking occur is uncertain, although since lesions of the reticular formation can produce a state resembling sleep, it is likely that sleep is brought about when this formation stops sending alerting impulses to the cortex

A. INSOMNIA
  1. Is rarely a sign of organic nervous disease and commonly a sign of psychoneurosis
  2. Some degree is normal in old age
  3. Occurs in many toxic and fever states
  4. In certain types of encephalitis, especially encephalitis lethargica, sleep rhythm is reversed and the patients are drowsy during the day and very wakeful and troublesome during the night

B. HYPERSOMNIA (increased sleepiness)
  1. Common in fatigue and while taking sedatives
  2. May become severe when there is a rise of intracranial pressure due to a cerebral tumor or brain swelling

C. NARCOLEPSY is a condition usually affecting rather plump adults in which they are suddenly overcome by irresistible sleep, often under embarrassing circumstances
  1. They may fall asleep in the middle of a meal or under monotonous conditions such as when they are driving their cars
  2. A touch will arouse them, and they do not feel sleepy in-between times
  3. Narcoleptics also suffer from 4 other conditions
    a. Sleep paralysis: here the mind wakes up before the body and for some time (minutes) after waking, the patient is unable to move his limbs. (This may occur in normal people at times)
    b. Dream hallucinations: the patient wakes, but a dream seems to continue
    c. Sonambulism: sleep-walking, when the body wakes up before the mind
    d. Cataplexy is a condition in which, on intense emotion, such as fright, a sudden outburst of laughter or of crying takes place. The patient's knees suddenly give way and he falls helplessly to the ground. This should not be confused with catalepsy, which is a psychogenic trancelike state, often affecting young females, lasting hours or days

D. TREATMENT: all of the above disorders are helped by regular doses of a stimulant such as amphetamine sulfate, taken in the morning and at midday. As the patient grows older, the attacks tend to ease up and even disappear, but the treatment may be needed for years

APPENDIX III.   EXAMINATION OF THE NERVOUS SYSTEM

One depends on inspection, palpation, occasionally percussion, and if vascular lesions are suspected, auscultation

   I. **History:** age, chief complaint, date and manner of onset of symptoms, description of symptoms, course of illness, and medical and social background

  II. **Neurologic examination**
   A. MENTAL STATE: state of awareness and degree of patient cooperation, memory and recall, intellectual functions (performance on math problems, ability to perform abstractions), assay of general fund of information, affect, personality, and speech
   B. CRANIAL NERVE FUNCTION (see Appendix IV)
   C. MOTOR STATUS (power and coordination)
      1. Are all muscles involved or only specific muscle groups?
      2. Check for fatigability, tenderness, atrophy, degree of resistance to passive movement, strength of muscles, and their ability to perform coordinated movements quickly
   D. REFLEXES: there are deep, superficial, pathologic, and meningeal reflexes
      1. A few of the major reflexes to be tested are
         a. Direct light reflex: if light is shined into one eye, the pupil will constrict
         b. Consensual light reflex: shine light as above and note constriction of opposite pupil
         c. Pupillary accommodation reflex: on near and distant objects of focus
         d. Corneal reflex: touch edge of cornea and get closure of eyelids (for 1st division of 5th cranial nerve)
         e. Abdominal reflex: stroke skin of upper abdomen toward midline and recti muscles contract
         f. Cremasteric reflex: stroke upper and inner thigh, causing elevation of testicle on same side
         g. Tendon reflexes: quickly tap muscle tendon, producing transient stretch with reflex muscle shortening. Biceps (C5-C6); triceps (C7-C8); knee jerk (L3-L4); Achilles tendon (L5, S1)
         h. Babinski reflex: stroke outer part of sole of foot from heel toward small toe and across ball of foot medially (tests pyramidal system)
         i. Trömmer or Hoffmann reflex: support forearm and stroke nail or middle finger. Normally, no movement; in pathology, thumb is flexed and adducted (test pyramidal system)
         j. Oppenheim's sign: draw fingernail along skin of shin from knee to medial malleolus. Dorsal flexion of big toe is pathognomonic (pyramidal system)
         k. Kernig's sign: extend leg while patient lies on back with thigh flexed. If limited by spasm and pain, is pathognomonic of meningeal irritation
         l. Brudzinski's sign: with patient supine and limbs extended, passively flex the neck. If it produces flexion of hips, is pathognomonic of meningeal irritation
   E. SENSORY SYSTEM: study pain, light touch, temperature, vibration, and position sense
   F. SYMPTOMS OF DISEASE OF THE NERVOUS SYSTEM
      1. Lower motor neuron syndrome: atrophy and fasciculation (twitching), flaccidity (limpness) of involved muscles, and loss or diminution of deep tendon reflexes
      2. Upper motor neuron syndrome: no atrophy of involved muscles except of disuse, spasticity of involved muscles, hyperactive deep tendon reflexes, pathologic reflexes, e.g., Babinski
      3. Extrapyramidal system syndrome: rigidity, changes in gait and stance, tremor, little if any change in deep tendon reflexes, and rarely any pathologic reflexes

4. Dysfunction of cerebrum: convulsions or their equivalents, mental and behavioral changes (organic type), aphasia (disorder in language and communication), hemiplegia, hemisensory defect, homonymous field defects
5. Dysfunction of basal ganglia: dyskinesias (such as tremors, choreas, athetoses, hemiballism, etc.), increased muscle tone, attitudinization, and alterations in posture and movement
6. Dysfunction of the cerebellum: unsteadiness in motor performances (adiadokinesis, asynergia, ataxia, unchecked rebound and pastpointing), tremor more prominent in movement than at rest, and hypotonicity of muscle
7. Dysfunction of brainstem (midbrain, pons, and medulla): bilateral pupillary changes, paralysis of conjugate eye movements (horizontal, vertical, or convergence), crossed syndromes (paralysis of eyes on one side and body on the other, paralysis of face on one side and body on the other side, paralysis of tongue on one side and body on the other, incoordination of one side and sensory defect on opposite side, loss of sensation in the face and on opposite side of body, etc.), defects in articulation, vocalization, phonation, swallowing, and respiration, and nystagmus (persistent) particularly on upward gaze
8. Dysfunction of spinal cord: bilateral involvement, particularly of the lower extremities, presence of sensory levels, Brown-Sequard syndrome (impairment of motor function on one side and pain-temperature sensations on other side), severe impairment of sense of vibration over the trunk and below and urinary and rectal sphincter disturbances in absence of mental changes
9. Dysfunction of spinal roots: sensory and motor changes in distribution of spinal roots, unilateral (such as herniated intervertebral disk syndrome) or bilateral (such as cauda equina root syndrome)
10. Dysfunction of peripheral nerves: mononeuropathy (sensorimotor defects in distribution of certain peripheral nerves), and polyneuropathy (sensory motor defects with bilateral symmetrical glove-and-stocking distribution)

G. CLASSIFICATION OF DISEASE
1. Symptomatic classification: neuralgia, convulsions, palsies, etc.
2. Anatomic classification: encephalopathy, myelopathy, radiculopathy, neuropathy, etc.
3. Etiologic classification: neoplastic, psychogenic, vascular, etc.

H. ETIOLOGIC CATEGORIES
1. Psychogenic: autonomic nervous system or somatic nervous system reactions
2. Infectious: bacterial, parasitic, viral, spirochetal, and fungal
3. Vascular: thrombosis, hemorrhage, embolism, spasm, aneurysms, and blood dyscrasias
4. Traumatic: concussion, contusion, fracture, and herniations
5. Toxic metabolic: drugs, metals, endocrine dysfunctions, biochemical changes, nutritional, and vitamin deficiencies
6. Neoplastic: benign, primary, malignant, metastatic
7. Degenerative: demyelinating, sclerotic, atrophic, and hypertrophic
8. Congenital and developmental anomalies: cysts, tumors, defects and malformations
9. Osseous and muscular: arthropathies and myopathies
10. Idiopathic: paroxysmal disorders without known cause, such as Ménière's syndrome, migraine, epilepsies, neuralgias, and tics

APPENDIX IV. CLINICAL TESTING OF CRANIAL NERVE FUNCTION*

Nerve	Fiber Component	Clinical Test
I	Olfactory (afferents)	Odor applied in one nostril at a time
II	Visual pathway (OD, OS)	Visual acuity (uncorrected and corrected for errors of refraction) using Snellen's chart, visual fields (tangent screen, confrontation, perimeter), fundi (ophthalmoscopy), light reflexes (direct and consensual)
III, IV, VI	Motor to extrinsic eye muscles	Range of ocular movements and convergence, lid movements, look for nystagmus, particularly lateral and upward gaze, look for ptosis of upper lid, look for diplopia
III	Parasympathetic	Pupillary response to light and accommodation; observe shape, size, and equality of pupils; test pupillary reflex to accommodation by having patient first look at near and then far object
V	Sensory fibers	Sensation of skin of face, corneal reflex
	Motor fibers	Muscles of mastication (close jaws forcibly and palpate muscles on both sides), masseter reflex
VII	Sensory fibers	Taste, anterior $2/3$ of tongue
	Motor fibers	Muscles of facial expression (observe asymmetry or wasting in facial muscles, test their action by having patient smile and show teeth, test for frontalis by having patient wrinkle forehead, test orbicularis oculi by having patient close his eyes against resistance), look for herpes in ear, observe degree of lacrimation, test corneal reflex
VIII	Auditory portion	Qualitative test for hearing (watch ticking), Rinne and Weber tuning fork tests (for air and bone conduction), audiometry, lateralization of sound, examine eardrum with otoscope for signs of middle ear infection
	Vestibular portion	Posture and coordination, caloric tests, oculonystagmography
IX, X	Motor fibers	Swallowing, gag reflex, movement of palate and uvula (IX), voice (X)
	Sensory fibers	Gag reflex, pharyngeal sensation, taste on posterior $1/3$ of tongue (IX), look at vocal cords if indicated by abnormal voice (X)
XI	Motor fibers	Test sternocleidomastoid and trapezius muscles
XII	Motor fibers	Tongue function (look for evidence of wasting or fibrillation), have patient stick out tongue and look for deviation to either side

*The commonly used tests at the bedside or in the office.

APPENDIX V.   FUNCTIONAL CLASSIFICATION OF
SYSTEMS AND REFLEXES

I. Neurons
A. AFFERENT (A): means the conduction current is traveling toward the brain or spinal cord. Used often synonymously with sensory, but sensory implies consciousness and not all afferent impulses reach consciousness
  1. Somatic (S) pertains to the framework of the body as distinguished from the viscera or organs. Embryologically is concerned with those parts derived from somatopleure. SA fibers carry information about changes in the external environment
     a. General (G) refers to impulses that begin at or near the body surface. In the GSA system are 3 sensory modalities of sensation: pain; temperature (both warm and cold); and touch (light touch and deep touch (synonymous with pressure)
     b. Special (S) receptors are highly specialized and are concentrated in relatively small areas or organs. Usually contained in a framework of nonnervous tissue and often associated with accessory structures (eye and ear)
        i. Vision—stimuli are short, high-frequency waves called light
        ii. Hearing—stimuli are long, low-frequency waves called sound
  2. Visceral (V) implies afferent impulses arising in or around the viscera. Are essential in the reflex regulation of all visceral activity
     a. General (G)
        i. Receptors are found in or on mucous membranes and in walls of most organs
        ii. The GVA system carries impulses arising as a result of physical or chemical composition of substances contained in the organ and the degree of distention of their walls
B. EFFERENT (E) impulses originate in the CNS and flow outward away from the brain or spinal cord. The term "motor" can be used since its conduction currents result in activity
  1. Somatic (S) tissue activated is skeletal muscle which traces its origin to embryonic somites
     a. General (G): there is no "special" category
        i. The terminations lie in motor end-plates of skeletal muscle
        ii. Most of the efferent fibers (GSE) originate in the ventral gray columns of the cord, pass via ventral roots, join spinal nerve trunks and run to the myoneural junction (motor end plate) of skeletal muscle
        iii. Fibers also originate in cranial nuclei of nerves III, IV, VI, and XII
  2. Visceral (V): system of efferent fibers that activate organs
     a. General (G) pass to all smooth muscle, to cardiac muscle, and to all glands
        i. The GVE system is concerned with body homeostasis: regulation of heart rate and blood pressure; temperature, glandular secretion, peristalsis, sphincter tension, and pupil size
     b. Special (S): relatively small distribution and tissue activated is typical striated muscle but not from somites but from the embryonic branchial arches
        i. Innervate muscles of facial expression, of mastication, of pharynx and larynx
        ii. Cells of origin of the SVE fibers are in groups or nuclei in the SVE column of the brainstem and use nerves V, VII, IX, X, and XI.
     c. Special (S): the receptors do answer, in part, the structural and functional requirements of organs of special sense. The stimuli are chemical, and the impulses have profound effects on the activity of certain organs
        i. The SVA category also includes
           (a) Taste: receptors are taste buds on the tongue, inside cheek and lips, palate, and pharynx
           (b) Smell: Olfactory receptors in the roof of the nasal cavity

C. Proprioception (P): although its receptors lie deep in the body tissue, this system is concerned with the position and movement of the body
  1. General (G)
      a. Receptors are widely dispersed throughout body: in muscles (together with their fasciae), in tendons, and in and around articular capsules
      b. The GP system processes receptors of the pressure or tension type activated by changes in tension of muscles or tendons and by movements at joints
      c. Activation of system is outside consciousness: impulses making possible normal walking, in part, balancing, regulation of head position, muscle contraction adjustment
  2. Special (S)
      a. Receptors are in the inner ear (ampullae of the semicircular canals, utricle, and saccule)
      b. SP system consists of a group of cells related to a fluid medium so that head movement causes current flow, stimulating the receptors

II. Reflexes
A. SOMATOSOMATIC: afferent, GSA or SSA; efferent, GSE
B. SOMATOVISCERAL: afferent, GSA or SSA; efferent, GVE or SVE
  1. General somatovisceral: afferent GSA or SSA; efferent, GVE
  2. Special somatovisceral: afferent GSA or SSA; efferent, SVE
C. VISCEROSOMATIC: afferent, GVA or SVA; efferent, GSE
D. VISCEROVISCERAL: afferent, GVA or SVA; efferent, GVE or SVE

APPENDIX VI. LANGUAGE

I. Introduction
A. Language is the only fundamental process in which man differs from the animals
B. Since no experimental animal, as yet known, has highly developed language skills, the study of language is difficult
C. There are no simple anatomic differences between the brains of man and other animals to account for language, yet subtle differences between the two hemispheres of man's brain do exist and are related to the fact that in adults, language functions occur predominantly in the left hemisphere

II. General
A. LANGUAGE IS SEPARABLE INTO TWO COMPONENTS: conceptualization and expression
  1. In forms of aphasia (language deficits not due to mental defects) related to conceptualization, the patient cannot understand spoken or written language, even though hearing and vision are not impaired
  2. In expressive aphasia: the patient is unable to carry out the coordinated oral and respiratory movements needed for language, even though he can move his tongue and lips, understand the spoken language, and knows what he wants to say. This is often associated with inability to write
B. Different cortical areas have been related to specific aspects of language
  1. Areas in the frontal lobe near the motor cortex are involved in the articulation of speech
  2. Areas in the parietal and temporal lobes are involved in sensory functions and language interpretations
  3. The cortical specializations are not present at birth but develop gradually in childhood during language acquisition
  4. Localization of language functions in the left hemisphere (in 96% of population) is not true during early childhood where both hemispheres function
    a. Damage to left hemisphere of a child results in no impediment of future language development—develops in intact right hemisphere
    b. Even if the left hemisphere is traumatized after language onset, it develops in the right hemisphere after transient periods of loss
    c. Prognosis becomes worse as the age at which the damage occurs increases so that after early teens, language is interfered with permanently
    d. The change in the possibility of establishing language in the teens is probably related to the fact that the brain reaches its final structural, biochemical, and functional maturity in that time
      i. With maturation of the brain, language functions are apparently irrevocably assigned and language utilization by the right hemisphere is no longer possible
C. Because language resides in the left hemisphere, only that hemisphere can communicate orally or in writing about its conscious experience
  1. If vision is limited so that only the part of the retina whose fibers pass to the right side is excited, the left hemisphere, which controls speech, is unaware of the visual experience and the patient cannot describe it orally or in writing
  2. If the portion of return projecting to the left hemisphere is activated, the patient can easily describe the experience
  3. The right silent hemisphere does understand language and has linguistic functions, but not those involved in graphic or vocal expression
D. The capacity for language is characteristic of man, and language develops provided there is some stimulation or opportunity
  1. The capacity is partially inherited although environmental circumstances are important and can limit language activity
  2. Behavioral characteristics and anatomic correlates of language can be identified, but mechanisms of neuronal operations resulting in language are unknown

APPENDIX VII. CONSCIOUSNESS

I. **Consciousness includes two distinct concepts:** states of consciousness and conscious experience (of which one is aware, such as thoughts, feelings, ideas, dreams, etc.)
A. A PERSON'S STATE OF CONSCIOUSNESS (awake, asleep, drowsy) is defined by his behavior, from coma to maximum attentiveness and by the pattern of brain activity that can be electrically recorded as the electroencephalogram or EEG (electric potential between 2 points on the scalp)
B. Wavelike patterns of the EEG change in frequency and amplitude as behavior changes
   1. The EEG is produced by the intermittent synchronization of the electric activity of small groups of neurons of the cerebral cortex
   2. The basic units of the electric activity are considered to be individual synaptic potentials (or groups of them) rather than action potentials
   3. The active synapses are driven by some subcortical center (possibly the thalamus)
C. The EEG is a useful tool clinically because the normal patterns are altered over the brain areas that are diseased or damaged or injured
D. The EEG is also useful in defining states of consciousness, but one does not know what function, if any, this electric activity serves in the processing of information

II. **States of consciousness**
A. THE WAKING STATE AND AROUSAL are far from homogeneous and comprise an infinite variety of things one can be doing
   1. The prominent EEG wave pattern of an awake relaxed adult with eyes closed is a slow oscillation of 8–13 Hz, known as the alpha rhythm
      a. Each person has a characteristic pattern of alpha rhythm and so does each region of the brain
      b. The alpha rhythms are always larger at the back of the head over the visual cortex and are also larger when a person has his eyes closed and is not thinking—associated with decreased levels of attention
   2. When a person is attentive to external stimuli, the alpha rhythm is replaced by lower, faster oscillations. The change is called arousal and is associated with the act of attending to stimuli rather than perception itself
B. SLEEP
   1. Sleep is an active process and not a mere absence of wakefulness
   2. Sleep is not a simple phenomenon. There are 2 states of sleep that are characterized by different EEG patterns and behavior patterns
   3. As a person becomes drowsy, the alpha rhythm is gradually replaced by irregular, low-voltage potential differences, and as sleep deepens, the EEG waves become slower, larger, and more irregular. This slow wave is periodically interrupted by episodes of paradoxical sleep during which the subject still seems asleep, but has an EEG pattern similar to that of EEG arousal or like an awake alert person
      a. In slow-wave sleep there is considerable tonus in postural muscles and only a small change in cardiovascular or respiratory activity. The person can be awakened, and rarely reports dreaming (thoughts rather than dreams)
      b. During paradoxical sleep the behavioral criteria are precise. At the onset, there is an abrupt and complete inhibition of tone in postural muscles, although there are periodic episodes of twitching of facial muscles and limbs, and rapid eye movements behind closed lids occur. This sleep is thus called rapid-eye movement or REM sleep
         i. Respiration and heart rate are irregular, and blood pressure may go up or down
         ii. When awakened from REM sleep, 80–90% of people report having had dreams

4. The two states of sleep follow a regular 30- to 90-minute cycle, each episode of paradoxical sleep lasting 10 to 15 minutes. Thus, slow-sleep takes up about 80% of total sleeping time and paradoxical about 20%
   a. Normally it is not possible to pass directly from the waking state to REM sleep, but it takes at least 30 minutes of slow-sleep
   b. If deprived of REM sleep for several nights, all subjects spend a greater portion of time in REM sleep the next time they sleep, thus, the total number of hours spent in REM sleep tends to remain constant
5. The brain, as a whole, does not rest during sleep, and there is no general inhibition of activity of cerebral neurons. In actuality, there is a considerable amount of neuronal activity during slow-wave sleep, and many areas of the brain are more active during REM sleep than during wakefulness. In addition, blood flow and oxygen consumption by the brain do not decrease in sleep
6. During sleep there is a change in distribution or reorganization of neuronal activity, some individual neurons being less active than during waking, although the brain as a whole remains relatively active
7. Sleep may not be a period of generalized rest but may represent a period of rest for certain specific elements during which time they can replenish substrates important for generation of potentials
   a. It has been suggested that the importance of sleep lies not in short-term recovery but in the relatively long-term chemical and structural changes that the brain must undergo to make learning and memory possible
   b. We really do not know for certain what the functional significance of sleep is
C. RETICULAR FORMATION AND CONSCIOUSNESS
   1. Consciousness is the result of interplay between three neuronal systems, one causing arousal and the other two sleep, and all three are parts of the reticular formation which lies in the central core of the brainstem in the center of the neural pathways ascending and descending between the brain and spinal cord
   2. Neural structures in the reticular formation have been found to be essential for the maintenance of a waking EEG since stimulation causes EEG arousal, and destruction results in coma and EEG characteristic of the sleeping state
D. Other areas of the brain, the cortex and certain hypothalamic areas, also play a role in wakefulness and alert state
E. Two neuronal systems control sleep by opposing the tonic activity of the reticular activating system
   1. One is in the center core of the brainstem
   2. The other is the sleep center in the pons whose activity induces paradoxic sleep
   3. The two centers interact with each other in a true feedback phenomenon, so that the cycling of sleep and wakefulness is due, in part, to the slow accumulation and dissipation of chemical transmitters
   4. Of the two basic components of the sleep-wake cycle, the waking mechanisms seem to be more easily activated than those causing sleep

APPENDIX VIII.   REFERENCES

Allen, D. J.: Scanning Electron Microscopy of Epiplexus Macrophages (Kolmer Cells) in the Dog. *J. Comp. Neurol.,* **161:**197–214, 1975.

Allen, D. J., and Low, F. N.: The Ependymal Surface of the Lateral Ventricle of the Dog as Revealed by Scanning Electron Microscopy. *Am. J. Anat.,* **137:**483–89, 1973.

Allen, D. J., and Low, F. N.: Scanning Electron Microscopy of the Subarachnoid Space in the Dog. III. Cranial Levels. *J. Comp. Neurol.,* **161:**515–40, 1975.

Allen, D. J.; Persky, B.; and Low, F. N.: Some Regional Variations in Ventricular Lining Material in Laboratory Mammals and Man. *Scanning Electron Microscopy,* **2:**45–253, 1978.

Barr, M. L.: *The Human Nervous System: An Anatomical Viewpoint,* 3rd ed. Harper and Row, Publishers, Inc., New York, 1979.

Bullock, T. H.; Orkand, R.; and Grinnell, A.: *Introduction to the Nervous System.* W. H. Freeman and Company, San Francisco, 1977.

Carpenter, M. B.: *Human Neuroanatomy,* 7th ed. The Williams & Wilkins Co., Baltimore, 1976.

Carpenter, M. B.: *Core Text of Neuroanatomy,* 2nd ed. The Williams & Wilkins Co., Baltimore, 1978.

Chusid, J. G.: *Correlative Neuroanatomy and Functional Neurology,* 17th ed. Lange Medical Publications, Los Altos, Calif., 1979.

Cooper, J. R.; Bloom, F. E.; and Roth, R. T.: *The Biochemical Basis of Neuropharmacology,* 3rd ed. Oxford University Press, New York, 1978.

Daube, J. R.; Sandok, B. A.; Reagan, T. J.; and Westmoreland, F. F.: *Medical Neurosciences.* Little, Brown & Company, Boston, 1978.

Dunkerley, G. B.: *A Basic Atlas of the Human Nervous System.* F. A. Davis Co., Philadelphia, 1975.

Escourocle, R., and Poirier, J.: *Manual of Basic Neuropathology,* 2nd ed. W. B. Saunders and Co., Philadelphia, 1978.

Globus, J. H.: *Practical Neuroanatomy.* William Wood and Company, 1937.

Guyton, A. C.: *Basic Human Physiology: Normal Function and Mechanisms of Disease,* 2nd ed. W. B. Saunders Company, Philadelphia, 1977.

Haymaker, W.: *Bing's Local Diagnoses in Neurological Diseases.* The C. V. Mosby Company, St. Louis, 1956.

House, E. L.; Pansky, B.; and Siegel, A.: *A Systematic Approach to Neuroscience,* 3rd ed. McGraw-Hill Book Co., New York, 1979.

Jensen, D.: *The Principles of Physiology.* Appleton-Century-Crofts, New York, 1976.

Kuffler, S. W., and Nicholls, J. G.: *A Cellular Approach to the Function of the Nervous System.* Sinauer Associates, Inc., Sunderland, Mass., 1976.

Lewis, A. J.: *Mechanisms of Neurological Disease.* Little, Brown & Company, Boston, 1976.

Noback, C. R., and Demarest, R. J.: *The Human Nervous System: Basic Principles of Neurobiology,* 2nd ed. McGraw-Hill Book Company, New York, 1977.

Otsuka, M., and Hall, Z. W. (eds.): *Neurobiology of Chemical Transmission.* John Wiley & Sons, Inc., New York, 1979.

Peele, T. L.: *The Neuroanatomic Basis for Clinical Neurology,* 3rd ed. McGraw-Hill Book Company, New York, 1977.

Robbins, S. L., and Cotran, R. S.: *Pathologic Basis of Disease,* 2nd ed. W. B. Saunders Co., Philadelphia, 1979.

Schade, J. P., and Ford, D. H.: *Basic Neurology,* 2nd rev. ed. Elsevier Scientific Publishing Company, New York, 1973.

Schmidt, R. F. (ed.): *Fundamentals of Neurophysiology,* 2nd ed. Springer-Verlag, New York, Heidelberg, Berlin, 1978.

Scientific American: The Brain, Vol. 241, No. 3, September, 1979.

Tenth International Anatomical Nomenclature Committee in Tokyo: *Nomina Anatomica,* 4th ed. Excerpta Medica Foundation, Amsterdam, 1977.

Thorek, P.: *Anatomy in Surgery,* J. B. Lippincott Co., Philadelphia, 1951.

Truex, R. C., and Carpenter, M. B.: *Strong and Elwyn's Human Neuroanatomy.* The Williams & Wilkins Co., Baltimore, 1964.

Vander, A. J.; Sherman, J. H.; and Luciano, D. S.: *Human Physiology: Mechanisms of Body Function,* 2nd ed. McGraw-Hill Book Co., New York, 1975.

Walton, J. N.: *Essentials of Neurology,* 4th ed. J. B. Lippincott Co., Philadelphia, 1975.

Warwick, R., and Williams, P. L.: *Gray's Anatomy,* 35th Brit. ed. W. B. Saunders Co., Philadelphia, 1973.

Watson, C.: *Basic Human Neuroanatomy: An Introductory Atlas,* 2nd ed. Little, Brown & Company, Boston, 1977.

Williams, P. L., and Warwick, R.: *Functional Neuroanatomy of Man.* W. B. Saunders Co., Philadelphia, 1975.

Willis, W. D., Jr., and Grossman, R. G.: *Medical Neurobiology,* 2nd ed. The C. V. Mosby Company, St. Louis, 1977.

INDEX

Figures in **boldface** type refer to pages on which illustrations appear.

genu, 198, **199**, 274, **275**
  posterior limb, 194, **195**, 196, **197**, 274, **275**
Caravan's disease, 466
Carbohydrate metabolism, disorders, 508
Carcinoma, metastatic, **537**
Carcinomatous neuropathy, 504
Carotid artery, internal, thrombosis of, 416, **417**
Carotid body, blood pressure regulation, 36
  development, 36, **37**
Carpal tunnel syndrome, 472, **473**, 566
Cataplexy, 581
Cataract, 386
  congenital, 42
Catechol-O-methyltransferase, **163**
Cations, 130
Cauda equina (tail of a horse), **61**, 566
Caudate nucleus, **97**, 272, **273**, 274, **275**, 566
Cavernous sinus, **227, 229**
Cell(s), acidophilic, 532
  amacrine, of retina, **391**, 392, 394
  anterior horn, 66, **67**
  ascending axon cells, 286, **287**
  astrocytes, 56, **59**, 565
  basket, **287, 319**, 320, **321**
  Betz, 286, **287**
  bipolar, of retina, **391**, 392, 394
  cerebellar, 320, **321**
  chromophobe, 532
  climbing, 320, **321**
  cortical, 284, **285**, 286, **287**, 288, **289**
  fusiform, 286, **287**
  ganglion, of retina, **391**, 394
  giant pyramidal, Betz, 286, **287**
  glia, 54, 56, 58
  Golgi, 54, **319**, 320, **321**
  granule, 286, **287**, 320, **321**
  horizontal cells, of Cajal, 286, **287**
    of retina, **391**, 392, 394
  internal granule, **355**
  Kolmer, **110, 111**
  Martinotti, 286, **287**
  microglia, **59**, 528, **571**
  mitral, **355**
  mossy, **319**, 320, **321**
  nerve (neuron), 8, **9**, 54, **55, 59**
  neuroepithelial, 6, **7**, 8, **9**
  neuroglia, 56–**59**
  neurogliaform, **287**
  olfactory, 354, **355**
  oligodendrocytes, 58, **59**, 572
  optic, 390, **391**
  polymorphic multiform, 286, **287**
  Purkinje, **319**, 320, **321**
  pyramidal, 194, **195**, 196, **197**, 286, **287, 319**
    of olfactory tract, **355**
  Renshaw, 204, **205, 209**
  Schwann, **55**, 146, **147**
  sensory, 54, **55**, 148, **149**
  stellate, 286, **287, 319**, 320, **321**

supraependymal, **110, 111**
sustentacular (supporting), **359, 361**
tufted, olfactory, 352, **353, 355**
Cell bodies, 54, 146, **147**
  somatic afferent pathways, 170–**85**
Cell permeability, 132, **133**
Central artery, of retina, 40, **41**
Central canal, **107, 109**
Central gyrus, **91**
Central lobule, **315, 316**
Central nervous system compartments, 114, **115**
Central scotoma, 402
Central sulcus, **89, 91**
Cephalgia. *See* Headaches
Cephalohematoma, parietal, **449**
Cerebral asthenia, **329**
  ataxia, 328, **329**
  corticonuclear projections, **323**
  degenerative disease, 484
  hypotonia, **329**
  nystagmus, **329**
  tremor, **329**
  wiring diagram, **321**
Cerebellar ataxia, 328, **329**, 462
Cerebellar connections, **329**
  afferent, 186, **187**, 188, **189**, 190, **191**, 324, **325, 326**
  efferent, 192, **193**, 324, **325, 327**
  general proprioception, 186, **187**
Cerebellar cortex, 318, **319, 325**
  cells, 318, **319**, 320, **321**
  layers, 318, **319**, 320, **321**
Cerebellar nuclei, **317**, 322, **323**
Cerebellar peduncles, 84, **85, 87**, 188, **189**, 190, **191**, 314, **315**, 324, **325, 326, 327**
Cerebellar plate, 20, **21**
  mantle layer, 20, **21**
  marginal layer, 20, **21**
  neuroepithelial layer, 20, **21**
Cerebellar vermis, **316, 317**
Cerebellar wiring diagram, **321**
Cerebellomedullary cistern, **109**, 112, **113**
Cerebellopontine tumor, 530
Cerebelloreticular fibers, **327**
Cerebellorubral tract, 192, **193**, 324, **327**
Cerebellospinal tract (uncinate bundle), 192, **193**
Cerebellovestibular control, 242, **243**
Cerebellovestibular tract, 192, **193, 243, 327, 383**
Cerebellum, 314–**29**
  absence of, 440
  afferent pathways, 186–**91**, 324, **325**
  anatomy, 314, **315**
  cortex, 314, 318, **319**, 320, **321, 325**
    cell and nerve types, 320, **321**
  corticonuclear projections, **323**
  development, 20, **21**
  dysfunction, 328, **329**
  efferent pathways, 192, **193**, 324, **325**
  flocculus, **95**
  function, **317**, 328, **329**

Fiber(s), adrenergic, 162, **163**
  arcuate, internal, 176, **177**
  association, cerebral cortex, 96, **97**
  auditory, 372, **373**
  bag, 206, **207, 209**
  beta, **207**
  cerebelloreticular, **327**
  chain, **209**
  cholinergic, 162, **163**
  climbing, cerebellar cortex, **319**, 320, **321**
  commissural, 64, **65**, 92, **93**, 96, **97, 257**
  corticobulbar, 194, **195**, 198, **199**, 274, **275**
  corticocollicular, 200, 372
  corticofugal, 192, **193**, 194, 372
  corticonigral, **201**
  corticonuclear, 192, **193**
  corticospinal, 68, **69**, 84, **85**, 274, **275**
  corticothalamic, 200, **201**
  cranial nerve, 204, **205**, 224-54
  dentatothalamic, **263**, 276, **277, 327**
  dorsal arcuate, 188, **189**
  extrafusal, 206, **207**
  flocculovestibular, **327**
  frontopontine, 190, **191, 201**
  gamma efferent, 206, **207**, 212
    dynamic fusimotor, 212, **213**
    static fusimotor, 212, **213**
  geniculocollicular, 392, **393**, 396, **399**
  hypothalamic, 266, **267**, 268, **269**
  intrafusal, 206, **207**
  mossy, **319**, 320, **321**
  motor, 194-**201**
  nucleofugal, 192
  olfactohypothalamic, 354, **355**
  olivocerebellar, **327**
  pallidohypothalamic, **263**, 276, **277**
  parallel, of cerebellum, 320, **321**
  parietotemporopontine, 274
  postcommissural, 344, **345**
  postganglionic, **303, 309**, 320
  potentials, 136-45
  precommissural, 344, **345**
  preganglionic, 302, **303**
  projection, 96, **97**
  pyramidal, **65**, 68, **69, 85, 87**, 194, **195**, 196, **197, 275**
  reticulocerebellar, 326
  reticulolenticular, 274, **275**
  reticulospinal, 334, **335**, 336, **337**
  rubrocerebellar, 326
  rubrothalamic, 276, **277, 337**
  spinocerebellar, **325** 326
  spiral sensory, 206, **207**
  sublenticular, 274, **275**
  subthalamic, 262, **263**
  sympathetic, 300, **301**, 302, **303**, 304, **305**
    postganglionic, 300, **301**, 302, **303**, 304, **305**
    preganglionic, 300, **301**, 302, **303**, 304, **305**
  thalamocortical, 260, **261**, 262, **263**
  thalamohypothalamic, 266, **267**

  ventral external arcuate, 190, **191**
  vestibular (archicerebellar), 188, **189**
  vestibulocerebellar, **327**
Fields of Forel, 262, **263**, 276, **277**
Filum terminale, 12, **13**, 60, **61**, 62, **63**
Fimbria, of fornix, 340, **341**, 568
  of hippocampus, **341**
Fissure(s), anterior median, of cord, 64, **65**
  calcarine, **89**, 92, **93, 95**
  cerebellar, 316
  collateral, **93, 95, 341**
  hippocampal, **95**
  lateral (sylvian), **89, 91, 95**
  longitudinal, **95**
  parieto-occipital (sulcus), **89**, 92, **93, 95**
  transverse cerebral, 92
Fits. *See* Epileptic seizures (fits)
Flaccid paralysis, 432, 568
Flexion reflex, 164, **165**, 212, **213**
Flexor muscles (innervation), 210, **211**, 212, **213**
  arm, 72, **73**
  forearm, 72, **73**
  hand, 72, **73**
Flexures. *See* Cervical flexure; Pontine flexure
Flocculonodular lobe, 316
Flocculovestibular tract, 192, **193**
Flocculus, of cerebellum, **95**, 316
Floor plates. *See* Plates
Flower-spray endings, **185**, 206, **207, 209**
Fluid, cerebrospinal. *See* Cerebrospinal fluid
Focal epilepsy, 488, 490, 569
Folia, 316
Folic acid deficiency, 500, 502
Folic acid disturbances, 494
Folium, **315**, 316
Footdrop, 472
Foramen(mina), 569
  cecum, **95**
  interventricular (of Monro), 92, **93**
  intervertebral, 60, **61**
  Luschka, 16, **17**, 106, **107, 315**, 569
  Magendie, 16, **17**, 106, **107, 315**, 569
  magnum, 569
  of Monro, 92, **93**, 106, **107**
  ovale, **47, 49**, 364, **365**
Forbes' disease, 508
Forceps major, 92, **93**, 569
Forceps minor, 92, **93**, 569
Forebrain, 4, **5, 15, 19**, 28, **29**
Forel's field, 262, **263**, 276, **277**
"Formes frustes," 466
Fornix, **79**, 92, **93**, 266, **267**, 338, **339**, 569
  body, 344, **345**
  columns, **79**, 92, **93**
  crura, **93**, 344, **345**
  fimbria, 344, **345**
  postcommissural, 344, **345**
  precommissural, 344, **345**
Fossa, interpeduncular 80, **81**, 569
  rhomboid, **315**

Fovea centralis, 384, **385, 389,** 569
Fractures, of skull, 426, 428
Free nerve endings, **149,** 170, **171,** 176, **177,** 180,
    181, 184, **185**
Frequency (Hz), 384
Friedreich's ataxia, 462, **463**
Frontal gyri, 90, **91,** 94, **95**
Frontopontine tract, 190, **275**
Fungal infections, 550
Fungiform papillae, 358, **359**
Funiculi, spinal cord, 64, **65**
Funiculus, 569
    anterior (ventral), 64, **65**
    dorsal, 64, **65**
    lateral, 64, **65**
    posterior, 176, **177**
Fusiform cells, 286, **287**

Galactosemia, 508
Galea aponeurotica, **101**
Gamma motor axons, 206, **207,** 212, **213**
Gamma reflex loop, 206, **207,** 212, **213**
Ganglia(on), 54, 569
    Auerbach, 306, **307**
    autonomic, 300, **301, 303, 305**
    basal, 272–**81,** 565
    celiac, **301,** 302, **303, 305**
    cervical, **301,** 302, **303, 309**
    ciliary, 254, **301, 307,** 308, **309**
    collateral, **301,** 302
    cranial, 300, **301**
    dorsal root, **57,** 70, **71, 301**
    gasserian, 230, **231, 233**
    geniculate, **237, 363**
    inferior, of glossopharyngeal, **245**
    inferior mesenteric, **301,** 302, **303**
    interpeduncular, **345, 349**
    jugular, **247**
    Meissner, 306, **307**
    mesenteric, 302, **303, 305**
    myenteric, 306, **307**
    nodose, **247, 363**
    optic, 390, **391**
    otic, **245,** 254, **301, 307,** 308, **309,** 363
    parasympathetic, 254, 300, **301,** 306, **307**
    paravertebral, 300, **301**
    petrosal, **363**
    prevertebral, 300, **301,** 302
    pterygopalatine, **237,** 254, **301, 307,** 308, **309**
    Scarpa's, 378, **379**
    semilunar (gasserian), 230, **231, 233**
    spiral, **239**
    statoacoustic, 238, **239,** 240, **241, 373, 375, 379,**
        383
    submandibular, **237,** 254, **301, 307,** 308, **309**
    submucosal, 306, **307**
    superior, of glossopharyngeal, **245**
    sympathetic, **57, 71,** 300, **301, 303, 305**
    terminal, 300, **301**

vestibular, 378, **379**
Ganglioneuromas, 530
Gangliosidoses (amaurotic idiocies), 508, **509**
Gate control (pain modulating), **175**
Gaucher's disease, 508
General somatic afferent (GSA) system, 170–**85,**
    230, **231**
    discriminative (fine) touch, from body, 176,
        177
        from face, 178, **179**
    light (crude) touch, from body, 180, **181**
        from face, 182, **183**
    pain, 170, **171,** 172, **173**
    proprioception, face and head, 184, **185**
    temperature, 170, **171,** 172, **173**
General somatic efferent fibers, 228
General visceral afferent group, 14, **15,** 16, **17,**
    18, **19,** 585, 586
General visceral efferent group, 14, **15,** 16, **17,**
    18, **19,** 585, 586
Generator potential (receptor activation), 150,
    151, 152, **153**
    adaptation, 152, **153**
    amplitude, 152, **153**
    intensity coding, 152, **153**
    summary, 152
Geniculate bodies, 81, **83, 85, 87,** 224, **225,** 258,
    259, 260, **261,** 392, **393,** 394, **395**
Geniculate herpes, 558
Geniculocalcarine tract, 392, **393,** 394, **395**
Geniculocortical radiations. *See* Auditory
    radiations
Genital corpuscles, 180, **181**
Genu, of corpus callosum, 92, **93, 97,** 569
Genu recurvatum, **549**
Gland(s), Bowman's, 352, **353**
    lacrimal, **233,** 234, 236, **237,** 306, **307,** 308, **309**
    nasal, 236, **237,** 308, **309**
    parotid, **233,** 244, **245,** 308, **309**
    pineal, 24, **25, 85,** 256, **257,** 573
    sublingual, 236, **237,** 300, **301, 305,** 306, **307,**
        308, **309**
    submandibular, 236, **237,** 300, **301, 305,** 306,
        307, 308, **309**
    suprarenal, 36, **37,** 300, **301, 305**
Glaucoma, 384, 386
    congenital, 42
Glia, 54, 56, **57,** 58, **59,** 569. *See also* Neuroglia
Glia limitans, **105**
Gliablasts, 8, **9**
Glioblastomas, **525,** 528, **529,** 569
Gliomas, 528, 569
Gliosis, 569
Globose nucleus, 314, **317,** 322, **323**
Globus pallidus, **79,** 272, **273,** 569
    connecting fibers, 272, 276, **277**
    efferent pathways, 276, **277**
Glomerulus, **321**
Glossopharyngeal nerve, 86, **87,** 94, **95,** 244, **245,**
    246, **247, 253.** *See also* Nerve(s)

Jacksonian seizures, 490, **495,** 570
Joint capsules, 206, **207**
Joint of Charcot, 548, 567
Joint organ, 206, **207**
Juxtarestiform body, 324, **325, 383**

Kaes-Bechteren, band of, **287**
Kayser-Fleischer ring, 506, **507**
Kemadrin, 482
Kernig's sign, 542, 570, 582
Kinky hair disease (Menkes' disease), 506
Kinocilia, 374, **375**
Klinefelter's syndrome, 32
Kolmer cells (supraependymal), **110, 111**
Konicortex, 570
Korsakoff's syndrome, 510
Krabbe's disease, 508
Krause's corpuscles, **149,** 170, **171**
Kuru, 560

Labyrinth, 364, **365,** 570
  bony, 364, **365**
  development, 44, **45**
  membranous, 364, **365**
  osseous, 364, **365**
Labyrinthine pathways, 372, **373**
Labyrinthitis, 580
Laceration, 428
Lacrimal apparatus, 384, **385**
Lacrimal gland, innervation, **233,** 234, 236, **237,**
  306–**309**
Lacrimal nucleus, **205**
Lafora's disease, 508
Lamina affixe, **259**
Lamina terminalis, **79,** 92, **93,** 257
Laminae, 68, **69**
  external medullary, 258, **259**
  internal medullary, 258, **259**
  pyramidal tract, 68, **69**
  Rexed's, 68, **69**
Landouzy-Déjerine dystrophy, 516
Lange curve, 458, 548, 570
Language, 587
Lasègue's sign, 570
Lateral popliteal nerve injury, 432
Layers (cerebral cortex), 286, **287,** 288, **289**
  external granular layer II, **287,** 288, **289**
  external pyramidal layer III, **287,** 288, **289**
  fusiform (multiform) layer VI, **287,** 288, **289**
  internal granular layer IV, **287,** 288, **289**
  internal pyramidal (ganglionic) layer V, **287,**
    288, **289**
  molecular layer I, **287,** 288, **289**
  multiform (polymorphous), **287,** 288, **289**
L-dopa, 482
Lead poisoning, 512, **513**
Leigh-Feigin disease, 508
Lemniscal decussation, 176, **177**

Lemniscus, 570
  lateral, **83, 85**
  medial, 82, **83, 87,** 176, **177,** 182, **183**
  nucleus, **85**
  trigeminal, 172, **173**
Lens, 384, **385,** 386, **387**
  development, 40, **41**
Lens placode, 38, **39, 41**
Lens refractive power, **387**
Lens vesicle, 38, **39, 41**
Lenticular nucleus, 272, **273,** 570
Lentiform nucleus, 272, **273**
Lentil-like neoplasia, 446
Leprosy, 548
Leptomeninges, 100, 570
Leptomeningitis, **545**
Leptospirosis, 550
Leukinosis, 506
Leukodystrophies, 452, **467,** 566, 570
Leukoencephalopathy, multifocal, 558
Light reflexes, 582
Light refraction, **387**
Lightning pains, 570
Limbic gyrus, 90, **91,** 338, **339**
Limbic lobe, 88, **89,** 338, **339**
Limbic system, 338, **339**
  primary roles, 338
  structures, 338, **339**
Limen insula, 90, **91,** 570
Lingual gyrus, 92, **93, 95, 339**
Lingula, **315**
Lipid metabolism, disorders, 508
Lipomas, 534
Lipoproteinoses, 508
Lissauer's tract, 68, **69,** 170
Little's disease, 448
Lobe(s), anterior of cerebellum, 314, 316
  archicerebellum, 314, **317**
  central (insular), 88, **89,** 90, **91**
  flocculonodular, 314, 316
  frontal, 88, **89,** 90, **91,** 94, **95**
  insular, 88, **89,** 90, **91**
  limbic, 88, **89,** 90, **91**
  neocerebellum, 314, **317**
  occipital, 88, **89,** 90, **91**
  paleocerebellum, 314, **317**
  parietal, 88, **89,** 90, **91**
  posterior of cerebellum, 314, 316
  temporal, 88, **89,** 90, **91**
Lobule, paracentral, **93**
Locus ceruleus, 84, **85, 87,** 571
Longitudinal fissure, 571
Longitudinal stria, **79, 97**
Loop, flexor reflex, **213**
  gamma reflex, **213**
  of Meyer, 402
Loudness (intensity), 366, **367**
Lower motor neuron syndrome, 582
Lumbar disk prolapse, 470
Lumbar punctures, 62

Lumbosacral plexus, 74, **75**
  lumbar plexus, 74, **75**
  sacral plexus, 74, **75**
Luminosity signals (intensity), visual, 394
Luschka's foramen, 16, **17**
Luys, subthalamic nucleus of, 262, **263**
Lysivane, 482

MacArdle's disease, 508
Macrocytic anemia, 494, 500
Macroglia, 571
Macrophages, suprapial, **105**
Macula, 571
Macula lutea, 384, **385, 389**
Macula sacculi, **241**, 374, **375, 377**
Macula utriculi, **241**, 374, **375, 377**
Magendie, foramen of, 16, **17**
Malleus, anatomy, 364, **365, 369**
  development, 48, **49**
Malta fever, 548
Mamillary bodies, 78, **79**, 94, **95, 264, 265, 339**
Mamillotegmental fasciculus, 266, **267**
Mamillothalamic tract, 266, **267**, 344, **345**
Mantle, layer, 6, **7**, 8, **9**
Maple sugar urine disease, 506
Marginal layer, 6, **7**, 8, **9**
Marie's ataxia, 462
Martinotti, cells of, 286, **287**
Masklike appearance, **279**, 482
Massa intermedia, 24, 78, **79**, 571
Massive congenital leukodystrophy, 466
Mastication, 230, **231**
Mastoid process, 48
Matrix layer, 8, **9**
Measles encephalitis, 552, 556
Meatal plug, 48, **49**
Meckel's cartilage, **49**
Medial forebrain bundle, 266, **267, 349**
Medial lemniscus, 82, **83**
Medial longitudinal fasciculus, 82, **83, 85**, 86, **87, 241, 379, 381**
Median (medial) eminence, 78, **79**, 84, **85, 87**, 264
Median nerve, 72, **73**
  injury, 432, 472, **473**
Medulla oblongata, 86, **87**, 571
  afferent pathways, 86, **87**
  development, 14, **15**, 16, **17**
Medullary lamina, **259**
Medullary plate, 4, **5**
Medullary velum, superior (anterior), **79**, 84, **85**, 186, **187, 315**, 316
Medulloblastoma, 530
Megalencephaly, 440
Meissner's corpuscle, **149**, 176, **177**, 178, 180, **181**
Melanomas, 530
Membrane permeability, 134, 136, **137**, 138, **139**
Membrane potentials, 126, **127**, 128, **129**, 130, **131**

chemical reactions, 126, **127**
electric charges, 126, **127**, 128, **129**, 130, **131**
Membrane pump, 134, **135**
Memory activities, disturbances, 478
Ménière's disease, 382, 580
Meninges, 98–**105**, 571
  arachnoid mater, 100, **101, 104**
  dura mater, 98, **99, 104**
  pia mater, 100, **101, 104**
  relation to vessels and nerves, **101**
  relation to spinal cord, **109**
  scanning electron microscopy, **104**
  venous dural sinuses, 98, **99**
Meningiomas, **527, 528, 529**, 536, 571
Meningism (meningismus), 542, 571
Meningitis, 100, 542, 544, **545**, 571
  bacterial, 546, **547**
  diagnosis, 544
  hemophilus, **543**
  influenzal, 542
  lymphocytic, 542
  meningococcal, 542
  pneumococcal, 542
  purulent, 542, **543**
Meningocele, 12, **13, 33**, 438, 571
Meningoencephalitis, 544, 552, 571
Meningoencephalocele, **33**
Meningohydroencephalocele, **33**
Meningomyelocele, 12, **13**, 571
Menkes' disease, 506
Mental retardation, 442
Mercury, 512
Merkel, tactile disks, **149**, 176, **177**, 180, **181**
Mesencephalic structures, 94, **95**
Mesencephalon, 4, **5**, 80–**83**, 88, 571
  alar plate, 22, **23**
  basal plate, 22, **23**
  colliculi, 80, **81**, 82, **83**
  crus cerebri (basis pedunculi), 80, **81**, 82, **83**
Mesocortex, 571
Mesoglia, 571
Metabolic diseases, neurologic complications, 506, **507**
Metabolism, inborn errors, 506
Metachromatic leukodystrophy, 466, 508
Metal metabolism, disorders, 506
Metastases, **411, 413**, 524, 536
Metathalamus, 571
Metencephalic structures, 84, **85**, 94, **95**
Metencephalon, 84, **85**, 88, **89**, 571
  alar plate, 18, **19**
  basal plate, 18, **19**
  cerebellum, formation, 20, **21**
  development, 18, **19**, 20, **21**
  pons, formation, 18, **19**
  rhombic lip, 18, **19**
Methoin, 494
Meynert, retroflex tract, 256, **257**
Microcephalus, **33**, 440, **441**
Microglia, 528, 571

Nerve(s) [*Cont.*]
  axillary, 72, **73**
  brachial plexus, 72, **73**
    nerves of arm, forearm, hand, 72, **73**
    small nerves, 72, **73**
  buccal, **233**
  cell, ionic concentrations, 130, **131**
  cervical plexus, 72, **73**
  chorda tympani, 235, **237**, **363**, 364
  cochlear, 84, **85**, **87**, **205**, 238, **239**
    auditory system, 238, **239**
    symptoms of disease, 238
  cranial, **205**, 224–55
    function, 252, **253**
    ganglia, **205**, 252, **253**
    nuclei, **205**, 252, **253**
    origin, **205**, 252, **253**
  cutaneous, 72, **73**
  endings, free, **149**
  facial, 84, **85**, 94, **95**, **205**, 234, **235**, 236, **237**,
    252
    components, 234, 252
    course, 234, **235**, 236, **237**
    motor nuclei, **205**, 234, **235**, 252
    symptoms of disease, 236
  femoral, 74, **75**
  fiber, diameter, 144
    type, 144
  glossopharyngeal, 86, **87**, 94, **95**, **205**, 244, **245**,
    246, **247**, 253, **363**
    components, 244, **245**, 253
    symptoms of disease, 246
    taste, **363**
  hypoglossal, 86, **87**, 94, **95**, **205**, 250, **251**
    components, 250
    course, 250, **251**
    nucleus, 86, **87**, **205**, 250, **251**
    symptoms of disease, 250
  impulses. *See* Action potentials
  inferior alveolar, **233**
  intermediate, **95**, **205**
  internal acoustic, **235**
  lateral popliteal injury, 432
  level of cord exit, 60, **61**
  maxillary, **231**, **233**
  median, 72, **73**
    injury, 432
  motor, 54, **55**, 148, **149**, **155**
  musculocutaneous, 72, **73**
  myelinated, cross section, **57**
  obturator, 74, **75**
  oculomotor, 80, **81**, 94, **95**, **205**, 226, **227**, 252
    course, 226, **227**
    eye movements, 226, **227**
    general somatic efferent, 226, 252
    general visceral efferent, 226, 252
    nucleus, **205**, 226, **227**
    symptoms of disease, 226
  olfactory, 224, **225**, 252
    components, 224, 252

    symptoms of disease, 224
  ophthalmic, **231**, **233**
  optic, **95**, 224, **225**, 252
    components, 224
    development, 40, **41**
    symptoms of disease, 224
  pelvic, **305**, 306, **307**
  peripheral, 72, **73**
  petrosal, **235**, **237**
  phrenic, 72, **73**, **307**
  potentials, 136–**41**, 150–**53**
  pudendal, **303**
  radial, 72, **73**
  sciatic, 74, **75**
    injury, 432
  spinal, 70, **71**, 146, **147**
    accessory, 86, **87**, 94, **95**, **205**
    cross section, **57**
    longitudinal section, **57**
    rami, 70, **71**
    roots, 70, **71**
    sympathetic chain ganglia, **57**, 300, **301**, 302,
      303, 304, **305**
    tracts, **69**, 70, 170–**201**. *See also* Tract(s)
  splanchnic, 302, **303**
  trigeminal, 84, **85**, 94, **95**, **205**, 230, **231**, 232,
    233, 252
    components, 230, 252
    main sensory nucleus, 84, **85**, **205**, 230, **231**
    mesencephalic, nucleus, 84, **85**, **205**, 230, **231**
    motor nucleus, 84, **85**, **205**, 230, **231**
    nucleus, 84, **85**, **205**, 230, **231**
    portio major, 84, **85**, 230, **231**
    portio minor, 84, **85**, 230, **231**
    spinal tract, 230, **231**
    symptoms of disease, 232
  trochlear, 80, **81**, 84, **85**, 94, **95**, **205**, 228, **229**,
    252
    eye movements, control, 228, **229**, 252
    nucleus, **205**, 228, **229**
    symptoms of disease, 228
  ulnar, 72, **73**
    injury, 432
  vagus, 86, **87**, 94, **95**, **205**, 244, **245**, 246, **247**,
    253, **363**
    branches, **245**, **247**, **253**
    cardiovascular, **247**
    components, 244, **245**, **253**
    description with glossopharyngeal nerve,
      244, **245**, 246, **247**
    fibers, **205**, 244, **245**, 246, **247**
    functions, **253**
    nucleus, 86, **87**, **205**, **253**
    pelvic, 302, **303**, 304, **305**, 306, **307**
    respiratory center, 244
    symptoms of disease, 246
    taste, 244, **245**, 362, **363**
  vertebral, **303**
  vestibular, 84, **85**, **205**, 240, **241**, 374, **375**
    components, 240, 374, **375**

Olfactory tract, 94, **95, 339,** 352, **353,** 354, **355**
Olfactory trigone, 94, **95,** 354, **355**
Olfactory tubercle, **339, 349,** 354, **355**
Olfactory vesicle, 352, **353**
Oligodendroglia, 58, **59,** 572
Oligodendroglioma, 528
Olivary nuclear complex, **87**
Olive, 86, **87**
Olivocerebellar tract, 190, **191**
Olivocochlear bundle, 372, **373**
Olivopontocerebellar degeneration, 462, 484
Olivospinal tract, **201**
Operculum, 572
Ophthalmic herpes, 558
Opisthotonos, 542, 572
Oppenheim's sign, 582
Opsin, 388
Optic atrophy, 402
Optic chiasma, 78, **79, 95,** 224, **225,** 392, **393**
  midline lesion, 402, **403**
Optic cortex, **225**
Optic cup, 38, **39**
Optic disk, 38, **39,** 384, **385, 389**
  swelling of, 524
Optic nerve and tract, **95,** 224, **225,** 252, 392, **393**
  development, 40, **41**
  interruption, 402, **403**
Optic neuritis, 456
Optic pathways, lesions, 402, **403**
Optic radiations, 224, **225,** 392, **393**
Optic tracts, 224, 392, **393**
Optic vesicles, 38, **39,** 40, **41**
Optics of vision, 386, **387**
Ora serrata, 384, **385, 387**
Oral contraceptives and thrombosis, 424, 512
Orbitofrontal cortex, 94, **95**
Organ of Corti, 364, **365**
  development, 44, **45**
Orthochromatic leukodystrophy, 466
Orthostatic hypotension, **433**
Osseous labyrinth, 364, **365**
Ossicles of ear, 364, **365, 369**
  development, 48, **49**
Ossification defect in skull, 446
Osteomas, 538
Osteosarcoma, 538
Otic ganglion, 254
Otic pit, 44, **45**
Otic placode, 44, **45**
Otic vesicles, 44, **45**
Otocephalus, 50, **51**
Otoconia, **241**
Otoliths, 364, **377**
Oval window, **47, 49,** 364, **365**
Oxycephaly, 446

Pacchionian bodies, 100, **101**
Pacemaker neurons, 164
Pachygyria, 440
Pachymeninx, 98, 572

Pacinian corpuscle, **149, 151,** 176, **177,** 180, **181,** 184, **185,** 186, **187**
Pain, from body, 170, **171**
  dull, **175**
  from face, 172, **173**
  modulating, **175**
  pathways, 170, **171,** 172, **173**
  sharp, **175**
  somatic, **174**
  spread, **174**
  visceral, **174**
Paleocerebellum, 314, **317, 327,** 572
Paleospinothalamic path (dull pain), **175**
Paleostriatum, 272, 276, 572
Pallidal atrophy, 484
Pallium, 28, **29,** 572
  archipallium, 30, **31**
  neopallium, 30, **31**
  paleopallium, 30, **31**
  paleostriatum, 276, **277**
Palpebrae (eyelids), 384, **385**
Panencephalitis, subacute sclerosing, **557**
Panurin, 494
Papillae of tongue, foliate, 358, **359**
  fungiform, 358, **359**
Papilledema, 524, **573**
  brain tumor, 524
Papoviruses, 558
Paracentral lobule, 92, **93**
Paraflocculus, 316
Paraldehyde, 494
Paralysis agitans (Parkinson's disease), 280, **281,** 480
Paralytic poliomyelitis, 552, 554
Paraparesis, progressive spastic, 546
Paraphysis, 504
Paraplegia, **61,** 470, **573**
  thoracic disk prolapse, 470
Parasitic infections, 550
Parasympathetic fibers, 300, **301,** 306, **307,** 308, **309**
Parasympathetic ganglia, 254, 300, **301,** 306, **307,** 308, **309**
Parasympathetic nervous system, 300, **301,** 306–**309**
  development, 34, **35**
Parasympathetic postganglionic fibers, 306, **307,** 308, **309**
Paraterminal body, **79**
Paraterminal gyrus, 92, **93, 339**
Paresis, 546
Parkinson state, **279**
Parkinson's disease and syndromes, **281,** 480, **481,** 482, **573**
Parolfactory area, **93, 339**
Parotid gland, innervation, 244, **245,** 308, **309**
Pars caeca retinae. *See* Optic cup
Pars intermedia, 26, **27**
Pars nervosa, 26, **27**
Pars optica retinae. *See* Optic cup
Pars tuberalis, 26, **27**

Trapezoid body, 84, **85, 373,** 576
Trapezoid nuclei, **373**
Trauma, 426–**33**
    concussion, 428
    consciousness, 428
    contusion, 428
    cord level damage, 432, **433**
    extradural hematoma, 426
    head injuries, 426, **429**
    laceration, 428
    neonatal, 448, **449**
    peripheral injury, 432
    posttraumatic events, 426, **427**
    spinal (neck), 424, 432, **433**
    subarachnoid bleeding, 426, **427**
    subdural hematoma, 426, **427**
Tremor, intention, 328
Treponema pallidum, 546
Trichinella spiralis, 550
Trichiniasis, 550
Trigeminal lemniscus, 172, **173, 174**
Trigeminal nerve. *See* Nerve(s)
Trigeminal neuralgia, 579
Trigeminocerebellar, 188, **189,** 326
Trigeminothalamic tract, 172, **173, 179, 183, 185**
Trigone, habenular, **85**
    hypoglossal, **85,** 86, **87**
    vagal, 86, **87**
Trisomy, **21,** 436
Trochlear nerve, 84, **85,** 94, **95,** 228, **229,** 252
    components, 252
    control of eye movements, 228, **229**
    nucleus, 82, **83,** 228, **229,** 252
    symptoms of disease, 228, **229**
Trömmer (Hoffmann) reflex, 582
Trousseau's sign (latent tetany), **505**
Trunk (brachial plexus), 72, **73**
    inferior, 72, **73**
    middle, 72, **73**
    superior, 72, **73**
Tuber, **315,** 316
Tuber cinereum, 78, **79,** 94, **95,** 264, **265**
Tubercles, cuneate, 81, **85,** 86, **87**
    gracilis, 81, **85,** 86, **87**
    olfactory, **339, 349,** 354, **355**
Tuberculosis of brain and spine, 546
    epidural tuberculous abscess, 546
    subdural tuberculous abscess, 546
Tuberculum cinereum, 81, **85,** 86, **87**
Tuberculum meningitis, 546
Tuberous sclerosis (Bourneville's disease), 446
Tumors, 524–**39**
    central nervous system, 524, **525**
    cerebral, 524–**35**
    effects on brain, 526
    metastatic, **537**
    spinal cord, 524, 536
Tunic (eye), inner, 384, **385**
    middle vascular, 384, **385**
    outer fibrous, 384, **385**
Turricephaly, 446

Tympanic cavity, 46, **47, 49,** 364, **365, 369**
    development, 48, **49**
Tympanic membrane (eardrum), 46, **47, 369**
Typhus, 550
Tyranosis, 506

Ulnar nerve, 72, **73**
    injury, 432
Uncinate bundle of Russell, 192, **193,** 324, **327,** 576
Uncinate fasciculus, 96, **97**
Uncus, **79,** 92, **93, 339,** 576
Unmyelinated neuron, **147**
Upper motor neuron syndrome, 582
Urinary expulsion, **307**
Urinary retention, **305**
Utricle, 364, **365, 374, 375,** 376, **377**
    development, **45,** 46, **47**
Uveal tract (uvea), 384, **385**
Uveomeningoencephalitides, 560
Uvula, **315,** 316, 576

Vagal trigone (ala cinerea), 86, **87**
Vagus nerve, 86, **87,** 94, **95,** 244, **245,** 246, **247, 253,** 576. *See also* Nerve(s)
Vallecula, **359,** 576
Varicella, 558
Vascular encephalopathies, 448
Vascular malformations, 424
Vein(s), anterior caudate, **123**
    basal, 122, **123**
    basilar, **103**
    central, **77**
    cerebellar, 122, **123**
    cerebral, 102, **103,** 122, **123**
    choroidal, **103,** 122, **123**
    cingulate, **123**
    emissary, 102, **103**
    epidural plexus, 76, **77**
    great anastomotic, 102, **123**
    great cerebral, 102, **103,** 122, **123**
    inferior striate, **123**
    insular, **123**
    intercostal, 76, **77**
    intervertebral, **77**
    lateral, 76, **77**
    lateral ventricular, **123**
    longitudinal caudate, **123**
    median, 76, **77**
    petrosal, **123**
    pontine, **123**
    posterior terminal, **103**
    segmental, 76, **77**
    septal, **103, 123**
    spinal, 76, **77**
    terminal, **123**
    thalamic, **123**
    thalamostriate, **123**
    thoracic, 76, **77**
    vasocoronal, 76, **77**